The Oxford Dictionary for Scientific Writers and Editors

The
Oxford Dictionary for Scientific Writers and Editors

CLARENDON PRESS • OXFORD
1991

Oxford University Press, Walton Street, Oxford OX2 6DP

Oxford New York Toronto
Delhi Bombay Calcutta Madras Karachi
Petaling Jaya Singapore Hong Kong Tokyo
Nairobi Dar es Salaam Cape Town
Melbourne Auckland

and associated companies in
Berlin Ibadan

Oxford is a trade mark of Oxford University Press

Published in the United States
by Oxford University Press, New York

British Library Cataloguing in Publication Data

A catalogue record for this book is
available from the British Library

Library of Congress Cataloging in Publication Data
The Oxford dictionary for scientific writers and editors.
1. Technical writing—Dictionaries. 2. Technical editing —Dictionaries. 3. English language—Technical English —Dictionaries.
T11.094 1991 808'.0666—dc20 91–350
ISBN 0–19–853920–7

Text prepared by
Market House Books Ltd, Aylesbury
Printed in Great Britain by
Bookcraft (Bath) Ltd
Midsomer Norton, Avon

Preface

This dictionary was devised as a companion volume to the *Oxford Dictionary for Writers and Editors*, which is itself a successor to the eleven editions of the *Authors' and Printers' Dictionary* first published (1905) under the editorship of F. Howard Collins.

The purpose of this dictionary is to provide scientists, science writers, and editors of scientific texts with a guide to the style for presenting scientific information most widely used within the scientific community. As far as possible this complies with the house style of the Oxford University Press; it also follows the recommendations of the International Union of Pure and Applied Physics and the International Union of Pure and Applied Chemistry.

The dictionary is not intended to be a dictionary of science; many of the entries have only a brief definition, no definition at all, or sometimes an identifying subject label. Full definitions for most of the terms found in this dictionary will be found in the *Concise Science Dictionary* (OUP, 2nd edn, 1991) or the *Concise Medical Dictionary* (OUP, 3rd edn, 1990).

The fields covered are primarily physics, chemistry, botany, zoology, biochemistry, genetics, immunology, microbiology, and astronomy, with some coverage of medicine, mathematics, and computer science.

Where a distinction exists between the usage or spelling common in the USA and that in the UK, the differences are shown in the dictionary. Both the British and American versions will be found as separate entries, for example:

> **oestrogen** US: **estrogen**
> **estrogen** US spelling of *oestrogen

Because this is primarily a dictionary for written English, no pronunciation guides have been given, except for a few acronyms, the pronunciation of which is not immediately apparent (e.g. CITES, IUPAC). For these pronunciations we have followed the respelling system used in *The Oxford Paperback Dictionary* (3rd edn, 1988).

As this is the first edition of an entirely new book, the publishers and compilers would welcome suggestions from readers for improvements or changes. For this purpose a card is provided.

A.I.
J.D.
E.M.

Editors

Alan Isaacs BSc, PhD
John Daintith BSc, PhD
Elizabeth Martin MA

(*Market House Books Ltd*)

Contributors and advisers

M. Barton BA
R. Cutler BSc
Robert S. Hine BSc, MSc
Alan Hughes MA
Valerie Illingworth BSc, MPhil
Anne Moorhead BSc
R. A. Prince MA
Ruth E. Taylor BSc, MIBiol, CBiol
Anthony L. Waddell BSc, PhD
J. W. Warren PhD

Acknowledgements

The editors wish to acknowledge the help and advice they have received from:

The Department of Zoology, Natural History Museum
The Royal Botanic Gardens, Kew
The Biochemical Society, London
The International Committee on Taxonomy of Viruses
Nature
Journal of Medical Microbiology

Note on proprietary status

Contents

Note

An asterisk before a word in an entry indicates that this word has its own entry in the dictionary and that additional information can be found there. However, not every word that appears as an entry in the dictionary has an asterisk placed before it when it is used in the text.

A

a Symbol for **1**. are. **2**. atto-. **3**. year. [from French *année*]

a Symbol (light ital.) for **1**. acceleration (in nonvector equations; *a*); in vector equations it is printed in bold italic (***a***). **2**. *Chem.* activity (a_B = activity of substance B). **3**. *Chem.* axial conformation (e.g. in cyclohexane derivatives). **4**. length (a_0 = Bohr radius). **5**. *Physics* specific activity. **6**. thermal diffusivity.

A 1. A blood group and its associated antigen (*see* ABO). **2**. Symbol for **i**. acid-catalysed (*see* reaction mechanisms). **ii**. adenine. **iii**. adenosine. **iv**. alanine. **v**. ampere. **vi**. androecium (in a *floral formula). **vii**. *Astron. See* spectral types.

Å Symbol for angstrom.

A Symbol for **1**. (light ital.) absorbance. **2**. (light ital.) activity. **3**. (light ital.) affinity (chemical). **4**. (light ital.) area. **5**. (light ital.) Helmholtz function (A_m = molar Helmholtz function). **6**. (bold ital.) magnetic vector potential. **7**. (light ital.) nucleon number. **8**. (light ital.) *Electronics* Richardson constant.

A_H (light ital. A) Symbol for Hall coefficient.

A_r (light ital. A) Symbol for relative atomic mass.

a- (an- before vowels and usually before h) Prefix denoting absence of, lacking, not (e.g. abiotic, *acellular, achromatic, asexual, anaerobic, anechoic, *anhydrous).

AAO Abbrev. for Anglo-Australian Observatory (Siding Spring, NSW).

AAS Abbrev. for atomic absorption spectroscopy.

AAT Abbrev. for Anglo-Australian Telescope (Siding Spring, NSW).

AAV Abbrev. for adeno-associated virus. *See* dependovirus.

AB A blood group of the *ABO system.

ab- Prefix denoting **1**. away from or opposite to (e.g. abaxial, aboral). **2**. *Obsolete* an electromagnetic unit in the cgs system (e.g. abampere).

ABA Abbrev. for abscisic acid.

Abbe, Cleveland (1838–1916) US meteorologist.

Abbe, Ernst (not Abbé) (1840–1905) German physicist.
Abbe condenser
Abbe criterion
Abbe number

Abegg, Richard (1869–1910) German physical chemist.
Abegg's rule

Abel, Sir Frederick Augustus (1827–1902) British chemist.

Abel, John Jacob (1857–1938) US biochemist.

Abel, Niels Henrik (1802–29) Norwegian mathematician.
Abelian group
Abel's theorem

Abell, George Ogden (1927–) British astronomer.
Abell catalogue

Abelson, Philip Hauge (1913–) US physical chemist.

Abney, Sir William de Wiveleslie (1844–1920) British chemist.
Abney level
Abney mounting

ABO The most important blood-group system, consisting of four groups (A, B, AB, and O) defined by the presence (or absence) of specific antigens on the red cells (see table).

Blood group	Antigen present	Antibody in serum
A	A	anti-B
B	B	anti-A
AB	A + B	–
O	–	anti-A + anti-B

abomasum (pl. abomasa) The fourth compartment of a ruminant's stomach. Adjectival form: **abomasal**.

abs. Abbrev. for absolute.

abscisic acid Abbrev.: ABA (no stops). A plant growth substance: promotes **abscission**.

abscissa (pl. abscissae; preferred to abscissas) *Maths.*

absolute activity Symbol: λ_B (Greek lamda) for a substance B or, when a complicated formula is to be written, $\lambda(B)$, as in $\lambda(H_2SO_4)$. A dimensionless physical quantity equal to:

$$\exp(\mu_B/RT),$$

where μ_B is the *chemical potential of substance B, R is the *molar gas constant and T the thermodynamic temperature.

absolute magnitude *See* magnitude.

absolute temperature Former name for thermodynamic temperature.

absorbance Symbol: A A dimensionless physical quantity indicating the reduction of intensity that occurs when electromagnetic radiation is passed through a substance. In chemistry it is defined as:

$$-\log_{10}(I/I_0) = \epsilon cl,$$

where I is the intensity of the radiation after transmission through the sample, I_0 is the initial intensity of the radiation, ϵ is the molar absorption coefficient, c is the sample concentration, and l is the path length of the sample. In physics, the absorbance is the *internal transmission density. *Compare* absorptance.

absorbed dose Symbol: D A physical quantity measuring irradiation; the mean energy per unit mass transferred to matter by ionizing radiation. The *SI unit is the gray (Gy), which has replaced the rad.

absorptance Symbol: α (Greek alpha). A dimensionless physical quantity equal to the ratio Φ_a/Φ_o, where Φ_o is the radiant or luminous flux incident on a body or substance and Φ_a is the flux absorbed by it. *Compare* absorbance.

absorption *Chem., physics* **1.** The process in which one substance, usually a gas or liquid, permeates or is dissolved by another liquid or solid substance (the **absorbent**). Adjectival form: **absorbent**. *Compare* adsorption. **2.** The conversion of energy, falling on a substance (an **absorber**), into some other form of energy within the substance. *See* absorbance.

absorption coefficient *See* acoustic absorption coefficient; linear absorption coefficient.

abysso- Prefix denoting the abyssal zones of oceans or lakes (e.g. abyssobenthic).

Ac 1. Symbol for actinium. **2.** Symbol often used to denote the acetyl (ethanoyl) group in chemical formulae, e.g. CH_3COCl can be written as AcCl.

Ac (ital.) Abbrev. for altocumulus.

AC Abbrev. for acyl-oxygen cleavage (*see* reaction mechanisms).

a.c. (stops, or **AC**) Abbrev. for alternating current.

ac- (ital., always hyphenated) Prefix denoting alicyclic. Often used in nomenclature for polycyclic compounds (e.g. *ac*-tetrahydro-2-naphthylamine).

7-ACA Abbrev. for 7-aminocephalosporanic acid.

acac (no stop) Abbrev. for the acetylacetonate ion often used in the formulae of coordination compounds, e.g. $[Co(acac)_2(py)_2]^{2+}$.

Acanthodii (cap. A; not Acanthodi) A class of extinct jawed fishes, sometimes regarded as a subclass of the Osteichthyes (bony fishes). Individual name and adjectival form: **acanthodian** (no cap.).

Acari (cap. A; not Acarii) An order of arachnids comprising the mites and ticks. Also called Acarina. Individual name and adjectival form: **acarid** (no cap.).

acaro- (or **acari-**) Prefix denoting mites and ticks, order *Acari (e.g. acarology, acarophily, acaricide).

acceleration Symbol: $\boldsymbol{a}$ A *vector quantity, the rate of change of *velocity, $d\boldsymbol{v}/dt$ or $d^2\boldsymbol{s}/dt^2$. The *SI unit is the metre per second per second ($m\ s^{-2}$). *See also* angular acceleration.

acceleration of free fall Symbol: g The acceleration of a body freely falling in a vacuum at a point close to the earth's surface. It varies with locality. The standard value, symbol g_n, is

$$9.806\ 65\ m\ s^{-2}.$$

Also called acceleration due to gravity. Former name: acceleration of gravity.

acceptor number density *See* number density.

acclimation *See* acclimatization.

acclimatization The adaptation of an organism to a change in its environment. The US form, **acclimation,** is becoming more widely used in Britain, but **acclimatize** is still the preferred British verb form; **acclimate** is restricted to US usage.

Ac-Ds (ital., hyphenated) *Genetics* Abbrev. for Activator-Dissociation transposable element. *See* transposon.

-aceae Noun suffix denoting a *family in plant classification (e.g. Cucurbitaceae, Rosaceae). Adjectival form (no initial cap.): **-aceous** (e.g. cucurbitaceous).

acellular *Biol.* Not divided into cells: used to describe multinucleate structures and organisms, such as *coenocytes, *plasmodia, and *syncytia, and uninucleate organisms, such as protozoans; in the latter context it is preferred to *unicellular.

ACES Abbrev. for *N*-(2-acetamido)-2-aminoethanesulphonic acid.

acetal $CH_3CH(OC_2H_5)_2$ The traditional name for 1,1-diethoxyethane.

acetaldehyde CH_3CHO The traditional name for ethanal.

acetamide CH_3CONH_2 The traditional name for ethanamide.

acetanilide $C_6H_5NHCOCH_3$ The traditional name for *N*-phenylethanamide.

acetate The traditional name for a salt or ester of acetic (ethanoic) acid. The recommended name is ethanoate.

acetbromamide $BrNHCOCH_3$ The traditional name for *N*-bromoethanamide.

acetic acid CH_3COOH The traditional name for ethanoic acid.

acetic anhydride $(CH_3CO)_2O$ The traditional name for ethanoic anhydride.

aceto- (often **acet-** before vowels) Prefix used in traditional chemical names to denote association with ethanoic (acetic) acid (e.g. acetoacetic acid, acetonitrile, acetamide).

acetoacetic acid CH_3COCH_2COOH The traditional name for 3-oxobutanoic acid.

acetoacetic ester $CH_3COCH_2COOC_2H_5$ The traditional name for ethyl 3-oxobutanoate.

acetoin $CH_3CH(OH)COCH_3$ The traditional name for 3-hydroxy-2-butanone.

acetol CH_3COCH_2OH The traditional name for hydroxypropanone.

acetone CH_3COCH_3 The traditional name for propanone. Acetone is acceptable in nonchemical contexts.

acetonitrile CH_3CN The traditional name for ethanenitrile.

acetophenone $C_6H_5COCH_3$ The traditional name for phenylethanone.

acetyl- *Prefix denoting the group CH_3CO- derived from ethanoic (acetic) acid (e.g. acetylacetone, acetyl azide). Ethanoyl- is recommended in all contexts.

acetylacetonate *See* acac.

acetylacetone $CH_3COCH_2COCH_3$ The traditional name for pentane-2,4-dione.

acetylation Use *ethanoylation.

acetyl chloride CH_3COCl The traditional name for ethanoyl chloride.

acetylcholine (one word) Abbrev.: ACh (no stops). *See also* cholinergic.

acetylcholine receptor Abbrev.: AChR (no stops).

acetylcholinesterase Abbrev.: AChE (no stops).

acetyl coenzyme A (not hyphenated) Abbrev.: acetyl CoA (no stops).

acetylene 1. C_2H_2 The traditional name for ethyne. Acetylene is used in nonchemical contexts, e.g. oxyacetylene welding. **2.** The traditional name for an *alkyne. In all chemical contexts use alkyne.

acetylene dichloride $CHCl=CHCl$ The traditional name for 1,2-dichloroethene.

acetylene tetrachloride $CHCl_2CHCl_2$ The traditional name for 1,1,2,2-tetrachloroethane.

ACh Abbrev. for acetylcholine.

AChE Abbrev. for acetylcholinesterase.

Achilles tendon (no apostrophe)

acholeplasma *See* mollicute.

AChR Abbrev. for acetylcholine receptor.

A chromosome (not hyphenated) Any of the chromosomes forming the normal complement of the species. *Compare* B chromosome.

acicula (pl. aciculae) A needle-shaped part: usually refers to needle-shaped leaves or crystals. Adjectival forms: **acicular, aciculate**. *Compare* aciculum.

aciculum (pl. acicula) A needle-like bristle in the parapodium of a polychaete worm. Adjectival form: **acicular**. *Compare* acicula.

acid anhydride (preferred to acyl anhydride) Any of a class of organic compounds that can be regarded as formed by the elimination of a water molecule from either a dicarboxylic acid molecule or two monocarboxylic acid molecules. Acid anhydride molecules contain the group –CO.O.CO–.

Acid anhydrides are systematically named as the anhydride of the appropriate carboxylic acid(s), e.g. ethanoic anhydride, $(CH_3CO)_2O$, ethanoic methanoic anhydride, $CH_3COOOCH$, and butanedioic anhydride, $(CH_2CO)_2O$.

In nonsystematic nomenclature, acid anhydrides are named as in systematic nomenclature but the trivial name(s) of the carboxylic acids are used, e.g. acetic anhydride, $(CH_3CO)_2O$.

acid–base balance (en dash, not hyphen)

acid dissociation constant *See* pH.

acid halide Use *acyl halide.

Acidiphilium (cap. A, ital.; not *Acidophilum*) A genus of acidophilic aerobic bacteria.

acidophilic (preferred to acidophil and acidophilous) **1.** Denoting microorganisms (known as **acidophiles**) that thrive in acid media. **2.** Denoting material that stains strongly with acid dyes.

ACK (or **ack**) *Computing, telecom* Abbrev. for acknowledgment.

acorn worm (two words) *See* Hemichordata.

acoustic absorption coefficient Symbol: α_a (Greek alpha). A dimensionless physical quantity given by $(1 - \rho)$, where ρ (Greek rho) is the reflection coefficient, P_r/P_0; P_0 and P_r are the *sound power (or more generally the acoustic power) incident on a body and reflected from it, respectively. The transmission coefficient, symbol τ (Greek tau), is the ratio P_{tr}/P_0; P_{tr} is the transmitted acoustic power. The dissipation factor, symbol ψ or δ (Greek psi or delta), is then equal to $(\alpha_a - \tau)$.

acousticolateralis system (one word)

acoustic power *See* sound power.

acoustic pressure *See* sound pressure.

acoustics 1. (takes a sing. form of verb) The science of sound, ultrasound, and infrasound. **2.** (takes a pl. form of verb) The characteristics of a room, hall, etc., with respect to audibility of sound. Adjectival form: **acoustic** (not acoustical).

ACP Abbrev. for acyl carrier protein, important in fatty-acid synthesis. It is qualified according to the fatty acid it binds to, e.g. malonyl ACP (not hyphenated) is malonic acid bound to ACP.

acquired immune deficiency syndrome (not immunodeficiency) Acronym: AIDS (no stops) or Aids.

Acrania Use *Cephalochordata.

Acrasiomycetes (cap. A) A class of the division *Myxomycota containing the cellular slime moulds. In some classifications these fungi are regarded as an order, Acrasiales, of the class *Myxomycetes (slime moulds). Individual name and adjectival form: **acrasiomycete** (no cap.).

acre A unit of area used in land measurement, equal to 4840 square yards, $^1/_{640}$ square mile. One acre = 4046.86 square metres, i.e. 0.404 686 hectare.

Acrilan (cap. A) A trade name for a synthetic acrylic fibre, poly(propenonitrile).

acro- Prefix denoting tip, apex, extremity (e.g. acropetal, acrosome).

acrolein $CH_2{=}CHCHO$ The traditional name for propenal.

acrylic acid $CH_2{=}CHCOOH$ The traditional name for propenoic acid.

acrylonitrile $CH_2{=}CHCN$ The traditional name for propenenitrile.

ACS Abbrev. for American Chemical Society.

ACTH Abbrev. and preferred form for *adrenocorticotrophic hormone.

actinides Use *actinoids.

actinium Symbol: Ac *See also* periodic table; actinoids; nuclide.

actino- (actin- before vowels) Prefix denoting **1.** a radial pattern or arrangement (e.g. actinomorphic). **2.** radiation (e.g. actinobiology, actinometer).

Actinobacillus (cap. A, ital.) A genus of bacteria of the family Pasteurellaceae. Individual name: **actinobacillus** (no cap., not ital.; pl. actinobacilli).

actinobacteria (sing. actinobacterium) A group of *actinomycete bacteria characterized by minimal or absent mycelial development. They include the genera *Actinomyces*, *Arthrobacter*, *Cellulomonas*, and *Micrococcus*. *Compare Actinobacterium*.

Actinobacterium (cap. A, ital.) A now obsolete genus of bacteria whose members have been transferred to the genus *Actinomyces*. *Compare* actinobacteria.

actinoids The recommended name for the group of elements with proton numbers from 89 (actinium) to 103 (lawrencium). The traditional name is actinides.

Actinomadura (cap. A, ital.) A genus of *maduromycete bacteria, some species of which cause Madura foot in humans. Individual name: **actinomadura** (no cap., not ital.; pl. actinomadurae).

Actinomyces (cap. A, ital.) A genus of *actinomycete bacteria. It includes bacteria formerly classified as the genus *Actinobacterium*.

actinomycete Any Gram-positive eubacterium of a diverse group characterized by the production of a mycelium. Note that the term includes, but is not restricted to, any member of the genus *Actinomyces*. Adjectival form: **actinomycete**.

actinomyosin Use actomyosin.

Actinoplanes (cap. A, ital.) A genus of *actinoplanete bacteria. The genus is considered by some authorities to be synonymous with *Amorphosporangium*.

actinoplanete Any actinomycete bacterium of a group that includes the genus **Actinoplanes* and closely related genera. The name 'actinoplanete' does not correspond with any particular taxon and is used for descriptive rather than taxonomic purposes. Adjectival form: **actinoplanete**.

Actinopterygii (cap. A; not Actinopterygi) A subclass of the Osteichthyes (bony fishes) comprising the ray-finned fishes. Individual name and adjectival form: **actinopterygian** (no cap.). letter).

Actinozoa Use *Anthozoa.

active power *See* power.

activity 1. Symbol: A A physical quantity associated with radioactive decay. It is the average number of spontaneous nuclear disintegrations in an amount of a radionuclide in a small time interval divided by that time interval. The *SI unit is the becquerel (Bq), which has replaced the curie and is dimensionally equivalent to the reciprocal of the second (s^{-1}). *See also* decay constant. **2.** Symbol: a_B (for a substance B) or, when a complicated formula is to be written, a(B), as in $a(H_2SO_4)$. A dimensionless physical quantity, a thermodynamic function that can be used in chemistry in place of *concentration. It is proportional to the *absolute activity. Also called relative activity.

activity coefficient A dimensionless physical quantity associated with a liquid or solid mixture. On a *mole fraction basis, the symbol is f_B (for a substance B) or, when a complicated formula is to be written, f(B), as in $f(H_2SO_4)$. For a liquid mixture, f_B is proportional to λ_B/x_B, where λ_B is the *absolute activity of substance B and x_B is the mole fraction of B. On a *molality basis, the symbol is γ_B (Greek gamma) or γ(B). For a solute B, γ_B is proportional to a_B/m_B, where a_B is the *activity of substance B and m_B is its molality.

actomyosin (not actinomyosin)

acyl- *Prefix denoting the group RCO–, where R is any alkyl or aryl group, derived from a carboxylic acid (e.g. acyl chloride).

acyl anhydride Use *acid anhydride.

acyl carrier protein Abbrev.: *ACP (no stops).

acyl halide Any of a class of organic compounds containing the group –COX, where X is a halogen atom. Also called acid halide.

Acyl halides are systematically named by replacing the suffix -oic of the corresponding carboxylic acid by -oyl followed by the name of the appropriate halide, e.g. ethanoyl chloride, CH_3COCl, and 2-methylpropanoyl bromide, $(CH_3)_2CHCOBr$.

In nonsystematic nomenclature the suffix -ic of the trivial name of the corresponding carboxylic acid is replaced by -yl followed by the name of the appropriate halide, e.g. acetyl chloride, CH_3COCl.

AD (small caps., no stops) Abbrev. (Latin) for *anno Domini*; always place before the year.

A/D (or **A-D, a/d, a-d**) Abbrev. for analog-to-digital.

ad- Prefix denoting to, towards, or near (e.g. adaxial, adsorb, advection).

Ada (cap. A) A registered trade mark (US government) for a programming language designed for embedded systems.

ADA Abbrev. for *N*-(2-acetamido)-iminodiacetic acid.

Adams, John Couch (1819–92) British astronomer.

Adams, Robert (1804–78) Irish physician.

Stokes–Adams attack or **syndrome** (en dash)

Adams, Roger (1889–1971) US organic chemist.

Adams, Walter Sydney (1876–1956) US astronomer.

adaptor Preferred to adapter in British (but not US) English for senses in electrical engineering, computing, etc.

ADC Abbrev. for analog-to-digital converter.

ADCC Abbrev. for antibody-dependent cellular cytotoxicity.

adelpho- Prefix denoting kinship, relatedness (e.g. adelphogamy).

-adelphous Adjectival suffix denoting a bundle or bundles of stamens (e.g. monadelphous).

adenine Symbol: A A purine base. *See also* base pair. *Compare* adenosine.

adeno- (aden- before vowels) Prefix denoting **1.** a gland or glandlike structure (e.g. adenohypophysis, adenophyllous, adenitis). **2.** adenine (e.g. adenosine, adenylation).

adeno-associated virus Abbrev.: AAV (no stops). *See* dependovirus.

adenosine Symbol: A A *nucleoside consisting of *adenine combined with D-ribose.

adenosine 5′-diphosphate Abbrev. and preferred form: ADP (no stops).

adenosine 3′,5′-phosphate Usually shortened to cyclic AMP; abbrev.: cAMP (no stops).

adenosine 5′-phosphate Abbrev. and preferred form: AMP (no stops).

adenosine 5′-triphosphate Abbrev. and preferred form: ATP (no stops).

adenovirus (one word) Any virus belonging to the family Adenoviridae. The serovar designations of the human adenoviruses consist of the letter 'h' and a number, e.g. h1, h2, etc. They are also commonly referred to by the label 'Ad' plus a number, e.g. Ad12. There are currently 42 human adenovirus serovars, grouped into six subgenera, denoted by the letters A to F. Adenovirus serovars of other host species are designated by a three-letter label, derived from the host's generic name, and a number: e.g. can1 (of dogs) [from *Canis*]; equ1 (of horses) [from *Equus*]. There are alternative systems of nomenclature for nonhuman adenoviruses: e.g. SAV-1 (simian adenovirus), BAV-1 (bovine adenovirus), etc.

ADH Abbrev. for antidiuretic hormone. *See also* vasopressin.

Adhemar, Alphonse Joseph (1797–1862) French mathematician.

adiabatic Without loss or gain of heat. *Compare* isothermal.

adipic acid $HOOC(CH_2)_4COOH$ The traditional name for hexanedioic acid.

adipo- Prefix denoting fatty tissue (e.g. adipocyte).

Adler, Alfred (1870–1937) Austrian psychiatrist. Adjectival form: **Adlerian.**

admittance Symbol: *Y* A physical quantity, the reciprocal of the *impedance, i.e. $1/Z$. Admittance may be written as a complex number:

$$Y = G + iB,$$

where *G*, the real part, is the *conductance and *B*, the imaginary part, is the *susceptance*; $i = \sqrt{-1}$. The *SI unit of admittance, conductance, and susceptance is the siemens.

ADP Abbrev. and preferred form for adenosine 5′-diphosphate.

adrenaline (not adrenalin) US: epinephrine. *See also* adrenergic.

adrenergic Denoting nerve fibres that release adrenaline or noradrenaline as a neurotransmitter. *Compare* cholinergic.
adrenergic receptor Abbrev.: AR (no stops). Often shortened to adrenoceptor. *See also* alpha receptor; beta receptor.

adreno- (**adren-** before vowels) Prefix denoting **1.** the adrenal gland (e.g. adrenocorticotrophic, adrenotropic). **2.** adrenaline (e.g. adrenergic).

adrenoceptor Abbrev.: AR (no stops). An acceptable shortened form of adrenergic receptor. *See also* alpha receptor; beta receptor.

adrenocorticotrophic hormone US: **adrenocorticotropic hormone**. Abbrev. and preferred form: ACTH (no stops). Also called adrenocorticotrophin or corticotrophin (US: adrenocorticotropin, corticotropin).

Adrian, Edgar Douglas, Lord Adrian (1889–1977) British neurophysiologist.

adsorption *Chem.* The taking up of one substance (the **adsorbate**) at the surface of another (the **adsorbent**). Adjectival form: **adsorbent**. *Compare* absorption.

advanced gas-cooled reactor Abbrev.: AGR (no stops).

Ae *Astron. See* spectral types.

AE *Photog.* Abbrev. for autoexposure.

AEA Abbrev. for Atomic Energy Authority.

Aegyptopithecus (cap. A, ital.) A genus of fossil apes, sometimes included in the genus **Propliopithecus*.

-aemia US: **-emia**. Noun suffix denoting blood (e.g. anaemia). Adjectival form: **-aemic** (US: **-emic**).

aeon *Geol.* Use *eon (aeon is used in non-scientific contexts).

aer- *See* aero-.

aerial *Telecom., etc.* *Antenna is now preferred in scientific literature although 'aerial' is still in general use.

aero- (sometimes **aer-** before vowels) Prefix denoting **1.** air or gas (e.g. aerobic, aerodynamics, aerotolerant, aerotropism). **2.** aircraft (e.g. aeroengine, aerofoil, aeronautics).

Aerococcus (cap. A, ital.) A genus of microaerophilic coccoid bacteria. Individual name: **aerococcus** (no cap., not ital.; pl. aerococci).

Aeromonas (cap. A, ital.) A genus of bacteria of the family Vibrionaceae. Individual name and adjective form: **aeromonad** (no cap., not ital.).

aestivation US: **estivation**. **1.** *Zool.* Dormancy during periods of drought or heat. *Compare* hibernation. **2.** *Bot.* The arrangement of flower parts in the bud. *Compare* vernation.

aether Use ether.

aetiology US: **etiology**. Adjectival form: **aetiological** (US: **etiologic**).

AF Abbrev. for **1.** (or **a.f.**) audiofrequency. **2.** *Photog.* autofocus.

a.f. (stops, or **AF**) Abbrev. for audiofrequency.

a factor (bold lower-case a) A pheromone secreted by yeast cells of mating type *MAT***a**. *Compare* alpha factor.

a.f.c. (stops, or **AFC**) Abbrev. for automatic frequency control.

afp (no stops) Abbrev. for alpha-fetoprotein.

affinity Symbol: *A* A physical quantity indicating the tendency of a chemical reaction to take place and expressed in terms of change of free energy. The *SI unit is the *joule.

AFRC Abbrev. for Agricultural and Food Research Council.

afterbirth (one word) The placenta and fetal membranes after expulsion from the uterus, following the birth of the fetus.

after-ripening (hyphenated) A period of dormancy in a mature seed before germination.

Ag Symbol for silver. [from Latin *argentum*]

agaric Any fungus of the family Agaricaceae; originally nearly all these fungi were placed in the genus *Agaricus* (cap. A, ital.), but this is no longer the case with changes in classification. The word is used both specifically, with the appropriate qualifier (e.g. fly agaric), and more generally, for any member of the family. It is also used loosely for any member of the order Agaricales. Adjectival forms: **agaricaceous** (referring to the family), **agaricalean** (referring to the order).

Agassiz, Alexander Emmanuel Rodolphe (1835–1910) US naturalist.
Agassiz trawl

Agassiz, Jean Louis Rodolphe (1807–73) Swiss-born US biologist and geologist.

a.g.c. (stops, or **AGC**) Abbrev. for automatic gain control.

ageing US: **aging**.

Agitococcus (cap. A, ital.) A genus of coccoid gliding bacteria. Individual name: **agitococcus** (no cap., not ital.; pl. agitococci).

Agnatha (cap. A) A superclass or class comprising the jawless vertebrates, including the lampreys. Individual name and adjectival form: **agnathan** (no cap.). *See also* Cyclostomata. *Compare* Gnathostomata.

-agogue Noun suffix denoting a substance that stimulates the secretion of something (e.g. cholagogue, sialagogue). Adjectival form: **-agogic**.

Agonomycetales (cap. A) A grouping of the *Deuteromycotina containing imperfect fungi that do not produce spores. Also called Mycelia Sterilia. This taxon is obsolete, as such fungi may be classified as basidiomycetes, ascomycetes, or deuteromycetes, but the individual name and adjectival form, **agonomycete** (no cap.), is still used for descriptive purposes; 'mycelia sterilia' (pl.; no caps.) is also used as a trivial name for these fungi.

AGR Abbrev. for advanced gas-cooled reactor.

Agricola, Georgius (1494–1555) German metallurgist. Also called Georg Bauer.

Agricultural and Food Research Council Abbrev.: AFRC (no stops).

agro- (or **agri-**) Prefix denoting agriculture (e.g. agrobiology, agribusiness).

Agrobacterium (cap. A, ital.) A genus of bacteria of the family Rhizobiaceae. Individual name: **agrobacterium** (no cap., not ital.; pl. agrobacteria).

A horizon A mixed mineral-organic *soil horizon at or near the surface. It was formerly designated the A_1 horizon; A_2 is now the *E horizon.

AI Abbrev. for **1.** artificial intelligence. **2.** artificial insemination.

AID Abbrev. for artificial insemination (by) donor, former name for *donor insemination.

AIDS (or **Aids**) Acronym for acquired immune deficiency syndrome.

AIH Abbrev. for artificial insemination (by) husband.

Aiken, Howard Hathaway (1900–73) US mathematician and computer engineer.

air bladder (two words) *Zool.*

air sac (two words) *Zool.*

Airy, Sir George Biddell (1801–92) British astronomer.
Airy disc
Airy isostasy hypothesis
Airy rings

AIV method [from *A. I.* **V*irtanen]

Akers, Sir Wallace Allen (1888–1954) British industrial chemist.

Al Symbol for aluminium.

Al (ital.) Symbol for Alfvén number.

AL Abbrev. for alkyl-oxygen cleavage (*see* reaction mechanisms).

-al 1. Noun suffix denoting an aldehyde (e.g. butanal, ethanal). **2.** Adjectival suffix denoting relationship with (e.g. atrial, fungal, tundral, viral).

Ala (or **ala**; no stop) Abbrev. for alanine. *See* amino acid.

alanine Abbrev.: Ala or ala (no stop). Symbol: A *See* amino acid.

al-Battani (*c.* 858–929) Arab astronomer. Also called Albategnius.

albino (pl. albinos) Derived noun: **albinism.**

Albinus, Bernhard Siegfried (1697–1770) German anatomist.

Albright, Arthur (1811–1900) British chemist and industrialist.

albumen Former name for **1.** egg white. **2.** *albumins, or mixtures of albumins and other proteins.

albumin Any of a class of water-soluble proteins, including serum albumin, lactalbumin, and ovalbumin. Adjectival form: **albuminous.** *Compare* albumen.

alcohol Any of a class of organic compounds containing the hydroxyl group –OH directly joined to a carbon atom.

Alcohols are systematically named by adding the suffix -ol to the name of the parent hydrocarbon, e.g. ethanol, CH_3CH_2OH, and 3-methylbutan-2-ol, $(CH_3)_2CHCHOHCH_3$.

Two nonsystematic methods of nomenclature are encountered. In one, alcohols are named according to the group attached to the hydroxyl group, e.g. ethyl alcohol, CH_3CH_2OH, and *t*-butyl alcohol, $(CH_3)_3COH$. In the other, alcohols are named as derivatives of carbinol (methanol, CH_3OH), e.g. triphenylcarbinol,

$$(C_6H_5)_3COH.$$

See also diol.

aldehyde Any of a class of organic compounds containing the group –CHO directly joined to a carbon atom.

Aldehydes are systematically named by adding the suffix -al to the name of the parent hydrocarbon, e.g. ethanal, CH_3CHO, and phenylethanal, $C_6H_5CH_2CHO$. An exception to this nomenclature is the aromatic member C_6H_5CHO, which can either retain the nonsystematic name benzaldehyde or be called benzenecarbaldehyde.

Lower members have nonsystematic names treating them as derivatives of carboxylic acids, e.g. formaldehyde, HCHO, and acetaldehyde, CH_3CHO. *See also* aldo-.

Alder, Kurt (1902–58) German organic chemist.

Diels–Alder reaction (en dash)

aldo- (**ald-** before vowels) Prefix denoting aldehyde (e.g. aldohexose, aldose, aldoxime).

aldol $CH_3CHOHCH_2CHO$ The traditional name for 3-hydroxybutanal.

Aldrovandi, Ulisse (1522–1604) Italian naturalist.

Alembert, Jean Le Rond d' Usually alphabetized as *d'Alembert.

aleph Symbol: ℵ First letter of the Hebrew alphabet, used mainly in the form $\aleph_0$, which denotes the cardinal number of the set of natural numbers.

-ales Noun suffix denoting an *order in plant classification (e.g. Graminales, Ranunculales).

Alfvén, Hannes Olof Gösta (1908–) Swedish physicist.

***Alfvén number**
Alfvén's theory
Alfvén waves

Alfvén number Symbol: *Al* (ital.). A dimensionless quantity equal to $v\sqrt{(\rho\mu)}/B$, where v is a characteristic speed, ρ is the density, μ the magnetic *permeability, and B the *magnetic flux density. *See also* parameter.

algae (pl. noun, no cap.; sing. alga) A group of primitive plants formerly regarded as comprising a single division, Algae (cap. A). These plants are now regarded as so dissimilar that a recent classification assigns them to four different subkingdoms. The name 'algae' now has little taxonomic significance but is retained for descriptive purposes and for use (in some cases) as a common name (with the appropriate qualifier) for plants now classified as separate

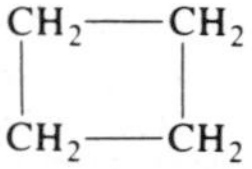

(1) cyclobutane (tetramethylene)

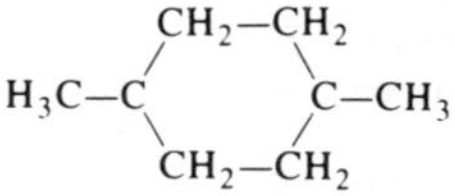

(2) 1,4-dimethylcyclohexane

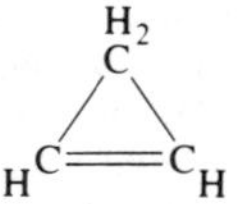

(3) cyclopropene

(4) cyclobutane

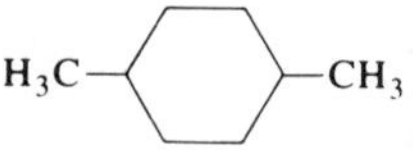

(5) 1,4-dimethylcyclohexane

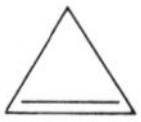

(6) cyclopropene

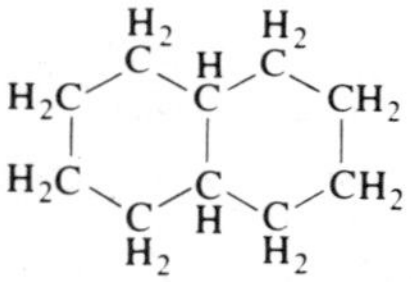

(7) bicyclo[4.4.0]decane (decalin, decahydronaphthalene)

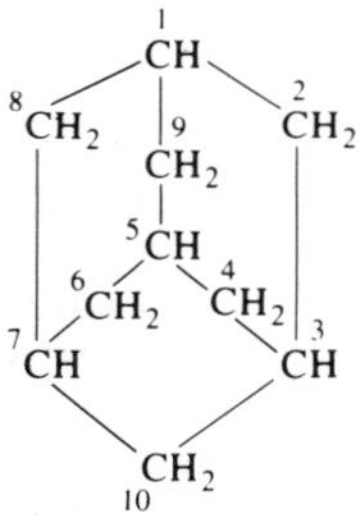

(8) tricyclo[3.3.1.$1^{3,7}$]decane (adamantane)

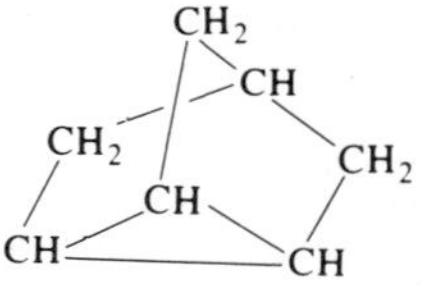

(9) tricyclo[2.2.1.$0^{2,6}$]heptane

taxonomic groups. These include: *Chrysophyta (golden-brown algae), *Xanthophyta (yellow-green algae), *Haptophyta, *Bacillariophyta (diatoms), *Chlorophyta (green algae), *Charophyta (stoneworts), *Euglenophyta (euglenoids), *Dinophyta (including dinoflagellates), *Cryptophyta, *Phaeophyta (brown algae), and *Rhodophyta (red algae). The prokaryotic organisms commonly known as blue-green algae were formerly classified as the division Cyanophyta but have now been reclassified as bacteria (*see* cyanobacteria). *See also* Thallophyta. Adjectival form: **algal**.

-algia Noun suffix denoting pain (e.g. neuralgia). Adjectival form: **-algic**. *See also* algo-.

algo- 1. (or **algi-**) Prefix denoting algae (e.g. algology, algicide, algicolous). 2. (or **alge-**, **algesi-**, **algio-**) Prefix denoting pain (e.g. algogenic, algefacient, algesimeter, algiomotor). *See also* -algia.

Algol 1. (or **ALGOL**) *Computing* Acronym for algorithmic language. 2. *Astron.* An eclipsing binary star in the constellation Perseus.

algorithm (preferred to algorism)

Alhazen (*c.* 965–1038) Arab scientist. Also called al Haytham.

alicyclic hydrocarbon Any of a class of hydrocarbons containing closed rings of carbon atoms with aliphatic rather than aromatic properties.

Monocyclic compounds are systematically named by adding the prefix cyclo- to the name of the open-chain hydrocarbon with the same number of carbon atoms as in the ring, e.g. cyclobutane (1), 1,4-dimethylcyclohexane (2), cyclopropene (3). Conventionally, alicyclic ring systems can be represented by the rings, without showing the carbon and hydrogen atoms, *see* (4), (5), and (6).

Polycyclic compounds contain linked rings and are named as follows. A numerical prefix, bicyclo-, tricyclo-, etc., denotes the number of linked rings present. The number of carbon atoms in the linkages connecting the tertiary carbon atoms is then indicated in brackets in descending order of magnitude (note that the numbers are separated by stops, not commas). The numbering of these polycyclic ring systems starts with a carbon atom forming a junction and proceeds along the longest linkage to the next junction, continues along the next longest path, and progresses in this fashion until the shortest linkage is specified.

Polycyclic compounds of three or more rings are named by choosing the largest ring and the main bridge and numbering the atoms. The compound is then named as for a bicyclic compound with sub-bridges indicated by pairs of superscript numbers separated by commas. Some examples are shown in the diagrams (7), (8), and (9).

Various nonsystematic methods of nomenclature have been used. For example, monocyclic compounds have been named as polymethylenes, e.g. tetramethylene (1). Alternatively, ring compounds can be named as hydrogenated aromatic compounds, e.g. decahydronaphthalene (7). A number have trivial names, e.g. (7) and (8).

alkali (pl. alkalis) Adjectival form: **alkaline**. Derived noun: **alkalinity**.

alkaline–earth metals The elements of group IIA of the *periodic table.

alkane Any of a class of aliphatic hydrocarbons of general formula C_nH_{2n+2}, where $n = 1, 2, 3, \ldots$.

The first four straight-chain alkanes have nonsystematic names, i.e. CH_4 (methane), CH_3CH_3 (ethane), $CH_3CH_2CH_3$ (propane) and $CH_3CH_2CH_2CH_3$ (butane); the higher members are named systematically with a numerical prefix to denote the number of carbon atoms (e.g. pent-, hex-, hept-, etc.) and the suffix -ane, e.g. eicosane, $CH_3(CH_2)_{18}CH_3$. The systematic names of branched alkanes use the longest continuous chain as the root. The substituent alkyl groups are assigned numbers corresponding to their positions on the chain, the direction of numbering chosen to give the lowest numbers possible and, if a series of numbers is required, the lowest first number; the order of the substituent groups is usually alphabetical. Examples are 2,2,3-trimethylbutane, $(CH_3)_3CCH(CH_3)_2$, and 5-(1-methylpropyl)-decane, $CH_3(CH_2)_3$-$(CH(CH_3)CH_2CH_3)(CH_2)_4CH_3$.

In nonsystematic nomenclature straight-chain alkanes have the same names as in systematic nomenclature but have the prefix *n*-, e.g. *n*-pentane, $CH_3CH_2CH_2CH_2$-CH_3. Branched-chain alkanes use the additional structural prefixes iso-, neo-, *s*- and *t*-, e.g. 4-*t*-butyl-4-isopropyldecane, $CH_3CH_2CH_2C(C(CH_3)_3)(CH(CH_3)_2)$-$CH_2CH_2CH_2CH_2CH_2CH_3$. The pentyl group ($C_5H_7-$) is sometimes named amyl, e.g. *n*-amyl, $CH_3CH_2CH_2CH_2CH_2-$, and isoamyl, $CH(CH_3)_2CH_2CH_2-$.

The name 'paraffin' for a member of this class of hydrocarbons is deprecated.

alkene Any of a class of aliphatic hydrocarbons containing one or more double bonds.

The general formula for alkenes containing one double bond is C_nH_{2n}, where $n = 1, 2, 3, \ldots$ The systematic nomenclature for alkenes containing one double bond employs the suffix -ene in place of the -ane used for the corresponding alkane, e.g. ethene, CH_2CH_2, and 2-methylpropene, $(CH_3)_2$-CCH_2. The chain is numbered so that the position of the first carbon atom of the double bond is indicated by the lowest possible number. Examples are eicos-1-ene, CH_3-$(CH_2)_{17}CH{=}CH_2$, and 3-ethylhept-1-ene, $CH_3(CH_2)_3CH(CH_2CH_3)CH{=}CH_2$.

Alkenes containing two or more double bonds are known as alkadienes, alkatrienes, alkatetraenes, etc., the suffix denoting the number of double bonds. The numbering of the root is such that the sum of the numbers for the first carbon atoms of each double bond is the lowest possible. Examples are buta-1,2-diene, $CH_2{=}C{=}CHCH_3$, and penta-1,3-diene, CH_3-$CH{=}CHCH{=}CH_2$.

In nonsystematic nomenclature the simpler alkenes have the suffix -ene appended to the name of the appropriate alkyl group, e.g. CH_2CH_2, $CH_3CH_2CH_2$, and $(CH_3)_2$-CCH_2 are named ethylene, propylene, and isobutylene, respectively (the corresponding systematic names are ethene, propene, and 2-methylpropene).

The name 'olefin' or 'olefine' for a member of this class of hydrocarbons is deprecated.

alkenyl- *Prefix denoting any alkene group after removal of a hydrogen atom (e.g. alkenylbenzene, alkenyl chloride).

al-Khwarizmi (*c.* 800–*c.* 847) Arab mathematician, astronomer, and geographer. Also called Ibn Musa.

alkoxide Any of a class or organic compounds in which the hydrogen atom belonging to an aliphatic alcohol has been replaced by a (strongly electropositive) metal.

Alkoxides are systematically named by replacing the suffix -al of the corresponding alcohol by -oxide and prefixing with the name of the metal, e.g. sodium ethoxide, $CH_3CH_2O^-Na^+$, and potassium dimethylethoxide, $(CH_3)_3CO^-K^+$.

In nonsystematic nomenclature alkoxides are named as in systematic nomenclature but the trivial names of the corresponding alkanols are used, e.g. potassium *t*-butoxide, $(CH_3)_3CO^-K^+$.

The corresponding phenol derivatives are named (systematically and nonsystematically) phenoxides, e.g. sodium phenoxide, $C_6H_5O^-Na^+$.

alkoxy- Prefix denoting the group RCH_2O-, where R is any alkyl group (e.g. alkoxybenzoic acid, alkoxyethanol).

alkyl- *Prefix denoting any alkane group after removal of a hydrogen atom (e.g. alkylbenzene, alkyl chloride).

alkyl halide The traditional name for a *haloalkane.

alkyne Any of a class of aliphatic hydrocarbons containing one or more triple bonds.

The general formula for alkynes containing one double bond is C_nH_{2n-2}, where $n = 1, 2, 3, \ldots$ The systematic nomenclature for alkynes containing one triple bond employs the suffix -yne in place of the -ane used for the corresponding alkane, e.g. ethyne, $CH{\equiv}CH$, and propyne, $CH_3C{\equiv}CH$. The chain is numbered so that the position of the first carbon atom of the triple bond is indicated by the lowest possible number. Examples are eicos-1-yne, $CH_3(CH_2)_{17}$-$C{\equiv}CH$, and 3-ethylhept-1-ene, CH_3-$(CH_2)_3CH(CH_2CH_3)C{\equiv}CH$.

Alkynes containing two or more triple bonds are known as alkadiynes, alkatriynes, alkatetraynes, etc., the suffix denoting the number of triple bonds. The numbering of the root chain is such that the sum of the numbers for the first carbon atoms of each triple bond is the lowest possible. Examples are penta-1,3-diene, $CH_3C{\equiv}CC{\equiv}CH$, and hexa-1,3,5-triene, $HC{\equiv}CC{\equiv}CC{\equiv}CH$.

Aliphatic hydrocarbons containing both double and triple bonds are named alkenynes, alkadienynes, alkendiynes, etc., according to the number of each type of multiple bond present: a double bond takes precedence over a triple bond when numbering the chain. Examples are butenyne, $CH_2{=}CHC{\equiv}CH$, and hex-1-en-3,5-diyne, $CH{\equiv}CC{\equiv}CCH{=}CH_2$.

The name 'acetylene' for a member of this class of hydrocarbons is deprecated, although the traditional name 'acetylene' is still used for ethyne, particularly in nonchemical contexts, e.g. oxyacetylene welding.

alkynyl- *Prefix denoting any alkyne group after removal of a hydrogen atom (e.g. alkynylbenzene, alkynyl chloride).

allele (preferred to allelomorph) One of the alternative forms of a gene. Adjectival form: **allelic**.

Symbols for alleles are printed in italic type. In classical genetics the simplest system uses a single letter (typically the initial letter of the characteristic determined by the dominant allele), with capitals for dominant alleles (e.g. *L* for *l*ong stem) and lower case for recessives (e.g. *l* for short stem). Normal alleles (which are typically dominant in a wild population) may be designated by a plus sign (*see* wild type), especially in species used in genetic research in which mutant recessives are common. For further information on the nomenclature of genetic loci, genes, and their alleles in different species, *see* gene.

allelomorph Use *allele.

allelopathy The production of a chemical by a plant that inhibits the growth of neighbouring plants. Not to be confused with any sense in genetics relating to *alleles. Adjectival form: **allelopathic**.

Allen, James Alfred Van Usually alphabetized as *Van Allen.

allene $CH_2=C=CH_2$ The traditional name for propadiene.

allergenic Causing an allergy; relating to or denoting a substance (**allergen**) that causes an allergy. Not to be confused with **allergic** (relating to or caused by an allergy).

allo- Prefix denoting **1**. difference, separation, disjunction (e.g. allobar, allochthonous, allogamy, allopatric, allotrope). **2**. *Chem.* a close (often isomeric) relationship between compounds (e.g. allocholesterol (coprostenol) is an isomer of cholesterol).

allochthonous Denoting rocks, deposits, etc., that did not originate where they are found. *Compare* autochthonous.

all-or-none (hyphenated) Characterized by a complete response or effect or by none at all.

alluvium (pl. alluvia; preferred to alluviums) Adjectival form: **alluvial**.

allyl- *Prefix denoting the group $CH_2=CHCH_2-$ (e.g. allylbenzene, allyl chloride). Propenyl- is recommended in all contexts.

allyl alcohol $CH_2=CHCH_2OH$ The traditional name for prop-2-en-1-ol.

allyl chloride $CH_2=CHCH_2Cl$ The traditional name for 3-chloroprop-1-ene.

Alnico (cap. A) A trade name for a series of alloys of aluminium, nickel, iron, cobalt, and copper.

alpha Greek letter, symbol: α (lower case), A (cap.).

α Symbol for **1**. absorptance. **2**. alpha particle. **3**. angular acceleration. **4**. brightest star in a constellation (*see* stellar nomenclature). **5**. electric polarizability of a molecule. **6**. expansion coefficient: α_V = cubic, α_l = linear. **7**. fine structure constant. **8**. heavy chain of IgA (*see* immunoglobulin). **9**. linear absorption coefficient (α_a = acoustic absorption coefficient). **10**. a plane angle. **11**. relative pressure coefficient. **12**. right ascension.

Alpha (cap. A, followed by the genitive form of a constellation name; e.g. Alpha Crucis or α Cru) Usually the brightest star in that constellation. *See* stellar nomenclature.

alpha factor (two words; usually written **α factor**) A pheromone secreted by yeast cells of mating type *MAT*α. *Compare* **a** factor.

alpha-fetoprotein (hyphenated) Abbrev.: afp (no stops).

alpha helix (two words; often written **α helix**) *Biochem.*

alphaherpesvirus (one word) A herpesvirus belonging to the subfamily Alphaherpesvirinae (vernacular name: herpes simplex virus group). *See also* herpes simplex virus.

alpha iron (two words; also written **α-iron** (hyphenated))

alphanumeric (one word; preferred to alphameric)

alpha particle (two words; also written **α-particle** (hyphenated)) A positively charged particle that is the nucleus of a helium-4 atom, $^4He^{2+}$. It is denoted α in nuclear reactions, etc. The nucleus of a helium-3 atom, $^3He^{2+}$, is called a helion. It is denoted h although previously the symbol τ was used; τ is now reserved for the *tauon.

alpha receptor (two words; usually written **α receptor**) Shortened and preferred form of alpha-adrenergic receptor (hyphenated; abbrev.: α-AR). There are two types: α_1 receptors and α_2 receptors (note subscript numbers).

Alphavirus (cap. A, ital.) A genus of *togaviruses. Individual name: **alphavirus** (no cap., not ital.).

Alpher, Ralph Asher (1921–) US physicist.

Alpher–Bethe–Gamow theory (en dashes) Also called alpha–beta–gamma theory or αβγ theory.

ALS Abbrev. for antilymphocyte serum.

alt (no stop) Short for altitude.

alternating current Abbrev.: a.c. (or sometimes AC). *See also* electric current.

alti- Prefix denoting height (e.g. altimeter).

alto- Prefix denoting high (e.g. altostratus).

altocumulus (pl. altocumuli) Abbrev.: *Ac* (ital., no stop).

altostratus (pl. altostratus) Abbrev.: *As* (ital., no stop).

ALU *Computing* Abbrev. for arithmetic logic unit.

alum $AlK(SO_4)_2{\cdot}12H_2O$ The traditional name for aluminium potassium sulphate-12-water.

alumina Al_2O_3 The traditional name for aluminium oxide.

aluminium US: **aluminum**. Symbol: Al *See also* periodic table; nuclide.

aluminium ammonium sulphate-12-water $AlNH_4(SO_4)_2{\cdot}12H_2O$ US: **aluminum ammonium sulfate-12-water.** The recommended name for the compound traditionally known as ammonium alum.

aluminium oxide Al_2O_3 The recommended name for the compound traditionally known as alumina.

aluminium potassium sulphate-12-water $AlK(SO_4)_2{\cdot}12H_2O$ US: **aluminum potassium sulfate-12-water.** The recommended name for the compound traditionally known as alum.

Alu sequence (cap. A, not ital.) A repeated base sequence found in the human genome, named after the *restriction enzyme *Alu*I, (*Alu* ital.), which cleaves such sequences. These sequences form the **Alu family**; the **Alu-equivalent family** is a family of similar sequences found in the Chinese hamster.

Alvarez, Luis Walter (1911–88) US physicist.

Alzheimer, Alois (1864–1915) German physician.

Alzheimer's disease

Am Symbol for americium.

AM (or **a.m.**) Abbrev. for amplitude modulation.

Ambartsumian, Viktor Amazaspovich (1908–) Soviet astrophysicist.

ambi- (or **ambo-**) Prefix denoting both (e.g. ambipolar, amboceptor).

ambo- *See* ambi-.

ameba US spelling of *amoeba.

amebo- US spelling of *amoebo-.

Amentiferae An obsolete plant taxon containing certain catkin-bearing families now separated in modern systems. It is equivalent to the subclass Hamamelidae plus the family Salicaceae.

American Standards Association Abbrev.: *ASA (no stops).

americium Symbol: Am *See also* periodic table; nuclide.

Amici, Giovanni Battista (1786–1863) Italian astronomer and instrument maker.

Amici prism

amide Any of a class of organic compounds containing the group $-CONH_2$.

Amides are systematically named by replacing the suffix -oic of the corresponding carboxylic acid by -amide, e.g. ethanamide, CH_3CONH_2, and 2-methylpropanamide, $(CH_3)_2CHCONH_2$. An exception to this nomenclature is the aromatic member $C_6H_5CONH_2$, which can either retain the nonsystematic name benzamide or be called benzenecarbamide.

In nonsystematic nomenclature the suffix -ic of the trivial name of the corresponding carboxylic acid is replaced by -amide, e.g. acetamide, CH_3CONH_2.

amido- Prefix formerly used to denote the amino group $-NH_2$ (e.g. amidol). *Amino- is now recommended in all contexts.

amine Any of a class of organic compounds that are analogues of ammonia.

Amines are systematically named by adding the suffix -amine to the name of the hydrocarbon group(s) attached to the nitrogen atom, e.g. ethylamine, $CH_3CH_2NH_2$, (dimethylethyl)amine, $(CH_3)_3CNH_2$, dimethylethylamine, $(CH_3)_2NCH_2CH_3$, and phenylamine, $C_6H_5NH_2$.

In nonsystematic nomenclature amines are named as in systematic nomenclature but the trivial names of the hydrocarbon groups are used, e.g. *t*-butylamine, $(CH_3)_3CNH_2$. Many benzene derivatives have acceptable trivial names, e.g. aniline, $C_6H_5NH_2$.

amino- Prefix denoting the group $-NH_2$ when joined to a carbon atom (e.g. aminoacyl, aminoazobenzene, aminopeptidase, aminotoluene). Formerly, amido- was sometimes used.

amino acid (two words) Abbreviations for the 20 or so amino acids commonly found in proteins are typically formed from the first three letters of the name, printed in lower case with no stops and with an initial capital or lower-case letter; for example, alanine is Ala or ala, glycine is Gly or gly (see table). An initial capital for the abbreviation is usually preferred. These abbreviations are used in depicting the sequences of proteins (*see* peptide).

aminoazobenzene $C_6H_5N{=}NC_6H_4$-NH_2 The traditional name for *(phenylazo)phenylamine.

4-aminobenzenesulphonic acid $H_2NC_6H_4SO_3H$ The recommended name for the compound traditionally known as sulphanilic acid.

aminobenzoic acid $H_2NC_6H_4COOH$ The recommended name for the isomer traditionally known as anthranilic acid is 2-aminobenzoic acid.

aminobutanedioic acid $HOOCCH_2$-$CH(NH_2)COOH$ The recommended name for the compound traditionally known as *aspartic acid.

γ-aminobutyric acid Abbrev.: *GABA.

aminoethanoic acid H_2NCH_2COOH The recommended name for the compound traditionally known as *glycine.

2-aminoethyl alcohol $HOCH_2CH_2$-NH_2 The traditional name for 2-hydroxyethylamine.

aminoethylsulphonic acid NH_2CH_2-CH_2SO_3H The recommended name for the compound traditionally known as taurine.

2-aminopentanedioic acid HOOC-$(CH_2)_2CH(NH_2)COOH$ The recommended name for the compound traditionally known as *glutamic acid.

aminophenol $H_2NC_6H_4OH$ The recommended name for the compound traditionally known as *o*-aminophenol is 2-aminophenol, etc.

aminosulphonic acid H_3NSO_3 The recommended name for the compound traditionally known as sulphamic acid.

ammocoete (not ammocoet) US: **ammocoete** or **ammocete**. The larva of a lamprey.

ammonium alum $AlNH_4(SO_4)_2{\cdot}12H_2O$ The traditional name for aluminium ammonium sulphate-12-water.

ammonium hydroxide NH_4OH The traditional name for aqueous ammonia.

amnion (pl. amnions; preferred to amnia) One of the extraembryonic membranes. Adjectival form: **amniotic**; not to be confused with *amniote.

amniote Any vertebrate belonging to the taxon Amniota, characterized by three membranes (amnion, allantois, and chorion) surrounding the yolk sac, i.e. reptiles, birds, and mammals. Adjectival form: **amniote**; not to be confused with amniotic (*see* amnion).

amoeba (no cap.; pl. amoebas or amoebae) US: **ameba** (pl. amebas). Any protozoan of the genus *Amoeba* (cap. A, ital.) or related genera (e.g. *Entamoeba*). Adjectival forms: **amoebic** (US: **amebic**), **amoeboid** (US: **ameboid**; resembling an amoeba).

amoebo- (**amoeb-** before vowels, **amoebi-**) US: **amebo-**, etc. Prefix denoting amoebae (e.g. amoebocyte, amoebiasis, amoebicide).

Amoebobacter (cap. A, ital.; not *Amebobacter*) A genus of purple sulphur bacteria.

Amorphosporangium (cap. A, ital.) *See Actinoplanes*.

amount of substance Symbol: *n*, or *ν* (Greek nu) if *n* is being used for number

The amino acids occurring in proteins

Amino acid	Abbreviation	Symbol	Formula
alanine	Ala	A	$CH_3{-}CH(NH_2){-}COOH$
arginine	Arg	R	$H_2N{-}C({=}NH_2^+){-}NH{-}CH_2{-}CH_2{-}CH_2{-}CH(NH_2){-}COOH$
asparagine	Asn	N	$H_2N{-}C({=}O){-}CH_2{-}CH(NH_2){-}COOH$
aspartic acid	Asp	D	$HOOC{-}CH_2{-}CH(NH_2){-}COOH$
cysteine	Cys	C	$HS{-}CH_2{-}CH(NH_2){-}COOH$
glutamic acid	Glu	E	$HOOC{-}CH_2{-}CH_2{-}CH(NH_2){-}COOH$
glutamine	Gln	Q	$H_2N{-}C({=}O){-}CH_2{-}CH_2{-}CH(NH_2){-}COOH$
glycine	Gly	G	$H{-}CH(NH_2){-}COOH$
histidine	His	H	$HC{=}C{-}CH_2{-}CH(NH_2){-}COOH$ (imidazole ring: HC=C, N, NH, CH)
isoleucine	Ile	I	$CH_3{-}CH_2{-}CH(CH_3){-}CH(NH_2){-}COOH$

Amino acid	Abbrev–iation	Symbol	Formula
leucine	Leu	L	$(H_3C)_2CH{-}CH_2{-}CH(NH_2){-}COOH$
lysine	Lys	K	$^{+}H_3N{-}CH_2{-}CH_2{-}CH_2{-}CH_2{-}CH(NH_2){-}COOH$
methionine	Met	M	$CH_3{-}S{-}CH_2{-}CH_2{-}CH(NH_2){-}COOH$
phenylalanine	Phe	F	$C_6H_5{-}CH_2{-}CH(NH_2){-}COOH$
proline	Pro	P	pyrrolidine ring ($H_2C{-}CH_2$, H_2C, $CH{-}COOH$, $N{-}H$) $\longrightarrow$ 4-hydroxyproline (Hyp): $H(OH)C{-}CH_2$, H_2C, $CH{-}COOH$, $N{-}H$
serine	Ser	S	$HO{-}CH_2{-}CH(NH_2){-}COOH$
threonine	Thr	T	$CH_3{-}CH(OH){-}CH(NH_2){-}COOH$
tryptophan	Trp	W	indole ring ($C{=}CH{-}NH$ fused to benzene) $C{-}CH_2{-}CH(NH_2){-}COOH$
tyrosine	Tyr	Y	$HO{-}C_6H_4{-}CH_2{-}CH(NH_2){-}COOH$
valine	Val	V	$(H_3C)_2CH{-}CH(NH_2){-}COOH$

density of particles. A fundamental physical quantity proportional to the number of specified elementary entities of that substance. The constant of proportionality is the reciprocal of the *Avogadro constant. The specified entity may be an atom, molecule, ion, ion pair, radical, electron, photon, etc., or any specified group of such entities (whether or not such a group has any separate existence); it may also be an equation. Examples include a mole of NaCl(s), a mole of C–C bonds, or a mole of the reaction $N_2O_4 \rightleftharpoons 2NO_2$. A symbol or formula must be stated rather than a name to avoid ambiguity. The *SI unit of amount of substance is the mole.

amount-of-substance concentration *See* concentration.

amp (no stop) Abbrev. for ampere, in nonscientific use.

AMP Abbrev. and preferred form for adenosine 5′-phosphate (adenosine monophosphate).

ampere (no cap., no accent) Symbol: A Abbrev. (in nonscientific writing): amp (no stop). The *SI unit of electric current. It is one of the seven base SI units, defined as that constant electric current that, if maintained in two straight parallel conductors of infinite length, of negligible circular cross section, and placed one metre apart in vacuum, would produce between these conductors a force equal to 2×10^{-7} newton per metre of length. Named after A. M. *Ampère.

Ampère, André Marie (1775–1836) French physicist and mathematician.
***ampere** (no accent)
Ampère balance
Ampère–Laplace law (en dash)
Ampère's law
Ampère's rule
Ampère's theorem

ampere-hour A unit of electric charge, equal to 3600 coulombs.

ampere-turn The product *NI*, where *N* is the total number of turns in a coil carrying a current of *I* amperes. The ampere-turn has been used as a unit of *magnetomotive force, although in *SI units this is usually expressed in amperes.

amphi- Prefix denoting both (e.g. amphimixis, amphistylic).

amphi- (ital., always hyphenated) Prefix sometimes used to denote 2,6-substitution on the naphthalene ring (e.g. *amphi*-dimethylnaphthalene). *Compare ana-*; *epi-*; *kata-*; *peri-*; *pros-*.

Amphibia (cap. A) A class of vertebrates containing the frogs, toads, newts, and salamanders. Individual name and adjectival form: **amphibian** (*compare* amphibious).

amphibious Living or designed to operate both on land and in water. The adjective 'amphibian' is restricted to the characteristics of the *Amphibia.

amphibole A mineral. Not to be confused with **amphibolite**, a rock rich in amphiboles.

amphioxus (no cap.; pl. amphioxi) A small burrowing marine animal of the genus *Branchiostoma* (originally named *Amphioxus*), subphylum Cephalochordata. Also called lancelet.

Amphipoda (cap. A) An order of crustaceans including the sandhoppers. Individual name: **amphipod** (no cap.). Adjectival form: **amphipod** or **amphipodan**.

amphoteric Denoting a compound that can display both acidic and basic properties. An amphoteric compound is sometimes called an **ampholyte**.

amplitude modulation Abbrev.: AM (or a.m.).

amu (or **a.m.u.**) Abbrev. for atomic mass unit.

amygdale (preferred to amygdule) *Geol.* A small oval mass of pale mineral within a volcanic rock. *See also* amygdaloid.

amygdalin A glucoside obtained from almonds. Not to be confused with the adjective **amygdaline** (relating to a tonsil).

amygdaloid (noun) A rock containing amygdales. Adjectival form: **amygdaloidal**.

amyl- *Prefix denoting the group C_5H_7– (e.g. amylbenzoic acid, amyl chloride). Pentyl- is recommended in all contexts.

amylase Any enzyme that catalyses the hydrolysis of starch or glycogen. The two forms are:
α-amylase, occurring in saliva (when it is known as ptyalin) and pancreatic juice;
β-amylase, occurring in germinating seeds, including barley; *see* diastase.

amylo- (**amyl-** before vowels) Prefix denoting starch (e.g. amylopectin, amylase).

an- *See* a-; ana-.

ana- (**an-** before vowels) Prefix denoting **1.** up (e.g. anabatic, anadromous, anaphoresis, anion). **2.** again, increased, renewed (e.g. anabolism, anaphase). **3.** inverted (e.g. anaplasia). *Compare* cata-.

ana- (ital., always hyphenated) Prefix sometimes used to denote 1,5-substitution on the naphthalene ring (e.g. *ana*-dinitronaphthalene). *Compare amphi-*; *epi-*; *kata-*; *peri-*; *pros-*.

anaemia US: **anemia**. Adjectival form: **anaemic** (US: **anemic**).

anaerobe US same. Adjectival form: **anaerobic**.

anaeroplasma *See* mollicute.

analog US spelling of *analogue, also preferred in computing and electronics.

analogous (not analagous) Noun form: **analogy**.

analog-to-digital (hyphenated) Abbrev.: A/D, A-D, a/d, or a-d.
analog-to-digital converter Abbrev.: ADC or A/D (or A-D) converter.

analogue US: **analog**. Although the US spelling is often used in scientific literature in Britain the British spelling is preferred except in computing and electronics, where analog is preferred.

analyse US: **analyze.** *See* -yse.

analysis (pl. analyses) Adjectival form: **analytical** or **analytic**. In general and chemical contexts 'analytical' is preferred (e.g. analytical philosophy, analytical chemistry) but in mathematics 'analytic' is often used (e.g. analytic function, analytic geometry). Derived nouns: **analyser** (US: **analyzer**), **analyst**.

analytical reagent *Chem.* Abbrev.: AR (no stops).

anatomy Adjectival form: **anatomical.** Derived noun: **anatomist.**
Names of anatomical parts are Latinized but not printed in italic type (e.g. vas deferens, not *vas deferens*). Plurals are also Latinized, and in two-word terms both words are pluralized (e.g. bronchus, pl. bronchi; cerebrum, pl. cerebra; foramen, pl. foramina; vena cava, pl. venae cavae).

Ancalomicrobium (cap. A, ital.) A genus of prosthecate bacteria. Individual name: **ancalomicrobium** (no cap., not ital.; pl. ancalomicrobia).

And (no stop) *Astron.* Abbrev. for Andromeda.

AND *See* logic symbols; electronics, graphical symbols.

Anderson, Carl David (1905–) US physicist.

Anderson, Philip Warren (1923–) US physicist.
Anderson model

andesite An igneous rock in which one of the main components is the feldspar mineral **andesine**. Named after the Andes.

Andrade, Edward Neville da Costa (1887–1971) British physicist.

Andreaeales *See* Andreaeidae.

Andreaeidae A subclass of mosses comprising the granite mosses. In some classifications it is reduced to an order, **Andreaeales**; in others it is elevated to a class, **Andreaeopsida**, within the superclass Bryatinae (*see* Bryopsida). Also called Andreaeobrya.

Andreaeobrya *See* Andreaeidae.

Andreaeopsida *See* Andreaeidae.

Andrews, Thomas (1813–85) Irish physical chemist.

andro- (**andr-** before vowels) Prefix denoting **1.** the male sex (e.g. androdioecious, androsterone). **2.** mankind (e.g. androphilous). **3.** a stamen or anther (e.g. androecium).

androecium (pl. androecia) US same. *Bot.*

Andromeda A constellation. Genitive form: Andromedae. Abbrev.: And (no stop). *See also* stellar nomenclature.

-androus Adjectival suffix denoting stamens (e.g. protandrous). Noun form: **-andry**.

-ane Noun suffix denoting **1.** a saturated hydrocarbon (e.g. ethane). *See* alkane. **2.** a saturated heterocyclic compound (e.g. furane).

anemia US spelling of *anaemia.

anemo- Prefix denoting the wind or air currents (e.g. anemometer, anemophily, anemotaxis).

anemone (not anenome) Generic name: *Anemone* (cap. A, ital.).

anestrus (or **anestrum**) US forms of *anoestrus.

aneurine (or **aneurin**) Obsolete name for *thiamin (vitamin B_1).

aneurysm (not aneurism) Adjectival form: **aneurysmal**.

Anfinsen, Christian Boehmer (1916–) US biochemist.

angio- (**angi-** before vowels) Prefix denoting **1.** blood or lymph vessels (e.g. angiogram, angiography, angioplasty, angiitis). **2.** a seed vessel (e.g. angiocarpous).

Angiospermae (cap. A) A class or subdivision comprising the flowering plants. Traditionally it is split into the subclasses *Monocotyledonae and *Dicotyledonae. In other classifications flowering plants are regarded as a division, Magnoliophyta; a subdivision, Magnoliophytina; or – more recently – as the class Magnoliopsida. Individual name and adjectival form: **angiosperm** (no cap.; not angiospermous).

angle Symbol: α, β, γ, ϕ, θ, etc. A dimensionless quantity indicating the inclination of two lines or planes. The *SI units are the radian and the degree. Also called plane angle. *See also* solid angle.

angle brackets *See* brackets.

Anglo-Australian Observatory Abbrev.: AAO (no stops). An observatory in Siding Spring, NSW, housing the Anglo-Australian Telescope (abbrev.: AAT).

angstrom (no cap., no accents) Symbol: Å A unit equal to 10^{-10} metre, i.e. 0.1 nanometre. Formerly used for wavelengths, intermolecular distances, etc., it has been largely replaced by the nanometre. Named after A. J. *Ångström. *See also* SI units.

Ångström, Anders Jonas (1814–74) Swedish physicist and astronomer.
***angstrom** (no accents)
Ångström pyrheliometer

angular acceleration Symbol: α (Greek alpha). A scalar physical quantity, equal to the rate of change of *angular velocity, $d\omega/dt$. It is a *pseudovector. The *SI unit is the radian per second per second (rad s^{-2}).

angular frequency Symbol: ω (Greek omega). A physical quantity equal to $2\pi f$, where f is the *frequency of a periodic phenomenon. The *SI unit is the radian per second (rad s^{-1}) or the reciprocal of the second (s^{-1}). Also called pulsatance; circular frequency.

angular impulse Symbol: $\boldsymbol{H}$ A *pseudovector quantity, equal to the time integral $\int \boldsymbol{M}\,dt$, where $\boldsymbol{M}$ is the *moment of force acting over time t. The *SI unit is the newton metre second (N m s).

angular momentum Symbol: $\boldsymbol{L}$ or $\boldsymbol{J}$ A *vector quantity, equal to the vector product of *position vector, $\boldsymbol{r}$, and *momentum, $\boldsymbol{p}$. For a rigid body rotating about a fixed axis OZ it is a *pseudovector quantity, L, given by:

$$I\omega = \omega\,\Sigma mr^2,$$

where ω is the *angular velocity of the body and I its *moment of inertia about OZ. The *SI unit is the kilogram metre squared per second (kg m^2 s^{-1}). Also called moment of momentum.

angular velocity Symbol: ω (Greek omega). A physical quantity, equal to the rate of rotation about an axis, i.e. the rate of change of angular displacement. It is a *pseudovector. The *SI unit is the radian per second. *See also* angular acceleration.

angular wavenumber Another name for circular wavenumber. *See* wavenumber.

anhyd. Abbrev. for anhydrous.

anhydro- (**anhyd-** before vowels) Prefix denoting abstraction of water (e.g. anhydroxyprogesterone).

anhydrous Abbrev.: anhyd. (stop). Containing no water. Used in chemical compounds to indicate lack of water of crystallization, e.g. anhydrous copper sulphate.

aniline $C_6H_5NH_2$ The traditional name for *phenylamine.

Animalia (cap. A) The kingdom comprising all the animals.

animal taxonomy *See* taxonomy.

aniso- (**anis-** before vowels) Prefix denoting dissimilarity or inequality (e.g. anisogamy).

Annelida (cap. A) A phylum comprising the segmented worms. Individual name and adjectival form: **annelid** (no cap.).

annulus (pl. annuli; preferred to annuluses for scientific and mathematical senses) Adjectival form: **annular**.

anoestrus US: **anestrus** or **anestrum**. Adjectival form: **anoestrous** (US: **anestrous**).

anomaly (not anomoly) **1.** Deviation from the normal or expected. Adjectival form: **anomalous** (e.g. an anomalous measurement). **2.** *Astron.* Any of three angles used to calculate the position of a body moving in an elliptical orbit. Adjectival form: **anomalistic** (e.g. anomalistic year).

Anoplura (cap. A) An order (or suborder: *see* Phthiraptera) of insects comprising the sucking lice. Former name: Siphunculata. Individual name and adjectival form: **anopluran** (no cap.).

anorexia Loss of appetite: the word is often used without qualification for the psychological illness **anorexia nervosa**. Derived noun and adjectival form: **anorectic** (preferred to anorexic).

anorthite A feldspar mineral of the plagioclase group. *Compare* anorthosite.

anorthosite A plutonic igneous rock composed almost entirely of plagioclase feldspar. *Compare* anorthite.

ANS Abbrev. for autonomic nervous system.

ANSI Acronym for American National Standards Institute.

Ant (no stop) *Astron.* Abbrev. for Antlia.

ant- *See* anti-.

ante- Prefix denoting preceding or in front of (e.g. antedorsal, antenatal).

antenna 1. (pl. antennae) A paired appendage on the head of an arthropod. **2.** (preferred to aerial; pl. antennas) Equipment for transmitting and receiving radio waves.

anthelmintic (not anthelminthic)

antherozoid (preferred to spermatozoid) *Bot.*

antho- (**anth-** before vowels) Prefix denoting flowers (e.g. anthochlor, anthocyanin, anthesis).

Anthoceratae *See* Anthocerotopsida.

Anthocerotopsida A class of bryophytes comprising the horned liverworts (or hornworts), contained within a single order, Anthocerotales. Also called Anthocerotae; this should not be confused with Anthoceratae (-atae, not -otae), an infradivision to which this class is elevated in a recent classification (*see* Bryophyta).

Anthozoa (cap. A) A class of coelenterates (phylum Cnidaria) comprising the corals and sea anemones. Former name: Actinozoa. Individual name and adjectival form: **anthozoan** (no cap.).

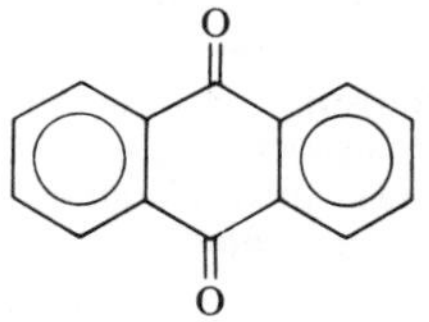

anthracene-9,10-dione
(9,10-anthraquinone)

anthracene-9,10-dione The recommended name for the compound traditionally known as 9,10-anthraquinone.

anthranilic acid $H_2NC_6H_4COOH$ The traditional name for 2-aminobenzoic acid. *See* aminobenzoic acid.

9,10-anthraquinone The traditional name for *anthracene-9,10-dione.

anthropo- Prefix denoting human beings (e.g. anthropocentric, anthropogenic).

anti- (**ant-** before a and h) Prefix denoting opposed to, opposite to, counteracting (e.g.

antibiotic, anticyclone, antiemetic, antimatter, antimetabolite, antitropism, antacid, antagonist).

anti- (ital., always hyphenated) *Chem. See trans-*.

anti-A (hyphenated) The antibody to the A antigen in the *ABO blood-group system.

anti-B (hyphenated) The antibody to the B antigen in the *ABO blood-group system.

antibody-dependent cellular cytotoxicity Abbrev.: ADCC (no stops).

anticodon The sequence of three bases carried by a tRNA molecule that is complementary to a specific *codon carried by the mRNA. The base sequence of an anticodon is conventionally written in the 5′ to 3′ direction, like the codon, although the sequences must run in opposite orientations during the matching process. Hence, for example, the codon AUC has its anticodon written as GAU; some authors place an arrow above the anticodon (e.g. $\overleftarrow{\text{GAU}}$) to indicate the actual orientation.

anti-D (hyphenated) The antibody to the most important rhesus antigen (called the D antigen). *See* rhesus factor.

antidiuretic hormone Abbrev.: ADH (no stops). Also called *vasopressin.

anti-inflammatory (hyphenated) In the pharmacological sense, used both as an adjective and as a noun (i.e. meaning an anti-inflammatory agent).

antilog (no stop) Abbrev. for antilogarithm.

antilymphocyte serum Abbrev.: ALS (no stops).

antimonic Denoting compounds in which antimony has an oxidation state of +5. The recommended system is to use oxidation numbers, e.g. antimonic oxide, Sb_2O_5, has the systematic name antimony(V) oxide.

antimonous Denoting compounds in which antimony has an oxidation state of +3. The recommended system is to use oxidation numbers, e.g. antimonous oxide, Sb_2O_3, has the systematic name antimony(III) oxide.

antimony Symbol: Sb *See also* periodic table; nuclide.

antimony(III) chloride oxide SbOCl The recommended name for the compound traditionally known as antimonyl chloride.

antimony(III) hydride SbH_3 The recommended name for the compound traditionally known as stibine.

antimony(III) oxide Sb_2O_3 The recommended name for the compound traditionally known as antimonous oxide or antimony trioxide.

antimony(V) oxide Sb_2O_5 The recommended name for the compound traditionally known as antimonic oxide or antimony pentoxide.

antimony(III) potassium 2,3-dihydroxybutanedioate oxide $KSbO(C_4H_4O_6)\cdot\frac{1}{2}H_2O$ The recommended name for the compound traditionally known as antimonyl potassium tartrate.

antimonyl Denoting compounds containing the ion SbO^+ or the group SbO. The recommended system is to use oxidation numbers, e.g. antimonyl chloride, SbOCl, has the systematic name antimony(III) chloride oxide.

antimonyl potassium tartrate $KSbO(C_4H_4O_6)$ The traditional name for antimony(III) potassium 2,3-dihydroxybutanedioate oxide.

antimony pentoxide Sb_2O_5 The traditional name for antimony(V) oxide.

antimony trioxide Sb_2O_3 The traditional name for antimony(III) oxide.

antiparticle (not hyphenated) A particle with the same mass as a given particle and with charge, isospin quantum number, and strangeness of identical magnitude and opposite sign. An antiparticle is indicated in nuclear reactions, etc., by means of a bar (or sometimes a tilde, ˜) above the symbol, as in $\bar{p}$ (antiproton), $\bar{n}$ (antineutron); for the positron, both $\bar{e}$ and e^+ are commonly used.

antirrhinum (not antirhinum; pl. antirrhinums) Generic name: *Antirrhinum* (cap. A, ital.).

Antlia A constellation. Genitive form: Antliae. Abbrev.: Ant (no stop). *See also* stellar nomenclature.

Antoniadi, Eugène Michael (1870–1944) Greek-born French astronomer.

Anura (cap. A) An order of amphibians comprising the frogs and toads. Individual name and adjectival form: **anuran** (no cap.). Not to be confused with 'anuria', a medical term meaning absence of urine production.

AO Abbrev. for atomic orbital.

aorta (pl. aortae; US: aortas) Adjectival form: **aortic**.

Ap *Astron. See* spectral types.

ap- *See* apo-.

6-APA Abbrev. for 6-aminopenicillanic acid.

APDC Abbrev. for the ammonium salt of 1-pyrrolidinecarbodithioic acid.

apex (pl. apexes in astronomy, mathematics, and general senses; apices in anatomy, zoology, and botany) Adjectival form: **apical**.

Aphaniptera Use *Siphonaptera.

aphelion (pl. aphelia) *Astron.*

aphid Any plant-eating insect of the family Aphididae. The plural, aphids, should not be confused with aphides, the plural of *aphis.

aphis (pl. aphides) Any *aphid of the genus *Aphis* (cap. A, ital.).

Aphyllophorales (cap. A) An order of basidiomycete fungi (*see* Basidiomycotina) containing the pore fungi. Also called Polyporales. Individual name and adjectival form: **aphyllophoralean** (no cap.).

Apiaceae *See* Umbelliferae.

apical *See* apex.

apo- (**ap-** before h) Prefix denoting **1**. separation (e.g. apocarpous, apocrine, apogee, apoinducer, aphelion). **2**. absence, lack (e.g. apochromatic, apoenzyme, apogamy, apomixis). **3**. derivation or relationship (e.g. apophysis).

Apoda (cap. A) An order of limbless tropical amphibians. Not to be confused with 'Apodes' (an obsolete order comprising the eels). Individual name and adjectival form: **apodan** (no cap.). *Compare* apodal.

apodal (or **apodous**) Describing any animal that lacks limbs or feet. Not to be confused with apodan (*see* Apoda) or 'apodid' (a bird of the swift family, Apodidae).

Apollonius of Perga (*c.* 262 BC – *c.* 190 BC) Greek mathematician.
Apollonius' theorem

a posteriori (two words; not ital.) *Logic* from effect to cause. [from Latin: from the latter] *Compare* a priori.

apothecaries' ounce *See* ounce.

apparatus (pl. apparatuses, not apparati; use an alternative where possible, e.g. 'appliances' or 'pieces of apparatus')

apparent magnitude *See* magnitude.

apparent power *See* power.

appendicular Relating to an appendage or appendages: used especially to denote the skeleton of the limbs (appendicular skeleton). Less commonly, the word is used as the adjectival form of *appendix.

appendix (pl. appendices for all senses; US: appendixes) In zoology and anatomy, it is often used without qualification to denote the vermiform appendix. *See also* appendicular.

Appert, Nicolas François (*c.* 1750–1841) French inventor.

Appleton, Sir Edward Victor (1892–1965) British physicist.
Appleton layer Also called F layer.

applications programmer (for applications in general, hence not 'application') Similarly **applications software**, **applications science**. 'Application' is often used for a specific task, as in **application package**.

a priori (two words; not ital.) *Logic* From cause to effect. [from Latin: from the previous] Noun form: **apriorism** (one word). *Compare* a posteriori.

Aps (no stop) *Astron.* Abbrev. for Apus.

apsis (pl. apsides) Also called apse (pl. apses). *Astron.*

APT Abbrev. for automatic picture transmission (from satellites).

Apus A constellation. Genitive form: Apodis. Abbrev.: Aps (no stop). *See also* stellar nomenclature.

aq (no stop; in parentheses, immediately following a chemical name, as in copper sulphate(aq)) Symbol for dissolved in water or, in quantitative work, at infinite dilution.

Aql (no stop) *Astron.* Abbrev. for Aquila.

Aqr (no stop) *Astron.* Abbrev. for Aquarius.

aqua- (or **aqui-**) Prefix denoting water (e.g. aqualung, aquiculture).

Aquarius A constellation. Genitive form: Aquarii. Abbrev.: Aqr (no stop). *See also* stellar nomenclature.
Delta Aquarids (meteor shower)
Eta Aquarids (meteor shower)

Aquaspirillum (cap. A, ital.) A genus of aerobic helical Gram-negative bacteria. Individual name: **aquaspirillum** (no cap., not ital.; pl. aquaspirilla).

aqueous Abbrev.: aq. (stop). *Chem.*

aqueous ammonia NH_4OH The recommended name for the solution traditionally known as ammonium hydroxide.

Aquila A constellation. Genitive form: Aquilae. Abbrev.: Aql (no stop). *See also* stellar nomenclature.

Ar 1. Symbol for argon. **2.** Symbol often used to denote an aryl group in chemical formulae, e.g. ArOH.

AR Abbrev. for **1.** analytical reagent. **2.** adrenergic receptor (or adrenoceptor). *See also* alpha receptor; beta receptor.

ar- (ital., always hyphenated) Prefix denoting aromatic. Often used in nomenclature for polycyclic compounds (e.g. *ar*-tetrahydro-1-naphthylamine).

Ara 1. A constellation. Genitive form: Arae. Abbrev.: Ara (no stop). *See also* stellar nomenclature. **2.** Abbrev. for arabinose.

arabinose Abbrev.: Ara (no stop). *See* sugars.

Araceae A family of herbaceous monocotyledons including the *arums. Individual name and adjectival form: **aroid** (preferred to araceous). *Compare* Arecaceae (*see* Palmae).

Arachnia (cap. A, ital.) A genus of Gram-positive colonial bacteria. Individual name: **arachnia** (no cap., not ital.; pl. arachniae).

Arachnida (cap. A) A class of arthropods including the spiders, scorpions, mites, and ticks. Individual name and adjectival form: **arachnid** (no cap.; not arachnidan); *compare* arachnoid.

arachno- (**arachn-** before vowels) Prefix denoting **1.** spiders (e.g. arachnophobia). **2.** the arachnoid membrane (e.g. arachnitis).

arachnoid (noun) One of the membranes (meninges) surrounding the vertebrate brain and spinal cord. Not to be confused with arachnid (*see* Arachnida), although arachnoid is sometimes used as an adjective or noun to describe or denote any invertebrate resembling the arachnids.

Arago, Dominique François Jean (1786–1853) French physicist.
Fresnel–Arago laws (en dash)

Araldite (cap. A) A trade name for an epoxy-resin adhesive.

Arber, Werner (1929–   ) Swiss microbiologist.

arbovirus Obsolete name for viruses now classified in the genera **Alphavirus* and *Flavivirus*. [from *ar*thropod-*bo*rne *virus*]

Arcanobacterium (cap. A, ital.) A genus of bacteria containing the single species *A. haemolyticum*, formerly classified as **Corynebacterium haemolyticum*.

arccos (no space) Symbol for inverse cosine. *See* cos.

arccosec (no space) Symbol for inverse cosecant. *See* cosec.

arccot (no space) Symbol for inverse cotangent. *See* cot.

arch Short for arcosh. *See* cosh.

Archaean (not Archean) **1.** (adjective) Denoting the earliest eon of geological time, i.e. before the *Proterozoic eon. **2.** (noun; preceded by 'the') The Archaean eon.

archaebacterium (pl. archaebacteria) Any of the so-called 'primitive' bacteria. A taxon, Archaeobacteria (cap. A, not ital., not Archaebacteria), of various ranks (e.g. kingdom, class) according to author, has been proposed to include several groups of primitive bacteria – the methanogens, halobacteria, and thermoacidophiles. However, the term 'archaebacteria' is used to refer collectively to the primitive bacteria without implying acceptance of the proposed taxon. *Compare* eubacterium.

archaeo- (**archae-** before vowels) US: **archeo-**. Prefix denoting ancient, primitive (e.g. archaeocyte, archaeophyte, archaeozoology).

Archaeopteryx (cap. A, ital.; not *Archeopteryx*) A genus of fossil birds.

arche- (**arch-** before vowels, **archi-**, **archo-**) Prefix denoting first, principal, primitive, ancestral (e.g. archegonium, archesporium, archenteron, archipallium, archipelago, archosaur).

Archimedes (287 BC–212 BC) Greek mathematician.
Archimedean screw
Archimedean solid
Archimedes' principle

archipelago (pl. archipelagos; preferred to archipelagoes) Adjectival forms: **archipelagic, archipelagian.**

Archosauria (cap. A) A subclass of reptiles including the dinosaurs, pterosaurs, and crocodiles, regarded by some authorities as a class comprising these reptiles and the birds. Individual name and adjectival form: **archosaur** (no cap.).

arc minute Symbol: Abbrev.: arcmin (no stop). A unit of angular measure equal to 1/60 of a degree, approx. 0.2909 milliradian. Also called minute of arc, minute.

arcosech Symbol for inverse hyperbolic secant. *See* cosech.

arcosh Symbol for inverse hyperbolic cosine. *See* cosh.

arcoth Symbol for inverse hyperbolic cotangent. *See* coth.

arcsec (no space) **1.** Symbol for inverse secant. *See* sec. **2.** Abbrev. for arc second.

arc second Symbol: ″ Abbrev.: arcsec (no stop). A unit of angular measure equal to 1/60 of an arc minute, i.e. 4.848 microradian. Also called second of arc, second.

arcsin (no space) Symbol for inverse sine. *See* sin.

arctan (no space) Symbol for inverse tangent. *See* tan.

Arctogaea US: **Arctogea**. A major zoogeographical region comprising the northern continents. Adjectival form: **Arctogaean** (US: **Arctogean**).

are Symbol: a A unit of area equal to 100 square metres or one square decametre. One are = 119.60 square yards, 0.0247 acre. *See also* hectare.

area Symbol: A or S A scalar physical quantity indicating extent in two dimensions. The *SI unit is the square metre or sometimes the hectare or are.

Arecaceae *See* Palmae. *Compare* Araceae.

Arecibo Radio Observatory An observatory in Puerto Rico.

arenavirus (one word) Any virus of the family Arenaviridae, which contains the single genus *Arenavirus* (cap. A, ital.).

arene Any of a class of hydrocarbons with aromatic properties, such as benzene, naphthalene, or anthracene.
Arenes are systematically named by choosing either a side chain or the ring system as the root and considering the remaining groups as substituents, e.g. methylbenzene, $C_6H_5CH_3$, and 1,2-diphenylethane, $C_6H_5CH_2CH_2C_6H_5$.
In nonsystematic nomenclature arenes are named as in systematic nomenclature but the trivial names of the roots and substituent groups are used, e.g. phenylacetylene, $C_6H_5C{\equiv}CH$, and *t*-butylbenzene, C_6H_5-$C(CH_3)_3$. Many benzene derivatives have acceptable trivial names, e.g. toluene, $C_6H_5CH_3$, phenol, C_6H_5OH, styrene, $C_6H_5CH=CH_2$. The traditional name for an arene is aromatic hydrocarbon.

arête (preferred to arete) A mountain ridge. [from French]

Arg (or **arg**; no stop) Abbrev. for arginine. *See* amino acid.

Argand, Jean Robert (1768–1822) French mathematician.
Argand diagram

Argelander, Friedrich Wilhelm August (1799–1875) German astronomer.

argentic Denoting compounds in which silver has an oxidation state of +2. The recommended system is to use oxidation numbers, e.g. argentic oxide, AgO, has the systematic name silver(II) oxide.

argentous Denoting compounds in which silver has an oxidation state of +1. The recommended system is to use oxidation numbers, e.g. argentous oxide, Ag_2O, has the systematic name silver(I) oxide.

arginine Abbrev.: Arg or arg (no stop). Symbol: R *See* amino acid.

argon Symbol Ar *See also* periodic table; nuclide.

Ari (no stop) *Astron.* Abbrev. for Aries.

Aries A constellation. Genitive form: Arietis. Abbrev.: Ari (no stop). *See also* stellar nomenclature.

Aristarchus of Samos (*c.* 320 BC–*c.* 250 BC) Greek astronomer.

Aristotle (384 BC–322 BC) Greek philosopher, logician, and scientist.
Aristotelianism
Aristotelian logic
Aristotelian mechanics

arithmetic (or **arithmetical**) Adjectives used interchangeably.

arithmetic logic unit (preferred to arithmetic and logic unit) Abbrev.: ALU (no stops).

arithmetic operations Addition and subtraction of two quantities are indicated respectively by the operators + and – (usually an en dash).

Multiplication of quantities may be indicated by a cross (×), a centred dot (·), parentheses, or by using a solid form; examples are:

$$5.2 \times 7.1, 2\pi \cdot 10^{-3}, (2.13)(-3.6)$$
$$a \times b, a \cdot b, xyz, mc^2, pV(T_1 + T_2).$$

The centred dot should not be used in numbers containing a decimal point.

Multiplication of two units is indicated by a thin space or centred dot between the symbols, as in

m s   m · s.

The symbols must not be joined in a solid form (ms could then represent either millisecond or metre second; *see also* SI units).

Division of one quantity by another is normally indicated by one of the following ways:

$$a \div b \qquad a/b \qquad ab^{-1} \qquad \frac{a}{b}$$

Division of units is indicated in the same way. Use of the horizontal bar causes spacing problems in running text and for this reason is usually avoided in preference to the solidus. When one or both of the quantities or units are products, quotients, sums or differences, parentheses (with square brackets and braces if required) should be used, as in

$$\{[z - (x + y)^2]^2 - 4\}/(2 - x).$$

It is usual to set a thin space or space on either side of the operators +, –, ×, and ÷ or ·.

Arizona (cap. A, ital.) A genus created for certain bacteria of the family *Enterobacteriaceae. In some taxonomic systems it is subsumed in the genus **Salmonella* as a 'subgenus'.

Arkwright, Sir Richard (1732–92) British inventor and industrialist.

Armco (cap. A) A trade name for a soft-iron material.

Armstrong, Edwin Howard (1890–1954) US electrical engineer.

Armstrong, Henry Edward (1848–1937) British chemist and teacher.

Armstrong, William George, Baron (1810–1900) British engineer and industrialist.

aroid **1.** (noun) Any plant of the family *Araceae. **2.** (adjective; not araceous) Relating to or describing the Araceae.

aromatic hydrocarbon The traditional name for an *arene.

Arrhenius, Svante August (1859–1927) Swedish physical chemist.
Arrhenius's equation

arrow worm (two words) *See* Chaetognatha.

arsech Symbol for inverse hyperbolic secant. *See* sech.

arsenate Denoting a compound containing the ion AsO_4^{3-}, e.g. sodium arsenate, K_3AsO_4. The recommended name is arsenate(V).

arsenate(III) Denoting a compound containing the ion AsO_3^{3-}, e.g. potassium arsenate(III), K_3AsO_3. The traditional name is arsenite.

arsenate(V) Denoting a compound containing the ion AsO_4^{3-}, e.g. sodium arsenate(V), Na_3AsO_4. The traditional name is arsenate.

arsenic 1. (noun) Symbol: As *See also* periodic table; nuclide. **2.** (adjective) Denoting compounds in which arsenic has an oxidation state of +5. The recommended system is to use oxidation numbers, e.g. arsenic chloride, $AsCl_5$, has the systematic name arsenic(V) chloride.

arsenic chloride $AsCl_5$ The traditional name for arsenic(V) chloride.

arsenic(III) chloride $AsCl_3$ The recommended name for the compound traditionally known as arsenious chloride or arsenic trichloride.

arsenic(V) chloride $AsCl_5$ The recommended name for the compound traditionally known as arsenic chloride or arsenic pentachloride.

arsenic(III) hydride AsH_3 The recommended name for the compound traditionally known as arsine or arsenurretted hydrogen.

arsenic pentachloride $AsCl_5$ The traditional name for arsenic(V) chloride.

arsenic trichloride $AsCl_3$ The traditional name for arsenic(III) chloride.

arsenious Denoting compounds in which arsenic has an oxidation state of +3. The recommended system is to use oxidation numbers, e.g. arsenous chloride, $AsCl_3$, has the systematic name arsenic(III) chloride.

arsenious chloride $AsCl_3$ The traditional name for arsenic(III) chloride.

arsenite Denoting a compound containing the ion AsO_3^{3-}, e.g potassium arsenite, K_3AsO_3. The recommended name is arsenate(III).

arsenurretted hydrogen AsH_3 The traditional name for arsenic(III) hydride.

arsh Short for arsinh. *See* sinh.

arsine AsH_3 The traditional name for arsenic(III) hydride.

arsinh Symbol for inverse hyperbolic sine. *See* sinh.

artanh Symbol for inverse hyperbolic tangent. *See* tanh.

artefact US: **artifact**.

arterio- (**arter-** or **arteri-** before vowels) Prefix denoting an artery (e.g. arteriography, arteriovenous, arteritis, arteriectomy).

arteriosclerosis (one word) Loosely, any of several conditions characterized by hardening of the artery walls: the term is most commonly used as a synonym (not in medical use) for **atherosclerosis**.

artesian well (not Artesian) Named after Artois, former French province.

arth Short for artanh. *See* tanh.

arthro- (**arthr-** before vowels) Prefix denoting a joint (e.g. arthropod, arthrospore).

Arthrobacter (cap. A, ital.) A genus of obligately aerobic bacteria. It includes the species formerly classified as *Corynebacterium ilicis*. Individual name: **arthrobacter** (no cap., not ital.). *See also Aureobacterium*.

Arthropoda (cap. A) A phylum comprising invertebrates with hard exoskeletons and jointed appendages (e.g. insects, crustaceans, etc.). Individual name and adjectival form: **arthropod** (no cap.; not arthropodan).

artificial insemination Abbrev.: AI (no stops).

artificial insemination (by) **donor** Abbrev.: AID (no stops). Former name for *donor insemination.

artificial insemination (by) **husband** Abbrev.: AIH (no stops).

artificial intelligence Abbrev.: AI (no stops).

Artiodactyla (cap. A) An order of hoofed mammals comprising the even-toed ungulates (e.g. pigs, antelopes, cattle, sheep). Individual name and adjectival form: **artiodactyl** (no cap.; not artiodactylous).

arum (pl. arums) Any of various plants of the family Araceae; the term is sometimes used in the plural for all members, though strictly it should be reserved for those of the genus *Arum* (cap. A, ital.), which includes cuckoopint (*A. maculatum*). The so-called 'arum lily' belongs to a different genus of this family, *Zantedeschia*.

aryl- *Prefix denoting any aromatic hydrocarbon residue after removal of a hydrogen atom (e.g. arylhydrazone, aryl methyl ketone).

As Symbol for arsenic.

As (ital.) Abbrev. for altostratus.

as- (ital., always hyphenated) *See unsym-*.

ASA Abbrev. for American Standards Association. ASA numbers used in classifying film speed have now been replaced by an *ISO classification.

aschelminth (no cap.) Any invertebrate animal formerly regarded as a member of the phylum Aschelminthes (cap. A), now reclassified in one of five separate phyla (including the Rotifera and Nematoda). The term aschelminth is still used for descriptive (rather than taxonomic) purposes.

ascidian A member of the Ascidiacea, the class of urochordates comprising the sea squirts. Adjectival form: **ascidian**. *Compare* ascidium.

ascidium (pl. ascidia) A pitcher-shaped leaf or part of a leaf. *Compare* ascidian.

ASCII *Computing* Acronym for American standard code for information interchange.

Asn (or **asn**; no stop) Abbrev. for asparagine. *See* amino acid.

asco- Prefix denoting an ascus (e.g. ascogenous, ascogonium).

ascomycete Any fungus belonging to the class Ascomycetes (cap. A). Authors using alternative classifications (*see* Ascomycotina) still use 'ascomycete' as the individual name for these fungi. Adjectival form: **ascomycete** (preferred to ascomycetous).

Ascomycotina A subdivision of true fungi (Eumycota) constituting the sac fungi. It was formerly divided into the classes Hemiascomycetes, Plectomycetes, Pyrenomycetes, Discomycetes, Laboulbeniomycetes, and Loculoascomycetes, but is now usually divided directly into 40–45 orders. In other classifications these fungi are regarded as a class, Ascomycetes, divided into the subclasses Hemiascomycetidae and Euascomycetidae; or a division, Ascomycota, of the kingdom *Mycobiota. Whichever classification is used, the term 'ascomycete' is acceptable as an individual name for these fungi.

ascorbic acid Permitted synonym: vitamin C.

ascus (pl. asci) The structure producing sexual spores (**ascospores**) in ascomycete fungi. In many ascomycetes it is produced in a fruiting body (**ascocarp**).

asdic (or **ASDIC**) Acronym for Allied Submarine Detection Investigation Committee: it usually denotes the echo-sounding device the committee developed, now known as *sonar.

-ase Noun suffix denoting an enzyme (e.g. amylase, ATPase, DNAase, lactase, nuclease). *See* enzyme nomenclature.

ASIC *Electronics* Acronym for application-specific integrated circuit.

Asp (or **asp**; no stop) Abbrev. for aspartic acid. *See* amino acid.

asparagine Abbrev.: Asn or asn (no stop). Symbol: N *See* amino acid.

aspartic acid (or **aspartate**) Abbrev.: Asp or asp (no stop). Symbol: D Recommended name: aminobutanedioic acid. *See* amino acid.

assimilation A process of incorporation. *Compare* simulation.

astatine Symbol: At *See also* periodic table; nuclide.

Astbury, William Thomas (1889–1961) British crystallographer and molecular biologist.

Asteraceae *See* Compositae. Adjectival form: **asteraceous**.

Asteridae A subclass of dicotyledons containing herbaceous plants whose flowers have fused petals. It is equivalent to the former taxon, Sympetalae.

asteroid 1. (noun) *Astron.* Use minor planet in scientific literature. However, **asteroid belt** is in technical usage. **2.** (noun and adjective) *Zool. See* Asteroidea.

Asteroidea (cap. A) A class of echinoderms comprising the starfishes. Individual name and adjectival form: **asteroid** (no cap.; not asterid or asteroidal).

Aston, Francis William (1877–1945) British chemist and physicist.
Aston dark space

astro- Prefix denoting **1.** *Astron.* stars or other celestial bodies (e.g. astrocompass, astrophysics). **2.** *Biol.* a star-shaped structure (e.g. astrocyte).

Astronomer Royal (initial capital letters; not hyphenated; pl. Astronomers Royal)

astronomical unit Abbrev.: AU (no stops). A unit of length used in astronomy. It was originally defined as (and is very nearly equal to) the semimajor axis of the earth's orbit. Its present value, adopted in 1976, is

$$1.495\ 978\ 70 \times 10^{11}\ \text{m}.$$

Although not an *SI unit, it is officially recognized because of its specialized usage and may be used with the SI units. It has no internationally agreed symbol. *See also* parsec.

ASV Abbrev. for avian sarcoma virus, another name for Rous sarcoma virus.

asymmetry Adjectival form: **asymmetric** (preferred to asymmetrical).

asymptote Adjectival form: **asymptotic.** Symbol for 'asymptotically equal to': $\cong$

At Symbol for astatine.

-ate 1. Noun suffix characteristic of salts and esters (e.g. ethanoate, phosphate, propanoate, sulphate). **2.** Adjectival suffix denoting possession of (e.g. ciliate, nucleate, septate).

ATEE Abbrev. for *N*-acetyl-L-tyrosine ethyl ester.

atelo- (**atel-** before vowels (except o), **ateli-** before o) Prefix denoting imperfect or incomplete development (e.g. atelocardia, atelectasis, ateliosis).

atherosclerosis (one word) Adjectival form: **atherosclerotic**. *See also* arteriosclerosis.

atm Symbol for atmosphere. *See* standard atmosphere.

atmo- Prefix denoting air or vapour (e.g. atmometer, atmosphere).

atmosphere *See* standard atmosphere.

atomic absorption spectroscopy Abbrev.: AAS (no stops). *Chem.*

Atomic Energy Authority Abbrev.: AEA (no stops). A UK government authority.

atomic mass constant Symbol: m_u A fundamental constant equal to 1 unified *atomic mass unit (u).

atomic mass unit Abbrev.: amu (or a.m.u.) A unit used in chemistry to express the mass of an isotope of an element. In 1961 it was redefined and took the name **unified atomic mass unit**, symbol u: it is the fraction 1/12 of the mass of an atom of carbon–12 in the ground state:

$$1\ \text{u} = 1.660\ 5402 \times 10^{-27}\ \text{kg}.$$

Although not an *SI unit, it is officially recognized because of its specialized usage and may be used with the SI units. *See also* dalton.

atomic number Another name for proton number.

atomic orbital Abbrev.: AO (no stops). *See* orbital.

atomic weight Abbrev.: at. wt. (stops). Former name for relative atomic mass.

ATP Abbrev. and preferred form for adenosine 5′-triphosphate.

ATPase Abbrev. and preferred form for adenosine triphosphatase, a protein that acts as a pump to move ions actively against a concentration gradient across a biological membrane. ATPases are designated according to the ions they move, e.g. Na^+K^+ ATPase is the sodium pump.

atrio- (**atri-** before vowels) Prefix denoting an atrium (e.g. atriopore, *atrioventricular).

atrioventricular (not auriculoventricular) Abbrev.: AV (no stops).
atrioventricular bundle (preferred to bundle of *His) Abbrev.: AV bundle.
atrioventricular node Abbrev.: AV node.

atrium (pl. atria, not atriums) *Biol.* A chamber, especially an upper chamber of the vertebrate heart (*compare* auricle). Adjectival form: **atrial.** *See also* atrio-.

atto- Symbol: a. A prefix to a unit of measurement that indicates 10^{-18} of that unit. *See also* SI units.

at. wt. Abbrev. for atomic weight.

Au Symbol for gold. [from Latin *aurum*]

AU Abbrev. for astronomical unit. The abbreviation of the French name is UA.

aubrietia (not aubretia; pl. aubrietias) Generic name: *Aubrietia* (cap. A, ital.).

audible (not audable) Derived noun: **audibility**.

audio- Prefix denoting sound or hearing (e.g. audiofrequency, audiometer).

audiofrequency (one word) Abbrev.: a.f. (stops) or AF (no stops).

Audubon, John James (1785–1851) US ornithologist and naturalist.

Auer, Karl, Baron von Welsbach (1858–1929) Austrian chemist.

Auger, Pierre Victor (1899–) French physicist.

Auger effect
Auger electron
Auger shower

Aur (no stop) *Astron.* Abbrev. for Auriga.

aural Relating to the ear or hearing. *Compare* oral.

Aureobacterium (cap. A, ital.) A genus of rod-shaped obligately aerobic bacteria. This genus, created in 1983, contains species previously assigned to other genera, including *Arthrobacter* (*A. flavescens*, *A. terregens*), *Curtobacterium* (*C. testaceum*, *C. saperdae*), *Microbacterium* (*M. liquefaciens*), and *Corynebacterium* (*C. barteri*). Individual name: **aureobacterium** (no cap., not ital.; pl. aureobacteria).

auricle *Biol.* Any ear-shaped lobe or process, especially the sac in the wall of the *atrium of the vertebrate heart. The use of the word as a synonym for the atrium as a whole is deprecated. Adjectival forms: **auricular, auriculate**.

Auriga A constellation. Genitive form: Aurigae. Abbrev.: Aur (no stop). *See also* stellar nomenclature.

aurora (pl. auroras or aurorae) An **aurora borealis** is in northern skies, an **aurora australis** in southern skies.

Australopithecus (cap. A, ital.) A genus of fossil hominids. It contains three (possibly four) species, now thought to have coexisted with each other:

A. afarensis: includes the 'Lucy' skeleton;

A. africanus;

A. boisei (originally named *Zinjanthropus boisei* and included by some authorities in the species *H. robustus*);

A. robustus (originally named *Paranthropus robustus*).

Individual name and adjectival form: **australopithecine** (not ital., no cap.).

auto- (sometimes **aut-** before vowels) Prefix denoting **1.** self or the individual (e.g. autoimmunity, autolysis, autecology, autoxidation). **2.** automatic, self-regulating (e.g. autodyne).

autochthonous 1. Denoting rocks, deposits, etc., whose constituent parts originated *in situ*. *Compare* allochthonous. **2.** Denoting physiological processes (such as heartbeat) that originate within an organ, rather than being triggered by external stimuli.

autoecious US: **autecious**. Denoting a parasitic fungus that passes through different stages of its life cycle in or on the same host. Noun form: **autoecism**. *Compare* autoicous.

autoexposure *Photog.* Abbrev.: AE (caps., no stops).

autofocus *Photog.* Abbrev.: AF (caps., no stops).

autoicous Having male and female inflorescences on the same plant (i.e. monoecious). *Compare* autoecious.

automatic frequency control (not hyphenated) Abbrev.: a.f.c. (stops) or AFC (no stops).

automatic gain control (not hyphenated) Abbrev.: a.g.c. (stops) or AGC (no stops).

automatic picture transmission *Space* Abbrev.: APT (no stops).

automatic volume control (not hyphenated) Abbrev.: a.v.c. (stops) or AVC (no stops).

automaton (pl. automatons or (when used collectively) automata)

autonomic nervous system Abbrev.: ANS (no stops).

autotrophic Describing organisms (**autotrophs**) that synthesize their organic requirements from inorganic precursors. Noun form: **autotrophism**. *Compare* auxotrophic.

auxiliary (not auxillary) Supplementary. *Compare* axillary.

auxo- (**aux-** before vowels) Prefix denoting increase, growth (e.g. auxotrophic, auxesis, auxin).

auxotrophic Describing mutant strains of microorganisms (**auxotrophs**) that are unable to synthesize a particular nutrient. *Compare* autotrophic.

AV Abbrev. for **1.** *atrioventricular. **2.** *Photog.* aperture value.

a.v.c. (stops, or **AVC**) Abbrev. for automatic volume control.

Averroës (1126–98) Spanish-Muslim physician, astronomer, and philosopher. Also called ibn-Rushd.

Avery, Oswald Theodore (1877–1955) US bacteriologist.

Aves (cap. A) A class of vertebrates comprising the birds. Adjectival form: **avian**.

avian sarcoma virus Abbrev.: ASV (no stops). Another name for Rous sarcoma virus.

Avicenna (980–1037) Persian physician and philosopher. Also called ibn-Sina.

Avipoxvirus (cap. A, ital.) Approved name for a genus of *poxviruses. Vernacular name: fowlpox subgroup. Individual name: **avipoxvirus** (no cap., not ital.).

Avogadro, Lorenzo Romano Amedeo Carlo, Count of Quaregna and Cerreto (1776–1856) Italian physicist and chemist.
***Avogadro constant** (not Avogadro number)
Avogadro's hypothesis

Avogadro constant Symbol: L, N_A A fundamental constant equal to

$$6.022\,1367 \times 10^{23}\ \text{mol}^{-1}.$$

This is a physical quantity, not a pure number, and the term 'Avogadro number' should not be used.

avoir Short for avoirdupois. *See* pound.

avoirdupois units *See* pound; ounce.

axial Of or relating to an axis (e.g. axial skeleton).

axil The angle between the upper surface of a leaf or branch and the stem that bears it. Adjectival form: **axillary**.

axile (not axial) Denoting a type of placentation in plants.

axilla (pl. axillae) The hollow area at the junction of the arm or the wing of a bird with the body. *See also* axillary.

axillary 1. *Bot.* Relating to or growing in an *axil (e.g. axillary buds). **2.** *Anat.* Relating to the armpit (axilla) (e.g. axillary lymph nodes). *Compare* auxiliary.
axillaries US: **axillars**. The feathers growing in a bird's axilla.

axis (pl. axes) Adjectival form: **axial**.
***x*-axis, *y*-axis, *z*-axis** (hyphenated)

aza- (**az-** before vowels) Prefix denoting a heterocyclic compound in which the hetero- atom is nitrogen (e.g. azabicycloheptane, azepine, aziridene).

azalea Any deciduous *rhododendron, formerly included in the genus *Azalea* (cap. A, ital.).

3′-azido-3′-deoxythymidine Abbrev. and preferred form: *AZT (no stops).

azimuth Adjectival form: **azimuthal**.

azine Any of a class of organic compounds containing the group C=N–N=C. Azines are systematically named by adding the word azine after the name of the corresponding aldehyde or ketone, e.g. propanone azine, $(CH_3)_2C{=}NN{=}C(CH_3)_2$. In nonsystematic nomenclature azines are named as in systematic nomenclature but the trivial names of the corresponding aldehydes or ketones are used, e.g. acetone azine, $(CH_3)_2C{=}NN{=}C(CH_3)_2$.

azo- Prefix denoting the group –N=N–. *See* azo compound.

azobenzene $C_6H_5N{=}NC_6H_5$ The traditional name for *phenylazobenzene.

azo compound Any of a class of organic compounds containing the group $-N{=}N-$ joined directly to two carbon atoms, or to one carbon atom and one hydrogen atom.
Azo compounds are systematically named by prefixing the name of the root hydrocarbon with azo-, preceded by the name of the substituent group, e.g. methylazobenzene, $CH_3N{=}NC_6H_5$, and phenylazobenzene, $C_6H_5N{=}NC_6H_5$.
In nonsystematic nomenclature azo compounds are named as in systematic nomenclature but the trivial names of the substituent and root are used; for members with both the root and substituent the same, only the root is named, e.g. azomethane, $CH_3N{=}NCH_3$, and azobenzene, $C_6H_5N{=}NC_6H_5$.

Azomonas (cap. A, ital.) A genus of bacteria of the family Azotobacteraceae (*see Azotobacter*). Individual name: **azomonad** (no cap., not ital.).

Azospirillum (cap. A, ital.) A genus of bacteria (*see* spirillum). Individual name: **azospirillum** (no cap., not ital.; pl. azospirilla).

Azotobacter (cap. A, ital.) A genus of bacteria belonging to the nitrogen-fixing family Azotobacteraceae. Use of the trivial name **azotobacter** (no cap., not ital.) may cause confusion between the genus and the family and should be avoided.

AZT Abbrev. and preferred form for 3′-azido-3′-deoxythymidine. The recommended name for this drug in the UK is zidovudine.

B

b Symbol (light ital.) for **1.** breadth. **2.** molality (b_B = molality of substance B).

B **1.** A blood group and its associated antigen (*see* ABO). **2.** Symbol for **i.** aspartic acid or asparagine (unspecified). **ii.** base-catalysed (*see* reaction mechanisms). **iii.** bel. **iv.** boron. **v.** guanosine, thymidine (or uridine), or cytidine (unspecified). **vi.** *Astron. See* spectral types.

B Symbol for **1.** (bold ital.) magnetic flux density. **2.** (light ital.) susceptance.
$\boldsymbol{B}_i$ (bold ital. B) Symbol for magnetic polarization.

***B*-** (ital., always hyphenated) Prefix denoting substitution on a boron atom in an organic compound (e.g. *B*,*B*,*B*-trimethylborazine).

Ba Symbol for barium.

BA Abbrev. for **1.** British Association (for the Advancement of Science). **2.** British Association screw thread.

Baade, Wilhelm Heinrich Walter (1893–1960) German-born US astronomer.

Babbage, Charles (1792–1871) British mathematician and inventor.

Babbitt, Isaac (1799–1862) US inventor.
babbitt metal (no cap.)

Babcock, Harold Delos (1882–1968) US astronomer.

Babcock, Horace Welcome (1912–) US astronomer.

Babinet, Jacques (1794–1872) French physicist.
Babinet compensator
Babinet's principle

Babo, Lambert Heinrich Clemens von (1818–99) German chemist.
Babo's law

Bache, Alexander Dallas (1806–67) US geophysicist.

Bacillariophyta A division of *algae comprising the diatoms. In some classifications it is regarded as a class, Bacillariophyceae, either of the division *Chrysophyta or, in a recent classification, of the division *Chromophyta.

bacille Calmette–Guérin (en dash) Abbrev. and preferred form: BCG (no stops). Named after Albert Léon Charles Calmette (1863–1933) and Camille Guérin (1872–1961).

bacillus (pl. bacilli) Any rod-shaped bacterium, including (but not restricted to) any bacterium of the genus *Bacillus* (cap. B, ital.). Adjectival forms: **bacillar, bacillary, bacilliform** (preferred when describing shape). *Compare* coccus.

Back, Ernst E. A. (1881–1959) German physicist.
Paschen–Back effect (en dash)

backbone (one word) Use vertebral column or spinal column in zoological or anatomical contexts.

backcross (noun and verb; one word) Derived noun: **backcrossing**.

back focal plane Abbrev.: bfp or BFP (no stops).

background (one word)

back scatter (noun; two words) *Physics, etc.* Adjectival form: **back-scattered**.

backup (noun; one word) *Computing, etc.* Verb form: **back up** (two words).

Bacon, Roger (*c.* 1220–92) English philosopher and alchemist.

bactericide (not bacteriocide) Any substance or agent that kills bacteria. Adjectival form: **bactericidal**. *See also* bactericidin. *Compare* bacteriostatic.

bactericidin A naturally produced substance, especially an antibody, that kills bacteria, i.e. it is a *bactericide.

bacterio- (or **bacteri-**) Prefix denoting bacteria (e.g. bacteriochlorophyll, bacteriophage, bactericide).

bacteriochlorophyll (one word) A form of chlorophyll occurring in bacteria. The various types are designated by a lower-case italic letter, e.g. bacteriochlorophyll *a*, *b*, etc.

bacteriophaeophytin US: **bacteriopheophytin**. A form of bacteriochlorophyll. The various types are designated by a lower-case italic letter, e.g. bacteriophaeophytin *a*, *b*, etc.

bacteriophage Often shortened to **phage**. A virus that infects a bacterium. For the classification of phages, *see* virus.
There is no standardized system for naming individual phages (i.e. species and isolates). Existing names employ various combinations of Roman or Greek letters, Arabic or Roman numerals, and superscript or subscript characters. Many names are prefixed with a capital P or a lower-case phi, e.g. PM2, ϕ6, ϕX, Pf1, etc. In an attempt to rationalize the nomenclature, H.-W. Ackermann (Chairman, ICTV Bacterial Virus Subcommittee) has proposed that 'new' phages be designated according to their host bacterium. The phage name would consist of the first two letters of the host genus name, plus the first two letters of the host species name, suffixed by other characters as necessary. Members of known species would be designated by the name of the type virus plus an initial representing the place of isolation. Hence T2Q would exemplify a phage belonging to the T-even phage group (type species T2) isolated in Quebec. This system has yet to gain support.

bacteriostatic Capable of inhibiting the growth and multiplication of bacteria. A substance with this ability is called a **bacteriostat** (*compare* bactericide). Noun form: **bacteriostasis**.

bacterium (pl. bacteria) The rules for bacterial classification and nomenclature follow those for higher organisms, but more emphasis is placed on subspecific taxonomic ranks (*see* binomial nomenclature; taxonomy). Adjectival form: **bacterial**.

bacteroid 1. (adjective) Resembling a bacterium. **2.** (noun) Any body that resembles a bacterium. The term should not be used to mean a member of the family Bacteroidaceae or the genus *Bacteroides* (*see* bacteroides).

bacteroides (pl. noun, no cap.) Bacteria belonging to the family Bacteroidaceae, which may be referred to as the 'Bacteroides group' (cap. B). Care should be taken to avoid confusion between members of the constituent genus *Bacteroides* (cap. B, ital.) and those of the family as a whole. *Compare* bacteroid.

Baekeland, Leo Hendrik (1863–1944) Belgian-born US industrial chemist. *See also* Bakelite.

baeocyte (not beocyte) A small cell formed inside the parent cell in certain cyanobacteria. Former name: endospore.

Baer, Karl Ernst von (1792–1876) Estonian-born German biologist, comparative anatomist, and embryologist.

Baeyer, Johann Friedrich Adolph von (1835–1917) German organic chemist.
Baeyer strain theory

Baily, Francis (1774–1844) British astronomer.
Baily's beads

Baird, John Logie (1888–1946) British inventor.

Baird, Spencer Fullerton (1823–87) US biologist.

Bakelite (capital B) A trade name for certain phenol–formaldehyde resins.

Baker, Sir Benjamin (1840–1907) British civil engineer.

BAL Abbrev. for British anti-Lewisite (2,3-dimercapto-1-propanol).

Balard, Antoine-Jérôme (1802–76) French chemist.

Balbiani, Edouard Gérard (1823–99) French embryologist.
Balbiani ring *Genetics*

Balfour, Francis Maitland (1851–82) British zoologist.

ball-and-socket (hyphenated when used as an adjective, three words as a noun)

Balmer, Johann Jakob (1825–98) Swiss mathematician.
Balmer lines
Balmer series

Baltimore, David (1938–) US molecular biologist.

balun *Telecom., etc.* Acronym for balanced unbalanced.

Banach, Stefan (1892–1945) Soviet mathematician.
Banach space

bandpass (or **band-pass**) **filter**

bandstop (or **band-stop**) **filter**

b&w (no spaces) *Photog., etc.* Abbrev. for black and white.

bandwidth (one word) *Physics, etc.*

Banks, Sir Joseph (1743–1820) British botanist.

banksia or, as generic name (*see* genus), ***Banksia***

Banting, Sir Frederick Grant (1891–1941) Canadian physiologist.

BAP Abbrev. for 6-benzylaminopurine.

BAPNA Abbrev. for *N*α-benzoyl-DL-arginine-*p*-nitroaniline.

bar (no symbol) A *cgs unit of pressure. Although not an *SI unit, it is officially recognized because of its specialized usage, for example in meteorology, in measuring fluid pressure. It may be used with the SI units and SI prefixes can be attached to it, as in millibar (mbar or mb). One bar = 10^5 pascal, 10^6 dynes cm^{-2}.

barchan (preferred to other forms) A crescent-shaped dune.

bar code (two words)

Bardeen, John (1908–91) US physicist. *See also* BCS theory.

Barfoed, Christen Thomsen (1815–99) Swedish physician.
Barfoed's reagent
Barfoed's test

Barger, George (1878–1939) British organic chemist.

barium Symbol: Ba *See also* periodic table; nuclide.

Barkhausen, Heinrich Georg (1881–1956) German physicist.
Barkhausen effect

Barkla, Charles Glover (1877–1944) British physicist.

barley stripe mosaic virus Abbrev.: BSMV (no stops). Type member of the **Hordeivirus* group (vernacular name: barley stripe mosaic virus group).

barley yellow dwarf virus Abbrev.: BYDV (no stops). Type member of the *Luteovirus* group (vernacular name: barley yellow dwarf virus group).

barn A unit of area equal to 10^{-28} square metre, used in atomic and nuclear physics to express cross sections.

Barnard, Edward Emerson (1857–1923) US astronomer.
Barnard's star

Barnard, Joseph Edwin (1870–1949) British physicist.

baro- Prefix denoting pressure (e.g. barograph, barotaxis).

baroreceptor (preferred to baroceptor)

Barr, Murray Llewellyn (1908–) Canadian anatomist.
Barr body *Genetics*

barretter (not baretter or barreter) *Elec. eng.*

Barringer, Daniel Moreau (1860–1929) US mining engineer and geologist.

Barrow, Isaac (1630–77) British mathematician.

Bartholin, Caspar Thomèson (1655–1738) Danish anatomist.
Bartholin's glands Also called greater vestibular glands.

Bartholin, Erasmus (1625–98) Danish mathematician.

Barton, Sir Derek Harold Richard (1918–) British chemist.

Bary, Heinrich Anton de Usually alphabetized as *de Bary.

bary- Prefix denoting mass or massive (e.g. baryon, barysphere).

barycentre US: **barycenter**. Adjectival form: **barycentric**.

barye An obsolete unit of pressure equal to one dyne per square centimetre, i.e. 0.1 pascal.

baryons *See* hadrons.

basal metabolic rate Abbrev.: BMR (no stops).

basalt Adjectival form: **basaltic**.

baseline (one word)

base pair A pair of complementary nitrogenous bases linked by hydrogen bonds in a nucleic acid molecule. Paired bases can be indicated in text by a centred dot, e.g. an adenine·thymine (or A·T) base pair. *See also* bp; kilobase.

base unit *See* SI units; coherent units.

basi- Prefix denoting base (e.g. basifixed, basipetal).

Basic (or **BASIC**) *Computing* Acronym for beginners' all-purpose symbolic instruction code.

basidiomycete Any fungus belonging to the class Basidiomycetes (cap. B). Authors using other classifications (*see* Basidiomycotina) still use 'basidiomycete' as the individual name for these fungi. Adjectival form: **basidiomycete** (preferred to basidiomycetous).

Basidiomycotina (cap. B) A subdivision of true fungi (Eumycota) characterized by the production of basidia. It is divided into four classes: *Urediniomycetes (rusts), *Ustilaginomycetes (smuts), *Hymenomycetes, and *Gasteromycetes. In other classifications these fungi are regarded as a class, Basidiomycetes, divided into the subclasses Heterobasidiomycetidae and Homobasidiomycetidae; or a division, Basidiomycota, of the kingdom *Mycobiota. Whichever classification is used, the term 'basidiomycete' is acceptable as an individual name for these fungi.

basidium (pl. basidia) The structure producing sexual spores (**basidiospores**) in basidiomycete fungi; it may be borne on a fruiting body (**basidiocarp**). *Compare* conidium.

basophil A type of white blood cell. This word can also be used as an adjective to describe cells that are readily stained by basic dyes, but the word **basophilic** or, less commonly, **basophile**, is preferred for this sense.

Basov, Nikolai Gennediyevich (1922–) Soviet physicist.

Bates, Henry Walter (1825–92) British naturalist and explorer.
Batesian mimicry

Bateson, William (1861–1926) British geneticist.

batho- *See* bathy-.

bathochromic Denoting the shift of an absorption maximum as a result of a chromophore to longer wavelength. *Compare* hypsochromic.

bathy- (or **batho-**) Prefix denoting **1.** depth (e.g. bathymetry, bathyplankton, batholith, bathophilous). **2.** longer wavelengths (e.g. bathochrome).

baud Symbol: Bd A unit of signal speed in a computer system or telecommunications system equal to the number of times per second that the signalling element changes state. When the signal is a sequence of *bits, one baud is equal to one bit per second (1 bps). Named after J. M. E. Baudot (1845–1903).

Bauer, Georg *See* Agricola, Georgius.

Baumé, Antoine (1728–1804) French chemist.
Baumé scale

Bayer, Johann (1572–1625) German astronomer.

Bayes, Thomas (1702–61) English mathematician and clergyman.
Bayesian inference
Bayes's theorem

Bayliss, Sir William Maddock (1860–1924) British physiologist.

Bazalgette, Sir Joseph William (1819–91) British civil engineer.

9-BBN Abbrev. for 9-borobicyclo[3.3.1]-nonane.

BBO Abbrev. for 2,5-bis(4-biphenylyl)-oxazole.

BBOT Abbrev. for 2,5-bis(5-*t*-butyl-2-benzoxazolyl)thiophene.

BC (small caps., no stops) Abbrev. for before Christ; should always be placed after the year or century.

BC *Astron.* Abbrev. for bolometric correction.

b.c.c. *Crystallog.* Abbrev. for body-centred cubic.

BCD (or **bcd**) *Computing* Abbrev. for binary-coded decimal.

B cell (or **B lymphocyte**; not hyphenated) A type of *lymphocyte responsible for humoral immunity. Named from the initial letter of bursa of Fabricius, from which these cells are derived in birds. *Compare* T cell.

BCF Abbrev. for bromochlorodifluoromethane.

BCG Abbrev. and preferred form for *bacille Calmette–Guérin.

B chromosome (not hyphenated) An accessory or supernumary chromosome. *Compare* A chromosome.

BCS Abbrev. for British Computer Society.

BCS theory *Superconductivity* [from J. **B*ardeen, L. N. **C*ooper, and J. R. **S*chrieffer]

Bd (no stop) Symbol for baud.

BD- *Prefix to a number, used to designate a star listed in the Bonner Durchmusterung (Bonn Star Catalogue).

BDCS Abbrev. for *t*-butyldimethylchlorosilane.

BDPA Abbrev. for α,γ-bisdiphenylene-β-phenylallyl.

b.d.v. (stops, or **BDV**) Abbrev. for breakdown voltage.

Be 1. Symbol for beryllium. 2. *Astron. See* spectral types.

BE Abbrev. for Bachelor of Engineering.

Beadle, George Wells (1903–89) US geneticist.

beat-frequency oscillator *Telecomm., etc.* Abbrev.: b.f.o. (or bfo, BFO).

Beaufort, Sir Francis (1774–1857) British hydrographer.
Beaufort scale

Beche, Sir Henry De La Usually alphabetized as *De La Beche.

Beckmann, Ernst Otto (1853–1923) German organic and physical chemist.
Beckmann rearrangement
Beckmann thermometer

becquerel (no cap.) Symbol: Bq The SI unit of *activity of a radionuclide.

$$1\ \text{Bq} = 1\ \text{s}^{-1}.$$

The becquerel has recently replaced the curie: one curie is equal to 3.7×10^{10} Bq. Named after A. H. *Becquerel.

Becquerel, Antoine Henri (1852–1908) French physicist.
***becquerel**

bedbug (one word) *See* Heteroptera.

Bednorz, George (1950–) German physicist.

Beer, Sir Gavin Rylands de Usually alphabetized as *de Beer.

Beggiatoa (cap. B, ital.) A genus of filamentous gliding bacteria. Individual name: **beggiatoa** (no cap., not ital.; pl. beggiatoas). The genus is placed by some authorities in the order Beggiatoales, and some authors apply the trivial name 'beggiatoa' to any member of this order, causing confusion. Therefore 'beggiatoan' is recommended as the trivial name for any member of the order, 'beggiatoa' being reserved for any member of the genus.

behaviour US: **behavior.** Adjectival form: **behavioural** (US: **behavioral**). Derived noun: **behaviourism** (US: **behaviorism**).

Behring, Emil Adolf von (1854–1917) German immunologist.

Beijerinck, Martinus Willem (1851–1931) Dutch microbiologist.

Beilstein, Friedrich Konrad (1838–1906) Russian organic chemist.
Beilstein's test

Békésy, Georg von (1899–1972) Hungarian-born US physicist and physiologist.

bel Symbol: B *See* decibel.

Bel, Joseph Achille Le Usually alphabetized as *Le Bel.

Bell, Alexander Graham (1847–1922) British inventor.
bel (unit)

Belon, Pierre (1517–64) French naturalist.

benchmark (one word)

Beneden, Edouard van (1846–1910) Belgian cytologist and embryologist.

Benedict, Stanley Rossiter (1884–1936) US chemist.
Benedict's reagent
Benedict's test

BEng (or **B.Eng.**) Abbrev. for Bachelor of Engineering.

Benguela current *Oceanog.* Named after Benguela, W Angola.

Benioff zone *Seismol.* Named after Hugo Benioff (1899–1968).

Bennettitales (double t) An extinct order of cycads (*see* Cycadopsida). Also called Cycadeoidales. Named after J. J. Bennett (1801–76).

Benson, Andrew Alm (1917–) US biochemist and plant physiologist.
Benson–Calvin–Bassham cycle (en dashes) Use Calvin cycle.

Bentham, George (1800–84) British botanist.

benthos The organisms, collectively, that live on the bottom of a sea or lake. Adjectival form: **benthic** (not benthonic).

Benz, Karl Friedrich (1844–1929) German engineer.

benzal- *Prefix denoting the group $C_6H_5CH=$ (e.g. benzalacetone, benzal chloride). It is preferred to the alternative benzylidene-.

benzalacetophenone $C_6H_5CH=CHCOC_6H_5$ The traditional name for 1,3-diphenyl-2-propen-1-one.

benzal chloride $C_6H_5CHCl_2$ The traditional name for (dichloromethyl)benzene.

benzenediamine $C_6H_4(NH_2)_2$ The recommended name for the compound traditionally known as phenylenediamine. The recommended name for *o*-phenylenediamine is benzene-1,2-diamine, etc.

benzene-1,4-dicarboxylic acid $C_6H_4(COOH)_2$ The recommended name for the compound traditionally known as terephthalic acid.

benzene-1,2-diol $C_6H_4(OH)_2$ The recommended name for the compound traditionally known as catechol.

benzene-1,3-diol $C_6H_4(OH)_2$ The recommended name for the compound traditionally known as resorcinol.

benzene-1,4-diol $C_6H_4(OH)_2$ The recommended name for the compound traditionally known as hydroquinone or quinol.

benzene hexachloride $C_6H_6Cl_6$ The traditional name for 1,2,3,4,5,6-hexachlorocyclohexane.

benzene-1,2,3-triol $C_6H_3(OH)_3$ The recommended name for the compound traditionally known as pyrogallol.

benzene-1,3,5-triol $C_6H_3(OH)_3$ The recommended name for the compound traditionally known as phloroglucinol.

Benzer, Seymour (1921–) US geneticist.

benzidine The traditional name for *biphenyl-4,4′-diamine.

benzil $C_6H_5COCOC_6H_5$ The traditional name for 1,2-diphenylethanedione.

benzo- (**benz-** before vowels) Prefix denoting **1.** the group C_6H_5C- (e.g. benzotrichloride, benzamide). **2.** a benzene ring attached to a parent cyclic compound (e.g. benzoquinoline, benzanthracene).

benzoin $C_6H_5COCH(OH)C_6H_5$ The traditional name for 2-hydroxy-1,2-diphenylethanone.

benzophenone $C_6H_5COC_6H_5$ The traditional name for diphenylmethanone.

1,4-benzoquinone The traditional name for *cyclohexadiene-1,4-dione.

benzotrichloride $C_6H_5CCl_3$ The traditional name for (trichloromethyl)benzene.

benzoyl- *Prefix denoting the group $C_6H_5C(O)-$ (e.g. benzoylbenzoic acid, benzoyl chloride).

benzyl- *Prefix denoting the group $C_6H_5CH_2-$ (e.g. benzylamine, benzyl chloride).

benzyl alcohol $C_6H_5CH_2OH$ The traditional name for phenylmethanol.

benzylamine $C_6H_5CH_2NH_2$ The traditional name for (phenylmethyl)amine.

6-benzylaminopurine (or **6-benzyladenine**) Abbrev.: BAP (no stops). A synthetic cytokinin.

benzyl chloride $C_6H_5CH_2Cl$ The traditional name for (chloromethyl)benzene.

benzylidene- *See* benzal-.

Berg, Paul (1926–) US molecular biologist.

Bergeron, Tor Harold Percival (1891–1977) Swedish meteorologist.
Bergeron–Findeisen theory (en dash)

Bergius, Friedrich Karl Rudolph (1884–1949) German industrial chemist.
Bergius process

Bergman, Torbern Olaf (1735–84) Swedish chemist.

Bergmann's rule *Biol.* Named after Carl Bergmann (19th century).

bergschrund (no cap.) *Glaciol.* [from German: mountain crack]

Bergström, Sune (1916–) Swedish biochemist.

beriberi (one word) A disease caused by thiamin (vitamin B_1) deficiency.

berkelium Symbol: Bk *See also* periodic table; nuclide.

Bernal, John Desmond (1901–71) British crystallographer.

Bernard, Claude (1813–78) French physiologist.

Bernoulli, Daniel (1700–82) Swiss mathematician, son of Johann Bernoulli.
Bernoulli equation
Bernoulli's principle
Bernoulli's theorem (in hydrodynamics)

Bernoulli, Jakob (or **Jacques**) (1654–1705) Swiss mathematician, brother of Johann Bernoulli.
Bernoulli numbers
Bernoulli's theorem (on probability)

Bernoulli, Johann (or **Jean**) (1667–1748) Swiss mathematician, brother of Jakob Bernoulli.
Bernoulli–L'Hospital rule (en dash)

Berthelot, Pierre Eugène Marcellin (1827–1907) French chemist.
Berthelot equation
Berthelot relation

Berthollet, Comte Claude-Louis (1748–1822) French chemist.
berthollide compound (no initial capital letter)

beryllium Symbol: Be *See also* periodic table; nuclide.

Berzelius, Jöns Jacob (1779–1848) Swedish chemist.
Berzelius theory of valency

BES Abbrev. for *N*,*N*-bis(2-hydroxyethyl)-2-aminoethanesulphonic acid.

Bessel, Friedrich Wilhelm (1784–1846) German astronomer and mathematician.
Bessel functions
Besselian year

Bessel's differential equation
Fourier–Bessel series (en dash)

Bessemer, Sir Henry (1813–98) British inventor and engineer.
Bessemer converter
Bessemer process

Best, Charles Herbert (1899–1978) US-born Canadian physiologist.

beta Greek letter, symbol: β (lower case), B (cap.).
β Symbol for **1.** (or β^-) electron. **2.** a plane angle. **3.** pressure coefficient. **4.** ratio of a velocity to the speed of light. **5.** second brightest star in a constellation (*see* stellar nomenclature).
β- (always hyphenated) Symbol used in **1.** the names of organic compounds to indicate a substituent attached to the second carbon atom along from the functional group (e.g. β-phenylethanol). **2.** nomenclature for steroids to indicate substituent groups above the plane of the nucleus (e.g. β-androstane, 5β-cholestane).

Beta (cap. B, followed by the genitive form of a constellation name; e.g. Beta Centauri or β Cen) Usually the second brightest star in that constellation. *See* stellar nomenclature.

Betabacterium (cap. B, ital.) A subgenus of **Lactobacillus*. Individual name: **betabacterium** (no cap., not ital.; pl. betabacteria).

beta blocker (two words; always written out in full) A drug that blocks beta receptors.

betaherpesvirus (one word) Any member of the subfamily Betaherpesvirinae (vernacular name: cytomegalovirus group). *See* cytomegalovirus.

beta iron (two words; also written **β-iron** (hyphenated))

beta-pleated sheet (usually written **β-pleated sheet**) The secondary and tertiary molecular structure of some proteins.

beta receptor (two words; usually written **β receptor**) Shortened and preferred form of beta-adrenergic receptor (hyphenated; abbrev.: β-AR). There are two types: β_1 receptors and β_2 receptors (note subscript numbers).

Bethe, Hans Albrecht (1906–) German-born US physicist.
Alpher–Bethe–Gamow theory (en dashes) Also called alpha–beta–gamma theory or αβγ theory.
Bethe–Weizsächer cycle (en dash) or **Bethe cycle** Also called carbon cycle.

béton *Civ. eng.* concrete. [from French]

BeV (US) Abbrev. for billion electronvolts. Use GeV (*see* giga-).

Bevan, Edward John (1856–1921) British industrial chemist.

B²FH theory *Astron.* [from G. and E. M. **B*urbidge, W. **F*owler, and F. **H*oyle]

b.f.o. (stops, or **BFO**) Abbrev. for beat-frequency oscillator.

B/H loop (B and H not bold ital.) A closed figure showing variation of magnetic flux density ***B*** in a magnetizable material against magnetic field strength ***H***. Also called hysteresis loop.

B horizon A subsurface *soil horizon characterized by illuviation of material from the *A horizon.

bhp (no stops) Abbrev. for brake horsepower.

BHT Abbrev. for butylated hydroxytoluene (2,6-di-*t*-butyl-4-methylphenol).

Bi Symbol for bismuth.

bi- 1. (sometimes **bin-** before vowels) Prefix denoting two, both, or double (e.g. biaxial, bicollateral, bicuspid, bimetallic, bistable, binaural). **2.** *Chem.* Prefix denoting **a.** the linking of two groups that together form the root of a structure (e.g. biphenyl-4,4′-dicarboxylic acid); **bis-** is used when an expression to be multiplied already contains a multiplicative prefix (e.g. bis (dimethylamine)) or to avoid ambiguity. **b.** an acid salt (e.g. sodium bisulphate). In strict chemical usage, the prefix *hydrogen- should be used (e.g. sodium hydrogensulphate). **3.** *See* bio-. *See also* di-.

Bial, Manfred (1870–1908) German physician.
Bial's reagent

biannual Occurring twice a year. *Compare* biennial.

bias Verb form: **biases, biasing, biased** (not -ass-).

bicarbonate Denoting a compound containing the ion HCO_3^-, e.g. sodium bicarbonate ($NaHCO_3$). The recommended name is hydrogencarbonate.

bi-CMOS (pronounced by-**see**-moss) *Electronics* Acronym for (merged) bipolar/CMOS. *See* CMOS.

bicuspid (noun) US name for a premolar tooth.

bicuspid valve (preferred to mitral valve)

Biela, Wilhelm von (1784–1856) Austrian astronomer.
Biela's comet
Bielids

biennial Lasting two years or occurring every two years. *Compare* biannual.

BIF *Geol.* Abbrev. for banded iron formation.

Biffen, Sir Rowland Harry (1874–1949) British geneticist and plant breeder.

Bifidobacterium (cap. B, ital.) A genus of anaerobic rod-shaped Gram-positive bacteria. Individual name: **bifidobacterium** (no cap., not ital.; pl. bifidobacteria).

bifilar (not bifiler) *Physics, etc.*

big bang (no cap.) *Astron.*

BIH Abbrev. for Bureau International de l'Heure. *See* TAI.

bile duct (two words)

bilharzia, bilharziasis Use *schistosome, schistosomiasis.

billion Abbrev.: bn (no stop). One thousand million (10^9). Formerly, in the UK, one million million (10^{12}); if the context allows any doubt give the decimal form, 10^{12} or 10^9, in brackets after the first use of the word.

bimolecular reaction *See* reaction mechanisms.

bin- *See* bi-.

binary-coded decimal (hyphenated) Abbrev.: BCD or bcd (no stops).

Binet, Alfred (1857–1911) French psychologist.
Stanford–Binet test (en dash)

Bingham, Eugene Cook (1878–1945) US scientist.
Bingham flow
Bingham plastic

Binnig, Gerd (1947–) German physicist.

binoculars (takes pl. form of a verb)

binomen A binomial name. *See* binomial nomenclature.

binomial coefficient *See* mathematical symbols.

binomial nomenclature The system of naming *species of organisms devised by Linnaeus and still used today. Each species has a two-part Latin name (binomial or binomen), printed in italic type without accents or other special marks, consisting of a generic name (initial letter capitalized; *see* genus) and a specific epithet (for plants) or specific name (for animals). For example, the tawny owl is *Strix aluco*; *Strix* is the generic name and *aluco* the specific name. The latter is never capitalized in zoological specific names, even when derived from the name of a person; e.g. Thomson's gazelle is *Gazella thomsoni*. However, botanical epithets are capitalized by some authors when they derive from personal or vernacular names: this is not proscribed by the Botanical Code. Unidentified species are printed in the form *Drosophila* sp. (i.e. an unidentified species of the genus *Drosophila*). Specific epithets and names are often adjectival in form, in which case the ending always agrees with the gender of the generic name; for example *Moticilla alba* (pied wagtail), but *Lamium album* (white deadnettle). In botanical nomenclature the binomial is often followed by the name, often abbreviated, of the person who validly published it; thus the common daisy, *Bellis perennis* L., was named by Linnaeus. In zoological nomenclature it is recommended that the author's name should not be abbreviated except when the abbreviation is clearly recognizable as the author's name. When the classification of a plant has been

revised, the name of the original author is given first, in brackets, followed by the name of the author who published the new name according to the new classification. For example, *Medicago arabica* (L.) Huds. indicates that this name was given by Hudson to a plant originally named otherwise by Linnaeus. On first being mentioned, the binomial should be spelled out in full; on second and subsequent occasions the generic name may be abbreviated to its initial letter followed by a full stop (e.g. *Homo sapiens* becomes *H. sapiens*). Genera sharing an initial letter may be distinguished by shortened forms of their names (e.g. *Staphylococcus* and *Streptococcus* become *Staph.* and *Strep.*, respectively).
Rules for the naming of species are specified by the International Code of Zoological Nomenclature (ICZN; for animals), the International Code of Botanical Nomenclature (ICBN; for wild plants, including fungi), the International Code of Nomenclature of Cultivated Plants (ICNCP), and the International Code of Nomenclature of Bacteria (ICNB). Viruses, the naming of which is specified by the International Code of Nomenclature of Viruses (ICNV), do not have specific epithets (*see* virus). *See also* cultivar; graft hybrid; hybrid; subspecies; variety.

binoxalate Denoting a compound containing the ion $HC_2O_4^-$, e.g. potassium binoxalate, HC_2O_4K. The recommended name is hydrogenethanedioate.

bio- (sometimes **bi-** before vowels) Prefix denoting life or living organisms (e.g. bioassay, bioengineering, bioluminescence, biomass, biosphere, biome, biota).

biocenose (or **biocenosis**) US forms of *biocoenosis.

biochemical oxygen demand (preferred to biological oxygen demand) Abbrev.: BOD (no stops).

biocoenosis (pl. biocoenoses) US: **biocenose** or **biocenosis**. **1.** A community of organisms occupying a uniform habitat. **2.** The relationship between the organisms within such a community. Adjectival form: **biocoenotic** (US: **biocenotic**). Derived noun: **biocoenology** (US: **biocenology**).

biogenetic 1. Relating to the theory of **biogenesis**, that living organisms arise only from other living organisms. **2.** Denoting Haeckel's law of recapitulation. *Compare* biogenic.

biogenic Produced by or from living organisms (e.g. biogenic sediments). *Compare* biogenetic.

biological oxygen demand Use *biochemical oxygen demand.

BIOS *Computing* Acronym for basic input-output system.

-biosis Noun suffix denoting a way of life (e.g. necrobiosis, symbiosis). Adjectival form: **-biotic**.

Biot, Jean Baptiste (1774–1862) French physicist.
Biot–Fourier equation (en dash)
Biot–Savart law (en dash)

biotype A unit of biological classification ranking below a species. *See also* biovar.

biovar Abbrev.: bv. (stop). A bio(logical) var(iety): an unofficial category of classification used in microbiology and ranking below subspecies. Biovars are strains distinguished by some biochemical or physiological character.

biphenyl-4,4′-diamine
(benzidine)

biphenyl-4,4′-diamine The recommended name for the compound traditionally known as benzidine.

BIPM Abbrev. for Bureau International des Poids et Mésures (International Bureau of Weights and Measures). An organization situated near Paris, managed by the *CIPM on behalf of the *CGPM.

bird's-nest fungi (hyphenated) *See* Gasteromycetes.

Birkeland, Kristian Olaf Bernhard (1867–1917) Norwegian physicist and chemist.
Birkeland–Eyde process (en dash)

Birkhoff, George David (1884–1944) US mathematician.

bis- *See* bi-.

bis(butanedione dioximato)nickel(II) $C_8H_{14}N_4NiO_4$ The recommended name for the compound traditionally known as nickel dimethylglyoxime.

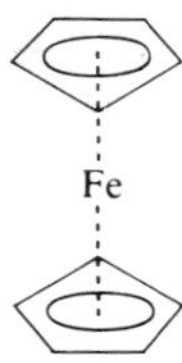

bis(h^5-cyclopentadienyl)iron (ferrocene)

bis(h^5-cyclopentadienyl)iron The recommended name for the compound traditionally known as ferrocene.

bismuth Symbol: Bi *See also* periodic table; nuclide.

bismuth(III) oxide chloride BiOCl The recommended name for the compound traditionally known as bismuthyl chloride.

bismuthyl chloride BiOCl The traditional name for bismuth(III) oxide chloride.

bisulphate US: **bisulfate.** Denoting a compound containing the ion HSO_4^-, e.g. sodium bisulphate $NaHSO_4$. The recommended name is hydrogensulphate.

bisulphite US: **bisulfite.** Denoting a compound containing the ion HSO_3^-, e.g. sodium bisulphite $NaHSO_3$. The recommended name is hydrogensulphite.

bit A unit of information derived from a choice between two equally probable events, conventionally represented by the presence of a 0 or a 1. The unit, short for binary digit, is much used in computing and telecommunications.

BITNET A computer network originally sponsored by IBM and later extended to other systems.

Bitter patterns *Magnetism* Named after Francis Bitter.

bivalve Any mollusc of the class Bivalvia (or Lamellibranchia), including the cockles, mussels, and oysters. Adjectival form: **bivalve.** Use 'bivalved' to refer to or describe any other organism with a two-valved shell, to avoid confusion with the Bivalvia.

Bjerknes, Jacob Aall Bonnevie (1897–1975) Norwegian meteorologist.

Bjerknes, Vilhelm Friman Koren (1862–1951) Norwegian meteorologist.

Bjerrum, Niels (1879–1958) Danish physical chemist.

Bk Symbol for berkelium.

Black, Sir James Whyte (1924–) British pharmacologist.

Black, Joseph (1728–99) British physician and chemist.

black body (two words) *Physics, etc.* Hyphenated when used adjectivally (e.g. black-body radiation).

black box (two words)

Blackett, Patrick Maynard Stuart (1897–1974) British physicist.

black hole (two words)

blackout (one word)

bladderworm (one word) A larval stage of many tapeworms. The *coenurus and *cysticercus are types of bladderworm.

-blast Noun suffix denoting an embryonic or formative cell (e.g. erythroblast, osteoblast). Adjectival form: **-blastic**.

blasto- Prefix denoting **1.** an embryo or germ cell (e.g. blastocyst, blastomere). **2.** budding or gemmation (e.g. blastogenesis, blastospore).

***Blastocaulis-Planctomyces* group** (generic names ital., with initial caps., and separated by a hyphen) A group, of indeterminate taxonomic status, containing all bacteria allocated to the genus *Planctomyces* or its synonym, *Blastocaulis*. (The genus *Planctomyces* was originally created for an aquatic 'fungus', later found to be identical to a bacterium in the genus *Blastocaulis*.) Until taxonomy of the group is clarified, members are formally classified as species of the genus *Planctomyces*, but additionally are assigned to one of five morphovars.

These are designated by Roman numerals, with subtypes designated by a lower-case letter: e.g. Ia, IIIc, etc. Hence, *Planctomyces bekefii* Gimesi 1924 (*Blastocaulis sphaerica* Henrici and Johnson 1935) is morphovar Ia of the *Blastocaulis-Planctomyces* group.

blastocoel (preferred to blastocoele) *See* blastula.

blastocyst *See* blastula.

blastoderm A type of *blastula resulting from cleavage of a very yolky fertilized egg cell, such as that of a bird. It originates as a small **blastodisc** on the yolk. Adjectival form: **blastodermic** (not blastodermatic).

blastodisc US: **blastodisk**. *See* blastoderm.

Blastomycetes (cap. B) A former class of imperfect fungi (*see* Deuteromycotina) containing the imperfect yeasts. They are now included in the class *Hyphomycetes. The individual name and adjectival form, **blastomycete** (no cap.), is still used for descriptive purposes.

blastula (pl. blastulas or blastulae) The product of cleavage of a fertilized animal egg cell, consisting of a ball of cells (**blastomeres**) around a central cavity (**blastocoel**). In mammals it is called a **blastocyst**. *See also* blastoderm.

Blenkinsop, John (1783–1831) British engineer.

Bloch, Felix (1905–83) Swiss-born US physicist.
Bloch functions
Bloch wall

Bloch, Konrad Emil (1912–) German-born US biochemist.

Bloembergen, Nicolaas (1920–) Dutch-born US physicist.

blood Two-word terms in which the first word is 'blood' (e.g. blood count, blood group, blood pressure, blood test) should not be hyphenated unless they are used adjectivally (e.g. blood-group system, blood-pressure measurement, blood-sugar level).

blood–brain barrier (en dash, not hyphen)

blood cell (two words; preferred to blood corpuscle) The two major types of blood cell are erythrocytes (red blood cells) and *leucocytes (white blood cells); *platelets are not blood cells.

blood corpuscle Use *blood cell.

blood group (two words) Blood-group antigens are typically designated by capital letters (sometimes combined with a lower-case letter or letters); e.g. A, B, Rh. Genes encoding these antigens are represented by the same letters printed in italic. The antibodies corresponding to the antigens are designated by the symbol for the antigen prefixed by 'anti-'; e.g. the antibody to the A antigen is anti-A. *See* ABO; rhesus factor. *See also* blood type.

bloodstream (one word)

blood type Another name for *blood group, used in certain contexts. For example, A is defined as a blood group (not a blood type), but blood is described as being type A (not group A).
blood-type (verb; hyphenated) To determine the type of blood or the blood group of a person. Noun form: **blood typing**.

blowhole (one word) *Eng., geol., zool.*

blowout (one word)

blue-green algae (hyphenated) Also called blue-green bacteria. *See* cyanobacteria.

blue vitriol $CuSO_4{\cdot}5H_2O$ A traditional name for copper(II) sulphate-5-water.

B lymphocyte (not hyphenated) *See* B cell.

BMR Abbrev. for basal metabolic rate.

BMV Abbrev. for *brome mosaic virus.

bn (no stop) Abbrev. for billion.

Bn (preferred to Bzl) Symbol sometimes used to denote the benzyl group in chemical formulae, e.g. $C_6H_5CH_2NH_2$ can be written as $BnNH_2$.

BNF *Computing* Abbrev. for Backus–Naur form (en dash). Named after John Backus and Peter Naur. Also called Backus normal form.

BN object *Astron.* Abbrev. for Becklin–Neugebauer object (en dash). Named after Eric Becklin and Gerald Neugebauer.

Board of Trade unit Former name for kilowatt hour.

Boc Symbol sometimes used to denote the *t*-butoxycarbonyl group in chemical formulae, e.g. $HONHCOOC(CH_3)_3$ can be written as HONHBoc.

BOC-ON Abbrev. for 2-(*t*-butoxycarbonyloxyimino)-2-phenylacetonitrile.

BOD Abbrev. for biochemical oxygen demand.

Bode, Johann Elert (1747–1826) German astronomer.
Bode's law or **Titius–Bode law** (en dash)

Bodenstein, Max Ernst August (1871–1942) German physical chemist.

body-centred cubic *Crystallog.* Abbrev.: b.c.c. (no caps., stops).

Bogoliubov, Nikolai Nikolaevich (1909–) Soviet mathematician and physicist.

Bohr, Aage Niels (1922–) Danish physicist.

Bohr, Christian (1855–1911) Danish physiologist.
Bohr effect (or **shift**)

Bohr, Niels Hendrik David (1885–1962) Danish physicist.
Bohr atom
***Bohr magneton**
***Bohr radius**
Bohr theory
Bohr–Sommerfeld theory (en dash)

Bohr magneton Symbol: μ_B (Greek mu). A fundamental constant equal to

$$9.274\,0154 \times 10^{-24}\,\mathrm{J\,T^{-1}}.$$

It is given by $eh/2\pi m_e$, where h is the Planck constant and e and m_e are the charge and mass of the electron, respectively. The nuclear magneton, symbol μ_N, is equal to

$$5.050\,7866 \times 10^{-27}\,\mathrm{J\,T^{-1}}.$$

It is given by $eh/2\pi m_p$, where m_p is the proton mass.

Bohr radius Symbol: a_0 A fundamental constant equal to

$$0.529\,177\,249 \times 10^{-10}\,\mathrm{m}.$$

It is given by $\alpha/4\pi R_\infty$, where α is the fine-structure constant and R_∞ is the Rydberg constant.

boiling point (not hyphenated) Abbrev.: b.p. (stops).

boiling-point constant (hyphenated; preferred to ebullioscopic constant) Symbol: K_b; recommended unit: kelvin kilogram per mole.

boiling-water reactor Abbrev.: BWR (no stops).

Bok, Bart Jan (1906–83) Dutch-born US astronomer.
Bok globules

bolometric correction *Astron.* Abbrev.: BC (no stops).

Boltwood, Bertram Borden (1870–1927) US chemist and physicist.

Boltzmann, Ludwig Edward (1844–1906) Austrian theoretical physicist.
***Boltzmann constant**
Boltzmann distribution Use Maxwell–Boltzmann distribution (en dash).
Boltzmann equation
Maxwell–Boltzmann distribution (en dash)
***Stefan–Boltzmann constant** (en dash)
Stefan–Boltzmann law (en dash)

Boltzmann constant Symbol: k A fundamental constant equal to

$$1.380\,658 \times 10^{-23}\,\mathrm{J\,K^{-1}}.$$

It is the ratio R/L, where R is the *molar gas constant and L is the *Avogadro constant. Former name: Boltzmann's constant. The ratio $1/kT$ (T is thermodynamic temperature) is given the symbol β.

bolus (pl. boluses, not boli) A mass of chewed food before it is swallowed.

Bolyai, János (1802–60) Hungarian mathematician.
Bolyai geometry

Bondi, Sir Hermann (1919–) Austrian-born British mathematician and cosmologist.

Bonne's projection *Cartog.*

Bonnet, Charles (1720–93) Swiss naturalist.

Boo (no stop, no accent) *Astron.* Abbrev. for Bootes.

Boole, George (1815–64) British mathematician.
Boolean algebra
Boolean operator

Boolean *See* logical.

Bootes (not Boötes) A constellation. Genitive form: Bootis. Abbrev.: Boo (no stop). *See also* stellar nomenclature.

bootstrap (one word) *Computing, electronics, etc.*

borax $Na_2B_4O_7{\cdot}10H_2O$ The traditional name for sodium heptaoxotetraborate(III)-10-water.

Bordet, Jules Jean Baptiste Vincent (1870–1961) Belgian bacteriologist and immunologist.
Bordetella

Bordetella (cap. B, ital.) A genus of aerobic mainly nonmotile bacteria that cause respiratory infections in humans and animals. Individual name: **bordetella** (no cap., not ital.; pl. bordetellae).

Borel, Émile (1871–1956) French mathematician.
Borel set

Borelli, Giovanni Alfonso (1608–79) Italian mathematician and physiologist.

boric Denoting compounds in which boron has an oxidation state of +3. The recommended system is to use oxidation numbers, e.g. boric oxide, B_2O_3, has the systematic name boron(III) oxide.

boric acid H_3BO_3 The traditional name for trioxoboric(III) acid. Condensed boric acids, $(HBO_2)_n$, form in concentrated solutions and have the systematic name polydioxoboric(III) acid.

Borlaug, Norman Ernest (1914–) US agronomist and plant breeder.

Born, Max (1882–1970) German-born British physicist.
Born–Oppenheimer approximation (en dash)

borohydride Denoting a compound containing the ion BH_4^-, e.g. sodium borohydride, $NaBH_4$. The recommended name is tetrahydridoborate(III).

boron Symbol: B *See also* periodic table; nuclide.

Borrelia (cap. B, ital.) A genus of *spirochaete bacteria. Individual name: **borrelia** (no cap., not ital.; pl. borreliae or borrelias).

Bosch, Carl (1874–1940) German industrial chemist.
Haber–Bosch process (en dash)

Bose, Sir Jagadis Chandra (1858–1937) Indian plant physiologist and physicist.

Bose, Satyendra Nath (1894–1974) Indian physicist.
Bose–Einstein statistics (en dash)
boson

Boss, Lewis (1846–1912) US astronomer.

botanical **1.** (adjective) Preferred to botanic in most contexts except *botanic garden and Royal Botanic Society. **2.** (noun) A substance, especially a pesticide or drug, derived from a plant.

botanic garden (not botanical) Also in proper names (e.g. Royal Botanic Gardens, Kew).

Bothe, Walther Wilhelm Georg (1891–1957) German atomic physicist.

boudinage *Geol.* A structure in sedimentary rocks. [from French]

bougainvillea (not bougainvillaea; pl. bougainvilleas) Generic name: *Bougainvillea* (cap. B, ital.).

Bouguer, Pierre (1698–1758) French physicist and mathematician.
Bouguer anomaly
Bouguer–Lambert–Beer law (en dashes) Use Lambert–Beer law (en dash).

Boulton, Matthew (1728–1809) British engineer.

Bourbaki, Nicolas Pseudonym for a group of 20th-century French mathematicians.

Bourdon, Eugène (1808–84) French hydraulic engineer.
Bourdon gauge

Boussingault, Jean Baptiste (1802–87) French agricultural chemist.

Boveri, Theodor Heinrich (1862–1915) German zoologist.

bovid Any hoofed mammal belonging to the family Bovidae, which includes cattle, sheep, goats, antelopes, etc. Adjectival form: **bovid.** *Compare* bovine.

bovine Any hoofed mammal belonging to the bovid tribe Bovini, which includes cattle. Adjectival form: **bovine**. *Compare* bovid.

bovine spongiform encephalopathy Abbrev.: BSE (no stops).

Bowman, Sir William (1816–92) British physician.
Bowman's capsule *Anat.*

Box–Jenkins model (en dash) *Maths.*

Boyer, Herbert Wayne (1936–) US biochemist.

Boyer, Paul Delos (1918–) US biochemist.

Boyle, Robert (1627–91) Irish chemist and physicist.
Boyle's law Called Mariotte's law in Europe.

bp (no stops) Symbol for base pair(s). It is used in molecular biology as a unit of length along a duplex polynucleotide, corresponding to the number of paired bases in a particular segment of DNA (or duplex RNA). *See also* kilobase.

BP (small caps., no stops) Abbrev. for before present; should be placed after the number of years, as in 20 000 BP, 3.7 Ma BP.

BP Abbrev. for **1.** blood pressure. **2.** British Pharmacopoeia.

b.p. Abbrev. for boiling point.

BPEA Abbrev. for 9,10-bis(phenylethenyl)anthracene.

BPEN Abbrev. for 5,10-bis(phenylethenyl)naphthacene.

bpi (no stops) Abbrev. for bits per inch, a measure of the maximum number of *bits that can be stored on one inch of a track of magnetic tape.

bps (no stops) Abbrev. for bits per second, i.e. the number of bits transmitted or transferred in a computer system or transmission line in one second. *See also* baud.

bpy (no stop) Abbrev. for bipyridine often used in the formulae of coordination compounds, e.g. $[Cr(bpy)_3]^{3+}$.

Bq Symbol for becquerel. *See also* SI units.

Br Symbol for bromine.

braces *See* brackets.

brachio- (**brachi-** before vowels) Prefix denoting an arm or armlike part (e.g. brachiocephalic, brachiopod).

Brachiopoda (cap. B) A phylum of bivalved marine invertebrates comprising the lamp shells. Individual name and adjectival form: **brachiopod** (no cap.). *Compare* Branchiopoda.

brachium (pl. brachia) The arm or an armlike part; commonly used in its adjectival form, **brachial** (e.g. brachial plexus); not to be confused with branchial (*see* branchia).

brachy- Prefix denoting shortness (e.g. brachycephalic, brachysclereid).

brackets A general name for parentheses, ( ), square brackets, [ ], braces, { }, and angle brackets ⟨ ⟩ (narrow) and < > (wide).
Parentheses, square brackets, and braces are used in mathematics, in that order, to define the extent of an expression or function, e.g.

$$\exp\{[(x^2 + y^2)^3 + z]/[2\pi(xy)^3]\}.$$

Pairs of brackets are also used for specific purposes. Parentheses are used to define the extent of a chemical group, as in $(C_2H_5)_3N$. Square brackets denote, for example, chemical *concentration, as in $[H_2SO_4]$, or a Fraunhofer line, as in [D], [H], [K]. Braces denote, for example, members of a set, as in {a,e,i,o,u}. Narrow angle brackets denote, for example, mean value of a quantity over a period of time, as in $\langle I\rangle$; wide angle brackets denote, for example, *greater than, *less than, etc. In computer programming, the different types of brackets have various specific uses, depending on the language.

Brackett, Frederick Sumner (1896–) US physicist.
Brackett series

Bradley, James (1693–1762) British astronomer.

brady- Prefix denoting slowness (e.g. bradycardia, bradytelic).

Bradyrhizobium (cap. B, ital.) A genus of nitrogen-fixing bacteria of the family Rhizobiaceae. Individual name: **bradyrhizobium** (no cap., not ital.; pl. bradyrhizobia). As this genus was formed by subdivision of the genus **Rhizobium*, some authors use 'rhizobia' to denote members of both genera, but this should be avoided.

Bragg, Sir (William) Lawrence (1890–1971) British physicist, son of William Henry Bragg.
Bragg angle
Bragg's law

Bragg, Sir William Henry (1862–1942) British physicist, father of Lawrence Bragg.

Brahe, Tycho (1546–1601) Danish astronomer.
Tycho's star
Tychonic system

Brahmagupta (*c.* 598–*c.* 665) Indian mathematician and astronomer.

brainstem (one word)

brake horsepower Abbrev.: bhp (no stops).

Bramah, Joseph (1748–1814) British engineer and inventor.

branchia (pl. branchiae) A gill; most commonly used in its adjectival form, **branchial** (e.g. branchial arch); not to be confused with brachial (*see* brachium) or bronchial (*see* bronchus).

branchio- Prefix denoting a gill (e.g. branchiomere, branchiopod).

Branchiopoda (capital B) A subclass of crustaceans including the brine shrimps, fairy shrimps, and water fleas. Individual name and adjectival form: **branchiopod** (no initial capital letter). *Compare* Brachiopoda.

Branchiostoma (cap. B, ital.) A genus of small burrowing marine cephalochordates. *See* amphioxus.

Brandt, Georg (1694–1768) Swedish chemist.

Branhamella (cap. B, ital.) A genus of bacteria closely related to the genus **Moraxella* and considered by some authorities to represent a subgenus of the latter. Individual name: **branhamella** (no cap., not ital.; pl. branhamellae).

Brans, Carl Henry (1935–) US mathematical physicist.

Brans–Dicke theory (en dash) Use Brans–Dicke–Jordan theory.

Brassica (cap. B, ital.) A genus of cruciferous plants including cabbages, cauliflower, Brussels sprouts, broccoli, etc. Individual name: **brassica** (no cap., not ital.; pl. brassicas).

Brassicaceae *See* Cruciferae. Adjectival form: **brassicaceous.**

Brattain, Walter Houser (1902–87) US physicist.

Braun, Karl Ferdinand (1850–1918) German physicist.
Braun tube Now called cathode-ray tube.

Braun, Wernher von Usually alphabetized as *von Braun.

Braun-Blanquet scale (hyphen) *Ecol.* Named after J. Braun-Blanquet.

Bravais, Auguste (1811–63) French physicist.
Bravais lattice

Brayton cycle *Eng.* Named after G. B. Brayton.

breadth *See* length.

breakdown (noun; one word)
breakdown voltage Abbrev.: b.d.v. (stops) or BDV (no stops).

breccia *Geol.* Adjectival form: **brecciated.**

Breit, Gregory (1899–1981) Russian-born US physicist.
Breit–Wigner formula (en dash)

bremsstrahlung (no cap.; not bremstrahlung) *Physics* [from German: braking radiation]

Brenner, Sydney (1927–) South African-born British molecular biologist.

Brevibacterium (cap. B, ital.) A genus of obligately aerobic bacteria. Individual name: **brevibacterium** (no cap, not ital.; pl. brevibacteria). Since the genus was redefined in 1980, a number of members have been reclassified. For example, *B. albidum*, *B. citreum*, *B. luteum*, and *B. pusillum* are now included in the genus **Curtobacterium*.

brewer's yeast (apostrophe) *See* Endomycetales.

Brewster, Sir David (1781–1868) British physicist.
Brewster angle
Brewster's law
Brewster window

Brianchon, Charles Julian (1783–1864) French mathematician.
Brianchon's theorem

Bridgman, Percy Williams (1882–1961) US physicist.

Briggs, Henry (1561–1630) English mathematician.
Briggsian logarithm Use common logarithm.

Brillouin, Léon (1889–1969) French physicist.
Brillouin scattering
Brillouin zone

Brindley, James (1716–72) British canal builder.

Brinell, Johann A. (1849–1925) Swedish metallurgist.
Brinell hardness
Brinell number

Bri-Nylon (initial caps.) *See* nylon.

bristleworm (one word) *See* Polychaeta.

British Summer Time Abbrev.: BST (no stops).

British thermal unit Any of several units of heat and *internal energy originally relating to the pound of water and degree Fahrenheit. The most important unit, symbol Btu, is now defined in terms of *SI units:
1 Btu/lb = 2.326 J/g (exactly),
thus 1 Btu = 1055.06 J.
The therm, a unit of heat and energy used in the gas industry, is equal to 100 000 Btu, i.e. 105.5 MJ.

brittlestar (one word) *See* Ophiuroidea.

Broca, Pierre Paul (1824–80) French physician and anthropologist.
Broca's aphasia *Med.*
Broca's area *Anat.*

Broglie, Prince Louis Victor de Usually alphabetized as *de Broglie.

bromal CBr_3CHO The traditional name for tribromoethanal.

bromate Denoting a compound containing the ion BrO_3^-, e.g. potassium bromate, $KBrO_3$. The recommended name is bromate(V).

bromate(I) Denoting a compound containing the ion BrO^-, e.g. sodium bromate(I), NaBrO. The traditional name is hypobromite.

bromate(V) Denoting a compound containing the ion BrO_3^-, e.g. potassium bromate(V), $KBrO_3$. The traditional name is bromate.

Bromeliaceae A family of monocotyledonous plants including pineapple. Individual name and adjectival form: **bromeliad** (preferred to bromeliaceous).

brome mosaic virus Abbrev.: BMV (no stops). Type member of the **Bromovirus* group (vernacular name: brome mosaic virus group).

bromic acid $HBrO_3$ The traditional name for bromic(V) acid.

bromic(I) acid HOBr The recommended name for the compound traditionally known as hypobromous acid.

bromic(V) acid $HBrO_3$ The recommended name for the compound traditionally known as bromic acid.

bromine Symbol: Br *See also* periodic table; nuclide.

bromo- Prefix denoting the bromine radical (e.g. bromobenzene, bromobutane, bromouracil).

bromoacetic acid $CH_2BrCOOH$ The traditional name for bromoethanoic acid.

bromoacetone CH_3COCH_2Br The traditional name for bromopropanone.

***N*-bromobutanedioic imide** The recommended name for the compound traditionally known as *N*-bromosuccinimide.

***N*-bromoethanamide** $BrNHCOCH_3$ The recommended name for the compound traditionally known as acetbromamide.

bromoethanoic acid $CH_2BrCOOH$ The recommended name for the compound traditionally known as bromoacetic acid.

bromoform $CHBr_3$ The traditional name for tribromomethane.

bromomethane CH_3Br The recommended name for the compound traditionally known as methyl bromide.

bromopropanone CH_3COCH_2Br The recommended name for the compound traditionally known as bromoacetone.

***N*-bromosuccinimide** The traditional name for *N*-bromobutanedioic imide.

Bromovirus (cap. B, ital.) Approved name for the *brome mosaic virus group. Individual name: **bromovirus** (no cap., not ital.). [from *brome mosaic virus*]

broncho- (**bronch-** before vowels) Prefix denoting the bronchi (e.g. bronchopulmonary, bronchitis).

bronchus (pl. bronchi) Any one of the branching air passages beyond the trachea that have cartilage and mucous glands in their walls. The smallest branches lead into **bronchioles**, which lack cartilage and mucous glands. The plural, bronchi, should not be confused with **bronchia**, the complete branching system of bronchi and bronchioles, more commonly known as the **bronchial tree**. Adjectival form: **bronchial** (not to be confused with branchial: *see* branchia).

Brongniart, Alexandre (1770–1847) French geologist and palaeontologist.

Brønsted, Johannes Nicolaus (1879–1947) Danish physical chemist.
Brønsted–Lowry theory (en dash)

brontosaurus (no cap., not ital.; pl. brontosauruses or brontosauri) A dinosaur of the genus *Apatosaurus* (originally named *Brontosaurus*).

Broom, Robert (1866–1951) British-born South African morphologist and palaeontologist.

Brouncker, William, Viscount (1620–85) English mathematician and experimental scientist.

Brouwer, Dirk (1902–66) Dutch-born US astronomer.

Brouwer, Luitzen Egbertus Jan (1881–1966) Dutch mathematician and philosopher of mathematics.

Brown, Alexander Crum Usually alphabetized as *Crum Brown.

Brown, Herbert Charles (1912–) US chemist.

Brown, Michael Stuart (1941–) US geneticist.

Brown, Robert (1773–1858) British botanist.
Brownian movement (or motion)

Brown, Robert Hanbury (1916–) British radio astronomer.

brown algae *See* Phaeophyta.

brown dwarf *See* dwarf.

Bruce, Sir David (1855–1931) British bacteriologist.
Brucella

Brucella (cap. B, ital.) A genus of nonmotile aerobic bacteria that cause brucellosis. Individual name: **brucella** (no cap., not ital.; pl. brucellae).

Brückner, Edouard (1862–1927) German meteorologist.
Brückner cycle

Brunel, Isambard Kingdom (1806–59) British civil and mechanical engineer, son of Marc Brunel.

Brunel, Sir Marc Isambard (1769–1849) French-born British engineer and inventor, father of Isambard Brunel.

Brunner, Johann Conrad von (1653–1727) Swiss anatomist.
Brunner's glands Also called duodenal glands.

Brunner, Sir John Tomlinson (1842–1919) British industrialist.

Bryales *See* Bryidae.

Bryatae *See* Bryophyta.

Bryatinae *See* Bryophyta.

Bryidae The largest subclass of mosses, comprising the true mosses. In some classifications it is reduced to an order, **Bryales**; in other classifications it is elevated to a class, *Bryopsida, which is subdivided into some 19 orders. Also called Eubrya.

bryo- Prefix denoting mosses and liverworts (e.g. bryokinin, bryology).

Bryophyta (cap. B) A division of nonvascular plants comprising the classes *Bryopsida (or Musci; mosses), *Hepaticopsida (or Hepaticae; liverworts), and *Anthocerotopsida (or Anthocerotae; horned liverworts or hornworts). Individual name and adjectival form: **bryophyte** (no cap.). In a recent classification the bryophytes comprise two infradivisions of the subdivision *Embryophytina (division Streptophyta): Bryatae, containing the superclasses Hepaticatinae (liverworts) and Bryatinae (mosses); and Anthoceratae (hornworts).

Bryopsida A class of bryophytes comprising the mosses, divided into the subclasses *Sphagnidae (bog or peat mosses), *Andreaeidae (granite mosses), and *Bryidae (true mosses). Also called Musci. In a recent classification mosses are classified as a superclass, Bryatinae (*see* Bryophyta); the subclasses mentioned above are then elevated to class status and named Sphagnopsida, Andreaeopsida, and Bryopsida, respectively.

Bryopsidophyceae A class of green algae (*see* Chlorophyta). Not to be confused with *Bryopsida.

Bryozoa (cap. B) A former phylum of aquatic invertebrate animals, also called Polyzoa, now split into two separate phyla, *Ectoprocta and *Entoprocta. The term is still used as a synonym for Ectoprocta. Individual name and adjectival form: **bryozoan** (no cap.; use only for ectoprocts).

BS Abbrev. for **1.** British Standard. **2.** Bachelor of Surgery. **3.** Bachelor of Science, from a US university.

BSA Abbrev. for bis(trimethylsilyl)-acetamide.

BSc (or **B.Sc.**) Abbrev. for Bachelor of Science.

BSE Abbrev. for bovine spongiform encephalopathy.

BSMV Abbrev. for barley stripe mosaic virus.

BST Abbrev. for British Summer Time.

BSTFA Abbrev. for bis(trimethylsilyl)-trifluoroacetamide.

BTech (or **B. Tech.**) Abbrev. for Bachelor of Technology.

BTEE Abbrev. for *N*-benzoyl-L-tyrosine ethyl ester.

BTM Abbrev. for bromotrifluoromethane.

BTMSA Abbrev. for bis(trimethylsilyl)-acetylene.

Btu Symbol for British thermal unit.

BTU Former symbol for **1.** Board of Trade Unit. **2.** British thermal unit (*see* Btu).

BTX Abbrev. for low-boiling-point mixture of benzene, toluene, and xylenes.

bu Symbol for US bushel. *See* bushel.

Bu Symbol often used to denote the butyl group in chemical formulae, e.g. $CH_3(CH_2)_2CH_2OH$ can be written as BuOH. The following symbols are similarly used:
$\mathbf{Bu^i}$ (not *i*-Bu, iBu, or isoBu) for isobutyl group, e.g. Bu^iOH is $(CH_3)_2CHCH_2OH$;
$\mathbf{Bu^n}$ (not *n*-Bu or nBu) for *n*-butyl group, e.g. Bu^nOH is $CH_3(CH_2)_2CH_2OH$;
$\mathbf{Bu^s}$ (not *s*-Bu or sBu) for *s*-butyl group, e.g. Bu^sOH is $CH_3CH_2CH(OH)CH_3$;
$\mathbf{Bu^t}$ (not *t*-Bu or tBu) for *t*-butyl group, e.g. Bu^tOH is $(CH_3)_3COH$.

buccopharyngeal (one word)

Buch, Christian Leopold von Usually alphabetized as *von Buch.

Buchner, Eduard (1860–1917) German organic chemist and biochemist.
Buchner funnel

Buchner, Hans Ernst Angass (1850–1902) German bacteriologist.

buddleia (not buddlea or buddlia) Generic name: *Buddleia* (cap. B, ital.). Named after A. Buddle (died 1715).

buffalopox (one word) *See also* poxvirus.

Buffon, Georges Louis Leclerc, Comte de (1707–88) French naturalist.
Buffon's needle *Maths.*

bug In zoology, usage should be restricted to insects of the order *Hemiptera (plant bugs, water bugs, bedbug, etc.) – the so-called 'true bugs'. The term is used loosely for any insect-like animal or pathogenic microorganism.

bulbourethral glands (preferred to Cowper's glands)

bulk modulus Symbol: K A physical quantity, the ratio of pressure, p, applied to a body and the resulting volume *strain. The isothermal bulk modulus is given by:

$$K = -V(\partial p/\partial V)_T.$$

Constant entropy rather than constant temperature gives the adiabatic bulk modulus, symbol K_S. The *SI unit is the pascal. Bulk modulus is the reciprocal of the *compressibility. Also called modulus of rigidity.

bulk strain Another name for volume strain. *See* strain.

Bullard, Sir Edward Crisp (1907–80) British geophysicist.

Bullen, Keith Edward (1906–76) Australian applied mathematician and geophysicist.

bundle of *His Use *atrioventricular bundle.

Bunsen, Robert Wilhelm (1811–99) German chemist.
Bunsen burner
Bunsen photometer Also called grease-spot photometer.

bunyavirus (one word) Any member of the genus *Bunyavirus* (cap. B, ital.). To avoid confusion, authors should not extend usage to include other genera of the family Bunyaviridae. Named after Bunyamwera, the Ugandan location where the type species was first isolated.

buoyancy (not bouyancy) Adjectival form: **buoyant**.

Burbank, Luther (1849–1926) US plant breeder.
Burbank potato

Burbidge, Geoffrey (1925–) British astrophysicist, husband of Margaret Burbidge. *See also* B^2FH theory.

Burbidge, (Eleanor) Margaret (1922–) British astronomer, née Peachey, wife of Geoffrey Burbidge. *See also* B^2FH theory.

Burnet, Sir Frank Macfarlane (1899–1985) Australian virologist.

burnout (noun; one word) *Elec. eng., electronics, etc.*

burnup (noun; one word) *Nuc. eng.*

bursa of *Fabricius

bushel In the US, a unit of capacity, symbol bu, used for dry measure only; it is equal to 2150.42 cubic inches, 35.2391 cubic decimetres. (The US *gallon can be used for liquid measure.)
1 bushel = 4 pecks = 32 dry quarts = 64 dry pints
The US bushel is being displaced by the cubic metre, cubic decimetre, etc.
In the UK, the bushel, equal to 8 UK gallons, and its submultiple the peck, equal to 2 UK gallons, are obsolete.

butadiene $CH_2CHCHCH_2$ The traditional name for buta-1,3-diene.

buta-1,3-diene $CH_2CH{=}CHCH_2$ The recommended name for the compound traditionally known as butadiene.

buta-1,3-diyne $CH{\equiv}CC{\equiv}CH$ The recommended name for the compound traditionally known as diacetylene.

butane $CH_3CH_2CH_2CH_3$ The recommended name for the compound traditionally known as *n*-butane.

butanedial $OHC(CH_2)_2CHO$ The recommended name for the compound traditionally known as succinaldehyde.

butanedioic acid $HOOC(CH_2)_2COOH$ The recommended name for the compound traditionally known as succinic acid.

```
CH2CO
 |    \
 |     O
 |    /
CH2CO
```

butanedioic anhydride
(succinic anhydride)

butanedioic anhydride The recommended name for the compound traditionally known as succinic anhydride.

butanedione $CH_3COCOCH_3$ The recommended name for the compound traditionally known as diacetyl.

butanedione dioxime $CH_3C(NOH)$-$C(NOH)CH_3$ The recommended name for the compound traditionally known as dimethylglyoxime.

butanoic acid $CH_3CH_2CH_2COOH$ The recommended name for the compound traditionally known as *n*-butanoic acid or butyric acid.

butan-1-ol $CH_3CH_2CH_2CH_2OH$ The recommended name for the compound traditionally known as *n*-butyl alcohol.

butan-2-ol $CH_3CH_2CH(CH_3)OH$ The recommended name for the compound traditionally known as *s*-butyl alcohol.

butanone $CH_3COCH_2CH_3$ The recommended name for the compound traditionally known as methyl ethyl ketone.

butanoyl- (preferred to butyryl-) *Prefix denoting either of the groups $CH_3(CH_2)_2CO$- (*n*-butanoyl-) and $(CH_3)_2CHCO$-(isobutanoyl) derived from the respective butanoic acids (e.g. *n*-butanoylacetone, isobutanoyl chloride).

buten-2-al $CH_3CH{=}CHCHO$ The recommended name for the compound traditionally known as crotonaldehyde.

Butenandt, Adolf Friedrich Johann (1903–) German organic chemist and biochemist.

butene $CH_3CH_2CH{=}CH_2$ The recommended name for the compound traditionally known as butylene. The recommended name for 1-butylene is but-1-ene, etc.

***cis*-butenedioic acid** $HCOOCH{=}CHCOOH$ The recommended name for the compound traditionally known as maleic acid.

***trans*-butenedioic acid** $HCOOCH{=}CHCOOH$ The recommended name for the compound traditionally known as fumaric acid.

```
HC — CO
||     \
||      O
||     /
HC — CO
```

butenedioic anhydride
(maleic anhydride)

butenedioic anhydride The recommended name for the compound traditionally known as maleic anhydride.

***cis*-but-2-enoic acid** $CH_3CH{=}CHCOOH$ The recommended name for the compound traditionally known as *cis*-crotonic acid.

***trans*-but-2-enoic acid** $CH_3CH{=}CHCOOH$ The recommended name for the compound traditionally known as *trans*-crotonic acid.

but-2-en-1-ol $CH_3CH{=}CHCH_2OH$ The recommended name for the compound traditionally known as crotyl alcohol.

Butlerov, Aleksandr Mikhailovich (1828–86) Russian chemist.

butyl- *Prefix denoting the group C_4H_9–. Isomers are *n*-butyl, $CH_3(CH_2)_2CH_2$–; isobutyl, $(CH_3)_2CHCH_2$–; *s*-butyl, $CH_3CH_2CH(CH_3)$–; and *t*-butyl, $(CH_3)_3C$–. Examples are *n*-butyl alcohol, isobutyl alcohol, *t*-butylbenzene.

***n*-butyl alcohol** $CH_3CH_2CH_2CH_2OH$ The traditional name for butan-1-ol.

***s*-butyl alcohol** $CH_3CH_2CH(CH_3)OH$ The traditional name for butan-2-ol.

***t*-butyl alcohol** $(CH_3)_3COH$ The traditional name for 2-methylpropan-2-ol.

***n*-butyraldehyde** $CH_3CH_2CH_2CHO$ The traditional name for butanal.

butylene $CH_3CH_2CHCH_2$ The traditional name for *butene.

***n*-butyric acid** $CH_3CH_2CH_2COOH$ The traditional name for butanoic acid.

butyro- (**butyr-** before vowels) Prefix denoting the group $CH_3(CH_2)_2C(O)$– (e.g. butyrophenone, butyrone).

butyryl- Use *butanoyl-.

Buys Ballot, Christoph Hendrik Diederik (1817–90) Dutch meteorologist.
Buys Ballot's law

bv. (stop) Abbrev. for biovar.

BWR Abbrev. for boiling-water reactor.

BYDV Abbrev. for barley yellow dwarf virus.

bypass (one word)

by-product (hyphenated)

Byrd, Richard Evelyn (1888–1957) US polar explorer.

byte A fixed number of *bits (now almost always 8 bits) that can be handled and stored as a single unit by a computer. It can for example be a character, such as a letter or digit, since characters are usually represented by an 8-bit code. The main store in a computer is divided into either byte or *word storage locations. A byte is shorter than a word.

bytownite (not bitownite) *Min.*

Bz Symbol often used to denote the benzoyl group in chemical formulae, e.g. C_6H_5-COCl can be written as BzCl.

Bzl Use *Bn.

C

c Symbol for centi-.

c Symbol (light ital.) for **1.** concentration (c_B concentration of substance B). **2.** specific heat capacity (c_p (ital. p) at constant pressure, c_V (ital. V) at constant volume). **3.** speed of light in vacuum; c_a speed of sound.

C 1. A programming language; C++ is an object-oriented version. **2.** Symbol for **i.** *carbon. **ii.** *complement. **iii.** constant region (of an *immunoglobulin chain). **iv.** corolla (in a *floral formula). **v.** coulomb. **vi.** cysteine. **vii.** cytidine. **viii.** cytosine. **3.** Abbrev. for **i.** Cambrian. **ii.** Carboniferous.
°C Symbol for degree Celsius.

C Symbol (light ital.) for **1.** capacitance. **2.** charm quantum number. **3.** Euler constant. **4.** heat capacity (C_p (ital. p) at constant pressure, C_V (ital. V) at constant volume, C_m molar heat capacity). **5.** molecular concentration (C_B of substance B).

c. (light ital.) Abbrev. for circa.

Ca Symbol for calcium.

cable *See* sea mile.

cable television Abbrev.: CATV (no stops).

cache (or **cache memory**) *Computing* [from French]

cactus (pl. cacti; not cactuses) Any flowering plant of the family Cactaceae. Use of the word as the trivial name for members of the genus *Cactus* (cap. C, ital.) of this family is confusing and should be avoided.

CAD Acronym for computer-aided design.

CADCAM (or **CAD/CAM**) Acronym for computer-aided design, computer-aided manufacturing.

CADMAT Acronym for computer-aided design, manufacturing, and testing.

cadmium Symbol: Cd *See also* periodic table; nuclide.

Cae (no stop) *Astron.* Abbrev. for Caelum.

CAE Abbrev. for computer-aided engineering.

caecum (pl. caeca) US: **cecum** (pl. ceca). *Anat., zool.* Adjectival form: **caecal** (US: **cecal**).

Caelum A constellation. Genitive form: Caeli. Abbrev.: Cae (no stop). *See also* stellar nomenclature.

caeno- (not caino-) US: **ceno-**. Prefix denoting recent or new (e.g. caenogenesis). *See also* ceno-.

Caesalpiniaceae *See* Leguminosae. Named after A. *Cesalpino.

Caesarean section US: **cesarean section**. *Obstet.* Named after Julius Caesar (100 BC–44 BC).

caesium US: **cesium**. Symbol: Cs *See also* periodic table; nuclide.

Cagniard De La Tour, Charles (1777–1859) French physicist.

Cahn–Ingold–Prelog sequence rules (en dashes) *Chem.*

Cahours, August André Thomas (1813–91) French organic chemist.

CAI Abbrev. for **1.** computer-aided (or -assisted) instruction. **2.** *Forestry* current annual increment.

Cailletet, Louis Paul (1832–1913) French physicist.
Cailletet process

caiman Use cayman as the common name, but note that the generic name is *Caiman* (cap. C, ital.).

caino- Use *caeno- or *ceno-.

cal (no stop) Symbol for *calorie.

CAL Acronym for computer-aided (or -assisted) learning.

Calamitales *See* Sphenopsida.

calc. Abbrev. for calculated.

calcareous (not calcarious) Containing or consisting of calcium carbonate or limestone. Not to be confused with 'calcarean', a sponge belonging to the class Calcarea.

calci- (**calc-** before vowels) Prefix denoting calcium, calcium salts, or lime (e.g. calcicole, calcifuge, calcite).

calciferol Use *ergocalciferol. *See also* vitamin D.

calcitonin (preferred to thyrocalcitonin)

calcium Symbol: Ca *See also* periodic table; nuclide.

calcium acetylide CaC_2 The traditional name for calcium dicarbide.

calcium bicarbonate $Ca(HCO_3)_2$ The traditional name for calcium hydrogencarbonate.

calcium carbide CaC_2 The traditional name for calcium dicarbide.

calcium dicarbide CaC_2 The recommended name for the compound traditionally known as calcium carbide or calcium acetylide.

calcium hydrogencarbonate $Ca(HCO_3)_2$ The recommended name for the compound traditionally known as calcium bicarbonate.

calcium hydroxide $Ca(OH)_2$ The recommended name for the compound traditionally known as slaked lime.

calcium octadecanoate $Ca(CH_3(CH_2)_{16}COO)_2$ The recommended name for the compound traditionally known as calcium stearate.

calcium oxide CaO The recommended name for the compound traditionally known as lime or quicklime.

calcium stearate $Ca(CH_3(CH_2)_{16}COO)_2$ The traditional name for calcium octadecanoate.

calcium sulphate-½-water $CaSO_4 \cdot \frac{1}{2}H_2O$ The recommended name for the compound traditionally known as plaster of Paris.

caldera (pl. calderas) *Geol.*

calendar A system for the reckoning of time. Adjectival forms: **calendrical**, **calendric**. *Compare* calender.

calender A machine for smoothing paper or cloth. *Compare* calendar.

calibre US: **caliber**.

californium Symbol: Cf *See also* periodic table; nuclide.

caliper US spelling of calliper.

Callendar, Hugh Longbourne (1863–1930) British physicist.

Callendar effect *Meteorol.* Named after G. S. Callendar.

calliper US: **caliper**.

callose A plant polysaccharide that forms a *callus in phloem sieve tubes.

callus (pl. calluses, not calli) A hard tissue mass that forms on rubbed skin (also called **callosity**), injured plant parts, or the fractured ends of bones. Adjectival form: **callous** (not to be confused with *callose).

calomel Hg_2Cl_2 The traditional name for dimercury(I) chloride.

calor- Prefix denoting heat (e.g. calorific, calorimeter).

calorie Symbol: cal (no stop) Any of several units of heat and *internal energy originally relating to the gram of water and the degree Celsius and now all deprecated. Three of these were used for precise measurements: the International Table calorie (symbol cal_{IT}), the thermochemical calorie (cal_{th}), and the 15°C calorie (cal_{15}):

1 cal_{IT} = 4.1868 joules (exactly)
1 cal_{th} = 4.1840 joules (exactly)
1 cal_{15} = 4.1855 joules (approx.).

The 'calorie' used in food science is in fact a kilocalorie, based on the cal_{15}, and is sometimes called a 'large calorie'. It is often written with a capital C to distinguish it from the 'small calorie'. All these units should be avoided in favour of the joule.

Caltech Short for California Institute of Technology.

Calvin, Melvin (1911–) US chemist and biochemist.
Calvin cycle (preferred to Benson–Calvin–Bassham cycle)

calyx (not calix; pl. calyces; preferred to calyxes) **1.** *Bot.* The sepals of a flower, collectively. **2.** *Anat., zool.* A cup-shaped part, especially a division of the kidney pelvis.

Cam (no stop) *Astron.* Abbrev. for Camelopardalis.

CAM Acronym for **1.** computer-aided manufacturing. **2.** content-addressable memory. **3.** crassulacean acid metabolism: often used adjectivally (e.g. CAM plants).

cambium (pl. cambia) *Bot.* Adjectival form: **cambial.**

Cambrian Abbrev.: C (no stop). **1.** (adjective) Denoting the first period of the Palaeozoic era. **2.** (noun; preceded by 'the') The Cambrian period. *See also* Precambrian.

camellia (not camelia; pl. camellias) Generic name: *Camellia* (cap. C, ital.).

Camelopardalis A constellation. Genitive form: Camelopardalis. Abbrev.: Cam (no stop). *See also* stellar nomenclature.

Camerarius, Rudolph Jacob (1665–1721) German botanist.

camomile (not chamomile)

cAMP (lower-case c) Abbrev. for cyclic AMP (adenosine 3′-5′-phosphate).

CaMV (lower-case a) Abbrev. for cauliflower mosaic virus.

Canada–France–Hawaii Telescope (en dashes) Abbrev.: *CFHT (no stops).

canary pox (two words) *See also* poxvirus.

Cancer A constellation. Genitive form: Cancri. Abbrev.: Cnc (no stop). *See also* stellar nomenclature.

candela Symbol: cd The *SI unit of *luminous intensity. It is one of the seven SI base units, defined since 1979 as the luminous intensity in a given direction of a source that emits monochromatic radiation of frequency 540×10^{12} hertz and of which the radiant intensity in that direction is 1/683 watt per steradian.

Candolle, Augustin Pyrame de (1778–1841) Swiss botanist.

Canes Venatici A constellation. Genitive form: Canum Venaticorum. Abbrev.: CVn (no stops). *See also* stellar nomenclature.

canine 1. A mammalian tooth between the incisors and premolars. **2.** Any mammal belonging to the dog family, Canidae, especially any member of the subfamily Caninae (which includes most of the family). Use 'canine' when the other two subfamilies are specifically excluded, but prefer 'canid' when referring to *all* members of the family. Adjectival form (for both senses): **canine.**

Canis Major A constellation. Genitive form: Canis Majoris. Abbrev.: CMa (no stops). *See also* stellar nomenclature.

Canis Minor A constellation. Genitive form: Canis Minoris. Abbrev.: CMi (no stops). *See also* stellar nomenclature.

Cannizzaro, Stanislao (1826–1910) Italian chemist.
Cannizzaro reaction

Cannon, Annie Jump (1863–1941) US astronomer.

canonical (not cannonical) *Physics, maths.*

Cantab (no stop) Abbrev. for Cantabrigiensis (Latin: of Cambridge), used with academic awards.

Cantor, Georg (1845–1918) German mathematician.
Cantor's continuum hypothesis
Cantor's paradox
Cantor's theory of sets

Cap (no stop) *Astron.* Abbrev. for Capricornus.

CAP Abbrev. for catabolite activator protein.

capacitance Symbol: C A physical quantity, the ability of a conductor or system to store *electric charge, given by the magnitude of the charge of one sign divided by *potential difference. The *SI unit is the farad. Former names: capacity, electrical capacity. Adjectival form: **capacitative.**

capacitor An electrical component having *capacitance. Former name: condenser.

capacity Former name for capacitance.

capillary 1. (adjective) Denoting a tube with a fine bore. Derived noun: **capillarity**. **2.** (noun) *Anat.* A small blood vessel.

capitellum (pl. capitella) The rounded articulating upper end of the humerus, i.e. the *capitulum of the humerus.

capitulum (pl. capitula) **1.** *Bot.* A type of inflorescence typical of family Compositae. **2.** *Anat.* The rounded articulating end of a bone. *Compare* capitellum.

Capricornus A constellation. Genitive form: Capricorni. Abbrev.: Cap (no stop). *See also* stellar nomenclature.
Capricornids (meteor shower)
Alpha Capricornids (meteor shower)

Capripoxvirus (cap. C, ital.) Approved name for a genus of *poxviruses. Vernacular name: sheep pox subgroup. Individual name: **capripoxvirus** (no cap., not ital.).

caproic acid $CH_3(CH_2)_4COOH$ The traditional name for hexanoic acid.

caprylic acid $CH_3(CH_2)_6COOH$ The traditional name for octanoic acid.

Car (no stop) *Astron.* Abbrev. for Carina.

carat 1. A measure of the quantity of gold in an alloy, expressed as parts of gold in 24 parts of the alloy: 24 carat gold is pure gold; 9 carat gold has 9 parts gold in 24 parts. **2.** *See* metric carat.

Carathéodory's principle *Thermodynamics* Named after Constantin Carathéodory (1873–1950).

carbamide $(H_2N)_2CO$ The recommended name for the compound traditionally known as urea.

carbaminohaemoglobin (one word) US: **carbaminohemoglobin**. The complex formed when haemoglobin combines with carbon dioxide. *Compare* carboxyhaemoglobin.

carbo- (**carb-** before vowels) Prefix denoting carbon (e.g. carbohydrate, carbamide).

carbobenzyloxy Abbrev.: CBZ (no stops).

carbolfuchsin (one word)

carbon Symbol: C It is recommended by IUPAC that the names of the different forms of carbon should be given as carbon (graphite), carbon (diamond), carbon (charcoal), carbon (wood charcoal), carbon (animal charcoal), etc. *See also* periodic table; nuclide.
The following conventions are used in chemistry where *n* and *m* are integers:
C_n denotes the number of atoms of carbon in a molecule.
C*n* (preferred to C-*n*) denotes the carbon atom at position number *n*.
$C_{n:m}$ denotes the number of atoms of carbon (*n*) and the number of double bonds (*m*) in a fatty acid.

carboniferous Bearing or yielding coal or carbon. *See also* Carboniferous.

Carboniferous (cap. C) Abbrev.: C (no stop). **1.** (adjective) Denoting the penultimate period of the Palaeozoic era, characterized by extensive deposits of coal (hence the name). **2.** (noun; preceded by 'the') The Carboniferous period. In European literature the period is divided into the Lower Carboniferous (or Dinantian) and the Upper Carboniferous (or Silesian) subperiods (initial caps.). In the USA the *Mississippian and *Pennsylvanian subperiods correspond approximately to the Lower and Upper Carboniferous, respectively.

carbon:nitrogen ratio *Ecol.* Abbrev.: C/N ratio.

carbon suboxide OCCCO The traditional name for tricarbon dioxide.

carbon tetrabromide CBr_4 The traditional name for tetrabromomethane.

carbon tetrachloride CCl_4 The traditional name for tetrachloromethane.

carbonyl- *Prefix denoting the group =CO in such compounds as aldehydes, ketones, and inorganic derivatives (e.g. carbonyl chloride, hexacarbonylcobalt).

carbonyl chloride Cl_2CO The recommended name for the compound traditionally known as phosgene.

carboxy- Prefix denoting an association with the carboxyl group –COOH (e.g. the enzyme carboxylase catalyses decarboxylation).

carboxyhaemoglobin (one word) US: **carboxyhemoglobin.** The complex formed when haemoglobin combines with carbon monoxide. *Compare* carbamino-haemoglobin.

carboxylic acid Any of a class of organic compounds containing the carboxyl group –COOH.

Carboxylic acids are systematically named either by adding the suffix -oic to the name of the parent hydrocarbon, e.g. ethanoic acid, CH_3COOH, and 3-phenylpropanoic acid, $C_6H_5CH_2CH_2COOH$, or by regarding the carboxyl group as a substituent, e.g. 3-phenylpropanecarboxylic acid, C_6H_5-$CH_2CH_2CH_2COOH$. The aromatic member C_6H_5COOH is anomalous as it can either retain its nonsystematic name, benzoic acid, or it can be called benzenecarboxylic acid.

Two nonsystematic methods of nomenclature are encountered. In one, the trivial names of carboxylic acids are used, e.g. α-methylbutyric acid, CH_3CH_2CH-$(CH_3)COOH$, and *n*-valeric acid, CH_3-$CH_2CH_2CH_2COOH$. In the other, carboxylic acids – with the exception of formic acid (methanoic acid, HCOOH) – are named as derivatives of acetic acid (ethanoic acid), e.g. methylacetic acid, CH_3CH_2-COOH, and *t*-butylacetic acid, $(CH_3)_3$-CCH_2COOH.

carboxymethylcellulose (one word) Abbrev. and preferred form: CM-cellulose (hyphenated).

carcino- (**carcin-** before vowels) Prefix denoting cancer (e.g. carcinogen, carcinoma).

carcinogen Any agent capable of causing cancer. Adjectival form: **carcinogenic.** *See also* carcinogenesis; carcinogenicity.

carcinogenesis The development of cancer from normal cells. *Compare* carcinogenicity.

carcinogenicity The extent to which a carcinogen will induce cancer. *Compare* carcinogenesis.

carcinoma (pl. carcinomata or (not in medical usage) carcinomas) Adjectival form: **carcinomatous.**

Cardano, Gerolamo (1501–76) Italian mathematician, physician, and astrologer. Anglicized name: Jerome Cardan.

Cardan's formula (preferred to Cardano's formula)

cardio- (**cardi-** before vowels) Prefix denoting the heart (e.g. cardiovascular, cardialgia).

cardo (pl. cardines) The hinge of a bivalve shell.

Carey–Foster bridge (en dash) *Elec. eng.*

Carina A constellation. Genitive form: Carinae. Abbrev: Car (no stop). *See also* stellar nomenclature.

carinate Describing birds having a keel (carina) to the sternum. Such birds were formerly classified on this basis as the subclass or superorder Carinatae, but the term carinate now has no taxonomic significance. *See also* neognathous. *Compare* ratite.

Carius method *Chem.*

Carlavirus (cap. C, ital.) Approved name for the *carnation latent virus group. Individual name: **carlavirus** (no cap., not ital.). [from *car*nation *la*tent *virus*]

carnation latent virus Abbrev.: CLV (no stops). Type member of the **Carlavirus* group (vernacular name: carnation latent virus group).

Carnivora (cap. C) An order of predominantly flesh-eating mammals including the cats, dogs, bears, raccoons, badgers, etc. Individual name: **carnivore** (no cap.), but as this word is also used for any flesh-eating animal, individuals are usually designated as 'a member of the Carnivora' (sing.) or 'the Carnivora' (pl.). This also avoids the apparent contradiction of describing pandas and other herbivorous Carnivora as 'carnivores'. The adjective 'carnivorous' is restricted to the flesh-eating habit.

carnivore 1. Any flesh-eating organism. Adjectival form: **carnivorous. 2.** *See* Carnivora.

carnivorous plant Any plant that supplements its mineral uptake by digesting small animals, including – but not restricted to – insects; hence 'carnivorous plant' is preferred to the synonym 'insectivorous plant'.

Carnot, Nicolas Leonard Sadi (1796–1832) French physicist.
Carnot–Clausius equation (en dash)
Carnot cycle
Carnot's theorem

Caro, Heinrich (1834–1910) German organic chemist.
***Caro's acid**

Caro's acid H_2SO_5 A traditional name for peroxosulphuric(VI) acid.

carotene (not carotin) Any one of a class of plant pigments; the principal types are **α-carotene** and **β-carotene** (hyphenated). *Compare* carotenoid.

carotenoid Any of a group of plant pigments, including the *carotenes and xanthophylls.

Carothers, Wallace Hume (1896–1937) US industrial chemist.

-carp Noun suffix denoting a fruit or fruiting body (e.g. ascocarp, endocarp, pericarp). Adjectival form: **-carpic**.

carpal 1. (adjective) Relating to the *carpus. **2.** (noun) A bone of the carpus; a carpal bone. *See also* carpo-. *Compare* carpel.

carpel The female reproductive structure in flowering plants. Adjectival form: **carpellary**. *See also* carpo-. *Compare* carpal.

-carpic *See* -carp; -carpy.

carpo- Prefix denoting **1.** fruit or a female reproductive structure in plants (e.g. carpogonium, carpospore). **2.** the carpus (e.g. carpometacarpus).

-carpous Adjectival suffix denoting carpels (e.g. apocarpous, syncarpous). Noun form: **-carpy**.

carpus (pl. carpi) The skeleton of the wrist or corresponding part, consisting of a number of small bones (**carpals** or **carpal bones**). Adjectival form: **carpal**.

-carpy Noun suffix denoting **1.** fruit production (e.g. parthenocarpy). Adjectival form: **-carpic**. **2.** carpals (*see* -carpous).

Cartan, Elie Joseph (1869–1951) French mathematician.

Cartesian Adjectival form of *Descartes.

Cartesian coordinates *See* coordinates.

Cartwright, Edmund (1743–1823) British inventor and industrialist.

caryo- (**cary-** before vowels) Prefix denoting a nut or nucleus; use *karyo- except in botanical senses denoting a nutlike fruit (e.g. caryopsis).

Cas (no stop) *Astron.* Abbrev. for Cassiopeia.

CAS Abbrev. for Chemical Abstracts Service.

CASE Acronym for computer-aided (or -assisted) software engineering.

Casimir, Hendrik Brugt Gerhard (1909–) Dutch physicist.
Casimir operator

Casparian strip (or **band**) (cap. C) *Bot.* Named after R. Caspary (19th century).

Caspersson, Torbjörn Oskar (1910–) Swedish cytochemist.

cassava latent virus Abbrev.: CLV (no stops).

Cassegrain, N. (*fl.* 1650–75) French telescope designer.
Cassegrain focus
Cassegrain telescope

cassia A spice resembling cinnamon, obtained from the bark of the tree *Cinnamomum cassia*. *Compare Cassia.*

Cassia (cap. C, ital.) A genus of plants, some species of which are the source of the laxative senna. *Compare* cassia.

Cassini, Giovanni Domenico (1625–1712) Italian-born French astronomer.
Cassini division

Cassiopeia A constellation. Genitive form: Cassiopeiae. Abbrev.: Cas (no stop). *See also* stellar nomenclature.

Castner, Hamilton Young (1858–98) US chemist.
Castner–Kellner process (en dash)

CAT Acronym for **1.** clear-air turbulence. **2.** computerized axial (or computer-assisted or -aided) tomography. Use CT (*see* computerized tomography). **3.** chloramphenicol acetyl transferase (*see* CAT assay).

cata- (**cat-** or **cath-** before vowels) Prefix denoting **1.** down or lower in position (e.g. catadromous, cataphyll, cation, cathode). **2.** breakdown or degeneration (e.g. catabolism, cataclasis, catalysis). **3.** reversal (e.g. catoptric). *See also* kata-. *Compare* ana-.

catabolism (not katabolism) Adjectival form: **catabolic**. Derived noun: **catabolite**.

catabolite activator protein Abbrev.: CAP (no stops). A protein required for the initiation of transcription by RNA polymerase of catabolite-dependent operons in *E. coli*. Also called cyclic AMP receptor protein (abbrev.: CRP).

catalogue equinox *See* equinox.

catalyse US: **catalyze**.

catalysis (pl. catalyses) Adjectival form: **catalytic**. Derived noun: **catalyst**.

catarrhine (not catarhine) A member of the Catarrhini, an infraorder of the Primates comprising the Old World monkeys. Adjectival form: **catarrhine**. *Compare* platyrrhine.

CAT assay Abbrev. and preferred form for chloramphenicol acetyl transferase assay, used to determine the activity of a particular eukaryote promoter gene.

catechol $C_6H_4(OH)_2$ The traditional name for benzene-1,2-diol.

catena- Prefix denoting a chain of atoms (e.g. catenasulphide).

cath- *See* cata-.

cathode ray (two words)
cathode-ray oscilloscope Abbrev.: CRO (no stops). Usually shortened to oscilloscope.
cathode-ray tube Abbrev.: CRT (no stops). Often shortened to tube.

CAT scanner Use CT scanner. *See* computerized tomography.

CATV Abbrev. for cable television.

Cauchy, Baron Augustin Louis (1789–1857) French mathematician.
Cauchy convergence test
Cauchy–Hadamard formula (en dash)
Cauchy integral
Cauchy–Riemann integral (en dash)
Cauchy sequence
Cauchy's integral theorem
Cauchy's residue theorem

cauliflower mosaic virus Abbrev.: CaMV (no stops). Type member of the **Caulimovirus* group (vernacular name: cauliflower mosaic virus group).

Caulimovirus (cap. C, ital.) Approved name for the *cauliflower mosaic virus group. Individual name: **caulimovirus** (no cap., not ital.). [from *cauli*flower *mo*saic *virus*]

Caulobacter (cap. C, ital.) A genus of prosthecate bacteria. Individual name: **caulobacter** (no cap., not ital.). Some authors use the term 'caulobacter' to include members not only of *Caulobacter* but also of *Asticcacaulis* and *Prosthecobacter*, genera containing species originally placed in the genus *Caulobacter*. Such usage can lead to confusion and should be avoided.

Cavalieri, Francesco Bonaventura (1598–1647) Italian mathematician.
Cavalieri's principle

Cavendish, Henry (1731–1810) English chemist and physicist.
Cavendish experiment

Caventou, Jean Bienaimé (1795–1877) French pharmacist and organic chemist.

Caxton, William (*c.* 1422–91) English printer.

Cayley, Arthur (1821–95) British mathematician.
Cayley–Hamilton theorem (en dash)
Cayley–Klein parameters (en dash)
Cayley table

Cayley, Sir George (1773–1857) British inventor

cayman (not *caiman) *See* Crocodilia.

Caytoniales An extinct order of cycads (*see* Cycadopsida).

Cb (ital.) Abbrev. for cumulonimbus.

C-banding (cap. C, hyphenated) Abbrev. and preferred form for centromeric banding, a staining technique that visualizes the centromeres of chromosomes.

CBiol Abbrev. for Chartered Biologist. Fellows and Members of the Institute of Biology have this additional designation, which precedes and is separate from the designations FIBiol and MIBiol.

CBZ Abbrev. for carbobenzyloxy.

cc Alternative symbol for cubic centimetre; cm^3 is preferred.

Cc (italic type) Abbrev. for cirrocumulus.

CCD *Electronics* Abbrev. for charge-coupled device.

CChem Abbrev. for Chartered Chemist. Fellows and Members of the Royal Society of Chemistry have this additional designation, which precedes and is separate from the desginations FRSC and MRSC.

CCIR Abbrev. for Comité Consultatif International des Radiocommunications (International Radio Consultative Committee).

CCITT Abbrev. for Comité Consultatif International Télégraphique et Téléphonique (International Telegraph and Telephone Consultative Committee).

CCK Abbrev. for *cholecystokinin.

c.c.p. Abbrev. for cubic close-packed.

CCTV Abbrev. for closed-circuit television.

cd Symbol for candela. *See also* SI units.

Cd Symbol for cadmium.

CD Abbrev. for **1.** compact disc. **2.** *Astron.* Córdoba Durchmusterung (Star Catalogue): used, followed by an Arabic numeral, to designate a star listed in this catalogue. **3.** *Immunol.* cluster of differentiation: used, followed by an Arabic numeral, to classify *T-cell differentiation antigens, i.e. CD1, CD2, CD3, etc.

CDF Abbrev. for Collider Detector at Fermilab.

cDNA (lower-case c) Abbrev. for complementary DNA.

CDP Abbrev. and preferred form for cytidine 5′-diphosphate.

CD-ROM (pronounced see-dee-**rom**) *Computing* Acronym for compact-disc read-only memory.

CDTA Abbrev. for 1,2-cyclohexylenedinitrotetraacetic acid.

Ce Symbol for cerium.

CE (small caps., no stops) Abbrev. for Common (or Christian) Era; always place before the year.

cecum US spelling of *caecum.

Ceefax (cap. C) A trade name for the BBC's teletext system.

-cele (not -coele) Noun suffix denoting a swelling or hernia (e.g. omphalocele). *Compare* -coel.

celio- US spelling of *coelio-.

cell cycle The series of events that occurs in a cell between one mitosis and the next. It is divided into phases, designated by capital letters: the M (mitotic) phase is followed successively by the G_1 phase (G = gap; note subscript number), S (synthesis) phase, and G_2 phase. *See also* restriction point.

Cellophane (cap. C) In the UK, a trade name for a cellulose-based transparent wrapping material. In the USA it is now a generic name rather than a trade name and thus not capitalized.

cellular slime moulds *See* Acrasiomycetes.

Cellulomonas (cap. C, ital.) A genus of rod-shaped Gram-positive bacteria. Individual name: **cellulomonad** (no cap., not ital.).

celom US variant spelling of *coelom.

Celon *See* nylon.

Celsius, Anders (1701–44) Swedish astronomer.
***Celsius temperature**
***degree Celsius**

Celsius temperature Symbol: t, or θ (Greek theta) if t is required as the symbol for time. A physical quantity, a *temperature, now defined in terms of *thermodynamic temperature. The *SI unit is the degree Celsius (formerly called degree centigrade), which is a special name for the kelvin used in expressing Celsius temperature. The relationship between Celsius temperature t and thermodynamic temperature T is:

$$t = T - T_0 = T - 273.15,$$

where T_0 is a thermodynamic temperature fixed as 0.01 K below the triple point of water. Originally, Celsius temperature was measured on a scale (the centigrade scale) in which the melting point of ice was designated 0 °C and the boiling point of water was 100 °C.

Cen (no stop) *Astron.* Abbrev. for Centaurus.

CEN Acronym for Comité Européen Normalisation (European Standardization Committee). CEN and *CENELEC are the official standards bodies of the European Community.

-cene Noun suffix denoting a recent geological epoch (e.g. Miocene).

CENELEC Acronym for Comité Européen Normalisation Electrotechnique (European Electrotechnical Standardization Committee). *See also* CEN.

CEng (or **C.Eng.**) Abbrev. for Chartered Engineer.

ceno- 1. (not caino-) Prefix denoting recent or new (e.g. Cenozoic). *See also* caeno-. **2.** US variant of *coeno-.

Cenozoic (preferred to Caenozoic, Cainozoic, and Kainozoic) Abbrev.: Cz (no stop). **1.** (adjective) Denoting an era of the geological time scale. **2.** (noun; preceded by 'the') The Cenozoic era.

cental A unit of mass equal to 100 pounds. In the US this is known as the short hundredweight.

Centaurus A constellation. Genitive form: Centauri. Abbrev.: Cen (no stop). *See also* stellar nomenclature.

center US spelling of centre.

centi- Symbol: c A prefix to a unit of measurement that indicates 10^{-2} times that unit, as in centimetre (cm). *See also* SI units.

centigrade *See* degree centigrade.

centimetre US: **centimeter**. Symbol: cm (no stop). The fundamental unit of length in the system of *cgs units. In *SI units, the metre, equal to 100 cm, is the base unit of length. The centimetre can therefore be used, as a submultiple of the metre, with SI units. One cm is equal to 0.393 701 inches, 0.032 808 feet.

centimorgan Abbrev.: cM (no stops; not cm). A unit of length used in chromosome mapping and equal to one *map unit.

Central European Time Abbrev.: CET (no stops).

central meridian Abbrev.: CM (no stops).

central nervous system Abbrev.: CNS (no stops). Latin names for parts of the CNS are printed in roman (upright) type, not italic (e.g. pars distalis of the pituitary gland).

central processing unit (not hyphenated) *Computing* Abbrev.: CPU or cpu (no stops). Now usually called central processor.

Central Standard Time Abbrev.: CST (no stops).

centre US: **center**. Verb form: **centring, centred** (US: **centering, centered**).

centri- (or **centro-**) Prefix denoting the centre (e.g. centrifuge, centripetal, centromere, centrosome).

centromeric banding *Genetics* Abbrev. and preferred form: *C-banding.

cep (no accent) Common name for any edible fungus of the genus *Boletus*, especially *B. edulis*. [from French *cèpe*]

Cep (no stop) *Astron.* Abbrev. for Cepheus.

-cephalic (or, less commonly, **-cephalous**) Adjectival suffix denoting the head (e.g. brachycephalic). Noun form: **-cephaly** or (less commonly) **-cephalism**.

cephalin Use phosphatidylethanolamine.

cephalo- (**cephal-** before vowels) Prefix denoting the head (e.g. cephalothorax).

Cephalochordata (cap. C) A subphylum of invertebrate chordate animals including *amphioxus. Former name: Acrania. Individual name and adjectival form: **cephalochordate** (no cap.).

Cephalopoda (cap. C) A class of marine molluscs including the squids, cuttlefishes, and octopus. Former name: Siphonopoda. Individual name and adjectival form: **cephalopod** (no cap.; not cephalopodous).

Cepheus A constellation. Genitive form: Cephei. Abbrev.: Cep (no stop). *See also* stellar nomenclature.
Cepheids (meteor shower)
Cepheid variable or **Cepheid**

cerato- (**cerat-** before vowels) Prefix denoting a horn, hornlike part, or horny tissue; use *kerato- except in some anatomical senses (e.g. ceratocricoid) and taxonomic names (e.g. *Ceratosaurus*).

cercaria (pl. cercariae) *Zool.* Adjectival form: **cercarial**.

cerco- (**cerc-** before vowels) Prefix denoting a tail (e.g. cercopithecoid, cercaria).

cercus (pl. cerci) *Zool.* Adjectival form: **cercal**.

cerebellum (pl. cerebella) Part of the hindbrain, concerned with coordinating movement and balance. Adjectival form: **cerebellar**. *Compare* cerebrum.

cerebro- Prefix denoting the brain (e.g. cerebrospinal, cerebrovascular).

cerebrospinal fluid Abbrev.: CSF (no stops).

cerebrum (pl. cerebra) The most highly developed part of the brain, responsible for rational thought, memory, etc.; consists of paired **cerebral hemispheres**. Adjectival form: **cerebral**. *Compare* cerebellum.

Cerenkov, Pavel Alekseyevich (preferred to Cherenkov) (1904–) Soviet physicist.
Cerenkov counter
Cerenkov radiation

ceric Denoting compounds in which cerium has an oxidation state of +4. The recommended system is to use oxidation numbers, e.g. ceric oxide, CeO_2, has the systematic name cerium(IV) oxide.

cerium Symbol: Ce *See also* periodic table; nuclide.

cermet Acronym formed from cer(amic and) met(al).

CERN Acronym for Conseil Européen (later Organisation Européenne) pour la Recherche Nucléaire, now the European Laboratory for Particle Physics.

cerous Denoting compounds in which cerium has an oxidation state of +3. The recommended system is to use oxidation numbers, e.g. cerous oxide, Ce_2O_3, has the systematic name cerium(III) oxide.

Cerro Tololo Interamerican Observatory An observatory in La Serena, Chile.

cervical 1. Relating to the neck (e.g. cervical vertebrae). **2.** Relating to the cervix uteri (e.g. cervical smear).

cervico- (**cervic-** before vowels) Prefix denoting **1.** a neck (e.g. cervicodorsal). **2.** the cervix uteri (e.g. cervicitis).

CerVit (cap. C and V) A trade name for a type of glass–ceramic material little affected by temperature changes.

cervix (pl. cervices) A neck or necklike part; often used without qualification to denote the neck of the uterus (cervix uteri). Adjectival form: ***cervical**.

Cesalpino, Andrea (1519–1603) Italian physician and botanist. Also called Andreas Caesalpinus.
Caesalpinia Generic name (*see* genus).
Caesalpiniaceae *See* Leguminosae.

cesium US spelling of caesium.

Cestoda (cap. C) A class of parasitic platyhelminths comprising the tapeworms. Individual name: **cestode** (no cap.). Adjectival forms: **cestode**, **cestoid** (describing any animal resembling a tapeworm in form).

Cet (no stop) *Astron.* Abbrev. for Cetus.

CET Abbrev. for Central European Time.

Cetacea (cap. C) An order of marine mammals containing the whales, porpoises, and dolphins. Individual name and adjectival form: **cetacean** (no cap.; not cetaceous).

cetane $CH_3(CH_2)_{14}CH_3$ The traditional name for hexadecane.

Cetus A constellation. Genitive form: Ceti. Abbrev.: Cet (no stop). *See also* stellar nomenclature.

cetyl alcohol $CH_3(CH_2)_{14}CH_2OH$ The traditional name for hexadecan-1-ol.

Cf Symbol for californium.

CFC (pl. CFCs) Abbrev. for chlorofluorocarbon.

CFHT Abbrev. for Canada–France–Hawaii Telescope (Mauna Kea, Hawaii).

CFSE Abbrev. for crystal-field stabilization energy.

CF theory Abbrev. for crystal-field theory.

CFU *Microbiol.* Abbrev. for colony-forming unit.

CGA *Computing* Abbrev. for colour graphics adapter.

CGPM Abbrev. for Conférence Générale des Poids et Mésures (General Conference of Weights and Measures) convened periodically and now concerned with all scientific measurements. *See also* CIPM; BIPM.

cgs (or **c.g.s.**) **units** A metric system of units in which the centimetre, gram, and second are the units of length, mass, and time. The dyne and erg are the units of force and energy. When electrical and magnetic properties are included a fourth fundamental property is required. The choice of this property led to several different forms of the cgs system. In the most popular, the electromagnetic system of Weber, magnetic permeability, μ, was selected and the unit chosen, the permeability of empty space, μ_0, was made numerically equal to unity. The electromagnetic units (emu) used to measure magnetic properties were the maxwell, gauss, oersted, and gilbert, and were of a convenient size. The emu for electrical measurements were, however, either extremely large or extremely small. The practical units – the ohm, volt, ampere, coulomb, farad, and henry – were therefore introduced. The practical units were a power of 10 times smaller or larger than their electromagnetic counterparts. A combination of the practical units and the magnetic emu gave the practical electrical system. This hybrid system was widely used in the late 19th and early 20th centuries but has been largely displaced by *SI units. *See also* mks units.

ch *Maths.* Short for cosh.

Cha (no stops) *Astron.* Abbrev. for Chamaeleon.

Chadwick, Sir James (1891–1974) British physicist.

chaeta (pl. chaetae) A bristle-like structure in some annelid worms. It is also called a *seta, but note that not all structures called setae are also known as chaetae.

chaeto- (or **chaeti-**) Prefix denoting a bristle or hair (e.g. chaetoplankton, chaetopod, chaetiferous).

Chaetognatha (cap. C) A phylum of small planktonic invertebrates comprising the arrow worms. Individual name and adjectival form: **chaetognath** (no cap.).

Chagas, Carlos (1879–1934) Brazilian physician.
Chagas' disease

chain *See* yard.

Chain, Sir Ernst Boris (1906–79) German-born British biochemist.

Chainia (cap. C, ital.) A former genus of actinomycete bacteria now subsumed in the genus **Streptomyces*.

chalco- (**chalc-** before vowels) Prefix denoting copper or a copper alloy (e.g. chalcolithic, chalcopyrite, chalcanthite).

chalcogens The elements of group VIB of the *periodic table.

chamae- Prefix denoting low or the ground (e.g. chamaephyte, chamaecephalic; but note chameleon).

chamaeleon *Zool.* Use chameleon.

Chamaeleon A constellation. Genitive form: Chamaeleontis. Abbrev.: Cha (no stop). *See also* stellar nomenclature.

Chamberlain, Owen (1920–) US physicist.

chameleon (not chamaeleon) *Zool.*

chamomile Use camomile.

Chance, Britton (1913–) US biophysicist.

Chandler, Seth Carlo (1846–1913) US astronomer.
Chandler wobble

Chandrasekhar, Subrahmanyan (1910–) Indian-born US astrophysicist.
Chandrasekhar limit (preferred to Schönberg–Chandrasekhar limit)

Chang Heng (AD 78–142) Chinese astronomer, mathematician, and instrument maker.

Chapman, Sydney (1888–1970) British mathematician and geophysicist.
Chapman–Enskogg theory (en dash)
Chapman–Ferraro theory (en dash)

CHAPS Abbrev. for 3-[(3-cholamidopropyl)dimethylammonio]-1-propanesulphonate.

Chaptal, Jean Antoine Claude (1756–1832) French chemist.
chaptalization (no cap.)

characteristic temperature Symbol: Θ (Greek cap. theta). A physical quantity used in statistical physics, etc. The *SI unit is the kelvin.

characters per inch Abbrev.: cpi (no stops).

characters per second Abbrev.: *cps (no stops).

charcoal *See* carbon.

Chardonnet, Louis Marie Hilaire Bernigaud, Comte de (1839–1924) French chemist.

Chargaff, Erwin (1905–) Austrian-born US biochemist.
Chargaff's rules

charge *See* electric charge.

charge-coupled device (hyphenated) Abbrev.: CCD (no stops).

charge density Symbol: ρ (Greek rho). A physical quantity, *electric charge divided by volume. It is more accurately called volume density of charge. The *SI unit is the coulomb per cubic metre ($C\ m^{-3}$). The surface density of charge, symbol σ (Greek sigma), is charge divided by surface area. The SI unit is $C\ m^{-2}$.

charge number of an ion Symbol: z or z_B (for a specific ion B). A number, positive for cations and negative for anions, indicating ionization state.

charge-transfer device (hyphenated) Abbrev.: CTD (no stops).

Charles, Jacques Alexandre César (1746–1823) French physicist and physical chemist.
Charles's law Called Gay-Lussac's law in France.
Charles's law of pressures

charm quantum number Symbol: C A quantum number associated with elementary particles. It takes zero or integral values.

Charnley, Sir John (1911–82) British orthopaedic surgeon.
Charnley clamps

Charophyta (cap. C) A division of algae commonly known as stoneworts. In some classifications this group is included in the division *Chlorophyta. In a recent classification it is treated as a class, Charophyceae, in the subdivision Charophytina (division *Streptophyta). Individual name and adjectival form: **charophyte** (no cap.).

Charpentier, Jean de (1786–1885) Swiss geologist and glaciologist.

chasma (not ital.; pl. chasmata) *Astron.* A deep steep-sided valley. The word, with an initial capital, is used in the approved Latin name of such valleys on the surface of a planet or satellite, as in Coprates Chasma on Mars and Artemis Chasma on Venus. *See also* vallis.

chasmo- Prefix denoting a crack, fissure, or opening (e.g. chasmogamy, chasmophyte).

chasmogamy The production of flowers that open at maturity. Adjectival form: **chasmogamous** (not chasmogamic). *Compare* cleistogamy.

Chatelier, Henri Louis Le Usually alphabetized as *Le Chatelier.

Chebyshev, Pafnuty Lvovich (not Tchebyshev) (1821–94) Russian mathematician.
Chebyshev inequality
Chebyshev polynomial

cheilo- (**cheil-** before vowels) Prefix denoting a lip or lips (e.g. cheiloplasty, cheilanthifolious).

cheiro- Prefix denoting the hand or hands; use *chiro-except in some medical senses (e.g. cheiromegaly, cheiropompholyx).

Cheiroptera Use *Chiroptera.

chela (pl. chelae) A paired pincer-like appendage in arthropods: forms the large claw of crabs and lobsters. Adjectival form: **chelate**. *Compare* chelicera.

chelate 1. (noun) *Chem.* An inorganic complex in which a metal atom or ion is coordinated to the same ligand at two or more different points. **2.** (adjective) *Zool.* Possessing chelae.

chelicera (pl. chelicerae) A paired feeding appendage of arachnids and horseshoe crabs (*compare* chela). Adjectival form: **cheliceral**; not to be confused with chelicerate (*see* Chelicerata).

Chelicerata (cap. C) A subphylum of arthropods containing the arachnids and horseshoe crabs, which possess chelicerae (*see* chelicera). Individual name and adjectival form: **chelicerate** (no cap.).

Chelonia (cap. C) An order of reptiles comprising the tortoises and turtles. Former name: Testudinata. Individual name and adjectival form: **chelonian** (no cap.).

chemi- Prefix denoting chemicals or chemical reactions (e.g. chemiluminescence, chemiosmosis, chemisorption). *See also* chemo-.

Chemical Abstracts Service Abbrev.: CAS.

chemical formulae There are various conventions for writing chemical formulae. In *organic chemical nomenclature, it is usual to write a formula as a series of groups, e.g. $CH_3COC_2H_5$. If it is necessary to divide off the groups (for the purpose of explanation), centred dots are used, e.g. $CH_3{\cdot}CO{\cdot}C_2H_5$. Double bonds should be indicated by equals signs, e.g. $CH_3CH{=}CH_2$ (preferred to colons, $CH_3CH{:}CH_2$). Triple bonds should be indicated by $CH_3C{\equiv}CH$ (preferred to three dots, $CH_3C{\vdots}CH$).

In *inorganic chemical nomenclature, the relationship between subscripts and superscripts is important, and these should be staggered so as to indicate the ions present, e.g. $Na^+{}_2CO_3{}^{2-}$ (sodium carbonate) but $Hg_2{}^{2+}Cl^-{}_2$ (dimercury(I) chloride). Coordinated species should be indicated by a centred dot, e.g. $Cu_2SO_4{\cdot}5H_2O$. Square brackets are preferred to indicate complex ions, e.g. $K^+{}_3[FeCl_6]^{3-}$.

chemically pure Abbrev.: CP (no stops).

chemical nomenclature *See* inorganic chemical nomenclature; organic chemical nomenclature.

chemical oxygen demand (not hyphenated) Abbrev.: COD (no stops).

chemical potential Symbol: μ_B (Greek mu) for a substance B or, when a complicated formula is to be written, μ(B), as in $\mu(H_2SO_4)$. A physical quantity that in a mixture of components B, C, . . . is defined as $\partial G/\partial n_B$ at constant temperature, pressure, and *amounts of substance of other components; n_B is the amount of substance of B and G is the *Gibbs function. The *SI unit is the joule per mole.

chemical shift Symbol: δ (Greek delta). A physical quantity, the difference between two absorption peaks in nuclear magnetic resonance spectroscopy. One peak is usually from the reference substance tetramethyl silane. The unit is parts per million (ppm). The chemical shift parameter τ (Greek tau) may be encountered and is defined as

$$\tau = 10 - \delta \text{ ppm}.$$

chemiosmosis (one word) Adjectival form: **chemiosmotic**.

chemisorption *See* adsorption.

chemo- Prefix denoting chemicals or chemical reactions (e.g. chemolithotroph(ic), chemoreceptor, chemotaxis, chemotaxonomy, chemotherapy, chemotropism). *See also* chemi-.

chemolithoautotrophic (one word) Noun form: **chemolithoautotroph**.

chemolithoheterotrophic (one word) Noun form: **chemolithoheterotroph**.

chemoorganoheterotrophic (one word) Noun form: **chemoorganoheterotroph**.

chemoorganotrophic (one word) Noun form: **chemoorganotroph**.

chemoreceptor (preferred to chemoceptor)

Chenopodiaceae A family of dicotyledonous plants, commonly known as the goosefoot family. Individual name and adjectival form: **chenopod** (preferred to chenopodiaceous).

Cherenkov Use *Cerenkov.

CHES Abbrev. for 2-(cyclohexylamino)-ethanesulphonic acid.

cheval vapeur French name for metric horsepower. *See* horsepower.

Chevreul, Michel Eugène (1786–1889) French organic chemist.

Cheyne, John (1777–1836) British physician.
Cheyne–Stokes respiration (en dash)

chi Greek letter, symbol: χ (lower case), X (cap.).
χ Symbol for **1.** (or χ_e) electric susceptibility. **2.** (or χ_m) magnetic susceptibility.

chiasma (pl. chiasmata) **1.** *Genetics* The cross shape formed by separating homologous chromosomes during meiosis at points where *crossing over has occurred. **2.** (US: **chiasm**; pl. chiasms) *Anat.* An X-shaped anatomical part (e.g. the optic chiasma).

chickenpox (one word) Medical name: varicella.

Child's law *Electronics* Use Langmuir–Child law (en dash).

Chilopoda (cap. C) A class of arthropods comprising the centipedes. Individual name and adjectival form: **chilopod** (no cap.; not chilopodan or chilopodous). *See also* myriapod.

chimaera 1. US: **chimera**. Any cartilaginous fish of the subclass Holocephali, especially any of the genus *Chimaera*. **2.** *Genetics* Use *chimera.

chimera (pl. chimeras) An organism composed of tissues of two or more genetically distinct types. Originally the US spelling, 'chimera' is now widely used in British English for the genetic sense, though *chimaera is retained for the fish.

chimno- (or **chimo-**) Prefix denoting winter (e.g. chimnophilous, chimopelagic).

***chi* mutant** (ital. *chi*) The mutant site found in lambda phage whose product stimulates recombination in *E. coli*. The name is derived from *c*rossover *h*otspot *i*nstigator; not to be confused with *chi structure.

chirality *Chem., maths.* Derived adjective: **chiral.**

chiro- Prefix denoting the hands (e.g. chiropractic, chiropterophilous). *See also* cheiro-.

Chiroptera (cap. C) An order of mammals comprising the bats. Individual name: **chiropteran** or (esp. US) **chiropter** (no caps.). Adjectival form: **chiropteran.**

chi-square (or **chi-squared**) *Stats.* Symbol: X^2 (cap. chi).
chi-square distribution
chi-square test

chi structure (not ital.) The four-armed structure obtained by cleavage following reciprocal recombination between homologous duplex DNA circles. It is named because of its resemblance to the Greek letter χ; not to be confused with **chi* mutant.

chitin A structural polysaccharide occurring in invertebrates and fungi. Adjectival form: **chitinous.** *Compare* chiton.

chiton Any mollusc of the genus *Chiton* (cap. C, ital.). *Compare* chitin.

Chittenden, Russell Henry (1856–1943) US physiologist and biochemist.

Chl Abbrev. for *chlorophyll: used in designating a particular type of chlorophyll (e.g. Chl *a*, Chl *b*).

Chladni, Ernst Florens Friedrich (1756–1827) German physicist.
Chladni figures

chlamydia (pl. chlamydiae or chlamydias) Any of the microorganisms (usually regarded as Gram-negative bacteria) belonging to the order Chlamydiales, which currently contains the single genus *Chlamydia* (cap. C, ital.). Adjectival form: **chlamydial.**

chloracetic acid The traditional name for *chloroethanoic acid.

chloral CCl_3CHO The traditional name for trichloroethanal.

chloramphenicol acetyl transferase Acronym: CAT (no stops). *See also* CAT assay.

chlorate Denoting a compound containing the ion ClO_3^-, e.g. potassium chlorate, $KClO_3$. The recommended name is chlorate(v).

chlorate(I) Denoting a compound containing the ion ClO^-, e.g. sodium chlorate(I), NaClO. The traditional name is hypochlorite.

chlorate(III) Denoting a compound containing the ion ClO_2^-, e.g. sodium chlorate(III), $NaClO_2$. The traditional name is chlorite.

chlorate(V) Denoting a compound containing the ion ClO_3^-, e.g. potassium chlorate(V), $KClO_3$. The traditional name is chlorate.

chloric acid $HClO_3$ The traditional name for chloric(V) acid.

chloric(I) acid HOCl The recommended name for the compound traditionally known as hypochlorous acid.

chloric(III) acid $HClO_2$ The recommended name for the compound traditionally known as chlorous acid.

chloric(V) acid $HClO_3$ The recommended name for the compound traditionally known as chloric acid.

chlorine Symbol: Cl *See also* periodic table; nuclide.

chlorine monoxide Cl_2O The traditional name for dichlorine oxide.

chlorite Denoting a compound containing the ion ClO_2^-, e.g. sodium chlorite, $NaClO_2$. The recommended name is chlorate(III).

chloro- (sometimes **chlor-** before vowels) Prefix denoting **1.** chlorine (e.g. chloroacetic, chlorohydrin, chlorate, chloride). **2.** a green colour (e.g. chlorocruorin, chlorophyll).

chloroacetic acid The traditional name for a *chloroethanoic acid.

chloroacetone $CH_2ClCOCH_3$ The traditional name for chloropropanone.

Chlorobionta In a recent classification, a subkingdom of plants that contains all the green plants, including green algae. *See also* Phytobiota.

chlorobium vesicle Obsolete term for chlorosome, a photosynthetic organelle of the green bacteria.

2-chlorobuta-1,3-diene $CH_2{=}CCl{-}CH{=}CH_2$ The recommended name for the compound traditionally known as chloroprene.

chloroethane CH_3CH_2Cl The recommended name for the compound traditionally known as ethyl chloride.

chloroethanoic acid Any of three acids: monochloroethanoic acid, $CH_2ClCOOH$; dichloroethanoic acid, $CHCl_2COOH$; and trichloroethanoic acid, CCl_3COOH. The traditional name for a chloroethanoic acid is chloracetic acid or chloroacetic acid.

2-chloroethanol $ClCH_2CH_2OH$ The recommended name for the compound traditionally known as ethylene chlorohydrin.

chloroethene $CH_2{=}CHCl$ The recommended name for the compound traditionally known as vinyl chloride.

chlorofluorocarbon (one word) Abbrev.: CFC (no stops; pl. CFCs).

chloroform $CHCl_3$ The traditional name for trichloromethane.

chloromethane CH_3Cl The recommended name for the compound traditionally known as methyl chloride.

(chloromethyl)benzene $C_6H_5CH_2Cl$ The recommended name for the compound traditionally known as benzyl chloride.

chloromethylbenzene $ClC_6H_4CH_3$ The recommended name for the compound traditionally known as chlorotoluene. The recommended name for *o*-chlorotoluene is chloro-2-methylbenzene, etc.

Chlorophyceae A class of algae in the division *Chlorophyta.

chlorophyll A photosynthetic pigment occurring in all plants and some bacteria (*see* bacteriochlorophyll). There are four groups, each designated by a lower-case italic letter, i.e. chlorophyll (or Chl) *a*, Chl *b*, Chl *c*, and Chl *d*. Chl *a* has several forms with different light absorption peaks (*see* photosystem).

Chlorophyta (cap. C) A large division comprising the green algae. Former name: Isokontae (*see* isokont). It contains the classes Chlorophyceae, Oedogoniophyceae, Bryopsidophyceae, and Zygnematophyceae. In classification systems that regard the stoneworts (*see* Charophyta) as green algae, there are two classes: Charophyceae (stoneworts) and Chlorophyceae. Recently green algae have been placed in the subkingdom *Chlorobionta. The classification of green algae is still controversial. Individual name and adjectival form: **chlorophyte** (no cap.).

chloropicrin CCl_3NO_2 The traditional name for trichloronitromethane.

chloroprene $CH_2CClCHCH_2$ The traditional name for 2-chlorobuta-1,3-diene.

chloropropanone $CH_2ClCOCH_3$ The recommended name for the compound traditionally known as chloroacetone.

3-chloroprop-1-ene $CH_2{=}CHCH_2Cl$ The recommended name for the compound traditionally known as allyl chloride.

chlorosulphonyl isocyanate Abbrev.: CSI (no stops).

chlorotoluene $ClC_6H_4CH_3$ The traditional name for chloromethylbenzene.

chlorous acid $HClO_2$ The traditional name for chloric(III) acid.

chlor-zinc iodide Abbrev.: CZI (no stops). *See* Schultze's solution.

choana (pl. choanae; usually referred to in the pl.) Either of the two openings of the nasal cavity into the mouth. Also called internal *nares. Adjectival form: **choanate** (possessing choanae: denoting lungfishes, amphibians, reptiles, and mammals, formerly classified on this basis in the taxon Choanata).

Choanichthyes (cap. C) A subclass of bony fishes that contains the crossopterygians and lungfishes. Also called Sarcopterygii (the definitive usage is currently under debate). Individual name and adjectival form: **choanichthyan** (no cap.).

choano- (**choan-** before vowels) Prefix denoting a funnel (e.g. choanocyte).

chole- (**chol-** before vowels) Prefix denoting bile (e.g. cholesterol, cholagogue).

cholecalciferol Synonyms: calciol, vitamin D_3.

cholecystokinin Abbrev.: CCK (no stops). A hormone, secreted by the duodenum, that stimulates gall-bladder contraction and pancreatic enzyme secretion. Originally, the latter effect was believed to be caused by a separate hormone, pancreozymin, but both effects are now known to be caused by a single hormone.

5-cholesten-3β-ol The systematic chemical name for the compound traditionally known as cholesterol.

cholesterol The traditional name for *5-cholesten-3β-ol.

cholinergic Denoting nerve fibres that release acetylcholine as a neurotransmitter. *Compare* adrenergic.

Chondrichthyes (cap. C) A class of vertebrates comprising the cartilaginous fishes (sharks, rays, etc.). Individual name and adjectival form: **chondrichthyan** (no cap.), although the common name is more widely used for individuals.

chondriosome Obsolete name for *mitochondrion.

chondro- (**chondr-** before vowels) Prefix denoting **1.** cartilage (e.g. chondroblast, chondrocranium). **2.** grain or granular (e.g. chondrodite, chondrite).

Chordata (cap. C) A phylum of animals possessing a notochord; includes all the vertebrates and a few invertebrates (*see* protochordate). Individual name and adjectival form: **chordate** (no cap.).

chordo- (**chord-** before vowels) Prefix denoting **1.** a cord (e.g. chordotonal, chorditis). **2.** (or **chorda-**) the notochord (e.g. chordate, chordamesoderm).

-chore Noun suffix denoting **1.** a plant distributed by some agency (e.g. anemochore, zoochore). Adjectival form: **-chorous**. **2.** space or volume (e.g. isochore). Adjectival form: **-choric**.

chori- Prefix denoting distinct or separate (e.g. choripetalous, choriphyllous).

C horizon A subsurface *soil horizon consisting of parent material.

Christian, Henry Asbury (1876–1951) US physician.

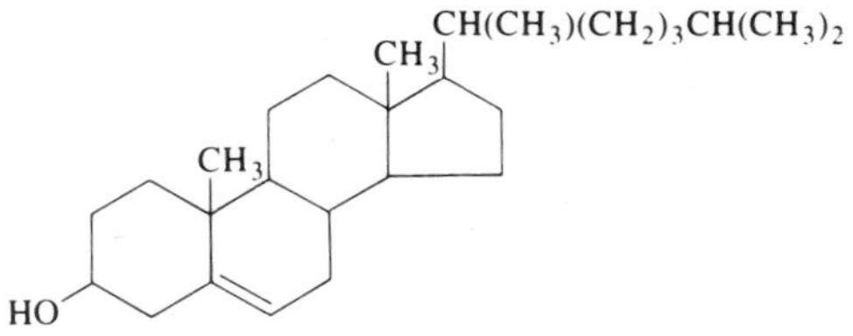

$CH(CH_3)(CH_2)_3CH(CH_3)_2$

5-cholesten-3*β*-ol (cholesterol)

Hand–Schüller–Christian disease or **Schüller–Christian disease** (en dashes)

Christoffel, Elwin Bruno (1829–1900) German mathematician.
Christoffel symbols
Riemann–Christoffel tensor (en dash)

chromate Denoting a compound containing the ion CrO_4^{2-}, e.g. potassium chromate, K_2CrO_4. The recommended name is chromate(VI).

chromato- (**chromat-** before vowels) Prefix denoting colour (e.g. chromatography, chromatophore).

chromatophore A pigment-containing cell or other structure in animals or plants. *Compare* chromophore.

chrome alum $KCr(SO_4)_2{\cdot}12H_2O$ The traditional name for chromium(III) potassium sulphate-12-water.

chromic Denoting compounds in which chromium has a high oxidation state. The recommended system is to use oxidation numbers, e.g. chromic chloride, $CrCl_3$, has the systematic name chromium(III) chloride.

chromium Symbol: Cr *See also* periodic table; nuclide.

chromium dichloride $CrCl_2$ The traditional name for chromium(II) chloride.

chromium(VI) dichloride dioxide CrO_2Cl_2 The recommended name for the compound traditionally known as chromyl chloride.

chromium(III) potassium sulphate-12-water $KCr(SO_4)_2{\cdot}12H_2O$ US: **chromium(III) potassium sulfate-12-water.** The recommended name for the compound traditionally known as chrome alum.

chromium sesquioxide Cr_2O_3 The traditional name for chromium(III) oxide.

chromium trichloride $CrCl_3$ The traditional name for chromium(III) chloride.

chromium trioxide CrO_3 The traditional name for chromium(VI) oxide.

chromo- (**chrom-** before vowels) Prefix denoting **1.** colour (e.g. *chromophore, chromoplast). **2.** chromium (e.g. chromate).

Chromobionta In a recent classification, a subkingdom of plants that includes many algae, including the brown and golden-brown algae and diatoms. *See also* Chromophyta; Phytobiota.

chromophore A group of atoms in a chemical compound that is responsible for its colour. *Compare* chromatophore.

Chromophyta A division of plants in the subkingdom *Chromobionta. It includes the classes Chrysophyceae (golden-brown algae), Xanthophyceae (yellow-green algae), *Oomycetes, Phaeophyceae (brown algae), and Bacillariophyceae (diatoms).

chromosome nomenclature *See* cytogenetics nomenclature.

chromous Denoting compounds in which chromium has a low oxidation state. The recommended system is to use oxidation numbers, e.g. chromous chloride, $CrCl_2$, has the systematic name chromium(II) chloride.

chromyl Denoting compounds containing the group CrO_2 or the ion CrO_2^{2+}. The recommended system is to use oxidation numbers, e.g. chromyl chloride, CrO_2Cl_2, has the systematic name chromium(VI) dichloride dioxide.

chromyl chloride CrO_2Cl_2 The traditional name for chromium(VI) dichloride dioxide.

chrono- (**chron-** before vowels) Prefix denoting time (e.g. chronostratigraphy, chronotaxis).

Chroococcales An order of *cyanobacteria. Adjectival form: **chroococcalean**.

chrysalis (pl. chrysalides; preferred to chrysalises) Also called chrysalid (pl. chrysalids).

chryso- (**chrys-** before vowels) Prefix denoting gold or golden (e.g. chrysoberyl, chrysolite, chrysotherapy).

Chrysophyta A division comprising the golden-brown algae. In some classifications it also includes the yellow-green algae (Xanthophyceae), diatoms (Bacillariophyceae), and Haptophyceae (*see* Haptophyta). In a recent classification this division is reduced to a class, Chrysophyceae, in the division *Chromophyta, subkingdom *Chromobionta.

Church, Alonzo (1903–) US mathematician and philosopher.
Church–Rosser theorem (en dash) Also named after J. B. Rosser (1907–).
Church's hypothesis

Chu Shih-Chieh (hyphen) (*fl.* 1300) Chinese mathematician.

chyle Lymph containing fat absorbed from the small intestine. *Compare* chyme.

chylo- (**chyl-** before vowels) Prefix denoting chyle (e.g. chylomicron, chyluria).

chyme The contents of the stomach and small intestine: food mixed with digestive enzymes. *Compare* chyle.

chymosin Use *rennin.

Chytridiales (cap. C) An order of fungi in the class *Chytridiomycetes. Individual name and adjectival form: **chytrid** (no cap.).

Chytridiomycetes A class of fungi in the subdivision *Mastigomycotina, mainly microscopic parasites of freshwater algae and animals. In a recent classification they are regarded as a division, Chytridiomycota.

Ci Symbol for curie.

Ci (ital.) Abbrev. for cirrus.

CI Abbrev. for **1.** Colour Index. **2.** *Photog.* contrast index.

-cide (usually preceded by i) Noun suffix denoting **1.** something that kills (e.g. insecticide, bactericide). **2.** killing (e.g. infanticide). Adjectival form: **-cidal**.

CIE Abbrev. for Commission International de l'Éclairage. Also called ICI, abbrev. for the English form, International Commission on Illumination.

ciliary 1. Relating to cilia. **2.** Designating or relating to the ciliary body, part of the vertebrate eye.

Ciliata (cap. C) A class of protozoans characterized by the possession of cilia. Regarded by some authorities as a subphylum, Ciliophora; this taxon is elevated to the status of phylum if these organisms are regarded as protists. Individual name and adjectival form: ***ciliate** (no cap.).

ciliate 1. (adjective) Relating to or describing the *Ciliata. **2.** (noun) Any ciliate protozoan. **3.** (adjective) Having cilia. The term 'ciliated' is preferred for this sense, which can apply to various cells and tissues, to avoid confusion with the Ciliata.

Ciliophora *See* Ciliata.

cilium (pl. cilia) A short hairlike locomotory appendage. Adjectival forms: ***ciliary**, **ciliated** (preferred to *ciliate).

CIM Acronym for computer-integrated manufacturing.

cineraria A widely cultivated ornamental plant, *Senecio* (formerly *Cineraria*) *cruentus*.

cinnamic acid $C_6H_5CH = CHCOOH$ The traditional name for 3-phenylpropenoic acid.

cipher (not cypher)

CIPM Abbrev. for Commission Internationale des Poids et Mésures (International Committee on Weights and Measures). A committee that meets annually and is concerned with international uniformity in standards of measurement throughout science. *See also* CGPM; BIPM.

CIPW classification *Geol.* Named after US petrologists *C*ross, *I*ddings, *P*irsson, and *W*ashington.

Cir (no stops) *Astron.* Abbrev. for Circinus.

Circinus A constellation. Genitive form: Circini. Abbrev.: Cir (no stop). *See also* stellar nomenclature.

circle of *Willis *Anat.*

circular frequency Another name for angular frequency.

circular wavenumber *See* wavenumber.

circum- Prefix denoting around or surrounding (e.g. circumnutation, circumoral, circumpolar).

cirque *Glaciol.*

cirri- *Zool.* Prefix denoting cirri (*see* cirrus) (e.g. cirripede).

Cirripedia (cap. C) A subclass of crustaceans comprising the barnacles. Individual name and adjectival form: **cirripede** (US: **cirriped**; no caps.).

cirro- Prefix denoting cirrus clouds (e.g. cirrocumulus).

cirrocumulus (pl. cirrocumuli) Abbrev.: *Cc* (cap. C, ital.).

cirrostratus (pl. cirrostrati) Abbrev.: *Cs* (cap. C, ital.).

cirrus (pl. cirri) **1.** *Meteorol.* Abbrev.: *Ci* (cap. C, ital.). A type of cloud. *See also* cirro-. **2.** *Biol.* A tentacle or tendril. *See also* cirri-.

cis Symbol for the function (cos + i sin); cis $x = e^{ix}$.

cis (ital.) Relating to or occurring on the same chromosome of a DNA molecule. A *cis* configuration is one in which loci occur on the same chromosome. A locus that exercises its effect on the same DNA molecule is described as *cis*-acting or *cis*-dominant (hyphenated). *Compare trans.*

cis- Prefix denoting on the same side or on this side (e.g. cislunar).

cis- (ital., always hyphenated) Prefix denoting an isomer in which **1.** two substituents are located on the same side of a double bond or a cyclic structure, e.g. *cis*-decalin, *cis*-dichloro-1,2-propadiene. **2.** two substituents on an atom are adjacent in a square-planar or octahedral inorganic complex, e.g. *cis*-dibromodichlorotitanium(IV). The alternative *syn*-is not recommended. *Compare trans-.*

***cis-trans* complementation test** (*cis-trans* ital. and hyphenated) A mating test used in genetics to determine whether two mutants belong to the same complementation group (*see cis*; *trans*).

CITES (pronounced **sy**-teez) Acronym for Convention on International Trade in Endangered Species.

citric acid The traditional name for *2-hydroxypropane-1,2,3-tricarboxylic acid. Citric acid is used in many nonchemical contexts.

citric acid cycle Use *TCA cycle.

Cl Symbol for chlorine.

cladistics A method of classification in which organisms are grouped into taxonomic ranks (**clades**) according to shared characteristics; relationships are indicated in a diagram called a **cladogram**. Proponents of the system are called **cladists**.

clado- (**clad-** before vowels) Prefix denoting branches or, by extension, cladistics (e.g. cladogram, cladoptosis).

Clairaut, Alexis-Claude (hyphen) (1713–65) French mathematician and physicist.
Clairault's equation

Claisen, Ludwig (1851–1930) German organic chemist.
Claisen condensation
Claisen flask
Claisen–Schmidt condensation (en dash)

Clapeyron, B. P. E. (1799–1864) French physicist.
Clausius–Clapeyron equation (en dash)

Clark, Alvan Graham (1832–97) US astronomer and instrument-maker.

Clark, Thomas (1801–67) British chemist.
clarking (no initial capital letter)
Clark process

Clark, Sir Wilfred Edward Le Gros Usually alphabetized as *Le Gros Clark.

Clark cell *Physics.* Named after Hosiah Clark (died 1898).

-clase *Mineral.* Noun suffix denoting breakdown or cleavage (e.g. orthoclase, plagioclase). Adjectival form: **-clastic**.

class A category used in biological classification (*see* taxonomy) consisting of a number of similar orders (sometimes only one order). Names of classes are printed in roman (not italic) type with an initial capital letter. Algal classes typically end in -phyceae (e.g. Chlorophyceae), fungal classes in -mycetes (e.g. Zygomycetes), and higher plant classes in -opsida (e.g. Lycopsida), although these endings are not universally recognized for the ranks indicated. Animal classes have a variety of endings (e.g. Insecta, Elasmobranchii, Reptilia, Aves). *See also* subclass.

classification, biological *See* taxonomy.

-clast *Biol.* Noun suffix denoting breakdown or resorption (e.g. osteoclast). Adjectival form: **-clastic**. Derived noun form: **-clasis**.

-clastic *See* -clase; -clast.

Claude, Albert (1898–1983) Belgian-born US cell biologist.

Claude, Georges (1870–1960) French chemist.
Claude process

Clausius, Rudolf Julius Emmanuel (1822–88) German physicist.
Carnot–Clausius equation (en dash)
Clausius–Clapeyron equation (en dash)
Clausius–Mosotti relation (en dash)
Clausius's equation
Clausius's theorem
Clausius virial

CLB Abbrev. for *Cytophaga*-like bacteria. *See Cytophaga.*

clear-air turbulence Abbrev.: CAT (no stops).

cleistogamy The production of flowers that do not open at maturity. Adjectival form: **cleistogamous** (not cleistogamic). *Compare* chasmogamy.

Clemence, Gerald Maurice (1908–74) US astronomer.

Clemmensen reduction *Organic chem.*

Cleve, Per Teodor (1840–1905) Swedish chemist.
Cleve's acids

-cline Noun suffix denoting a gradation or slope (e.g. anticline, monocline, thermocline). Adjectival form: **-clinal**.

clinostat (preferred to klinostat)

clitellum (pl. clitella) *Zool.* Adjectival form: **clitellar**.

cloaca (pl. cloacae) *Zool.* Adjectival form: **cloacal**.

closed-circuit television (hyphenated) Abbrev.: CCT (no stops).

Clostridium (cap. C, ital.) A genus of anaerobic spore-forming rod-shaped bacteria. Individual name: **clostridium** (no cap., not ital.; pl. clostridia). Adjectival form: **clostridial**.
C. botulinum: the causal agent of botulism. Strains are divided into seven types, according to the antigenic characteristics of their toxin, designated by the roman (upright) capital letters A, B, C, D, E, F, and G.
C. novyi: strains are designated as type A or B (roman caps.).
C. perfringens (formerly *C. welchii*): strains are designated A, B, C, D, or E according to their production of major lethal toxins (of which there are four: alpha, beta, gamma, and delta).

clubmoss (preferred to club moss) Any pteridophyte plant of the genera *Lycopodium* or *Selaginella*. *See* Lycopsida.

Clusius column *Physics.*

cluster of differentiation Abbrev. and preferred form: *CD (no stops). *See also* T-cell differentiation antigens.

CLV Abbrev. for **1.** carnation latent virus. **2.** cassava latent virus.

cm (no stop) Symbol for centimetre (or centimetres). *See* metre.

cM Abbrev. for *centimorgan.

Cm Symbol for curium.

CM Abbrev. for central meridian.

CMa *Astron.* Abbrev. for Canis Major.

CM-cellulose (hyphenated) Abbrev. and preferred form for carboxymethylcellulose.

CMi (no stops) *Astron.* Abbrev. for Canis Minor.

CMI Abbrev. for computer-managed instruction.

CMOS (or **C/MOS**) *Electronics* Acronym for complementary *MOS.

CMP Abbrev. and preferred form for cytidine 5′-phosphate (cytidine monophosphate).

CMV Abbrev. for **1.** cytomegalovirus. **2.** cucumber mosaic virus.

Cnc (no stop) *Astron.* Abbrev. for Cancer.

Cnidaria (cap. C) A phylum of *coelenterates including the jellyfishes, sea anemones, and corals. Individual name and adjectival form: **cnidarian** (no cap.).

C/N ratio *Ecol.* Abbrev. for carbon: nitrogen ratio.

CNS Abbrev. for central nervous system.

Co Symbol for cobalt.

co- Prefix denoting **1.** with, together, jointly or joint, similar, same (e.g. codominance, coefficient, coenzyme, cofactor, coherence, coordinate, covalent). **2.** the complement of an angle (e.g. cosecant, colatitude). *See also* com-; con-.

CoA (no stops) Abbrev. for coenzyme A.

coagulase reacting factor Abbrev.: CRF (no stops).

coagulation factor Any of a group of substances, present in blood plasma, that are necessary for blood clotting. Human coagulation factors are now designated by Roman numerals (I–XIII), but the eponymous or descriptive names originally applied to them are still permitted. It is recommended that both designations are given when a factor is first mentioned in text. *See also* factor VIII; factor IX.

coal-tar fuels (hyphenated) Abbrev.: CTF (no stops).

coax (no stop) Abbrev. for coaxial cable.

cobalamin (not cobalamine) *See* cyanocobalamin.

cobalt Symbol: Co *See also* periodic table; nuclide.

cobaltic Denoting compounds in which cobalt has an oxidation state of +3. The recommended system is to use oxidation numbers, e.g. cobaltic oxide, Co_2O_3, has the systematic name cobalt(III) oxide.

cobalticyanide Denoting a compound containing the ion $(Co(CN)_6)^{3-}$, e.g. potassium cobalticyanide, $K_3Co(CN)_6$. The recommended name is hexacyanocobaltate(III).

cobaltous Denoting compounds in which cobalt has an oxidation state of +2. The recommended system is to use oxidation numbers, e.g. cobaltous oxide, CoO, has the systematic name cobalt(II) oxide.

Cobol (or **COBOL**) *Computing* Acronym for common business-oriented language.

cocco- (**cocc-** before vowels) Prefix denoting spherical shape (e.g. coccobacillus, coccolith).

coccobacillus (pl. coccobacilli) Any bacterium having a shape intermediate between spherical (as in a coccus) and rod-shaped (as in a bacillus).

coccus (pl. cocci) Any spherical bacterium. The word is often used in combination, to denote spherical bacteria associated with particular diseases (e.g. gonococcus, pneumococcus) or specific genera of cocci (e.g. **Staphylococcus*, **Streptococcus*). Adjectival forms: **coccal**, **coccoid** (preferred for describing shape); coccid and coccous are rarely used. *Compare* bacillus.

-coccus Noun suffix denoting spherical bacteria (*see* coccus). Adjectival form: **-coccal**; -coccic and -coccous are rarely used.

coccygo- (**coccyo-**; **coccyg-** or **coccy-** before vowels) Prefix denoting the coccyx (e.g. coccygodynia, coccygectomy).

coccyx (pl. coccyges; preferred to coccyxes) The terminal element of the backbone. Adjectival form: **coccygeal**.

cochlea (pl. cochleae) Part of the inner ear of mammals, birds, and some reptiles. Adjectival form: **cochlear**.

Cockcroft, Sir John Douglas (1897–1967) British physicist.
Cockcroft–Walton generator (en dash)

COD Abbrev. for chemical oxygen demand.

codon Any of the 64 base triplets that constitute the genetic code. Each codon is designated by the three capital letters corresponding to the sequence of constituent bases, e.g. CUU, CUC, CUA, etc. *Compare* anticodon.

coefficient Abbrev.: coeff. (stop). The constant of proportionality, k, in an equation of the form $A = kB$, where A and B are physical quantities with different dimensions. An example is diffusion coefficient. The quantity k is also called a modulus, as in the Young modulus. If A and B have the same dimensions then k is known as a factor or index, as in coupling factor and refractive index.

-coel (not -coele) Noun suffix denoting a cavity (e.g. blastocoel, haemocoel). *Compare* -cele.

coel- *See* coelo-.

coelacanth (not coelocanth) A bony fish belonging to the crossopterygian suborder Coelacanthina. The only living genus is *Latimeria*.

coelenterate An invertebrate animal formerly (and sometimes still) regarded as a member of the phylum Coelenterata. Coelenterates are now usually classified as two separate phyla, the *Cnidaria and the *Ctenophora. Adjectival form: **coelenterate**.

coeliac US: **celiac**. Relating to the abdominal region (e.g. coeliac artery, coeliac disease). Not to be confused with coelomic (*see* coelom).

coelio- (**coeli-** before vowels) US: **celio-**, **celi-**. Prefix denoting the abdomen (e.g. coeliac).

coelo- (**coel-** before vowels) Prefix denoting **1.** the sky (e.g. coelostat). **2.** a body cavity or hollow (e.g. coelodont, coelenterate, coelenteron).

coelom (not coelome; pl. coeloms; preferred to coelomata) US: **coelom** or **celom**. The body cavity of many animals. Adjectival form: **coelomic** (*see also* coelomate). *Compare* coeliac.

coelomate Any animal that possesses a coelom. Such animals were formerly classified in the taxon Coelomata, but the term coelomate now has no taxonomic significance. Adjectival form: **coelomate**.

Coelomycetes (cap. C) A class of imperfect fungi (*see* Deuteromycotina) that possess recognizable fruiting bodies. Individual name: **coelomycete** (no cap.).

coeno- (**coen-** before vowels) US: **ceno-**, **cen-**. Prefix denoting a common characteristic or property (e.g. coenobium, *coenocyte, coenospecies, coenurus).

coenocyte *Bot.* A structure consisting of a mass of cytoplasm containing many nuclei and bounded by a cell wall. One or more coenocytes form the plant body of certain algae and fungi. The adjectival form, **coenocytic**, may be used in a wider sense to describe any *acellular multinucleate plant or plant structure. *Compare* plasmodium; syncytium.

coenurus (pl. coenuri) The larval stage of certain tapeworms: a type of *bladderworm.

coenzyme (one word)
coenzyme A Abbrev.: CoA (no stops).
coenzyme Q Former name for ubiquinone.

Cohen, Seymour Stanley (1917–) US biochemist.

Cohen, Stanley (1922–) US biochemist.

coherent units A system of units of measurement in which the quotient or product of any two quantities has a unit equal to the quotient or product of the units of these quantities. The **base units** of a coherent system are arbitrarily defined. All **derived units** in the system are formed from the base units without the introduction of factors of proportionality. *SI units are coherent units.

Cohn, Ferdinand Julius (1828–98) German botanist and bacteriologist.

Col (no stop) **1.** *Astron.* Abbrev. for Columba. **2.** *See* Col plasmid.

Colby bars *Surveying*

-cole *See* -colous.

coleo- Prefix denoting a sheath or annulus (e.g. *Coleoptera, coleoptile).

Coleoptera (cap. C) An order of insects comprising the beetles and weevils. Individual name and adjectival form: **coleopteran** (no cap.; not coleopterous).

coleorhiza (not coleorrhiza; pl. coleorhizae) A sheath covering a grass radicle.

coli- Prefix denoting the bacterium *Escherichia coli* (e.g. coliform, coliphage).

colicin plasmid *See* Col plasmid.

coliphage λ Usually shortened to *phage λ.

coliphage T2 Usually shortened to *phage T2.

Collider Detector at Fermilab Abbrev.: CDF (no stops).

colony-forming unit *Microbiol.* Abbrev.: CFU (no stops).

colony-stimulating factor *Genetics* Abbrev.: *CSF (no stops).

color US spelling of colour.

colorant (not colourant)

coloration (not colouration)

colorimeter (not colourimeter)

colour US: **color**.

colour graphics adapter Abbrev.: CGA (no stops).

Colour Index Abbrev.: CI (no stops). The definitive register of dyestuffs and pigments. Each compound is assigned a number and is listed with relevant information. For example, acid yellow 73 (fluorescein) is listed as CI 45350.

-colous Adjectival suffix denoting growing or living on (e.g. arenicolous, lignicolous). Noun form: **-cole**.

Colpitts oscillator *Elec. eng.*

Col plasmid (cap. C) Abbrev. and preferred form for colicin plasmid. The numerous types are generally designated by a capital letter, some with an additional Arabic numeral, e.g. ColE1, ColE2, ColK, etc. (no intervening space).

Columba A constellation. Genitive form: Columbae. Abbrev.: Col (no stop). *See also* stellar nomenclature.

columbium Former name for niobium.

Com (no stop) *Astron.* Abbrev. for Coma Berenices.

COM Abbrev. for computer output on microfiche.

com- Prefix denoting with, together, jointly or joint, similar, same (e.g. combustion, commensal). *See also* co-; con-.

Coma Berenices A constellation. Genitive form: Comae Berenices. Abbrev.: Com (no stop). *See also* stellar nomenclature.

comet nomenclature When a newly discovered comet has been confirmed, an interim designation is given by the IAU: year of discovery followed immediately by the letter a, b, c, . . . , assigned in order of discovery. Permanent designations are given later: year of perihelion passage followed (after a thin space) by a Roman numeral assigned in order of date of perihelion passage, as in 1988 VI, 1989 I. In addition, a comet is generally named after its discoverer(s) or the person who computed its orbit, as with comet Kohoutek, comet IRAS–Iraki–Alcock, and Halley's comet. Comet Kohoutek, 1973 XII, was first designated 1973f. A periodic comet is indicated by the letter P after the designation.

Commonwealth Scientific and Industrial Research Organization Abbrev.: CSIRO (no stops).

Comovirus (cap. C, ital.) Approved name for the *cowpea mosaic virus group. Individual name: **comovirus** (no cap., not ital.). [from *co*wpea *mo*saic *virus*]

compact disc US: **compact disk**. Abbrev.: CD (no stops).

compander *Telecom.* Short for compressor expander.

compass points Use abbreviated forms, no stops, as in N, NNE, NE, etc., except when denoting a particular region or place name, as in 'magnetic north', 'South Pole', 'South America', 'North Sea gas', 'in the West'. A compass bearing is usually given in degrees, 0° to 360°, measured clockwise from magnetic or geographical north.

complement A group of serum proteins that assists (complements) the action of antibodies and is involved in other aspects of humoral immunity. Its components, which react in sequence, are designated by a capital C followed by an Arabic numeral; they are (in order of reaction): C1 (having the subcomponents C1q, C1r, and C1s), C4, C2, C3, and C5–C9. Intermediate complexes are preceded by the symbols E (erythrocyte), A (antibody), and C (complement) preceding numerals designating the components that have already reacted; e.g. EAC142 is the intermediate produced by reaction of the first three components. In an alternative pathway for activation of complement, the components are designated C3, B, D, P, H, and I; fragments of proteins produced during this pathway are designated by the lower-case letters a and b; e.g. C3a, C3b, Ba, Bb.

complementary DNA Abbrev.: cDNA (lower-case c; no stops).

complexity *Genetics* The DNA content of a given sample. It is important to distinguish between **chemical complexity**, which is chemically determined and usually measured in picograms (pg); and **kinetic complexity**, which is determined from the reassociation kinetics of the DNA and is usually measured in base pairs (*see* bp).

complex *See* coordination compound.

complex numbers Quantities of the form

$$z = x + \mathrm{i}y,$$

where x and y are real numbers and i (roman type, also written j) is defined by:

$$\mathrm{i} = \sqrt{(-1)} \qquad \mathrm{i}^2 = -1.$$

The real part of z (i.e. x) and the imaginary part of z (i.e. y) are denoted, respectively:

$$\mathrm{Re}\, z \qquad \mathrm{Im}\, z.$$

The modulus of z is denoted $|z|$, where

$$z = |z| \exp{(\mathrm{i}\phi)};$$

ϕ (Greek phi) is called the phase or the argument of z, also denoted $\arg z$. The complex conjugate of z, i.e. $x - \mathrm{i}y$, is denoted z^* or $\bar{z}$.

In the case of physical quantities, the complex representation is often denoted using primes; for example for the dielectric constant,

$$\epsilon = \epsilon' + \epsilon''.$$

Compositae (cap. C) A large family of dicotyledonous plants, commonly known as the daisy family. Alternative name: Asteraceae. Individual name and adjectival form: **composite** (no cap.).

compound Abbrev.: cpd. (stop)

compressibility Symbol: κ_T or κ (Greek kappa). A physical quantity relating to the ease with which a body of volume V can be compressed at a constant temperature T, given by:

$$\kappa = -(1/V)(\partial V/\partial p)_T,$$

where p is the pressure. This is also called **isothermal compressibility**. Constant entropy rather than constant temperature gives the **adiabatic compressibility**, symbol: κ_S. The *SI unit is the reciprocal of the pascal (Pa^{-1}). Compressibility is the reciprocal of *bulk modulus.

Compton, Arthur Holly (1892–1962) US physicist.

Compton effect or, in continental Europe, **Compton–Debye effect** (en dash)

Compton telescope

***Compton wavelength**

Compton wavelength Symbol: λ_C (Greek lambda). A fundamental constant equal to

$$2.426\,310\,58 \times 10^{-12}\,\mathrm{m}.$$

It is given by $h/m_e c$, where h is the Planck constant, m_e the electron mass, and c the speed of light in vacuum. The Compton wavelengths of the proton and neutron are given by $h/m_p c$ and $h/m_n c$, where m_p and m_n are the proton and neutron mass, respectively; the symbols are then $\lambda_{C,p}$ and $\lambda_{C,n}$.

computer-aided (or **computer-assisted**; hyphenated)

computer-aided design Abbrev.: CAD

computer-aided design, manufacturing, and testing Abbrev.: CADMAT

computer-aided engineering Abbrev.: CAE

computer-aided instruction Abbrev.: CAI

computer-aided learning Abbrev.: CAL

computer-aided manufacturing Abbrev.: CAM

computer-aided software engineering Abbrev.: CASE

computer-integrated manufacturing (hyphenated) Abbrev.: CIM (no stops).

computerized tomography US: **computed tomography**. Abbrev. and preferred form: CT (no stops). Former name: **computerized axial tomography** (abbrev.: CAT), superseded as the technique can now be applied in any plane, not only the axial plane.

computer-managed instruction (hyphenated) Abbrev.: CMI (no stops).

con- Prefix denoting with, together, jointly or joint, similar, same (e.g. concentric, conglomerate, conjugation, consociation). *See also* co-; com-.

Con A (cap. C, no stop) Abbrev. for concanavalin A.

conc. Abbrev. for concentrated.

concanavalin A Abbrev.: Con A (cap. C, no stop).

concentrated Abbrev.: conc. (stop).

concentration Symbol: c_B for a solute B or, where a complicated formula is to be written, [B] or c(B), as in $[(NH_4)_2CO_3]$. A physical quantity, the *amount of substance of the solute divided by the volume of the solution. The *SI unit is the mole per cubic metre or the mole per litre. Also called amount-of-substance concentration.

The **mass concentration**, symbol ρ_B (Greek rho), is the mass of a substance B divided by the volume of the mixture. The SI unit is the kilogram per cubic metre or kilogram per litre. The full term, mass concentration, must be used if it is meant, to avoid confusion with the term concentration.

See also molality; mole fraction; molecular concentration.

Condamine, Charles-Marie de La Usually alphabetized as *La Condamine.

condenser (not condensor) 1. An apparatus for cooling a gas to form a liquid or solid. 2. A lens system to concentrate a beam of light. 3. Former name for capacitor.

Condon, Edward Uhler (1902–74) US physicist.

Franck–Condon principle (en dash)

conductance Symbol: G A physical quantity, the reciprocal of *resistance. The *SI unit is the siemens, formerly called a reciprocal ohm or mho.

conductivity (or **electrical conductivity**) Symbol: γ or σ (Greek gamma, sigma); κ (Greek kappa) is used for electrolytic conductivity. A physical quantity, the reciprocal of *resistivity. The *SI unit is the siemens per metre. *See also* thermal conductivity.

conductor (not conducter)

conidium (pl. conidia) An asexual spore (also called conidiospore) produced by certain fungi on a structure called a **conidiophore**. Note that the term does not denote a spore-producing body (*compare* basidium).

conifer Any plant of the order *Coniferales. Adjectival form: **coniferous.**

Coniferales (cap. C) An order of plants in the class *Coniferopsida that includes pine, fir, spruce, larch, etc. Also called Pinales. Individual name: **conifer** (no cap.). Adjectival form: **coniferous**.

Coniferophytina A synonym for *Gymnospermae used by some authors. Note that the trivial name **coniferophyte** is used for members of the *Pinicae, a subdivision in a classification that regards gymnosperms as a division, *Pinophyta.

Coniferopsida A class of gymnosperms containing the orders Ginkgoales (*see* ginkgo), *Coniferales (conifers), and *Taxales (e.g. yews). In a recent classification, Ginkgoales is raised to the status of a class, Ginkgoopsida; the remaining orders then constitute the class Pinopsida.

conjunctiva (pl. conjunctivae) The mucous membrane that covers the cornea and the inner eyelids. Adjectival form: **conjunctival.**

Conon of Samos (*fl.* 245 BC) Greek mathematician and astronomer.

consensus sequence *Genetics* An idealized sequence of bases that represents the base most likely to occur at each position in the sequence. For example, the base sequence TATAAT, known as the *Pribnow box, can be represented in the form $T_{80}A_{95}t_{45}A_{60}a_{50}T_{96}$; the subscript numbers refer to the percentage occurrence of the bases, and a lower-case letter is used to represent a base whose occurrence is less than 54% but still significantly greater than random. A site in which no particular base occurs with statistical significance is denoted by N.

const. Abbrev. for constant.

content-addressable memory (hyphenated) *Computing* Abbrev.: CAM (no stops).

continuous-wave radar (hyphenated) Abbrev.: CW radar.

continuous-wave spectroscopy (hyphenated) Abbrev.: CWS (no stops).

continuum (pl. continua)

contra- Prefix denoting against or opposite (e.g. contralateral).

contrast index *Photog.* Abbrev.: CI (no stops).

Control of Substances Hazardous to Health Abbrev.: CSHH (no stops).

convolvulus (pl. convolvuluses; not convolvuli) Generic name: *Convolvulus* (cap. C, ital.).

Conybeare, William Daniel (1787–1857) British geologist.

Cook, James (1728–78) British navigator and explorer.

Coombs, Robert Royston Amos (1921–) British immunologist.
Coombs test

Cooper, Leon Neil (1930–) US physicist.
Cooper pair *See also* BCS theory.

Coordinated Universal Time *See* TAI.

coordinates Space coordinates, defining the position of a point relative to a frame of reference, are usually denoted:
(x, y, z) Cartesian coordinates (Fig. 1)
r, θ, z cylindrical polar coordinates (Fig. 2)
(r, θ, ϕ) spherical polar coordinates (Fig. 3)
In two-dimensional problems only two coordinates need be specified: (x, y) in Cartesian coordinates, (r, θ) in plane polar coordinates.

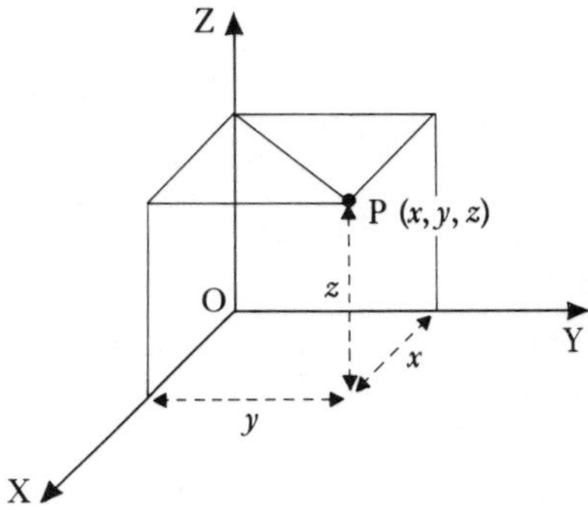

Fig.1 Cartesian coordinates

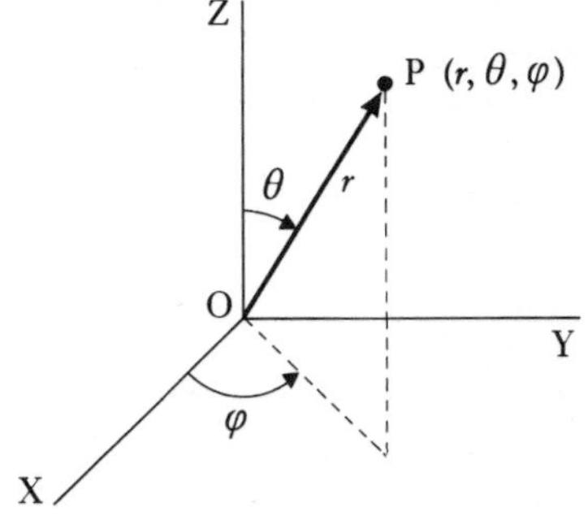

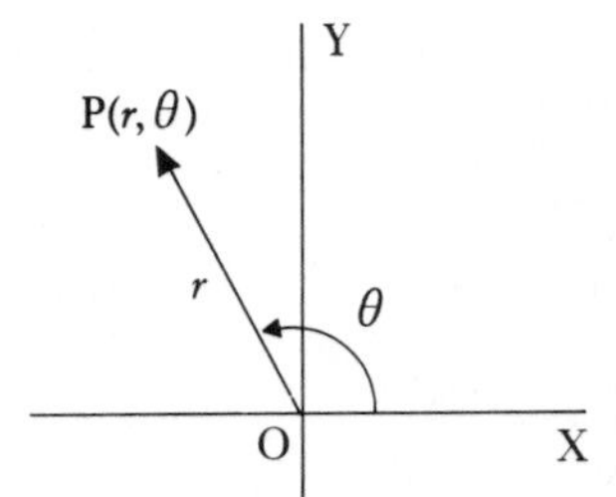

Fig.2 Spherical (above) and plane polar coordinates

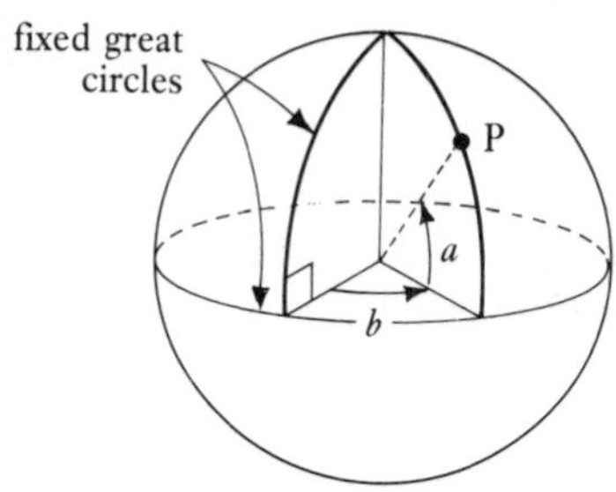

Fig.3 Coordinates of point on surface of sphere

The position of a point on the surface of a sphere can be specified by two coordinates (a, b) that are angles measured relative to two fixed great circles on the sphere (Fig. 3). A point on the earth is located by its latitude and longitude (ϕ, L). A celestial object on the celestial sphere is located, for example, by its *right ascension and *declination (α, δ).

Relativistic coordinates, defining an event, may be denoted:

(x_0, x_1, x_2, x_3) or (x_1, x_2, x_3, x_4)

where $x_0 = ct$, $x_1 = x$, $x_2 = y$, $x_3 = z$, $x_4 = \mathrm{i}ct$.

See also generalized coordinates; graphs.

coordination compound A type of chemical compound in which molecules or ions (ligands) form coordinate bonds with a central atom or ion.

Coordination compounds are systematically named by giving the ligands in alphabetical order (disregarding any prefixes indicating the number of each) followed by the name of the central atom or ion and its oxidation number (in Roman numerals), together with the name of any associated cationic or anionic species if the complex is charged, e.g. potassium hexacyanoferrate(III), $K_3[Fe(CN)_6]$, pentaamminecobalt(III) chloride, $[Co(NH_3)_5]Cl_3$, and tetracarbonylnickel(0), $[Ni(CO)_4]$. An alternative to using Roman numerals to indicate the oxidation number of the central atom or ion (the Stock system) is the Ewins–Bassett system, which uses Arabic numerals followed by a plus or a minus sign, indicating a cationic or an anionic complex respectively, to show the overall charge on a complex, e.g. potassium hexacyanoferrate (4-), $K_3[Fe(CN)_6]$.

In nonsystematic nomenclature coordination compounds are named by using the anionic or cationic form of the name of the central atom or ion as appropriate, e.g. potassium ferricyanide, $K_3[Fe(CN)_6]$, and pentaamminecobaltic chloride, $[Co(NH_3)_5]Cl_3$; neutral complexes are named by naming the central species followed by the name(s) of the ligands, e.g. nickel carbonyl, $[Ni(CO)_4]$. Also (until fairly recently), it was the practice to name negative ligands first, out of alphabetical sequence, e.g. dichlorodiammineplatinum(II), $[Pt(NH_3)_2Cl_2]$.

See also ligand.

Cope, Edward Drinker (1840–97) US vertebrate palaeontologist and comparative anatomist.

Copepoda (cap. C) A subclass of planktonic crustaceans. Individual name and adjectival form: **copepod** (no cap.).

Copernicus, Nicolaus (1473–1543) Polish astronomer.

Copernican system

***copia* element** (ital. *copia*) A transposable element (*see* transposon) found in the genome of *Drosophila melanogaster*. The name derives from the Latin *copia* (abundance) because *copia* elements code for abundant mRNAs.

copper Symbol: Cu *See also* periodic table; nuclide.

copper(I) dicarbide Cu_2C_2 The recommended name for the compound traditionally known as cuprous acetylide.

copper(I) oxide Cu_2O The recommended name for the compound traditionally known as cuprous oxide.

copper(II) oxide CuO The recommended name for the compound traditionally known as cupric oxide.

copper(II) sulphate-5-water $CuSO_4{\cdot}5H_2O$ US: **copper(II) sulfate-5-water**. The recommended name for the compound traditionally known as blue vitriol.

copro- (**copr-** before vowels) Prefix denoting faeces (e.g. coprophyte, coprozoic).

Coprococcus (cap. C, ital.) A genus of coccoid fermentative bacteria. Individual name: **coprococcus** (no cap., not ital.; pl. coprococci).

cord (not chord) *Anat., zool.* Any long thin structure, e.g. nerve cord, spinal cord, vocal cord. Note that the 'h' is retained in *notochord.

Cordaitales An extinct order of gymnosperms (*see* Pinicae).

Corey, Elias James (1928–) US chemist.

Cori ester Named after C. F. Cori (1896–1984) and his wife, G. T. R. Cori (1896–1957).

Coriolis, Gustave-Gaspard (1792–1843) French physicist.
Coriolis acceleration
Coriolis effect
Coriolis force
Coriolis parameter

Cormack, Allan Macleod (1924–) South African physicist.

Cormatae An infradivision in the division *Streptophyta that contains all the vascular plants.

cornea (pl. corneas; preferred to corneae) The transparent part of the front of the eyeball. Adjectival form: **corneal**.

Cornforth, Sir John Warcup (1917–) Australian chemist.

Cornu spiral *Maths.* Named after M. A. Cornu.

corona (pl. coronae (anatomical and sometimes astronomical senses) or coronas) Adjectival forms: **coronal** (e.g. coronal suture, coronal discharge, coronal holes), **coronary** (e.g. coronary artery).

Corona Australis A constellation. Genitive form: Coronae Australis. Abbrev.: CrA (no stops). *See also* stellar nomenclature.
Corona Australids (meteor shower)

Corona Borealis A constellation. Genitive form: Coronae Borealis. Abbrev.: CrB (no stops). *See also* stellar nomenclature.

coronavirus (one word) Any virus belonging to the family Coronaviridae.

corpus allatum (pl. corpora allata; usually referred to in the pl.) A paired endocrine gland in insects.

corpus callosum (pl. corpora callosa) The tissue connecting the two halves of the cerebrum.

corpus cardiacum (pl. corpora cardiaca; usually referred to in the pl.) A paired endocrine gland in insects.

corpuscle Use *blood cell.

corpus luteum (pl. corpora lutea) A temporary endocrine gland that forms in the ovary after ovulation. Adjectival form: **luteal**.

corpus striatum (pl. corpora striata) A part of the brain.

corr. Abbrev. for **1.** corrected. **2.** corresponding.

correlation (not corelation)

Correns, Karl Erich (1864–1933) German botanist and geneticist.

corresponding Abbrev.: corr. (stop).

Cort, Henry (1740–1800) British metallurgist and inventor.

cortex (pl. cortices) The outer zone of many organs or structures (e.g. of the kidney, adrenal gland, cerebrum, plant stems and roots, etc.). Adjectival form: **cortical**.

Corti, Alfonso Giacomo Gaspare (1822–88) Italian anatomist and histologist.
organ of Corti Also called spiral organ.

corticotrophin US: **corticotropin**. Another name for adrenocorticotrophic hormone (ACTH).
corticotrophin-releasing factor Abbrev.: CRF (no stops).

Corvus A constellation. Genitive form: Corvi. Abbrev.: Crv (no stop). *See also* stellar nomenclature.

Corynebacterium (cap. C, ital.) A genus of facultatively anaerobic rod- or club-shaped bacteria. Individual name:

corynebacterium (no cap., not ital.; pl. corynebacteria).
C. diphtheriae: the cause of diphtheria. Strains are assigned to one of three cultural types: *gravis*, *intermedius*, and *mitis* (ital.).
C. haemolyticum: now transferred to the genus *Arcanobacterium*.
C. pseudotuberculosis: the cause of lymphatic infections in domestic animals, often erroneously referred to as '*C. ovis*'.
C. pyogenes: now transferred to the genus *Actinomyces*.
None of the species causing plant diseases is now regarded as a true member of the genus. Some have already been assigned to other genera; others await reclassification. *See* *Arthrobacter*; *Aureobacterium*; *Curtobacterium*; *Erwinia*; *Propionibacterium*; *Rhodococcus*.

coryneform Describing any club-shaped bacterium. The term may also be used as a noun, to mean any coryneform bacterium. Note that usage is not restricted to members of the genus **Corynebacterium*. *See also Propionibacterium*.

cos Symbol for cosine, written with a space or thin space before an angle or variable:
$\cos 30°$, $\cos(-\theta)$, $\cos x$, $\cos(a + b)$.
The inverse function of $y = \cos x$ is denoted:
$$x = \arccos y$$
or
$$x = \cos^{-1}y \text{ (no space).}$$

cosec Symbol for cosecant (reciprocal of sine), written with a space or thin space before an angle or variable:
$\operatorname{cosec} 30°$, $\operatorname{cosec}(-\theta)$, $\operatorname{cosec} x$, $\operatorname{cosec}(a + b)$.
The symbol csc can also be used.
The inverse function of $y = \operatorname{cosec} x$ is denoted:
$$x = \operatorname{arccosec} y$$
or
$$x = \operatorname{cosec}^{-1}y \text{ (no space).}$$

cosech Symbol for hyperbolic cosecant, written with a space or thin space before a variable as in $\operatorname{cosech} x$, $\operatorname{cosech}(a + b)$. The inverse function of $y = \operatorname{cosech} x$ is denoted using the prefix ar-:
$$x = \operatorname{arcosech} y$$
or
$$x = \operatorname{cosech}^{-1}y \text{ (no space).}$$

cosh Symbol for hyperbolic cosine, written with a space or thin space before a variable as in $\cosh x$, $\cosh(a + b)$. The shortened form ch is also permitted.
The inverse function of $y = \cosh x$ is denoted using the prefix ar-:
$$x = \operatorname{arcosh} y \text{ (or arch } y\text{)}$$
or
$$x = \cosh^{-1}y \text{ (no space).}$$

COSHH Abbrev. for Control of Substances Hazardous to Health.

cosmo- (**cosm-** before vowels) Prefix denoting the world, universe, or space (e.g. cosmology, cosmonaut, cosmic).

cot 1. Symbol for cotangent (reciprocal of tangent), written with a space or thin space before an angle or variable:
$\cot 30°$, $\cot(-\theta)$, $\cot x$, $\cot(a + b)$.
The symbol ctg can also be used.
The inverse function of $y = \cot x$ is denoted:
$$x = \operatorname{arccot} y$$
or
$$x = \cot^{-1}y \text{ (no space).}$$
2. Abbrev. for cyclooctatetraene often used in the formulae of coordination compounds, e.g. $[Ti_2(cot)_3]$.

coth Symbol for hyperbolic cotangent, written with a space or thin space before a variable as in $\coth x$, $\coth(a + b)$. The inverse function of $y = \coth x$ is denoted using the prefix ar-:
$$x = \operatorname{arcoth} y$$
or
$$x = \coth^{-1}y \text{ (no space).}$$

Cotton, Aimé (1869–1951) French physicist.
Cotton balance
Cotton–Mouton effect (en dash) Also named after H. Mouton.

Cottrell, Sir Alan Howard (1919–) British physicist and metallurgist.

Cottrell precipitator *Chem. eng.* Also called electrostatic precipitator. Named after F. G. Cottrell.

Cot value (cap. C) A measure of DNA kinetic *complexity based on hybridization of single-stranded DNA. The notation is a modification of C_0t (subscript zero), where

C_0 is the initial DNA concentration and *t* is time. $Cot_{1/2}$ (subscript ½) is the value representing half complete reassociation. *Compare* Rot value.

cotylo- (**cotyl-** before vowels) Prefix denoting a cup or cup-shaped hollow (e.g. cotylosaur, cotyloid).

coudé mounting (lower-case c; acute accent) *Astron.* [from French: elbow]

coulomb (no cap.) Symbol: C The *SI unit of electric charge.

$$1\,C = 1\,A\,s.$$

Named after C. A. de *Coulomb.

Coulomb, Charles Augustin de (1736–1806) French physicist.
***coulomb**
Coulomb excitation
Coulomb field
Coulomb force
***coulombmeter**
Coulomb scattering
Coulomb's law
coulombmetric analysis

coulombmeter (preferred to coulometer) Former name: voltameter.

Coulson, Charles Alfred (1910–74) British theoretical chemist, physicist, and mathematician.

counter- Prefix denoting against, opposite, complementary (e.g. countercurrent, countershading).

counterstain (one word)

Couper, Archibald Scott (1831–92) British organic chemist.

coupling constant Symbol: J A physical quantity, the separation between two absorption peaks in nuclear magnetic resonance spectroscopy due to spin–spin coupling. The unit is the hertz (Hz).

Cousteau, Jacques-Yves (hyphen) (1910–) French oceanographer.

COV *Genetics* Abbrev. for crossover value.

coverslip (one word)

cowpea mosaic virus Type member of the **Comovirus* group (vernacular name: cowpea mosaic virus group).

Cowper, William (1666–1709) English surgeon.

Cowper's glands Another name for bulbourethral glands.

coxa (pl. coxae) The basal segment of an insect's leg. Adjectival form: **coxal**.

coxsackievirus (one word, no cap.; not cocksackie) Any of a subgroup of human *enteroviruses. They are divided into two categories: A (containing 24 serovars) and B (containing 16 serovars); individual serovars are designated coxsackievirus A3, coxsackievirus B6, etc. Named after the village in New York State where the virus was first reported.

cp (no stop) Abbrev. for the cyclopentadienyl radical often used in the formulae of coordination compounds, e.g. $[cp_2TiNCO]$.

CP Abbrev. for chemically pure.

cpd. Abbrev. for compound.

cpi (no stops) Abbrev. for characters per inch.

CP invariance (CP not ital.) *Nucl. physics* CP is the product of the charge-conjugation operator (symbol: C) and space-inversion operator (symbol: P). In the case of **CPT invariance**, T is the time-reversal operator (symbol: T).

C_3 plant (subscript number) A plant that produces the 3-carbon compound phosphoglyceric acid as the first stage of photosynthesis.

C_4 plant (subscript number) A plant that produces the 4-carbon compound oxaloethanoic (oxaloacetic) acid as the first stage of photosynthesis.

CPM Abbrev. for critical path method.

CPPO Abbrev. for bis(2-carbopentyloxy-3,5,6-trichlorophenyl) oxalate.

cps (no stops) Abbrev. for **1.** characters per second, a measure of the speed of a printer in a computer system. **2.** cycles per second, an obsolete measure of the rate of a periodic phenomenon, such as frequency. It has been replaced by the hertz, an *SI unit.

CPU (or **cpu**) *Computing* Abbrev. for central processing unit.

Cr Symbol for chromium.

CrA (no stops) *Astron.* Abbrev. for Corona Australis.

Crafts, James Mason (1839–1917) US chemist.
Friedel–Crafts reaction (en dash)

Cram, Donald James (1919–   ) US chemist.

Cramer's rule *Linear algebra* Named after Gabriel Cramer (1704–52).

Craniata (cap. C) The subphylum of the Chordata comprising all animals possessing a cranium and a vertebral column. Also called Vertebrata. Individual name and adjectival form: **craniate** (no cap.), although *vertebrate is more widely used.

craniate Possessing a cranium. *See also* Craniata.

cranio- (**crani-** before vowels) Prefix denoting the skull (e.g. craniometry).

cranium (pl. crania) The skull. The word is used in combination to denote anatomically distinct parts of the skull (e.g. neurocranium, splanchnocranium). Adjectival forms: **cranial**, ***craniate**.

crassulacean acid metabolism Acronym: CAM (no stops).

Crater A constellation. Genitive form: Crateris. Abbrev.: Crt (no stop). *See also* stellar nomenclature.

craters *Astron.* Craters on planets and satellites are usually named after famous people, as with Copernicus and Tycho on the moon. The fact that it is a crater is not indicated in the name, unlike other surface features, such as maria.

CrB (no stops) *Astron.* Abbrev. for Corona Borealis.

cresol $CH_3C_6H_4OH$ The traditional name for *methylphenol.

cretaceous Consisting of or resembling chalk; chalky. *See also* Cretaceous.

Cretaceous (cap. C) Abbrev.: K (no stop). **1.** (adjective) Denoting the most recent period of the Mesozoic era, characterized by the deposition of chalk (hence the name). **2.** (noun; preceded by 'the') The Cretaceous period. The period is divided into the Lower (or Early) Cretaceous and the Upper (or Late) Cretaceous epochs (initial caps.); abbrevs. (respectively): K_1 and K_2 (subscript numerals).

Creutzfeldt–Jacob disease (en dash) Named after Hans Gerhard Creutzfeldt (1885–1964) and Alfons Maria Jacob (1884–1931).

CRF Abbrev. for **1.** corticotrophin-releasing factor. **2.** coagulase reacting factor.

Crick, Francis Harry Compton (1916–   ) British molecular biologist.
Watson–Crick model (en dash)

Crinoidea (cap. C) A class of echinoderms containing the sea lilies and featherstars. Individual name and adjectival form: **crinoid** (no cap.).

crista (pl. cristae) **1.** A sensory structure in a semicircular canal of the inner ear. **2.** A fold on the inner membrane of a mitochondrion. Adjectival form: **cristate**.

Cristispira (cap. C, ital.) A genus of *spirochaete bacteria. Individual name: **cristispire** (no cap., not ital.).

crit. Abbrev. for critical.

criterion (pl. criteria)

critical path method Abbrev.: CPM (no stops).

CRO Abbrev. for cathode-ray oscilloscope.

Crocco's equation *Fluid mech.*

Crocodilia (cap. C) An order of reptiles containing the crocodiles, alligators, and caymans. Individual name and adjectival form: **crocodilian** (no cap.); the word 'crocodile' should be reserved for crocodilians of the family Crocodylidae.

crocus (pl. crocuses; not croci) Generic name: *Crocus* (cap. C, ital.).

Cro-Magnon man Common name for a type of fossil hominid (*see Homo*). Named after the Cro-Magnon caves near Les Eyzies, France, where the first fossils were found in 1868.

Crompton, Samuel (1753–1827) British inventor.

Cronin, James Watson (1931–   ) US physicist.

Cronstedt, Axel Frederic (1722–65) Swedish chemist and mineralogist.

Crookes, Sir William (1832–1919) British chemist and physicist.

Crookes dark space Also called cathode dark space, Hittorf dark space.
Crookes radiometer
Crookes tube

cross- Prefix denoting action, movement, or location between or across. It may or may not be hyphenated; for examples, see individual entries.

crossbar (one word)

cross-bedding (hyphenated) *Geol.*

crossbreed (verb and noun; one word) Derived noun: **crossbreeding**.

cross-check (hyphenated)

cross compiler (two words) *Computing*

cross correlation (two words) *Maths.*

cross-fertilization (hyphenated) Derived verb: **cross-fertilize**.

crossing over (two words) *Genetics.* The exchange of chromatid sections between paired homologous chromosomes during meiosis. *See also* chiasma.

cross-linkage (hyphenated) Derived verb: **cross-link**.

Crossopterygii (cap. C; not Crossopterygi) An order (sometimes regarded as a subclass) of bony fishes comprising the lobe-finned fishes. Individual name and adjectival form: **crossopterygian** (no cap.).

crossover (one word)

crossover value *Genetics* Abbrev.: COV (no stops).

cross-pollination (hyphenated) Derived verb: **cross-pollinate**.

cross product Another name for vector product.

cross section (two words) Symbol: σ (Greek sigma). A physical quantity relating to a specified atomic, molecular, or nuclear reaction or process at a specified target entity. The reaction or process is produced by charged or uncharged particles of specified type and energy incident on the target. If an incident particle travels distance dx in a medium with N target particles per square volume, the probability of interaction equals $\sigma N dx$. The *SI unit is the square metre, but in nuclear physics the *barn is sometimes used.

crosstalk (one word) *Telecom.*

croton (pl. crotons) A popular cultivated plant of the genus *Codiaeum* (not **Croton*), usually *Codiaeum variegatum*.

Croton (cap. C, ital.) A genus of mainly tropical plants of the family Euphorbiaceae. The species *C. tiglium* is a source of **croton oil**. Not to be confused with *croton, a house plant of the same family.

crotonaldehyde $CH_3CH=CHCHO$ The traditional name for buten-2-al.

crotonic acid $CH_3CH=CHCOOH$ The traditional name for butenoic acid.

crotyl alcohol $CH_3CH=CHCH_2OH$ The traditional name for but-2-en-1-ol.

CRP Abbrev. for cyclic AMP receptor protein. *See* catabolite activator protein.

Crt (no stop) *Astron.* Abbrev. for Crater.

CRT Abbrev. for cathode-ray tube.

Cru (no stop) *Astron.* Abbrev. for Crux.

Cruciferae (cap. C) A large family of dicotyledonous plants, commonly known as the mustard family. Alternative name: Brassicaceae. Individual name: **crucifer** (no cap.). Adjectival form: **cruciferous**.

Crum Brown, Alexander (1838–1922) British organic chemist.
Crum Brown's rule

Crustacea (cap. C) A class of arthropods including the shrimps, crabs, water fleas, barnacles, and woodlice. Individual name and adjectival form: **crustacean** (no cap.; not crustaceous).

Crux A constellation. Genitive form: Crucis. Abbrev.: Cru (no stop). *See also* stellar nomenclature.

Crv (no stop) *Astron.* Abbrev. for Corvus.

cryo- Prefix denoting cold or low temperature (e.g. cryogenics, cryophilic).

cryoscopic constant *See* freezing point.

crypto- (**crypt-** before vowels) Prefix denoting hidden (e.g. cryptography, cryptophyte, cryptanalysis, cryptorchid).

cryptogam In former plant classification schemes, any plant without obvious reproductive structures (e.g. an alga, bryophyte, or pteridophyte), placed in the taxon Cryptogamia. The term now has no taxonomic significance. *Compare* phanerogam.

Cryptophyta A division of algae that usually lack cell walls and are regarded as protozoans by some authorities. In a recent classification it is placed in the subkingdom Protistobionta (*see* Protista).

cryst. Abbrev. for crystalline.

crystal-field stabilization energy Abbrev.: CFSE (no stops). *Chem.*

crystal-field theory Abbrev.: CF theory. *Chem.*

Cs Symbol for caesium.

Cs (ital.) Abbrev. for cirrostratus.

csc Another symbol for cosecant. *See* cosec.

CSF Abbrev. for **1**. cerebrospinal fluid. **2**. *Genetics* colony-stimulating factor. The various types are designated by a hyphenated prefix: e.g. M-CSF (for macrophage CSF); GM-CSF (for granulocyte/monocyte CSF); and G-CSF (for granulocyte CSF).

CSI Abbrev. for chlorosulphonyl isocyanate.

CSIRO Abbrev. for Commonwealth Scientific and Industrial Research Organization, an Australian government organization.

CST Abbrev. for Central Standard Time (in the USA).

CT Abbrev. and preferred form for *computerized tomography.
CT scanner

CTAB Abbrev. for cetyltrimethylammonium bromide.

CTD *Electronics* Abbrev. for charge-transfer device.

ctDNA Abbrev. for chloroplast DNA.

ctenidium (pl. ctenidia) One of the comblike gills of some molluscs. Adjectival form: **ctenidial**.

Ctenophora (cap. C) A phylum of *coelenterates comprising the comb jellies. Individual name: **ctenophore** (no cap.). Adjectival form: **ctenophoran**.

CTF Abbrev. for coal-tar fuels.

ctg Another symbol for cotangent. *See* cot.

CTP Abbrev. and preferred form for cytidine 5′-triphosphate.

Cu Symbol for copper.

Cu (ital.) Abbrev. for cumulus.

cube root *See* square root.

cubic close-packed *Crystallog.* Abbrev.: c.c.p. (no caps., stops).

cubic expansion coefficient *See* linear expansion coefficient.

cubic metre US: **cubic meter**. Symbol: m^3 The derived unit of volume and capacity in *SI units. The SI prefixes permit smaller or larger volumes and capacities, such as the cubic decimetre (dm^3), cubic centimetre (cm^3), and cubic millimetre (mm^3), to be expressed:
$1\ dm^3 = (0.1\ m)^3 = 10^{-3}\ m^3$
$1\ cm^3 = (0.01\ m)^3 = 10^{-6}\ m^3$
$1\ mm^3 = (0.001\ m)^3 = 10^{-9}\ m^3$.
The *litre is now exactly the same as the cubic decimetre. One cubic metre = 1.307 95 cubic yards, 35.3147 cubic feet, 219.255 UK gallons, 264.172 US gallons.

cucumber mosaic virus Abbrev.: CMV (no stops). Type member of the **Cucumovirus* group (vernacular name: cucumber mosaic virus group).

Cucumovirus (cap. C, ital.) Approved name for the *cucumber mosaic virus group. Individual name: **cucumovirus** (no cap., not ital.). [from *cucu*mber *mo*saic *virus*]

cuesta *Geol.* A low ridge. [from Spanish: shoulder]

Cugnot, Nicolas-Joseph (hyphen) (1725–1804) French engineer.

Culpeper, Nicholas (1616–54) English medical writer and herbalist.

cultivar Abbrev.: cv. (stop). A culti(vated) var(iety): a plant maintained by horticultural or agricultural techniques. The epithet of a cultivar is printed in roman (rather than italic) type, with an initial capital letter, the whole name being printed in the form *Cornus controversa* Variegata or *C. controversa* 'Variegata'.

cumene $C_6H_5CH(CH_3)_2$ The traditional name for (1-methylethyl)benzene.

cumulonimbus (pl. cumulonimbi) *Meteorol.* Abbrev.: *Cb* (ital., no stop).

cumulus (pl. cumuli) *Meteorol.* Abbrev.: *Cu* (ital., no stop).

cumyl- Prefix denoting the group $C_6H_5C(CH_3)_2-$ (e.g. cumyl alcohol). Isopropylbenzyl is preferred in all contexts.

cup fungi Common name for fungi of the orders *Helotiales and *Pezizales.

cuprammonium Denoting compounds containing the ion $[Cu(NH_3)_4]^{2+}$. The recommended system is to use oxidation numbers (*see* cuprammonium sulphate).

cuprammonium sulphate $Cu(NH_3)_4SO_4$ US: **cuprammonium sulfate.** The traditional name for tetraamminecopper(II) sulphate.

cupric Denoting compounds in which copper has an oxidation state of +2. The recommended system is to use oxidation numbers, e.g. cupric oxide, CuO, has the systematic name copper(II) oxide.

cupro- (or **cupri-**) Prefix denoting copper (e.g. cuprophyte, cupriferous). *See also* cupric; cuprous.

cuprous Denoting compounds in which copper has an oxidation state of +1. The recommended system is to use oxidation numbers, e.g. cuprous oxide, Cu_2O, has the systematic name copper(I) oxide.

cuprous acetylide Cu_2C_2 The traditional name for copper(I) dicarbide.

cuprous potassium cyanide $K_2Cu(CN)_4$ The traditional name for potassium tetracyanocuprate(I).

curie Symbol: Ci A unit of *activity of a radioactive nuclide, equal to 3.7×10^{10} disintegrations per second; in practice the millicurie (mCi) is used. The curie is loosely used as a unit of quantity of any radioactive substance in which there is this degree of activity. The curie has been displaced by the becquerel, an *SI unit: 1 Ci = 3.7×10^{10} Bq. Named after Pierre *Curie (not Marie Curie).

Curie, Marie Sklodowska (1867–1934) Polish-born French chemist, wife of Pierre Curie.
curium

Curie, Pierre (1859–1906) French physicist, husband of Marie Curie.
***curie**
Curie's law
Curie temperature (or **point**) Symbol: T_C
Curie–Weiss law (en dash)
curium

curium Symbol: Cm *See also* periodic table; nuclide.

curl *See* vector.

curly dee Symbol: ∂ Used to denote a partial derivative. *See* mathematical symbols.

current *See* electric current.

current annual increment *Forestry* Abbrev.: CAI (no stops).

current density *See* electric current density.

Curtius, Theodor (1857–1928) German organic chemist.
Curtius transformation

Curtobacterium (cap. C, ital.) A genus of obligately aerobic rod-shaped or coccoid bacteria. Several members were formerly assigned to other genera, including **Brevibacterium*. The bacteria formerly classified as *Corynebacterium betae*, *C. oortii*, and *C. poinsettiae* are now regarded as pathovars of *Curtobacterium flaccumfaciens* (formerly *Corynebacterium flaccumfaciens*). Individual name: **curtobacterium** (no cap., not ital.; pl. curtobacteria).

Cushing, Harvey William (1869–1939) US surgeon.
Cushing's syndrome

cutoff (noun; one word) Verb form: **cut off** (two words).

cutout (noun; one word) Verb form: **cut out** (two words).

Cuvier, Georges Léopold Chrétien Frédéric Dagobert, Baron (1769–1832) French comparative anatomist, palaeontologist, and taxonomist.
Cuvierian duct US: **cuvierian duct.** Also called common cardinal vein, ductus Cuvieri (US: cuvieri).
Cuvierian organs

cv. (stop) Abbrev. for cultivar.

C value (cap. C) The total amount of DNA in the haploid genome of any particular species, usually measured in picograms (pg). The diploid amount is signified as 2C (no space), the tetraploid amount as 4C, etc.

CVn (no stops) *Astron.* Abbrev. for Canes Venatici.

CW radar Abbrev. for continuous-wave radar.

CWS Abbrev. for continuous-wave spectroscopy.

cwt Symbol for hundredweight. *See* pound.

cyano- (sometimes **cyan-** before vowels or h) Prefix denoting **1.** the cyanide group, –CN (e.g. cyanobenzamide, cyanoethanoic, cyanogenesis, cyanamide). **2.** a blue colour (e.g. cyanobacteria, cyanosis).

cyanobacteria (no cap.; sing. cyanobacterium) Preferred name for the photosynthetic bacteria traditionally known as the blue-green algae. Adjectival form: **cyanobacterial**.

Formerly classified as algae, these organisms have now been firmly established as eubacteria, and the term 'cyanobacteria' is widely accepted. However the taxonomy and nomenclature of the cyanobacteria are still confused. Botanists have been describing taxa for the blue-green algae according to the International Code of Botanical Nomenclature, while bacteriologists have been employing the different criteria of the International Code of Nomenclature of Bacteria. Hence, there is currently no generally agreed Approved List of names for the cyanobacteria and it is imperative that the authors of cyanobacterial names be cited wherever there is a risk of confusion. In spite of changed or uncertain status, former generic (and specific) names generally retain their italicized form; *see*, for example, *Pleurocapsa*.

cyanocobalamin (not cyanocobalamine) Permitted synonym: vitamin B_{12}; **cobalamin** is the form of the vitamin with coenzyme activity.

Cyanophyta (cap. C) Former name for the *cyanobacteria (blue-green algae), when these were classified as a division of algae. Its use is discouraged since the suffix -phyta implies some affiliation with plants.

***Cyanothece* group** (cap. C, ital.) A provisional assemblage of strains of *cyanobacteria previously assigned to various botanical genera, including *Cyanothece*.

cycad Any plant of class *Cycadopsida, especially any member of the extant order, Cycadales.

Cycadeoidales *See* Bennettitales.

Cycadicae A subdivision of the *Pinophyta, corresponding to the class Cycadopsida. Individual name: **cycadophyte**.

Cycadofilicales *See* Pteridospermales.

cycadophyte *See* Cycadicae.

Cycadopsida (cap. C) A class of primitive gymnosperm plants, including the extant order Cycadales and three fossil orders: *Pteridospermales, Caytoniales, and *Bennettitales. Individual name and adjectival form: **cycad** (no cap.; not cycadopsid).

cycles per second *See* cps.

-cyclic Adjectival suffix denoting **1.** a circle or ring (e.g. alicyclic, polycyclic). **2.** a cycle of activity (e.g. multicyclic).

cyclic AMP Short for adenosine 3′,5′-phosphate. Abbrev.: cAMP (lower-case c; no stops).

cyclo- Prefix denoting **1.** *Chem.* a cyclic compound (e.g. cycloalkane, cyclohexane, cyclosilicate). **2.** (**cycl-** before vowels) a circle or circular (e.g. cyclostome, cyclotron, cyclosis).

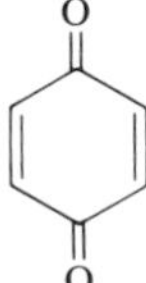

cyclohexadiene-1,4-dione
(quinone or 1,4-benzoquinone)

cyclohexadiene-1,4-dione The recommended name for the compound traditionally known as quinone or 1,4-benzoquinone.

```
H     OH
  \  /
   C
  / \
H2C   CH2
 |     |
H2C   CH2
  \  /
   C
   H2
```

cyclohexanol
(hexahydrophenol)

cyclohexanol The recommended name for the compound traditionally known as hexahydrophenol.

cyclooctatetraene *See* cot.

cyclopentadienyl *See* cp.

Cyclostomata (cap. C) A class or subclass comprising the living Agnatha, i.e. lampreys and hagfishes. Individual name and adjectival form: **cyclostome** (no cap.).

Cyg (no stop) *Astron.* Abbrev. for Cygnus.

Cygnus A constellation. Genitive form: Cygni. Abbrev.: Cyg (no stop). *See also* stellar nomenclature.

Cygnids (meteor shower)

Cyperaceae A large family of monocotyledonous plants, commonly known as the sedge family. There is no common name for individual members; 'sedge' is restricted to members of the genus *Carex*. Adjectival form: **cyperaceous**.

cypher Use cipher.

Cypovirus (cap. C, ital.) A genus of viruses belonging to the family Reoviridae. Individual name: **cypovirus** (no cap., not ital.). [from *cy*toplasmic *po*lyhedrosis *virus*]

Cys (or **cys**; no stop) Abbrev. for cysteine (*see* amino acid). The abbrev. for *cystine is Cys or Cys.

cyst- *See* cysto-.

-cyst Noun suffix denoting a bladder or sac (e.g. otocyst). Adjectival form: **-cystic**.

cysteine Abbrev.: Cys or cys (no stop). Symbol: C An *amino acid; not to be confused with *cystine.

cysticercus (pl. cysticerci) The larval form of certain tapeworms; often used synonymously with *bladderworm, but not all bladderworms are cysticerci.

cystine Abbrev.: Cys or Cys (no stop). A compound formed from two linked *cysteine molecules.

cysto- (**cyst-** before vowels) Prefix denoting a bladder or sac (e.g. cystocarp, cystolith).

-cyte Noun suffix denoting a cell (e.g. leucocyte, spermatocyte). Adjectival form: **-cytic**.

cytidine Symbol: C A *nucleoside consisting of *cytosine combined with D-ribose.

cytidine 5′-diphosphate Abbrev. and preferred form: CDP (no stops).

cytidine 5′-phosphate Abbrev. and preferred form: CMP (no stops).

cytidine 5′-triphosphate Abbrev. and preferred form: CTP (no stops).

cyto- (**cyt-** before vowels) Prefix denoting a cell (e.g. cytology, cytoplasm, cytotoxic).

cytochrome A conjugated protein forming part of the electron transport chain. There are five known cytochromes, each designated by an italic lower-case letter and, in some cases, by a subscript number, i.e. (in order in the chain) b, c, c_1, a, and a_3. The two terminal cytochromes form a complex, **cytochrome oxidase** (also called cytochrome aa_3).

cytogenetics nomenclature A standardized system of nomenclature for describing human chromosomes was agreed by the 1971 Paris Conference on Standardization in Human Cytogenetics. This employs symbols and abbreviations to describe the banding of chromosome arms (as visualized by various staining techniques) and uses the bands as landmarks to pinpoint particular loci and to describe rearrangements, deletions, or duplications of chromosomal material.

Each of the 22 human autosomes has an identifying number; the sex chromosomes are designated X and Y. The short arm of any chromosome is denoted 'p', the long arm 'q', and the bands on each arm are numbered in a standard sequence starting from the centromere outwards; individual bands may comprise numbered sub-bands. For example, a locus on the long arm of chromosome 14 situated at sub-band 3 of band 32 is designated 14q32.3 (no spaces, on-line point). The notation follows the prescribed order: chromosome number; arm; band; sub-band. A deletion is designated by 'del', a duplication by 'dup'. For example, del(1)(q32→q34) indicates a deletion of bands 32 to 34 on the long arm of chromosome 1. Duplication of the same region is denoted dup(1)(q32→q34). An arrow is used to denote the range of bands affected. The end of a chromosome is denoted by the abbreviation 'ter' (from 'terminal'). Hence, del(1)q32→qter) indicates a deletion from band 32 to the end of the q arm of chromosome 1.

Translocations are indicated by the abbreviation 't'; e.g. t(14;18)(q32.3;q21.3) denotes a reciprocal translocation between chromosomes 14 and 18 (the semicolons indicate structural rearrangements involving more than one chromosome) with breakpoints at band 32.3 on the long arm of chromosome 14 and at band 21.3 on the long arm of chromosome 18. The full description of a karyotype having the above translocation would be, in a male, 46,XY,t(14;18)(q32.3;q21.3): the translocation description is prefixed by the total number of chromosomes (i.e. 46) and the sex chromosome constitution (i.e. XY); note commas, no spaces between any characters.

A gain or loss of any particular chromosome is denoted by, respectively, a plus (+) or minus (−) sign before the chromosome number. Hence, +2 (no space) denotes an additional chromosome 2; −15 a missing chromosome 15. Enlargement or reduction in size is denoted by + or − after the character. For example, q− denotes a reduced long arm; h+ an enlarged secondary constriction (h from *h*eterochromatic region).

Nomenclature for the fruit fly *Drosophila melanogaster* and for mice is similar, but symbols for chromosome aberrations differ and are italicized (e.g. *Df* for deficiency, *Dp* for duplication, *T* for translocation); identifying numbers for autosomes and letters for sex chromosomes are printed in italic type for *Drosophila* and in bold roman type for mice.

cytomegalovirus (one word) Abbrev.: CMV (no stops). Any *herpesvirus belonging to the subfamily Betaherpesvirinae (vernacular name: cytomegalovirus group). There are two established member species: human (beta) herpesvirus 5 and murid (beta) herpesvirus 1; their vernacular names are human cytomegalovirus and mouse cytomegalovirus, respectively.

Cytophaga (cap. C, ital.) A genus of gliding bacteria. Individual name: **cytophaga** (no cap., not ital.; pl. cytophagas) Because definition of the genus remains imprecise, the term '*Cytophaga*-like bacteria' (abbrev.: CLB) has been suggested for all putative members. The trivial name 'cytophaga' is used by some authors not only for members of the genus *Cytophaga* but also other members of the family Cytophagaceae (the '*Cytophaga* group'). Care should be taken to avoid confusion.

cytosine Symbol: C A pyrimidine base. *See also* base pair. *Compare* cytidine.

Cz Abbrev. for Cenozoic.

CZI Abbrev. for chlor-zinc iodide. *See* Schultze's solution.

D

d Symbol for **1**. day. **2**. deci-. **3**. 2′-deoxyribonucleoside (preceding the *nucleoside symbol). **4**. a 2′-deoxy sugar (preceding the symbol for the sugar). **5**. deuteron. **6**. doublet (in nuclear magnetic resonance spectroscopy). **7**. the electron state $l = 2$ (*see* orbital angular momentum quantum number).

d Symbol (light ital.) for **1**. diameter. **2**. relative density. **3**. thickness.

D **1**. A rhesus antigen (*see* rhesus factor). **2**. Symbol for **i**. aspartic acid. **ii**. darcy. **iii**. deuterium. **iv**. diversity region (of an *immunoglobulin chain). **v**. D meson (D^+, D^0). **vi**. guanosine, adenosine, or thymidine (or uridine) (unspecifed). **3**. Abbrev. for **i.** Devonian. **ii.** dimension or dimensional (preceded by a number).

D Symbol for **1.** (light ital.) absorbed dose. **2.** (light ital.) diameter. **3.** (light ital.) diffusion coefficient. **4.** (bold ital.) electric flux density.

D_i Symbol for **1.** (bold ital. D) electric polarization. **2.** (light ital. D) internal transmission density (former name: optical density).

2,4-D Abbrev. and preferred form for 2,4-dichlorophenoxyacetic acid, a synthetic auxin.

***d*-** (ital., always hyphenated) *See* dextrorotatory.

D- (small cap., always hyphenated) Prefix denoting a structural relationship to dextrorotatory glyceraldehyde (D-(+)-glyceraldehyde) (e.g. D-glucose, D-threose). A D-compound is not necessarily dextrorotatory. *Compare* L-.

da Symbol for deca-. *See also* SI units.

Da Symbol for *dalton.

DA Abbrev. for *dopamine.

D/A (or **D-A, d/a, d-a**) Abbrev. for digital-to-analog.

DABCO Abbrev. for 1,4-diazobicyclo-[2.2.2]octane.

DAC Abbrev. for digital-to-analog converter.

Dacron The US equivalent of *Terylene.

dactylo- (**dactyl-** before vowels) Prefix denoting fingers or toes (e.g. dactylomegaly).

-dactyly Noun suffix denoting fingers or toes (e.g. polydactyly, syndactyly). Adjectival form: **-dactylous**.

Daguerre, Louis-Jacques-Mandé (hyphens) (1789–1851) French physicist, inventor, and painter.
daguerreotype

Daimler, Gottlieb Wilhelm (1834–1900) German engineer and inventor.

Dainton, Sir Frederick Sydney (1914–) British physical chemist and scientific administrator.

Dakin, Henry Drysdale (1880–1952) British chemist.
Dakin's solution

d'Alembert, Jean Le Rond (1717–83) French mathematician, encyclopedist, and philosopher.
d'Alembert's principle
d'Alembert's ratio test

Dalén, Nils Gustaf (1869–1937) Swedish engineer.

Dalitz plot *Particle physics* Named after Richard H. Dalitz (1925–).

DALR *Meteorol.* Abbrev. for dry adiabatic lapse rate.

dalton Symbol: Da The former name for the *atomic mass unit, still used in biochemistry to express the relative molecular masses of proteins. Named after John *Dalton.

Dalton, John (1766–1844) British chemist and physicist.
***dalton**
daltonism *Ophthalmol.*
Dalton's atomic theory
Dalton's law of partial pressures

Dam, Carl Peter Henrik (1895–1976) Danish biochemist.

DAMN Abbrev. for diaminomaleonitrile.

damping coefficient Symbol: δ or λ (Greek delta or lambda). A parameter relating to a decreasing periodic function $f(t)$, defined by the equation:

$$f(t) = \exp(-\delta t) \sin \omega t,$$

where ω is *angular frequency. The *SI unit is the reciprocal of the second (s^{-1}). The product $T\delta$ is the logarithmic decrement, symbol Λ (Greek cap. lambda), where T is the period. *Compare* growth rate; relaxation time.

Dana, James Dwight (1813–95) US geologist, mineralogist, and zoologist.

Dandelin sphere *Maths.* Named after Pierre Dandelin (1794–1847).

Daniell, John Frederic (1790–1845) British chemist and meteorologist.
Daniell cell

DAPI Abbrev. for 4′,6-diamidino-2-phenylindole.

Darby, Abraham (*c.* 1678–1717) British metallurgist.

darcy Symbol: D A *cgs unit of *permeability coefficient, now discouraged. The *SI unit of permeability coefficient is the square metre: 1 D = 0.9868×10^{-12} m^2. Named after H. P. G. *Darcy.

Darcy, Henri-Philibert-Gaspard (1803–58) French hydraulic engineer.
***darcy**
Darcy's law

Darlington, Cyril Dean (1903–81) British geneticist.

Darlington pair *Electronics.*

D arm (cap. D) The arm of a tRNA molecule that contains the unusual nucleoside dihydrouridine (D). Also called DHU loop. *Compare* D loop.

d'Arsonval, Jacques A. (1831–1940) French physicist.
d'Arsonval galvanometer

Dart, Raymond Arthur (1893–1988) Australian anatomist.

Darwin, Charles Robert (1809–82) British naturalist. Adjectival form: **Darwinian.**
Darwinism
Darwin's finches
neo-Darwinism

Darwin, Erasmus (1731–1802) British physician, grandfather of Charles Darwin.

DAST Abbrev. for diethylaminosulphur trifluoride.

DAT Acronym for digital audio tape.

data (pl. noun; sing. datum) Information prepared for a specific purpose. Strictly speaking, 'data' should take a plural form of verb. However, it is now widely used as a collective noun, taking a singular form of verb (as in 'the data has been fed into the computer').

database (one word) *Computing*
database management system Abbrev.: DBMS (no stops).

dataflow (one word) *Computing*

datagram (one word) *Computing*

data processing (two words) Abbrev.: DP or dp (no stops).

data type (two words) Often shortened to type.

dates Specific dates should be written using the style '2 Dec, 1943, . . . '. Note that the day precedes the month and the year has commas around it if followed by text. Abbreviations are used (no stops) for some months as follows: Jan, Feb, Mar, Apr, May, Jun, Jul, Aug, Sept, Oct, Nov, Dec (sometimes June and July are written out in full). In US usage, the month precedes the day (Dec 2, 1943).
In writing years, the abbreviations AD and BC are used (small caps., no stops). AD precedes the year (e.g. AD 1600) and BC follows the year (e.g. 600 BC). Sometimes CE (for Common (or Christian) Era) and BP (for Before Present) are used. CE is placed before the date (e.g. CE 600); BP is placed after the date (e.g. 2000 BP).

datum An item of *data.

Daubrée, Gabriel Auguste (1814–96) French geologist.

Dausset, Jean Baptiste Gabriel (1916–) French physician and immunologist.

Davaine, Casimir Joseph (1812–82) French physician and microbiologist.

Davenport, Charles Benedict (1866–1944) US zoologist and geneticist.

Davis, William Morris (1850–1934) US physical geographer.
Davisian systems

Davisson, Clinton Joseph (1881–1958) US physicist.
Davisson–Germer experiment (en dash)

Davy, Sir Humphry (1778–1829) British chemist.
Davy lamp

Dawes, William Rutter (1799–1868) British astronomer.
Dawes limit

day Symbol: d A unit of time equal to 24 hours, i.e. 86 400 seconds. Although not an *SI unit, it may be used with the SI units. The *sidereal day*, determined astronomically, is about 236.1 s shorter than the 24-hour day.

daylight exposure *Photog.* Abbrev.: *DX (no stops).

day-neutral plant (hyphenated)

dB Symbol for decibel.

2,4-DB Abbrev. for 4-(2,4-dichlorophenoxy)butyric acid.

DBMS Abbrev. for database management system.

DBN Abbrev. for 1,5-diazabicyclo[4.3.0]-non-5-ene.

DBS Abbrev. for direct broadcast by satellite.

DBU Abbrev. for 1,8-diazabicyclo[5.4.0]-undec-7-ene.

d.c. (stops, or **DC**) Abbrev. for direct current.

DCC Abbrev. for 1,3-dicyclohexylcarbodiimide.

DCU Abbrev. for *N,N*-dichlorourethane.

DDB Abbrev. for 2,3-dimethoxy-1,4-bis-(dimethylamino)butane.

DDD Abbrev. for dichlorodiphenyldichloroethane (a mixture of the isomers 1-(2-chlorophenyl)-1-(4-chlorophenyl)-2,2-dichloroethane and 2,2-bis(4-chlorophenyl)-1,1-dichloroethane).

(*E,E*)-8,10-DDDA Abbrev. for *trans*-8,*trans*-10-dodecadien-1-yl acetate.

(*E,Z*)-7,9-DDDA Abbrev. for *trans*-7,*cis*-9-dodecadien-1-yl acetate.

(*E,E*)-8,10-DDDOL Abbrev. for *trans*-8,*trans*-10-dodecadien-1-ol.

(*Z*)-7-DDOL Abbrev. for *cis*-7-dodecen-1-ol.

DDP Abbrev. for dichlorodiammine-platinum(II).

DDQ Abbrev. for 2,3-dichloro-5,6-dicyano-1,4-benzoquinone.

DDT Abbrev. for dichlorodiphenyltrichloroethane (a mixture of the isomers 1-(2-chlorophenyl)-1-(4-chlorophenyl)-2,2,2-trichloroethane and 1,1-bis(4-chlorophenyl)-2,2,2-trichloroethane).

de- (sometimes **des-** before vowels in chemical compounds) Prefix denoting removal, loss, or reversal (e.g. deamination, debug, decompose, de-energize (hyphenated), dehydration, dehydrogenation, deionize (hyphenated), deoxidize, deoxyribose, desoxyephedrine).

DEAD Abbrev. for diethyl azodicarboxylate.

DEAE-cellulose (hyphenated) Abbrev. and preferred form for diethylaminoethylcellulose.

de Bary, Heinrich Anton (1831–88) German botanist.

de Beer, Sir Gavin Rylands (1899–1972) British zoologist.

Debierne, André Louis (1874–1949) French chemist.

De Bort, Léon Teisserenc Usually alphabetized as *Teisserenc De Bort.

de Broglie, Prince Louis Victor Pierre Raymond (1892–1987) French physicist.

de Broglie wavelength

de Broglie wave

de Buffon, Comte Georges Usually alphabetized as *Buffon.

debye A *cgs unit of *electric dipole moment, now discouraged. The *SI unit of dipole moment is the coulomb metre: 1 debye = 3.336×10^{-30} C m. Named after P. J. W. *Debye.

Debye, Peter Joseph William (1884–1966) Dutch–US physicist and physical chemist.

Compton–Debye effect Use Compton effect.

***debye**

Debye–Hückel theory (en dash). Also named after Erick Hückel (1895–).

Debye length Symbol: λ_D (Greek lambda)

Debye temperature Symbol: Θ_D (Greek cap. theta)

Debye theory of specific heats (of solids)

dec (no stop) *Astron.* Abbrev. for declination.

deca- 1. Symbol: da A prefix to a unit of measurement that indicates 10 times that unit. *See also* SI units. **2.** (**dec-** before vowels) A general prefix denoting ten (e.g. decapod, decapetalous).

de Candolle, Augustin Usually alphabetized as *Candolle.

decanedioic acid $HOOC(CH_2)_8COOH$ The recommended name for the compound traditionally known as sebacic acid.

Decapoda (cap. D) **1.** An order of crustaceans containing the crabs, lobsters, prawns, shrimps, etc. **2.** An order of cephalopod molluscs including the cuttlefishes and squids. Individual name and adjectival form (for both senses): **decapod** (no cap.; not decapodan or decapodal).

decay constant Symbol: λ (Greek lambda). A physical quantity associated with radioactive decay. Also called disintegration constant. It is a constant for a particular radionuclide. The *SI unit is the reciprocal of the second (s^{-1}).

For exponential decay,

$$-dN/dt = \lambda N,$$

where N is the number of radioactive atoms present at time t and $-dN/dt$ is the *activity.

The mean life, symbol τ or τ_m (Greek tau), is equal to $1/\lambda$. It is the average time taken (in seconds) for the number N to be reduced to N/e.

The half-life, symbol $T_{1/2}$ or $\tau_{1/2}$, is the average time taken (in seconds) for N to be reduced by a half; it is equal to $(\log_e 2)/\lambda$.

de Chardonnet, Comte Usually alphabetized as *Chardonnet.

deci- Symbol: d A prefix to a unit of measurement that indicates one tenth of that unit, as in decibel (dB). *See also* SI units.

decibel Symbol: dB A dimensionless unit defined on a logarithmic basis and is used, especially in acoustics and telecommunications, to express *sound pressure levels or *power level differences. It is a more practical unit than the *bel*, which is equal to 10 decibels. It is related to the neper (Np):

$$1\ \text{dB} = (\log_e 10)/20\ \text{Np} = 0.115\ 129\ \text{Np}.$$

decimal sign In the UK and the USA, the decimal sign is set as a dot on the line (.), and is called the decimal point. The centred dot (·) should not be used as a decimal point in scientific writing. In most of Europe, including the USSR, the decimal sign is a comma on the line (,). This sign is preferred by *ISO. There should be at least one digit both before and after the decimal sign. For a number less than unity, the decimal sign should be preceded by a zero, as in 0.007.

declination Symbol: δ (Greek delta). A coordinate used with *right ascension to give the position of an astronomical object with respect to the celestial equator. It is the object's angular distance (from 0° to 90°) north (counted positive) or south (counted negative) of the equator; it is the equivalent of terrestrial latitude.

decomp. Abbrev. for decomposition.

de Coulomb, Charles Usually alphabetized as *Coulomb.

Dedekind, (Julius Wilhelm) Richard (1831–1916) German mathematician.

Cantor–Dedekind hypothesis (en dash)

Dedekind cut (or **section**)

de Duve, Christian René (1917–) Belgian biochemist.

de-emphasis (hyphenated) *Telecom.*

de-energize (hyphenated) *Elec. eng.*

Deep Space Network Abbrev.: *DSN (no stops).

defecate (not defaecate) Noun form: **defecation.**

de Fermat, Pierre Usually alphabetized as *Fermat.

de Ferranti, Sebastian Usually alphabetized as *Ferranti.

De Forest, Lee (1873–1961) US physicist and inventor.

degauss (degaussing, degaussed) [from the name *Gauss]

de Geer, Gerard Jacob Usually alphabetized as *Geer.

degree 1. Symbol: ° A unit of angle equal to $\pi/180$ radian, i.e.

$$1° = 0.174\ 533\ \text{rad},\ 360° = 2\pi\ \text{rad}.$$

Although not an *SI unit, the degree can be used with the SI units. Decimal subdivisions of degree are preferred to the minute (1/60 degree) or second (1/3600 degree), as in 60.75°.

2. A unit used in expressing a temperature interval or temperature difference on the Celsius, Fahrenheit, or other temperature scale. *See also* degree Celsius; kelvin.

degree Celsius (cap. C) Symbol: °C (there is either a thin space or no space between temperature value and symbol, as in 10 °C or 10°C; 10° C is incorrect). An *SI unit used in expressing Celsius temperatures or temperature differences. It is identical in magnitude to the kelvin. Named after Anders *Celsius. Former name: degree centigrade.

degree centigrade Another name for degree Celsius, which replaced it in 1948.

degree Fahrenheit Symbol: °F A unit used in expressing Fahrenheit temperatures and temperature differences. Fahrenheit temperature *t* is related to Celsius temperature θ and thermodynamic temperature *T* by the equations:

$$t = {}^{9}/_{5}\theta + 32 = {}^{9}/_{5}T - 459.67.$$

When considering temperature difference,

$$1\ {}^{\circ}\mathrm{F} = {}^{5}/_{9}\ {}^{\circ}\mathrm{C} = {}^{5}/_{9}\ \mathrm{K}.$$

The Fahrenheit scale is now generally calibrated in terms of the *IPTS: thermodynamic Fahrenheit temperatures are commonly used in the USA for scientific purposes. Originally the scale was calibrated at the ice and steam points. Named after G. D. *Fahrenheit.

degree Kelvin *See* kelvin.

degree Rankine Symbol: °R A unit formerly used to express thermodynamic temperature or temperature difference: a temperature of 0 °R is absolute zero. Rankine temperature *r* is related to Fahrenheit temperature *t* and thermodynamic temperature *T* (in kelvin) by the equations:

$$r = t + 459.67 = {}^{9}/_{5}T.$$

When considering temperature difference, 1 °R = 1 °F. Named after W. J. M. *Rankine.

de Haas–van Alphen effect *Physics* (en dash) Also called van Alphen effect. Named after W. J. de Haas and P. M. van Alphen.

De Havilland, Sir Geoffrey (1882–1965) British aeronautical engineer.

dehydrogenase Systematic name: oxidoreductase. *See also* enzyme nomenclature.

del (preferred to nabla) Symbols: ∇, $\partial/\partial \boldsymbol{r}$ The differential operator

$$\boldsymbol{i}(\partial/\partial x) + \boldsymbol{j}(\partial/\partial y) + \boldsymbol{k}(\partial/\partial z),$$

where ***i***, ***j***, ***k*** are unit vectors along the *x*, *y*, *z* axes respectively. Also called nabla.

The second differential of the operator

$$\partial^2/\partial x^2 + \partial^2/\partial y^2 + \partial^2/\partial z^2$$

has the symbol ∇^2, pronounced 'del squared'.

Del (no stop) *Astron.* Abbrev. for Delphinus. *See also* stellar nomenclature.

De La Beche, Sir Henry Thomas (1796–1855) British geologist.

de La Blache, Paul Vidal Usually alphabetized as *Vidal de La Blache.

de Laplace Usually alphabetized as *Laplace.

De La Tour, Charles *See* Cagniard De La Tour, Charles.

Delbrück, Max (1906–81) German biophysicist.

de L'Hospital, Marquis Guillaume Usually alphabetized as *L'Hospital.

D'Elhuyar, Fausto (1755–1833) Spanish chemist and mineralogist.

deliquescence Adjectival form: **deliquescent**. Verb form: **deliquesce**.

Dellinger fadeout (or **effect**) *Telecomm.*

Delphinus A constellation. Genitive form: Delphini. Abbrev.: Del (no stop). *See also* stellar nomenclature.

delta Greek letter, symbol: δ (lower case), Δ (cap.).

δ Symbol for **1.** chemical shift. **2.** damping coefficient. **3.** *Astron.* declination. **4.** fourth brightest star in a constellation (*see* stellar nomenclature). **5.** heavy chain of IgD (*see* immunoglobulin). **6.** thickness. **7.** *Maths.* variation of (followed by a variable or function).

δ- (always hyphenated) Symbol used in the names of organic compounds to indicate a substituent attached to the fourth carbon atom along from the functional group (e.g. δ-chloro-n-valeric acid).

$\boldsymbol{\delta}_{ij}$ Symbol for Kronecker delta.

$\boldsymbol{\delta}(\boldsymbol{r})$ (ital. r) Symbol for Dirac delta function.

Δ Symbol for **1.** *Maths.* finite increase in (followed by a variable). **2.** mass excess.

Δ- (always hyphenated) Symbol used to indicate the position of a double bond in an organic compound. A superscript numeral denotes the first double-bonded carbon atom in the compound. Examples are $\Delta^{1,3}$-butadiene, Δ^{2}-butene, Δ^{3}-isopentyl alcohol.

delta iron Also written **δ-iron**.

de Luc, Jean André (1727–1817) Swiss geologist and meteorologist.

Demarçay, Eugene Anatole (1852–1904) French chemist.

de Marignac, Jean Usually alphabetized as *Marignac.

Demerec, Milislav (1895–1966) Yugoslavian-born US geneticist.

demi- Prefix denoting a half (e.g. demifacet).

demo- Prefix denoting people or population (e.g. demography).

De Moivre, Abraham (1667–1754) French mathematician.
De Moivre–Laplace theorem (en dash)
De Moivre's theorem

De Morgan, Augustus (1806–71) British mathematician and logician.
De Morgan's laws

Dempster, Arthur Jeffrey (1886–1950) Canadian-born US physicist.

D.En. Abbrev. for Department of Energy, a UK government department.

dendrite 1. (not dendron) Any of the short branching processes of the cell body of a neurone, forming the input region of the neurone. **2.** A branching structure in some minerals and rocks. Adjectival form: **dendritic.**

dendro- (**dendr-** (before vowels), **dendri-**) Prefix denoting a tree or branching structure (e.g. dendrogram, dendrology, dendrite, dendriform).

denier A unit of linear density used in the textile industry, equal to 1 gram per 9 kilometres or $0.111\ 111 \times 10^{-6}$ kg m^{-1}.

density 1. Symbol: ρ (Greek rho). A physical quantity, mass divided by volume, m/V. The *SI unit is the *kilogram per *cubic metre (kg m^{-3}). This quantity is sometimes called mass density to differentiate it from linear density (mass divided by length, symbol ρ_l) and surface density (mass divided by area, symbol ρ_A). These are measured in kg m^{-1} and kg m^{-2} respectively. *See also* relative density. **2.** *See* internal transmission density. **3.** A word placed after a scalar physical quantity to indicate that the quantity is expressed per unit volume (e.g. energy density, charge density). **4.** A word placed after a vector quantity to denote a flow through unit area (e.g. current density, magnetic flux density).

density of heat flow rate *See* heat flow rate.

dental formula A graphic representation of the dentition of an animal. The dental formula of the rabbit, for example, could be represented as follows:

$$\frac{2\quad 0\quad 3\quad 3}{1\quad 0\quad 2\quad 3}$$

The numbers above (and below) the line represent (from left to right) the numbers of incisors, canines, premolars, and molars in each half of the upper (and lower) jaws, indicating – in the example shown – a total of 28 teeth.

denti- (sometimes **dent-** before vowels, **dento-**) Prefix denoting teeth (e.g. dentirostral, dentate, dentoalveolar).

dentine US: **dentin.**

deoxyribonuclease Abbrev.: *DNAase (no stops).

deoxyribonucleic acid (preferred to desoxyribonucleic acid) Abbrev. and preferred form: *DNA (no stops).

deoxyribose Abbrev.: dRib (no stops). *See* sugars.

DEP Abbrev. for diethyl pyrocarbonate.

Department of Education and Science Abbrev.: DES (no stops).

Department of Energy Abbrev.: D.En. (stops).

Department of the Environment Abbrev.: DoE (lower-case o, no stops).

dependovirus (one word) Any member of the genus *Dependovirus* (cap. D, ital.). Vernacular name: adeno-associated virus (abbrev.: AAV).

derivative (of a function *f*). *See* mathematical symbols.

derived unit *See* SI units; coherent units.

-derm Noun suffix denoting **1.** the skin (e.g. pachyderm). Adjectival form: **-dermatous. 2.** a germ layer (e.g. blastoderm, ectoderm). Adjectival forms: **-dermal** (e.g. ectodermal), **-dermic** (e.g. blastodermic).

Dermaptera (cap. D) An order of insects comprising the earwigs. Individual name and adjectival form: **dermapteran** (no cap.). *Compare* Dermoptera.

dermato- (dermat- before vowels) Prefix denoting the skin (e.g. dermatophyte, dermatitis, dermatome).

Dermatophilus (cap. D, ital.) A genus of actinomycete bacteria. Individual name: **dermatophilus** (no cap., not ital.; pl. dermatophili).

dermis (preferred to corium, derm, and derma) The inner layer of the skin, lying beneath the epidermis. Adjectival forms: **dermal, dermoid** (these also relate to the whole of the skin).

Dermoptera (cap. D) An order of mammals comprising the flying lemurs. Individual name and adjectival form: **dermopteran** (no cap.). *Compare* Dermaptera.

DES Abbrev. for Department of Education and Science, a UK government department.

Desargues, Girard (1591–1661) French mathematician and engineer.
Desargues's theorem

de Saussure, Horace Usually alphabetized as *Saussure.

Descartes, René du Perron (1596–1650) French mathematician, philosopher, and scientist. Adjectival form: **Cartesian.**
Cartesian coordinates
Cartesian geometry
Cartesian ovals
Descartes's rule of signs

Descemet, Jean (1732–1810) French anatomist.
Descemet's membrane

desert (not dessert) *Geog.* Derived noun: **desertification.**

desiccate (not dessicate) Noun form: **desiccation.** Derived nouns: **desiccator, desiccant.**

de Sitter, Willem (1872–1934) Dutch astronomer and mathematician.
de Sitter universe
Einstein–de Sitter universe (en dash)

desktop publishing (desktop one word) Abbrev.: DTP (no stops).

Desmarest, Nicolas (1725–1815) French geologist.

desmo- Prefix denoting a bond or chain (e.g. desmognathous, desmosome, desmotropism).

desorption The reverse process to *adsorption.

desoxyribonucleic acid Use deoxyribonucleic acid. *See* DNA.

Destriau effect *Physics* Named after G. Destriau.

Desulfococcus (cap. D, ital.; not *Desulphococcus*) A genus of sulphur-reducing bacteria; not to be confused with **Desulfurococcus*. Individual name: **desulfococcus** (no cap., not ital.; pl. desulfococci).

Desulfomonas (cap. D, ital.; not *Desulphomonas*) A genus of sulphur-reducing bacteria; not to be confused with *Desulfuromonas*. Individual name: **desulfomonad** (no cap., not ital.).

Desulfonema (cap. D, ital.; not *Desulphonema*) A genus of filamentous gliding sulphur-reducing bacteria.

Desulfurococcus (cap. D, ital.; not *Desulphurococcus*) A genus of extremely thermoacidophilic bacteria; not to be confused with **Desulfococcus*. Individual name: **desulfurococcus** (no cap., not ital.; pl. desulfurococci).

deuterate Preferred to deuteriate.

deuterium *See* hydrogen.

deutero- (deuter- (before vowels), **deuto-**) Prefix denoting **1.** second or secondary (e.g. deuterostome, deutonymph, deutoplasm). **2.** deuterium (e.g. deuteron).

deuteromycete Any imperfect fungus, formerly assigned to the class Deuteromycetes. Such fungi are now included in the subdivision *Deuteromycotina, but the term 'deuteromycete' is still used for descriptive purposes.

Deuteromycotina An artificial subdivision comprising the imperfect fungi, which lack any form of sexual reproduction. Also called Fungi Imperfecti. The grouping is purely biological and does not imply any taxonomic affinities between members. Two classes are recognized: *Hyphomycetes and *Coelomycetes; there is no separation into orders within these classes.

deuteron A nucleus of an atom of deuterium, $^2H^+$, denoted by d in nuclear reactions, etc.

De Vaucouleurs, Gerard Henri (1918–) French-born US astronomer.

Deville, Henri Étienne Sainte-Claire (hyphen) (1818–81) French chemist.

Devonian Abbrev.: D (no stop). **1.** (adjective) Denoting the fourth period of the Palaeozoic era. **2.** (noun; preceded by 'the') The Devonian period. The period is divided into the Lower (or Early) Devonian, Middle Devonian, and Upper (or Late) Devonian epochs (initial caps.; abbrevs. (respectively): D_1, D_2, and D_3 (subscript numerals)).

de Vries, Hugo (1848–1935) Dutch plant physiologist and geneticist.

Dewar, Sir James (1842–1923) British chemist and physicist.
Dewar flask Also called vacuum flask. Avoid Thermos flask (trade mark).

Dewar, Michael James Stewart (1918–) British-born US chemist.
Dewar benzene
Dewar structure

dew point (two words)

dextro- (dextr- before vowels) Prefix denoting on or towards the right (e.g. *dextrorotatory, dextrorse).

dextro- (ital., always hyphenated) *See* dextrorotatory.

dextrorotatory US: **dextrorotary.** Dextrorotatory compounds are denoted by the prefix (+)- (parentheses, always hyphenated), e.g. (+)-tartaric acid. The alternatives *d*- and *dextro*- are not recommended. Noun form: **dextrorotation.** *See also* D-. *Compare* laevorotatory.

dextrose $C_6H_{12}O_6$ The traditional name for (+)-glucose.

DFP Abbrev. for diisopropyl fluorophosphate.

DHBA Abbrev. for 3,4-dihydroxybenzylamine.

d'Herelle, Félix (1873–1949) French–Canadian bacteriologist.

DHU loop *See* D arm.

DI Abbrev. for *donor insemination.

di- Prefix denoting **1.** two or double (e.g. dibranchiate, dicotyledon, dimorphic, dipole). **2.** *Chem.* two separate identical atoms, ions, or groups (e.g. diethylamine, dioxide, diphenylmethane). **3.** *See* dia-. *See also* bi-.

dia- (di- before vowels) Prefix denoting **1.** through, during, throughout, or completely (e.g. diachronous, diagonal, diaphragm, dielectric, dioptre, diurnal). **2.** apart, distinct, in opposite directions (e.g. diakinesis, dialysis, diamagnetism, diapsid). **3.** across or at right angles (e.g. diatropism).

diacetone alcohol $HOC(CH_3)_2CH_2COCH_3$ The traditional name for 4-hydroxy-4-methylpentan-2-one.

diacetyl $CH_3COCOCH_3$ The traditional name for butanedione.

diacetylene $CH{\equiv}CC{\equiv}CH$ The traditional name for buta-1,3-diyne.

DIAD Abbrev. for diisopropyl azodicarboxylate.

diag- (ital., always hyphenated) Prefix denoting geometric isomerization in square-pyramidal inorganic complexes having a diagonal configuration of ligands (e.g. *diag*-dibromodicarbonylcyclopentadienylrhenium(III)). *Compare lat-*.

dialyse US: **dialyze.** Noun form: **dialysis** (pl. dialyses).

diameter *See* length.

diaminazobenzene $C_6H_5N{=}NNHC_6H_5$ The traditional name for *N*-(phenylazo)phenylamine.

1,4-diaminobutane $H_2N(CH_2)_4NH_2$ The recommended name for the compound traditionally known as putrescine.

diaminomaleonitrile Abbrev.: DAMN (no stops).

diamond *See* carbon.

diaphragm (not diaphram) Adjectival form: **diaphagmatic**.

diaphysis (pl. diaphyses) The shaft of a long bone. Adjectival form: **diaphyseal** (preferred to diaphysial). *Compare* epiphysis.

diarrhoea US: **diarrhea**.

diars (no stop) Abbrev. for *o*-phenylenebis(dimethylarsine) often used in the formulae of coordination compounds, e.g. $[Pd(diars)_2]^{2+}$.

diastase The name originally given to the active principle of malt extract, which includes β-amylase: sometimes used as a synonym for this enzyme in germinating barley.

diatom *Bot. See* Bacillariophyta. Adjectival form: **diatomaceous**; *compare* diatomic.

diatomic *Chem.* Denoting a molecule that contains two atoms. *Compare* diatom.

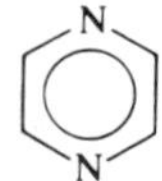

1,4-diazine
(pyrazine)

1,4-diazine The recommended name for the compound traditionally known as pyrazine.

diazo compound Any of a number of classes of organic compounds containing two linked nitrogen atoms, e.g. *azo compounds and *diazonium salts.

diazonium salt Any of a class of organic salts containing the cation $ArN_2{}^+$, where Ar is any aromatic group.
Diazonium salts are systematically named by adding the word diazonium followed by the name of the associated anion to the name of the parent hydrocarbon, e.g. benzenediazonium chloride, $C_6H_5N_2{}^+Cl^-$.
In nonsystematic nomenclature diazonium salts are named as in systematic nomenclature but the trivial names of the parent hydrocarbons are used, e.g. ρ-toluenediazonium chloride, $CH_3C_6H_4N^+Cl^-$.

DIBAL-H Abbrev. for diisobutylaluminium hydride.

1,1-dibromoethane CH_3CHBr_2 The recommended name for the compound traditionally known as ethylidene dibromide.

1,2-dibromoethane $BrCH_2CH_2Br$ The recommended name for the compound traditionally known as ethylene dibromide.

dichlorine oxide Cl_2O The recommended name for the compound traditionally known as chlorine monoxide.

dichlorobenzene $C_6H_4Cl_2$ The recommended name for the compound traditionally known as dichlorobenzene. The recommended name for *o*-dichlorobenzene is 1,2-dichlorobenzene, etc.

dichlorodiphenyltrichloroethane Abbrev. and preferred form: DDT (no stops).

dichloroethanoic acid $CHCl_2COOH$ *See* chloroethanoic acid.

1,2-dichloroethene $CHCl=CHCl$ The recommended name for the compound traditionally known as acetylene dichloride.

dichloromethane CH_2Cl_2 The recommended name for the compound traditionally known as methylene chloride.

(dichloromethyl)benzene C_6H_5-$CHCl_2$ The recommended name for the compound traditionally known as benzal chloride.

dichromate Denoting a compound containing the ion $Cr_2O_7{}^{2-}$, e.g. potassium dichromate, $K_2Cr_2O_7$. The recommended name is dichromate(VI).

Dicke, Robert Henry (1916–) US physicist.
Brans–Dicke theory (en dash) Use Brans–Dicke–Jordan theory.
Dicke radiometer

Dicotyledonae (cap. D) A subclass of flowering plants having embryos with two cotyledons. In some classifications these plants are regarded as a class, Magnoliopsida; in a recent classification they are regarded as a subclass, Magnoliidae, of the Magnoliopsida (*see* Angiospermae). Individual name: **dicotyledon** (no cap.; often shortened to **dicot**). Adjectival form: **dicotyledonous.** Note that this name and adjectival form apply irrespective of the classification used.

Dictyoptera (cap. D) An order of insects including the cockroaches and mantids, formerly classified in the Orthoptera. Individual name and adjectival form: **dictyopteran** (no cap.).

dieback (noun; one word) Verb form: **die back** (two words).

diecious US spelling of *dioecious.

dielectric constant Use relative *permittivity.

dielectric polarization Another name for electric polarization.

Diels, Otto Paul Hermann (1876–1954) German organic chemist.
dieldrin (no cap.)
Diels–Alder reaction (en dash)
Diels's hydrocarbon

dien (no stop) Abbrev. for diethylenetriamine often used in the formulae of coordination compounds, e.g. $[Co(dien)_2]^{2+}$.

Diesel, Rudolph Christian Carl (1858–1913) German engineer and inventor.
diesel (no cap.) (fuel)
Diesel cycle
diesel engine (no cap.)

diestrus (or **diestrum**) US forms of *dioestrus.

Dieterici equation *Chem.*

1,1-diethoxyethane $CH_3CH(OC_2H_5)_2$ The recommended name for the compound traditionally known as acetal.

diethylaminoethylcellulose (one word) Abbrev. and preferred form: DEAE-cellulose (hyphenated).

diethylenetriamine *See* dien.

diethyl ether $C_2H_5OC_2H_5$ The traditional name for ethoxyethane.

Dietl, Joseph (1804–78) Polish physician.
Dietl's crisis

differential scanning calorimetry *Chem.* Abbrev.: DSC (no stops).

differential thermal analysis *Chem.* Abbrev.: DTA (no stops).

diffusion coefficient Symbol: D A physical quantity used in statistical physics, solid-state physics, etc. The *SI unit is the square metre per second. In the theory of diffusion of thermal neutrons the term is commonly used for a similar quantity divided by the mean speed of the neutrons. The SI unit is then the metre.

diffusion pressure deficit Abbrev.: DPD (no stops). Formerly, the net force causing water to enter a plant cell; now expressed in terms of *water potential.

digital audio tape Abbrev.: DAT (no stops).

digital-to-analog (hyphenated) Abbrev.: D/A, D-A, d/a, or d-a.
digital-to-analog converter Abbrev.: DAC (no stops) or D/A (or D-A) converter.

digital voltmeter Abbrev.: DVM (no stops).

dihydrogen *See* hydrogen.

(±)-2,3-dihydroxybutanedioic acid HOOCCH(OH)CH(OH)COOH The recommended name for the compound traditionally known as racemic acid or *d,l*-tartaric acid.

3,4-dihydroxyphenylalanine Abbrev. and preferred form: *dopa.

diiodine hexachloride I_2Cl_6 The recommended name for the compound traditionally known as iodine trichloride.

dike *See* dyke.

DIL *Electronics* Acronym for dual in-line.

dil. Abbrev. for dilute.

dilatation In British (but not US) English, preferred to 'dilation' for the physiological sense of expansion of a hollow organ. However, **dilator** is preferred to 'dilatator' for a muscle, drug, or instrument that causes dilatation.

dilead(II) lead(IV) oxide Pb_3O_4 The recommended name for the compound traditionally known as red lead.

dilute Abbrev.: dil. (stop).

dimercury(I) chloride Hg_2Cl_2 The recommended name for the compound traditionally known as calomel.

dimethylbenzene $C_6H_4(CH_3)_2$ The recommended name for the compound traditionally known as oxylene. The recommended name for *o*-xylene is 1,2-dimethylbenzene, etc.

2,4-dimethylbenzoic acid C_6H_3-$(CH_3)_2COOH$ The recommended name for the compound traditionally known as xylic acid.

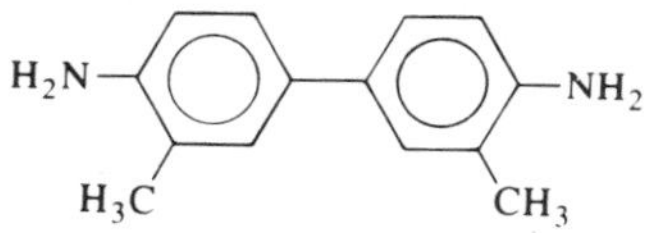

3,3′-dimethylbiphenyl-4,4′-diamine (*o*-tolidine)

3,3′-dimethylbiphenyl-4,4′-diamine The recommended name for the compound traditionally known as *o*-tolidine.

2,3-dimethylbutane-2,3-diol $(CH_3)_2C(OH)C(OH)(CH_3)_2$ The recommended name for the compound traditionally known as pinacol.

3,3-dimethyl-2-butanone CH_3COC-$(CH_3)_3$ The recommended name for the compound traditionally known as pinacolone.

dimethylglyoxime $CH_3C(NOH)$-$C(NOH)CH_3$ The traditional name for *butanedione dioxime.

2,2-dimethylpropanoic acid $(CH_3)_3$-$CCOOH$ The recommended name for the compound traditionally known as pivalic acid.

2,2-dimethylpropan-1-ol $(CH_3)_3$-CCH_2OH The recommended name for the compound traditionally known as neopentyl alcohol.

dimethyl sulphoxide Abbrev.: DMSO (no stops).

DIN Acronym for Deutsche Institut für Normung, the German national standards organization.

dinitrogen oxide N_2O The recommended name for the compound traditionally known as nitrous oxide.

dino- Prefix denoting **1.** terrible (e.g. dinosaur). **2.** whirling (e.g. dinoflagellate).

dinoflagellate (noun) Any alga of the class Dinophyceae. *See* Dinophyta.

Dinophyta A division of predominantly flagellate unicellular marine algae. Also called Pyrrophyta. It is divided into the classes Desmophyceae and Dinophyceae (dinoflagellates). In a recent classification the Dinophyta are included in the subkingdom Protistobionta (*see* Protista).

dinosaur An extinct reptile of the orders *Ornithischia or *Saurischia. It is a descriptive rather than a taxonomic term and applies only to terrestrial reptiles. Adjectival form: **dinosaurian.**

dioecious (not dioicous) US: **diecious.** *Bot.* Noun form: **dioecy** (US: **diecy**).

dioestrus US: **diestrus** or **diestrum.** *See* oestrus. Adjectival form: **dioestrous** (US: **diestrous**).

dioicous Use *dioecious.

diol Any of a class of organic compounds containing two hydroxyl groups, i.e. a dihydric alcohol.

Diols are systematically named by adding the suffix -diol to the name of the parent hydrocarbon, e.g. propane-1,2-diol, CH_3-$CHOHCH_2OH$ and pentane-1,5-diol, $HOCH_2CH_2CH_2CH_2CH_2OH$.

In nonsystematic nomenclature 1,2-diols are named by adding the word glycol to the trivial name of the parent alkene, e.g. propylene glycol, $CH_3CHOHCH_2OH$; other diols are named as polymethylene glycols, e.g. pentamethylene glycol, $HOCH_2CH_2$-$CH_2CH_2CH_2OH$.

Diophantus of Alexandria (*fl.* AD 250) Greek mathematician.

Diophantine equation

Diophantine problem

dioptre US: **diopter**. A unit of lens power, especially of a spectacle lens, equal to the reciprocal of the focal length of the lens in metres. Normally a converging lens is taken to have a positive value, a diverging lens a negative value. The dioptre is also used in measuring the curvature of a surface of a lens, a mirror, or wavefront.

Dioscorides, Pedanius (*c.* AD 40–*c.* AD 90) Greek physician.

dioxygen *See* oxygen.

DIP *Electronics* Acronym for dual in-line package.

1,2-diphenylethanedione C_6H_5-$COCOC_6H_5$ The recommended name for the compound traditionally known as benzil.

***trans*-1,2-diphenylethene** C_6H_5-$CHCHC_6H_5$ The recommended name for the compound traditionally known as stilbene.

***N,N′*-diphenylhydrazine** C_6H_5-$NHNHC_6H_5$ The recommended name for the compound traditionally known as hydrazobenzene.

diphenylmethanone $C_6H_5COC_6H_5$ The recommended name for the compound traditionally known as benzophenone.

1,3-diphenyl-2-propen-1-one $C_6H_5CH{=}CHCOC_6H_5$ The recommended name for the compound traditionally known as benzalacetophenone.

diphos (no stop) Abbrev. for 1,2-bis-(diphenylphosphino)ethane often used in the formulae of coordination compounds, e.g. $[Co(diphos)_2]^{2+}$.

diphosphopyridine nucleotide *See* DPN.

DIPI Abbrev. for 4′,6-di(2′-imidazolinyl-4*H*,5*H*)-2-phenylindole.

diplo- (**dipl-** before vowels) Prefix denoting double (e.g. diplobiont, diploblastic, diplacusis).

diploid Denoting a nucleus, cell, or organism having twice the *haploid number of chromosomes (*see also* genome; genotype). The diploid number is designated as $2n$; for example in humans $2n = 46$. Noun form: **diploidy**. *Compare* diplont.

diplont Any organism whose somatic cells have a *diploid chromosome set. Adjectival form: **diplontic**.

Diplopoda (cap. D) A class of arthropods comprising the millipedes. Individual name and adjectival form: **diplopod** (no cap.). *See also* myriapod.

Dipnoi (cap. D) An order of bony fishes comprising the lungfishes. Individual name and adjectival form: **dipnoan** (no cap.).

DipTech (or **Dip.Tech.**) Abbrev. for Diploma in Technology.

Diptera (cap. D) An order of insects comprising the two-winged flies. Individual name and adjectival form: **dipteran** (no cap.; not dipterous or dipteral).

Dirac, Paul Adrien Maurice (1902–84) British mathematician and physicist.
Dirac bra (or **ket**) **vector**
Dirac constant *See* Planck constant.
Dirac delta function
Dirac equation
Dirac notation
Fermi–Dirac–Sommerfeld law (en dashes)
Fermi–Dirac statistics (en dash)

direct broadcast by satellite Abbrev.: DBS (no stops).

direct current Abbrev.: d.c. (or sometimes DC). *See also* electric current.

direct memory access (not hyphenated) *Computing* Abbrev.: DMA (no stops).

directrix (pl. directrices) *Geom.*

Dirichlet, (Peter Gustav) Lejeune (1805–59) German mathematician.
Dirichlet boundary condition
Dirichlet problem
Dirichlet's theorem

dis- Prefix denoting separation, removal, lack, or reversal (e.g. disbud, disinfect, disoperation).

disc *See* disk.

Discomycetes (cap. D) A former class of ascomycete fungi (*see* Ascomycotina). The individual name and adjectival form, **discomycete** (no cap.), may still be used for descriptive purposes.

discrete (not discreet) Separate or distinct.

disintegration constant Another name for decay constant.

disk Originally the US spelling of **disc**, 'disk' is now widely used for most technical senses (e.g. **disk brake, disk drive, disk valve, magnetic disk**; but note **compact disc**). In biology and anatomy, however, 'disc' is used (e.g. **intervertebral disc, disc floret**).

disk operating system (not hyphenated) Abbrev.: DOS (no stops).

disodium hydrogen orthophosphate Na_2HPO_4 The traditional name for disodium hydrogenphosphate(V).

disodium hydrogenphosphate(V) Na_2HPO_4 The recommended name for the compound traditionally known as disodium hydrogen orthophosphate.

displacement 1. Symbol: ***s*** A *vector quantity expressing the difference in position of two points. The *SI unit is the metre. **2.** Another name for electric flux density.

dissipation factor *See* acoustic absorption coefficient.

distil (distilling, distilled) US: **distill.** Noun forms: **distillation, distillate**.

disulphate(IV) US: **disulfate(IV)**. Denoting a compound containing the ion $S_2O_5^{2-}$, e.g. sodium disulphate(IV). The traditional name is metabisulphite.

disulphide group US: **disulfide group**. Symbol: S–S (en dash).

disulphur dichloride S_2Cl_2 US: **disulfur dichloride.** The recommended name for the compound traditionally known as sulphur monochloride.

dithioerythritol Abbrev.: DTE (no stops).

div Symbol for divergence. *See* vector.

diverticulum (pl. diverticula) A pouch, often abnormal, formed in the walls of a hollow organ, especially the intestine. Adjectival form: **diverticular**.

division 1. *Bot.* A category used in plant classification (*see* taxonomy) consisting of a number of related classes (sometimes only one class). Names of divisions are printed in roman (not italic) type with an initial capital letter and typically end in -phyta, e.g. Bryophyta, Tracheophyta. Large divisions may be split into subdivisions. *Compare* phylum. **2.** *Maths. See* arithmetic operations.

***dl*-** (ital., always hyphenated) *Chem.* Use *DL-.

DL- (small caps., always hyphenated) Prefix denoting racemic (e.g. DL-lactic acid). The alternative (±)- (but not *dl*-) may be used, depending on the context (*see* racemic).

D layer (not hyphenated) *Physics* Also called D region.

D loop (cap. D) Abbrev. and preferred form for **1.** displacement loop – the displaced outer strand of DNA occurring at the initiation site shortly after the start of replication in some species of mitochondrial DNA. **2.** displaced loop – the displaced chain of DNA that forms at the initiation of recombination according to the double-strand break model. *Compare* D arm.

DMA *Computing* Abbrev. for direct memory access.

DMAP Abbrev. for dimethylaminopyridine.

DME Abbrev. for **1.** dropping-mercury electrode. **2.** 1,2-dimethoxyethane.

DMF Abbrev. for *N,N*-dimethylformamide.

DMI Abbrev. for 1,3-dimethyl-2-imidazolidinone.

DMPE Abbrev. for 1,2-bis(dimethylphosphino)ethane.

DMPO Abbrev. for 5,5-dimethyl-l-pyrroline-*N*-oxide.

DMPP Abbrev. for 1,1-dimethyl-4-phenylpiperazinium iodide.

DMPS Abbrev. for 2,3-dimercapto-1-propanesulphonic acid.

DMPU Abbrev. for N,N′-dimethylpropyleneurea.

DMSO Abbrev. for dimethyl sulphoxide.

DMT Abbrev. for dimethyl terephthalate.

dna (lower case, ital.) Any of a group of genes whose protein products control DNA replication; e.g. the products of *dnaA*, *dnaB*, and *dnaC* initiate unwinding of DNA before replication.

DNA Abbrev. and preferred form for deoxyribonucleic acid. When prefixed by one or more lower-case letters, it indicates a particular type of DNA. Thus, for example:
cDNA (complementary DNA)
ctDNA (chloroplast DNA)
dsDNA (double-stranded DNA)
mtDNA (mitochondrial DNA)
rDNA (ribosomal DNA)
ssDNA (single-stranded DNA)
Structural types of DNA are designated by hyphenated prefixed roman capital letters, i.e. A-DNA, B-DNA, C-DNA, Z-DNA (zigzagged DNA, describing the orientation of the phosphate groups).

DNAase Abbrev. and preferred form for deoxyribonuclease (note that this is the recommended form; DNase, though commonly used, is incorrect). The different types are designated by Roman numerals, e.g. DNAase I, DNAase II (note intervening space).

DNA polymerase An enzyme responsible for the synthesis of new DNA strands. In eukaryotes there are three classes, designated α, β, and γ. The three types found in *E. coli* are designated by Roman numerals: I, II, and III.

DNase Use *DNAase.

DNBP Abbrev. for 4,6-dinitro-2-*s*-butylphenol.

DNOC Abbrev. for 4,6-dinitro-*o*-cresol.

DNS Abbrev. for 5-dimethylamino-1-naphthalenesulphonic acid.

Döbereiner, Johann Wolfgang (1780–1849) German chemist.
Döbereiner's law of triads

Döbner–Miller process (en dash) *Organic chem.*

Dobzhansky, Theodosius (1900–75) Russian-born US geneticist.

dodeca- Prefix denoting twelve (e.g. dodecahedron, dodecanoic acid).

dodecanoic acid $CH_3(CH_2)_{10}COOH$ The recommended name for the compound traditionally known as lauric acid.

dodecan-1-ol $CH_3(CH_2)_{10}CH_2OH$ The recommended name for the compound traditionally known as lauryl alcohol.

DoE Abbrev. for Department of the Environment, a UK government department.

Doisy, Edward Adelbert (1893–1986) US biochemist.

Dolby, Ray Milton (1933–) US electrical engineer.
Dolbyed (UK trade mark)
Dolbyized (US trade mark)
Dolby system (trade mark) *Sound recording*

dolicho- (**dolich-** before vowels) Prefix denoting long (e.g. dolichocephalic).

Döllinger, Ignaz Christoph von (1770–1841) German physiologist.

Dollo, Louis Antoine Marie Joseph (1857–1931) Belgian palaeontologist.
Dollo's law of irreversibility

Dollond, John (1706–61) British optician.

dolomite 1. A mineral, systematic name: calcium magnesium biscarbonate. **2.** A sedimentary rock consisting principally of the mineral dolomite. Also called dolostone (to avoid confusion between the rock and the mineral). **3.** A basic refractory consisting of either calcined dolomite or a magnesian limestone without dolomite. Adjectival form: **dolomitic.** Derived noun: **dolomitization.**

Domagk, Gerhard (1895–1964) German biochemist.

Donati, Giovanni Battista (1826–73) Italian astronomer.
Donati's comet

Donnan, Frederick George (1879–1956) British chemist.
Donnan equilibrium

donor insemination Abbrev.: DI (no stops). Former name: artificial insemination by donor (abbrev.: AID).

donor number density *See* number density.

dopa (preferred to DOPA) Abbrev. and preferred form for 3,4-dihydroxyphenylalanine: a neurotransmitter and intermediate in the synthesis of *dopamine. *See also* L-dopa.

dopamine Abbrev.: DA (no stops). A neurotransmitter. Dopamine receptors are designated D_1, D_2, D_3, and D_4 (note subscript numbers).

D operator (not hyphenated) Short for differential operator. Symbol: d/d*x*

Doppler, Christian Johann (1803–53) Austrian physicist.
Doppler broadening
Doppler effect
Doppler radar
Doppler shift

DL-**DOPS** Abbrev. for DL-*threo*-3,4-dihydroxyphenylserine.

Dor (no stop) *Astron.* Abbrev. for Dorado. *See also* stellar nomenclature.

Dorado A constellation. Genitive form: Doradus (not Doradûs). Abbrev.: Dor (no stop). *See also* stellar nomenclature.

d orbital *See* orbital.

dormin Former name for *abscisic acid.

dorsiventral (one word) *Bot.* Describing a leaf or similar part with distinct upper and lower faces. *Compare* dorsoventral.

dorso- (or **dorsi-**) Prefix denoting the back (e.g. dorsomedial, dorsifixed, dorsiflexion).

dorsoventral (one word) *Anat.* Extending from the back (dorsal) to front (ventral) surfaces. *Compare* dorsiventral.

DOS *Computing* Abbrev. for disk operating system.

dose equivalent Symbol: *H* A physical quantity associated with ionizing radiation, equal to the product *DQN*, where *D* is the *absorbed dose, *Q* is the quality factor for a particular type of radiation – e.g. X-rays, α-particles, or medium-energy β-particles – and *N* is the product of any other modifying factors. The *SI unit is the sievert (Sv), which has replaced the rem.

dot product Another name for scalar product.

Douglas, David (1798–1834) British botanist.
Douglas fir (*Pseudotsuga menziesii*)

Douglas, Donald Wills (1892–1981) US aircraft engineer.

Douglas, James (1675–1742) British physician.
Douglas's pouch (or **pouch of Douglas**) *Anat.*

Dover, Thomas (1660–1742) British physician.
Dover's powder

Dow, Herbert Henry (1866–1930) US industrial chemist.
Dow process

Down, John Langdon Haydon (1828–96) British physician.
Down's syndrome (not mongolism)

DP (or **dp**) Abbrev. for data processing.

2,4-DP Abbrev. for 2-(2,4-dichlorophenoxy)propionic acid.

DPB Abbrev. for 1,4-diphenyl-1,3-butadiene.

DPD Abbrev. for *diffusion pressure deficit.

DPH Abbrev. for 1,6-diphenylhexatriene.

DPhil (or **D.Phil.**) Abbrev. for Doctor of Philosophy.

DPM Abbrev. for bis(diphenylphosphino)methane.

DPN Abbrev. for diphosphopyridine nucleotide, a former name for *NAD.

DPPA Abbrev. for diphenylphosphoryl azide.

DPPH Abbrev. for 2,2-di(4-t-octylphenyl)-1-picrylhydrazyl.

DPS Abbrev. for *trans*-4,4′-diphenylstilbene.

dr Symbol for dram.

Dra (no stop) *Astron.* Abbrev. for Draco. *See also* stellar nomenclature.

drachm US: **dram.** *See* ounce.

Draco A constellation. Genitive form: Draconis. Abbrev.: Dra (no stop). *See also* stellar nomenclature.

draft 1. An engineering drawing. **2.** A plan, sketch, or diagram. **3.** US spelling of *draught.
draftsman Use draughtsman.

dram **1.** Symbol: dr A unit of mass equal to $^1/_{16}$ ounce (in avoirdupois measure) or 1.771 85 grams. **2.** In the US, a unit of mass equal to $^1/_8$ apothecaries' ounce or 3.887 93 grams. In the UK this unit (spelt 'drachm') is no longer used.

DRAM *Electronics* Acronym for dynamic random-access memory.

Draper, Henry (1837–82) US astronomer.

Draper, John William (1811–82) British-born US chemist.

draught US: **draft. 1.** A flow of air. **2.** The pulling of a load. **3.** The depth of water displaced by a loaded vessel. *Compare* draft.

draughtsman (preferred to draftsman)

D region *See* D layer.

Dreyer, Johann Louis Emil (1852–1926) Danish astronomer.

dRib (no stops) Abbrev. for deoxyribose.

Driesch, Hans Adolf Eduard (1867–1941) German biologist.

dRNA Abbrev. for DNA-like RNA. Use hnRNA (heterogeneous nuclear RNA).

-dromous Adjectival suffix denoting speed or movement (e.g. anadromous, catadromous).

dropping-mercury electrode Abbrev.: DME (no stops). *Chem.*

Drosophila (cap. D, ital.) A genus of fruit flies. Individual name: **drosophila** (no cap., not ital.; pl. drosophilas).

D. melanogaster: the best known species, widely used in genetics research. For its genetic nomenclature, *see* gene.

dry adiabatic lapse rate Abbrev.: DALR (no stops). It has the value 9.76 °C km^{-1}.

Dryopithecus (cap. D, ital.) A genus of extinct apes. Fossils of this and similar genera are known as **dryopithecines** (not ital., no cap.).

DSc (or **D.Sc.**) Abbrev. for Doctor of Science.

DSC Abbrev. for **1.** differential scanning calorimetry. **2.** *N,N′*-disuccinimidyl carbonate.

dsDNA (lower-case d, s) Abbrev. for double-stranded DNA.

DSN Abbrev. for Deep Space Network, operated by *JPL for *NASA.

dT Symbol for thymidine. *See also* nucleoside.

DTA Abbrev. for differential thermal analysis.

dTDP Abbrev. and preferred form for thymidine 5′-diphosphate.

DTE Abbrev. for dithioerythritol.

dTMP Abbrev. and preferred form for thymidine 5′-phosphate.

DTNB Abbrev. for 5,5′-dithiobis-(2-nitrobenzoic acid).

DTP Abbrev. for desktop publishing.

DTT Abbrev. for dithiothreitol.

dTTP Abbrev. and preferred form for thymidine 5′-triphosphate.

dual in-line *Electronics* Acronym: DIL (no stops).

dual in-line package Acronym: DIP (no stops).

Du Bois-Reymond, Emil Heinrich (hyphen) (1818–96) German neurophysiologist.

Du Bois-Reymond law (hyphen)

Dubos, René Jules (1901–82) French-born US microbiologist.

Duchenne, Guillaume Benjamin Amand (1806–75) French neurologist.

Duchenne dystrophy

ductus (pl. ductus) An anatomical duct: usually qualified (e.g. ductus arteriosus, ductus deferens, ductus venosus).

Duggar, Benjamin Minge (1872–1956) US plant pathologist.

Duhamel Du Monceau, Henri-Louis (1700–82) French agriculturalist and technologist.

Duhamel's principle *Maths.* Named after J. M. C. Duhamel (1797–1872).

Duhem, Pierre (1861–1916) French physicist, mathematician, and philosopher of science.

Duhem theorem

Gibbs–Duhem equation (en dash)

Dujardin, Félix (1801–60) French biologist and cytologist.

Dulong, Pierre-Louis (1785–1838) French chemist and physicist.
Dulong and Petit's law or **Dulong–Petit law** (en dash) Also named after Alexis-Thérèse Petit (1791–1820).

Dumas, Jean Baptiste André (1800–84) French chemist.
Dumas method

Du Mont, Allen Balcom (1901–65) US electrical engineer.

Dunlop, John Boyd (1840–1921) British inventor.

Dupuytren, Baron Guillaume (1777–1835) French surgeon.
Dupuytren's contracture

durene $C_6H_2(CH_3)_4$ The traditional name for 1,2,4,5-tetramethylbenzene.

Dushman, Saul (1883–1954) US physicist.
Richardson–Dushman equation (en dash) Use Richardson's equation.

du Toit, Alexander Logie (1878–1949) South African geologist.

Dutrochet, René Joachim Henri (1776–1847) French biologist and physiologist.

Duve, Christian René de Usually alphabetized as *de Duve.

Du Vigneaud, Vincent (1901–78) US biochemist.

DVM Abbrev. for digital voltmeter.

dwarf *Astron.* Former name for a main-sequence star, although the term red dwarf is still sometimes used for a low-temperature main-sequence star. A white dwarf (a highly evolved collapsed star) and a brown dwarf (a low-mass object unable to become a star) are not main-sequence stars.

DX *Photog.* Abbrev. for daylight exposure, used to indicate that the speed of a particular film can be set automatically in a suitably equipped camera.

Dy Symbol for dysprosium.

dyestuff (one word)

dyke US: **dike**. The US spelling is sometimes used in UK scientific literature but 'dyke' is preferred.

dyn Symbol for dyne.

dynamic (preferred to dynamical)

dynamic equinox *See* equinox.

dynamic viscosity *See* viscosity.

dynamo- (**dynam-** before vowels) Prefix denoting power (e.g. dynamometer, dynamics).

dyne Symbol: dyn The *cgs unit of force, now discouraged. In *SI units, force is measured in newtons: 1 dyn = 10^{-5} N.

dys- Prefix denoting difficult, abnormal, or impaired (e.g. dysfunction, dystrophic).

Dyson, Sir Frank Watson (1868–1939) British astronomer.

Dyson, Freeman John (1923–) British-born US theoretical physicist.

Dyson notation *Chem.* Named after G. M. Dyson.

dysprosium Symbol: Dy *See also* periodic table; nuclide.

E

e Symbol for **1.** (or e^-) electron (e^+ or $\bar{e}$ = positron). **2.** the transcendental number 2.718 282 . . . .

e Symbol for **1.** (light ital.) eccentricity. **2.** (light ital.) electron (or proton) charge (e/m = electron specific charge). **3.** (light ital.) equatorial conformation (e.g. in cyclohexane derivatives). **4.** (bold ital.) unit coordinate vectors: $\boldsymbol{e}_x$, $\boldsymbol{e}_y$, $\boldsymbol{e}_z$.

E 1. Symbol for **i.** electromeric effect (*see* electron displacement). **ii.** elimination (*see* reaction mechanisms). **iii.** elliptical *galaxy (followed by a number). **iv.** exa-. **v.** glutamic acid. **2.** Abbrev. for **i.** *Elec. eng.* earth. **ii.** *east or eastern.

E Symbol (light ital.) for **1.** electric field strength (in nonvector equations; in vector equations it is printed in bold italic, ***E***). **2.** electrode potential. **3.** electromotive force; for an electrolytic cell, $E^{\ominus}$ = standard electrode potential (preferred to E_0). **4.** energy (E_p = potential energy, E_k = kinetic energy, E_F = Fermi energy). **5.** (or E_v) illuminance. **6.** (or E_e) irradiance. **7.** Young modulus.

E_A (light ital. E) Symbol for electron affinity.

(*E*)- (ital., parentheses, always hyphenated) Prefix denoting a geometric isomer in which the highest priority substituent groups are located on opposite sides of a double bond (e.g. (*E*)-3-methyl-2-pentenoic acid). The priority of the substituent groups is obtained using the Cahn–Ingold–Prelog sequence rules. *Compare* (*Z*)-.

Eads, John Buchanan (1820–87) US civil engineer.

-eae Suffix denoting a *tribe in plant taxonomy (e.g. Cichorieae).

EAN Abbrev. for effective atomic number.

eardrum (one word) *See also* tympanum.

EARN *Computing* Acronym for European Academic and Research Network.

earth 1. (or **Earth**; often preceded by 'the') The third planet from the sun. The mass of the earth, symbol $M_{\oplus}$, and its equatorial radius, symbol $R_{\oplus}$, have been adopted as astronomical constants:

$M_{\oplus} = 5.9742 \times 10^{24}$ kg

$R_{\oplus} = 6378.140$ km.

See also astronomical unit; month; year.

2. (not preceded by 'the') US: **ground.** *Elec. eng., etc.* The point held at zero potential in an electrical circuit. Adjectival form: **earthed** (US: **grounded**).

earthstar (one word) *See* Lycoperdales.

east Adjectival forms: **east, eastern.** Abbrev.: E (no stop). Use the abbrev. only descriptively, not in place names or concepts (e.g. E London, E Canada, but East Anglia, East Greenland Current, East Indies, East Pacific Basin).

Eastern Standard Time Abbrev.: EST (no stops).

Eastman, George (1854–1932) US inventor.

east-north-east (hyphenated) Adjectival form: **east-north-eastern.** Abbrev. (for both): ENE (no stops).

east-south-east (hyphenated) Adjectival form: **east-south-eastern.** Abbrev. (for both): ESE (no stops).

Ebashi, Setsuro (1922–) Japanese biochemist.

EBCDIC *Computing* Acronym for extended binary-coded decimal interchange code.

e-beam Short for electron beam.

ebullioscopic constant Use *boiling-point constant.

EBV (or **EB virus**) Abbrev. for *Epstein–Barr virus.

EC Abbrev. for Enzyme Commission: precedes the numerical designation of an enzyme. *See* enzyme nomenclature.

ec- Prefix denoting out of, outside, or away from (e.g. eccentric, ecdysis, ectopic).

eccentricity *Maths., astron.* Symbol: *e* Adjectival form: **eccentric.**

Eccles, Sir John Carew (1903–) Australian physiologist.

Eccles, William Henry (1875–1966) British physicist.

Eccles–Jordan circuit (en dash) Also named after F. W. Jordan.

ecdysis (pl. ecdyses) The periodic shedding of the cuticle of arthropods or the outer epidermal layer of reptiles; sometimes also used as a synonym for desquamation (the shedding of skin by scaling). Also called moulting, but note that the moulting of hair and feathers is not called ecdysis. Adjectival form: **ecdysial.**

ECG Abbrev. and preferred form for *electrocardiogram.

echelon (no accent) A diffraction grating. *Compare* etalon. [from French *échelon*: rung of ladder]

echino- (**echin-** before vowels) Prefix denoting spiny or prickly (e.g. echinococcus, echinoderm).

Echinodermata (cap. E) A phylum of marine invertebrates including the starfishes, sea urchins, sea cucumbers, brittlestars, etc. Individual name and adjectival form: **echinoderm** (no cap; not echinodermal or echinodermatous).

Echinoidea (cap. E) A class of echinoderms including the sea urchins. Individual name and adjectival form: **echinoid** (no cap.).

echo (pl. echoes) Verb form: **echoes, echoing, echoed.**

echolocation (one word)

echosounding (one word)

echovirus (one word) Any of a subgroup of human *enteroviruses associated with neurological disorders. There are 34 serovars, each designated by an Arabic numeral: echovirus 1, echovirus 2, etc. [from *e*nteric *c*ytopathic *h*uman *o*rphan *virus*]

Eckart, Carl (1902–71) US physicist and mathematician.

Eckert, John Presper (1919–) US electronics engineer.

ECL *Electronics* Abbrev. for emitter-coupled logic.

eco- Prefix denoting ecology or the environment (e.g. ecocline, ecosystem, ecotype).

E. coli *See Escherichia coli.*

ecto- Prefix denoting outer or external (e.g. ectoderm, ectoparasite).

ectocarp Use *exocarp.

ectoderm (preferred to ectoblast and exoderm) The outer germ layer of an animal embryo. Adjectival form: **ectodermal** (not ectodermic). *Compare* endoderm.

Ectoprocta (cap. E) A phylum of aquatic invertebrates including the moss animals and sea mats. Also called *Bryozoa. Individual name and adjectival form: **ectoproct** (no cap.). *Compare* Entoprocta.

ectothermic Denoting animals (called **ectotherms**) whose body temperature is determined largely by external heat sources. The term is often used synonymously with *poikilothermic, but this is incorrect as many ectotherms can maintain a relatively constant body temperature. Not to be confused with *exothermic. Noun form: **ectothermy**. *Compare* endothermic.

ED_{50} (subscript number) Abbrev. and preferred form for median effective dose.

edapho- (edaph- before vowels) Prefix denoting the soil (e.g. edaphology).

Eddington, Sir Arthur Stanley (1882–1944) British astrophysicist and mathematician.
Eddington limit

Edelman, Gerald Maurice (1929–) US biochemist.

edema US spelling of *oedema.

Edentata (cap. E) An order of virtually toothless mammals including the sloths, armadillos, and anteaters. Individual name and adjectival form: **edentate** (no cap.); the adjectives 'edentulate' and 'edentulous' are used in a more general sense to mean toothless.

Edison, Thomas Alva (1847–1931) US physicist and inventor.
Edison accumulator
Edison effect
Edison phonograph

Edlén, Bengt (1906–) Swedish physicist.

EDP *Computing* Abbrev. for electronic data processing.

Edsall, John Tileston (1902–) US biochemist.

edta (no stops) Abbrev. for ethylenediaminetetraacetic acid often used in the formulae of coordination compounds, e.g. $[Cr(edta)]^-$. Note that the form EDTA (caps.) should not be used in this context.

EDTN Abbrev. for 1-ethoxy-4-(dichloro-1,3,5-triazinyl)naphthalene.

EEDQ Abbrev. for 1-ethoxycarbonyl-2-ethoxy-1,2-dihydroquinoline.

EEG Abbrev. and preferred form for *electroencephalogram.

EEROM *Computing* Acronym for electrically erasable read-only memory.

EF *Genetics* Abbrev. for elongation factor, a protein that enables elongation of a polypeptide chain by delivering aminoacyl-tRNAs to the ribosome during translation. In *E. coli* it was originally obtained as a single fraction called the T factor (transfer factor);

this is now recognized as comprising two entities: EF-Tu (*u*nstable when heated) and EF-Ts (*s*table when heated). A third elongation factor, EF-G (*G*TP-requiring), participates in ribosome translocation along mRNA. Elongation factors in eukaryotes are denoted by Arabic numerals and prefixed by a lower-case e (for eukaryote), e.g. eEF1, eEF2.

EFA Abbrev. for essential fatty acid.

effective atomic number Abbrev.: EAN (no stops).

efficiency of plating *Virology* Abbrev.: EOP (no stops).

e.g. (stops) Abbrev. for *exempli gratia* (Latin: for example).

EGA *Computing* Abbrev. for enhanced graphics adapter.

EGF Abbrev. for epidermal growth factor.

EHF Abbrev. for extremely high frequency.

E horizon An e(luvial) mineral *soil horizon.

Ehrenberg, Christian Gottfried (1795–1876) German biologist and microscopist.

Ehrenfest, Paul (1880–1933) Dutch physicist.
Ehrenfest relations
Ehrenfest's theorem or **rule**

Ehrlich, Paul (1854–1915) German physician, bacteriologist, and chemist.
Ehrlich reaction

EHT (or **e.h.t.**) *Electronics* Abbrev. for extra-high tension.

EIA Abbrev. for Electronic Industries Association, a US organization.

Eichler, August Wilhelm (1839–87) German botanist.

EIEC Abbrev. for enteroinvasive *Escherichia coli*.

Eigen, Manfred (1927–) German physical chemist.

eigen- *Physics* Prefix denoting characteristic or natural (e.g. eigenfunction, eigenvalue). [from German: own]

Eijkman, Christiaan (1858–1930) Dutch physician.

Einstein, Albert (1879–1955) German-born US theoretical physicist.
Bose–Einstein statistics (en dash)
Einstein–de Haas effect (en dash) Also named after W. J. de Haas.
Einstein–de Sitter universe (en dash)
Einstein displacement
einsteinium
Einstein's equation ($E = mc^2$)
Einstein shift Also called gravitational redshift.
Einstein's photoelectric equation
Stark–Einstein equation (en dash)

einsteinium Symbol: Es *See also* periodic table; nuclide.

Einthoven, Willem (1860–1927) Dutch physiologist.
Einthoven galvanometer

Ekman, Vagn Walfrid (1874–1954) Swedish oceanographer.
Ekman flow
Ekman layer
Ekman spiral

Elasmobranchii (cap. E, not Elasmobranchi) A subclass of the Chondrichthyes (cartilaginous fishes) including the sharks, skates, and rays. Individual name and adjectival form: **elasmobranch** (no cap.).

E layer (not hyphenated) *Physics* Also called E region, Kennelly–Heaviside (or Heaviside) layer.

electric (or **electrical**) Although the words may be used interchangeably in some contexts, 'electric' is usually used to describe devices or concepts (e.g. electric charge, electric current, electric eel, electric field, electric fire, electric motor) while 'electrical' is used in a more general sense (e.g. electrical appliance, electrical energy, electrical engineering).

electric charge Symbol: Q or q A fundamental physical quantity, a property of certain subatomic particles including the *electron, which by convention has a negative charge, and the *proton, which by convention has a positive charge. The charge of a particle is indicated by adding a superscript +, −, or 0 after the particle symbol, as in π^+.

The *SI unit is the coulomb but the ampere hour (3600 coulombs) is also used in some contexts (e.g. the charge stored in an accumulator).

electric constant Another name for permittivity of vacuum. *See* permittivity.

electric current Symbol: *I* (*i* is often used for the instantaneous value of an alternating current). A fundamental physical quantity, a flow of *electric charge. The magnitude of a current is given by the amount of charge flowing in unit time. The direction is by convention the direction of flow of positive charge, i.e. the opposite direction to the flow of electrons. The *SI unit is the ampere.

electric current density Symbol: *J* or *j* A physical quantity, *electric current divided by the cross-sectional area of the current-carrying medium, the area being perpendicular to the direction of flow. The *SI unit is the ampere per square metre.

electric dipole moment Symbol: ***p*** A *vector quantity equal to the product of the distance between two equal but opposite electric charges and the magnitude of either charge. The *SI unit is the coulomb metre (C m).

electric displacement Another name for electric flux density.

electric field strength Symbol: ***E*** A *vector quantity, the *force exerted by an electric field on an electric point charge divided by the magnitude of the charge. The *SI unit is the volt per metre. Also called electric field.

electric flux Symbol: Ψ (Greek cap. psi). A physical quantity that is the scalar product $\mathbf{D}\cdot dA$, where ***D*** is the *electric flux density through a given element of area dA. The *SI unit is the coulomb.

electric flux density Symbol: ***D*** A *vector quantity, the product of *electric field strength, ***E***, and the *permittivity of the medium, ε. The direction at a given point is that of the field strength at that point. The *SI unit is the coulomb per square metre. The divergence of electric flux density, div ***D***, is equal to the volume density of charge (*see* charge density). Also called displacement, electric displacement.

electric intensity Former name for electric field strength.

electric polarization Symbol: ***P*** or $\mathbf{D}_i$ A physical quantity defined by the relation

$$\mathbf{D} - \epsilon_0 \mathbf{E},$$

where ***D*** is the *electric flux density, ***E*** the *electric field strength, and ϵ_0 the *permittivity of vacuum. The *SI unit is the coulomb per square metre. It resolves as a vector but does not add as a vector because of saturation effects. Also called dielectric polarization.

electric potential Symbol: *V* or ϕ (Greek phi). A physical quantity that is a *scalar for electrostatic fields. The *SI unit is the volt. The negative gradient of electric potential is equal to the *electric field strength, ***E***. *See also* potential difference.

electric susceptibility *See* permittivity.

electro- Prefix denoting electric charge, current, potential, or field (e.g. electrolysis, electromagnetic, electro-optical (hyphenated), electroreceptor, electrotaxis).

electrocardiogram (one word) Abbrev. and preferred form: ECG (no stops). A tracing of the electrical activity of the heart, recorded by an apparatus called an **electrocardiograph** in the technique of **electrocardiography**.

electroencephalogram (one word) Abbrev. and preferred form: EEG (no stops). A tracing of the electrical activity of the brain, recorded by an apparatus called an **electroencephalograph** in the technique of **electroencephalography**.

electrolyse US: **electrolyze**. Derived nouns: **electrolysis, electrolyte**. Adjectival form: **electrolytic**.

electromagnetic Abbrev.: EM, e.m. (stops), or em (no stops).

electromagnetic radiation Abbrev.: emr or EMR (no stops). See Appendix 1.

electromagnetic unit Abbrev.: emu (no stops) or e.m.u. (stops). *See* cgs units.

electromeric effect *See* electron displacement.

electromotive force Symbol: E Abbrev.: emf (or e.m.f.). A physical quantity; the rate at which work is done electrically upon a circuit (power) divided by the current. If the power is in watts and the current in amperes, the *SI unit is the volt. Electromotive force must be distinguished from *potential difference. In a circuit, potential difference is rate of energy dissipation divided by current, and is inherently irreversible.

electron A negatively charged elementary particle denoted e (or sometimes β) in nuclear reactions, etc.; the negative charge may be indicated by using the form e^- (or β^-). The antiparticle of the electron, the positron, is denoted e^+ or $\bar{e}$ (or sometimes β^+ or $\bar{\beta}$).

The magnitude of the charge of the electron is a fundamental constant, usually called the elementary charge. It has the symbol e (italic) and is equal to

$$1.602\,177\,33 \times 10^{-19}\,\text{C}.$$

The rest mass of the electron is also a fundamental constant, usually called the electron mass. It has the symbol m_e and is equal to

$$9.109\,3897 \times 10^{-31}\,\text{kg}$$
$$5.485\,799\,03 \times 10^{-4}\,\text{u}.$$

This is equivalent to

$$0.510\,999\,06\,\text{MeV}.$$

The electron specific charge is the ratio $-e/m_e$ and is equal to

$$-1.758\,819 \times 10^{11}\,\text{C kg}^{-1}.$$

Two- or three-word terms in which the first word is 'electron' are not hyphenated (e.g. electron beam, electron microscope, electron spin resonance).

electron displacement Various effects are distinguished in molecules in which electrons are displaced away from or towards a specified atom or group. The symbols used are as follows:

E – electromeric effect;

I – inductive effect (sometimes I_s);

I_d – inductomeric effect;

M – mesomeric effect;

R – resonance effect;

T – tautomeric effect, use mesomeric effect.

In addition, a plus sign before the symbol indicates that electron displacement is away from the atom or group (e.g. +I); a minus sign indicates displacement towards an atom or group (e.g. −I).

electronic configuration *See* orbital.

electronic data processing *Computing* Abbrev.: EDP (no stops).

Electronic Industries Association Abbrev.: EIA (no stops).

electronics, graphical symbols Figures, each with a defined meaning, conventionally used on electronics diagrams (including logic diagrams), on electric power diagrams, and on telecommunications diagrams, to represent a device or concept. There are over 2000 graphical symbols (given in a series of British Standards documents, BS 3939, IEC-approved).

The basic symbols are simple figures. General symbols, common to a whole family of items, are usually simple. More complex symbols are formed by the addition of so-called qualifying symbols to a simple symbol. The qualifying symbols provide additional information; for example they may indicate the kind of current or voltage, variability, direction of flow, type of material (solid, liquid, gas, insulating, etc.), type of effect (thermal, electromagnetic, etc.), incidence or emission of radiation, or signal waveform. A selection of qualifying symbols and the more commonly used graphical symbols is given in Appendix 2.

electronics, letter symbols The basic letter symbols, which are set in italic type, are as follows: I or i (current), U or u or V or v (potential difference or electromotive force), P or p (power). The symbols can have one or more subscripts, capital or lower case, set in roman (upright) type. The use of either capital or lower-case letters for the basic letters and the subscripts is summarized in Appendix 3, Table 3.1. Recommended general subscripts are given in Appendix 3, Table 3.2. If more than one subscript is used, these should be all capital or all lower-case, where both styles exist.

Many other subscripts are recommended for various fields in electronics. For example, subscripts for the base, collector, and emitter terminals of bipolar transistors are respectively B,b,C,c, and E,e. Subscripts for the source, drain, and gate terminals of field-effect transistors are respectively S,s,D,d, and G,g.

Examples of these letter symbols, applying to a transistor base current, are as follows

i_b	instantaneous value, varying component
i_B	instantaneous total value
I_b	r.m.s. value, varying component
I_B	continuous (d.c.) value (without signal)
I_{bav}	average value, varying component
I_{bm}	maximum value, varying component
I_{BAV}	average total value
I_{BM}	maximum total value

In the case of currents, if it is necessary to indicate the terminal carrying the current, this should be done using the first subscript. In the case of voltages, if it is necessary to indicate the points between which a voltage is measured, this should be done using the first two subscripts, as in V_{BE}. The first subscript indicates one terminal point of the device, the second the reference terminal or the circuit node; when there is no possibility of confusion the second subscript can be omitted. Supply voltages or supply currents are indicated by repeating the appropriate terminal subscript, as in I_{EE}; a reference terminal can be indicated, when necessary, by a third subscript, as in V_{CCE}.

If a device has more than one terminal of the same kind, the subscript is formed from the appropriate letter for the terminal followed by a number, as in I_{B2} (d.c. flowing into the second base terminal); hyphens can be used in multiple-letter subscripts, as in V_{B2-E} (d.c. voltage between second base and emitter). For multiple-unit devices, subscripts are modified by a number preceding the letter subscript, as in I_{3C} (d.c. flowing into collector of third unit); hyphens can be used in the case of multiple-letter subscripts, as in V_{1C-2C} (d.c. voltage between collectors of first and second unit).

Certain additional symbols for electrical parameters are used. These are set in italic type, the most important being:
C (capacitance), *L* (inductance), *Z* or *z* (impedance), *R* or *r* (resistance), *X* or *x* (reactance), *Y* or *y* (admittance), *G* or *g* (conductance), and *B* or *b* (susceptance).

Capital letters are used for the representation of parameters of external circuits and of circuits in which the device forms only a part. Lower-case letters are used for parameters inherent in the device. Symbols for capacitances and inductances, however, are always capitals.

Some recommended general subscripts for electrical parameters are listed in Table 3.3 (Appendix 3). Capital subscripts designate static (d.c.) values while lower-case subscripts designate small-signal values. If more than one subscript is used, these should be all capital or all lower case, where both styles exist. The numerical subscripts are applicable to all electrical parameters except four-pole matrix parameters.

The symbol for the commonly used hybrid four-pole (or quadripole) matrix parameter is *H* or *h*. Each element of the four-pole matrix is identified as follows. The first subscript can be a double numerical subscript or a single letter subscript: 11 or i (input), 22 or o (output), 12 or r (reverse transfer), 21 or f (forward transfer); examples are h_{11} or h_i, h_{21} or h_f. A further subscript is used for the identification of the circuit configuration, as in h_{21E} or h_{21e} (common emitter forward transfer).

electronic states A molecule in which the spin multiplicity $2S + 1$ is one, i.e. all the electrons are paired in orbitals, is in the singlet state, designated S. If the molecule contains two unpaired electrons, *S* is 1 and the spin multiplicity is three and the molecule is in the triplet state, symbolized by T. The ground state of a molecule is designated by a right subscript zero, e.g. S_0 or T_0. Similarly, excited states are denoted by right subscript numbers, e.g. the first excited singlet state is designated S_1 and the second excited triplet state is designated T_2.

electronic structure (of atoms) *See* orbital angular momentum quantum number.

electron mass *See* electron.

electron microscope Abbrev.: EM (no stops) or e.m. (stops).

electron number density *See* number density.

electron paramagnetic resonance Abbrev.: EPR (no stops). Use *electron spin resonance.

electron-probe microanalysis (hyphenated) *Chem.* Abbrev.: EPM (no stops).

electron specific charge *See* electron.

electron spectroscopy for chemical analysis Abbrev.: ESCA (no stops). Use *X-ray photo-electron spectroscopy (XPS).

electron spin resonance (preferred to electron paramagnetic resonance) Abbrev.: ESR (no stops).

electronvolt (preferred to electron-volt) Symbol: eV A unit of energy in atomic and nuclear physics and accelerator technology. Although not an *SI unit, it may be used with the SI units and SI prefixes may be attached to it, as in megaelectronvolt (MeV).

1 eV = $1.602\,177\,33 \times 10^{-19}$ joule.

electrophilic *See* reaction mechanisms.

electrostatic unit Abbrev.: esu (no stops) or e.s.u. (stops). *See* cgs units.

element *See* nuclide; periodic table.

element 104 Due to a conflicting claim to discovery between a US and a Soviet group, it has no official name. The proposed US name is rutherfordium, symbol Rf. The proposed Soviet name is kurchatovium, symbol Ku. The systematic name unnilquadium, symbol Unq, has also been suggested.

element 105 Due to a conflicting claim to discovery between a US and a Soviet group, it has no official name. The proposed US name is hahnium, symbol Ha. The proposed Soviet name is nielsbohrium. The systematic name unnilpentium, symbol Unp, has also been proposed.

element 106 Due to nearly simultaneous discoveries by a US and a Soviet group, no name has been proposed by either group. The systematic name unnilhexium, symbol Unh, has been suggested. The symbol '106' is also used.

elementary charge *See* electron.

elementary particles Another name for fundamental particles.

Elhuyar, Fausto D' Usually alphabetized as *D'Elhuyar.

elinvar (no cap.) A nickel–chromium steel alloy.

Elion, Gertrude Belle (1918–) US biochemist and pharmacologist.

ELISA Acronym and preferred form for enzyme-linked immunosorbent assay.

ellipse Adjectival form: **elliptical** or **elliptic**. Derived noun: **ellipsoid**, adjectival form: **ellipsoidal**.

An ellipse is a plane curve, an ellipsoid is a curved surface or solid body. The adjective elliptical is often used in optics, astronomy, etc., to describe an ellipsoidal surface (as in elliptical mirror, elliptical antenna), since it is the elliptical sections of the surface that are relevant.

El Niño southern oscillation (or **El Niño**; note tilde) An extensive prolonged warming of the eastern tropical Pacific occurring every few years with major meteorological effects. Originally El Niño was a weak warm ocean current flowing south along the Ecuador/Peruvian coast at Christmas time. [from Spanish: The Child, i.e. Christ]

elongation factor *Genetics* Abbrev.: *EF (no stops).

Elsasser, Walter Maurice (1904–) German-born US geophysicist.

Elton, Charles Sutherland (1900–91) British ecologist.

Elvehjem, Conrad Arnold (1901–62) US biochemist.

elytron (preferred to elytrum; pl. elytra; usually referred to in the pl.) Either of the leathery forewings of a beetle. *Compare* hemelytron.

em *See* en dash; point.

EM (or **e.m.**) Abbrev. for **1.** electron microscope. **2.** (or **em**) electromagnetic.

em- *See* en-.

email (or **e-mail, E-mail**) Short for electronic mail.

Embden, Gustav George (1874–1933) German physiologist.
Embden–Meyerhof pathway (en dash) Another name for glycolysis.

embryo (pl. embryos) An organism at any stage of development before birth, hatching, or germination. In mammals it denotes the earlier stages, before the main organs have developed (*compare* fetus). Adjectival form: **embryonic**.

Embryobionta *See* Embryophyta.

Embryophyta (cap. E) One of two subkingdoms into which plants are divided in certain classifications. It contains the bryophytes, pteridophytes, and spermatophytes. Also called Embryobionta. Not to be confused with *Embryophytina. Individual name and adjectival form: **embryophyte** (no cap.). *Compare* Thallophyta.

Embryophytina In a recent classification of plants, a subdivision of the division *Streptophyta, containing the bryophytes, pteridophytes, and spermatophytes. *Compare* Embryophyta.

EMC virus Abbrev. and preferred form for encephalomyocarditis virus – the type species of the genus *Cardiovirus* (vernacular name: EMC virus group).

Emerson effect *Bot.* Also called enhancement effect. Named after Robert Emerson.

emf (or **e.m.f.**) Abbrev. for electromotive force.

-emia US spelling of *-aemia.

emissivity Symbol: ϵ (Greek epsilon). A dimensionless physical quantity, the ratio M/M_B, where M and M_B are the radiant exitance of a particular surface and that of a black-body radiator at the same temperature.

emitter-coupled logic (hyphenated) *Electronics* Abbrev.: ECL (no stops).

Empedocles of Acragas (*c.* 490 BC–*c.* 430 BC) Greek philosopher.

emr (or **EMR**; no stops) Abbrev. for electromagnetic radiation.

emu (or **e.m.u.**) Abbrev. for electromagnetic unit. *See* cgs units.

emulation *Computing* The imitation of all or part of one computer system by another so that both execute the same programs and produce identical results from the same input. *Compare* simulation.

en (no stop) **1.** Abbrev. for ethylenediamine often used in the formulae of coordination compounds, e.g. $[Cr(en)_3]^{3+}$. **2.** *See* en dash.

en- (**em-** before b, m, and p) Prefix denoting **1.** in, into, or inside (e.g. endemic, enthalpy, empirical). **2.** put in, surround or cover by, make or become (e.g. encode, encyst, embed).

encephalin Use *enkephalin.

encephalo- (**encephal-** before vowels) Prefix denoting the brain (e.g. encephalopathy, encephalitis).

-enchyma Noun suffix denoting cellular tissue (e.g. collenchyma, parenchyma, sclerenchyma). Adjectival form: **-enchymatous**.

Encke, Johann Franz (1791–1865) German astronomer.
Encke division
Encke's comet

en dash (or **en rule**) A dash, –, half of one em in length (*see* point). Use an en dash, rather than a hyphen, for the following:
1. To separate a span of dates, numbers, etc.; e.g. 1800–56, 12–18 leaflets.
2. To denote a minus sign.
3. To separate parenthetical text. A space precedes and follows the dash; e.g. "A discovery – of great scientific significance – was made last week." An em dash can also be used for this purpose, in which case it is usual to omit the spaces.
4. To separate entities, individuals, etc., linked in a common context; e.g. acid–base balance, Joule–Thomson effect, urea–formaldehyde resin.

endemic 1. (adjective) *Med.* Describing a disease that is habitually present in a certain locality or population. The term should be reserved for human diseases; the equivalent term for animal diseases is enzootic. *Compare* epidemic; pandemic. **2.** (adjective) *Biol.* Describing a species of plant or animal that is restricted to a certain region, where it probably originated. *Compare* exotic. **3.** (noun) An endemic disease, plant, or animal.

Enders, John Franklin (1897–1985) US microbiologist.

endo- (**end-** before vowels) Prefix denoting **1.** within or inner (e.g. endocarp, endoparasite, endoskeleton, endarch). **2.** taking or passing into (e.g. endothermic, endergonic, endosmosis). *See also* ento-. *Compare* exo-.

endo- (ital., always hyphenated) *Chem.* Prefix denoting that a substituent is oriented to be on the concave side of a molecule (often a polycyclic compound) (e.g. *endo*-2-chlorobicyclo[2.2.1]heptane). *Compare* *exo-*.

endoblast Use *endoderm.

endoderm (preferred to endoblast and entoderm) The inner germ layer of an animal embryo. Adjectival form: **endodermal** (not endodermic). *Compare* endodermis.

endodermis The cell layer between the cortex and vascular tissue of plants. Adjectival form: **endodermal**. *Compare* endoderm.

endoergic (preferred to endothermic in nuclear reactions)

end of file *Computing* Abbrev.: EOF (no stops).

Endomycetales An order of ascomycete fungi (*see* Ascomycotina) containing the family Saccharomycetaceae, which includes the most important yeasts, e.g. brewer's yeast (*Saccharomyces cerevisiae*).

endoplasmic reticulum Abbrev.: ER (no stops).

Endoprocta Use *Entoprocta.

Endopterygota (cap. E) A division of winged insects in which complete metamorphosis occurs. Also called Holometabola. Individual name and adjectival form: **endopterygote** (no cap.). *Compare* Exopterygota.

endorphin Any one of a group of polypeptides with morphine-like activity, found in the central nervous system. They include the *enkephalins.

endothermic 1. (preferred to endothermal) Denoting animals (called **endotherms**) whose body temperature is determined by heat energy generated internally. The term is often used synonymously with *homoiothermic, but this is incorrect as not all homoiotherms are necessarily endothermic. Noun form: **endothermy**. *Compare* ectothermic. **2.** Denoting a chemical reaction that takes in heat from the surroundings. In the context of a nuclear reaction 'endoergic' is preferred.

endplate (one word) The area of muscle cell membrane immediately beneath a motor nerve ending.

endplate potential Abbrev.: e.p.p. (stops; pl. e.p.ps) or EPP (no stops; pl. EPPs).

ENE (no stops) Abbrev. and preferred form for east-north-east or east-north-eastern.

-ene Noun suffix denoting a carbon-to-carbon double bond (e.g. alkene, butadiene).

energy Symbol: E A physical quantity indicating the capacity of a body or system to do *work. The *SI unit is the joule; the kilowatt hour and electronvolt may also be used.

A body or system has **potential energy** by virtue of its position; the symbols E_p, V, or Φ (Greek cap. phi) are used.

The **kinetic energy** of a body or system arises from its motion; the symbols E_k, K, or T are used.

Care must be taken not to describe a physical entity, such as light or electricity, as energy; energy is the capacity that electromagnetic radiation or an electric current have for doing work.

energy fluence *See* particle fluence.

energy flux density Another name for energy fluence. *See* particle fluence.

energy imparted Symbol: ϵ (Greek epsilon). A physical quantity associated with the energy delivered to a particular volume of matter by all the directly and indirectly ionizing particles (i.e. charged and uncharged) entering that volume. The *SI unit is the joule.

The **mean energy imparted**, symbol $\bar{\epsilon}$, is the average energy imparted to irradiated matter. It is related to the *absorbed dose, D:

$$\bar{\epsilon} = \int D\mathrm{d}m.$$

The **specific energy imparted**, symbol z, is the energy imparted to an element of irradiated matter divided by the mass, dm, of that element:

$$\epsilon = \int z\mathrm{d}m.$$

The SI unit is the gray (Gy), which has replaced the rad.

Engelmann, George (1809–84) US botanist.

Engelmann spruce (*Picea engelmannii*)

Engler, (Heinrich Gustav) Adolf (1844–1930) German botanist.

enhanced graphics adapter *Computing* Abbrev.: EGA (no stops).

Enkalon (cap. E) *See* nylon.

enkephalin (preferred to encephalin) Either of two neuropeptides, **Leu-enkephalin** and **Met-enkephalin**, differing in a single amino acid (leucine and methionine, respectively). *See* endorphin.

entero- (**enter-** before vowels) Prefix denoting the intestine (e.g. enterovirus, enterozoon, enteritis).

Enterobacter (cap. E, ital.) Type genus of the family *Enterobacteriaceae. Individual name: **enterobacter** (no cap., not ital.). *Compare* enterobacterium.

Enterobacteriaceae (cap. E) A family of Gram-negative rod-shaped bacteria. This spelling is currently valid because of long-established usage, in spite of conflicting with strict rules of nomenclature which would require the spelling Enterobacteraceae (no i). Members of the family are commonly referred to as 'the enterobacteriaceae' (no cap.); this is preferred to 'enterobacteria' (*see* enterobacterium). *Compare Enterobacter.*

enterobacterium (pl. enterobacteria) Any bacterium occurring in or associated with the intestine. The term is commonly used to mean any member of the family *Enterobacteriaceae, with the family collectively called 'the enterobacteria', but this is confusing since not all members of the family are associated with the intestine. The term 'enterobacteriaceae' is preferred for members of this family.

enterococcus (pl. enterococci) Any coccoid bacterium inhabiting the intestine. The trivial name 'enterococci' has been given to a group of *Streptococcus* species. *Enterococcus* (cap. E, ital.) has been proposed (but not approved) as a new genus to include the species *Streptococcus faecalis*, *S. faecium*, *S. avium*, and *S. gallinarum*, found in the intestine. *See Streptococcus.*

enterovirus (one word) Any member of the genus *Enterovirus* (cap. E, ital.). Human viruses of this genus were formerly allocated to the subgroups *polioviruses, *coxsackieviruses, and *echoviruses. This classification is still used, but newly described types are designated as human enteroviruses with a specific type number, e.g. human enterovirus 70. Enteroviruses of other species are designated as simian enterovirus 18, porcine enterovirus 1, etc.

enthalpy Symbol: H A physical quantity, a thermodynamic property of a system defined as:

$$U + pV,$$

where U is the *internal energy, p the pressure, and V the volume. The *SI unit is the joule. By convention, the change in H, ΔH, due to an exothermic reaction is taken to be negative. **Molar enthalpy**, symbol H_{m}, is enthalpy divided by *amount of substance; it is measured in joules per mole. The quantity **molar enthalpy change**, symbol ΔH_{m}, is meaningless unless the entity concerned is specified unambiguously, as it refers to the enthalpy change per mole of the reaction as written.

ento- Prefix denoting within. *Endo- is preferred in most cases; exceptions include *Entoprocta.

entoderm Use *endoderm.

entomo- (**entom-** before vowels) Prefix denoting insects (e.g. entomology, entomophagous, entomophily).

entomopoxvirus (one word) A *poxvirus belonging to the subfamily Entomopoxvirinae, i.e. the poxviruses of insects.

Entoprocta (cap. E; not Endoprocta) A phylum of minute aquatic animals formerly included in the phylum *Bryozoa. Individual name and adjectival form: **entoproct** (no cap.). *Compare* Ectoprocta.

entropy Symbol: S A physical quantity, a thermodynamic property indicating the amount of disorder in a system. When a small quantity of heat dQ enters a system at temperature T, the entropy of that system is increased by dQ/T, provided no irreversible change takes place. The *SI unit is joule per kelvin.

Specific entropy, symbol s, is entropy divided by mass; it is measured in J/(K kg). **Molar entropy**, symbol S_m, is entropy divided by *amount of substance; it is measured in J/(K mol).

Environmentally Sensitive Area Abbrev.: ESA (no stops).

enzootic 1. (adjective) Describing a disease of animals that is habitually present in a certain locality or population. *See also* endemic. **2.** (noun) An enzootic disease. *Compare* epizootic; panzootic.

enzymatic (preferred to enzymic)

enzyme-linked immunosorbent assay (not immunoabsorbent) Acronym and preferred form: ELISA (no stops).

enzyme nomenclature Most enzyme trivial names are based on the name of the type of reaction they catalyse and the name of the substrate or product with which they are associated. Most have the ending -ase. For example, glutamate dehydrogenase catalyses the oxidative deamination of glutamate; phosphofructokinase catalyses the phosphorylation of fructose 6-phosphate to fructose 1,6-bisphosphate. Exceptions without the -ase ending include the digestive enzymes trypsin and pepsin.

A systematic nomenclature for enzymes has been devised by the International Union of Biochemistry. Any enzyme is assigned to one of six major numbered groups according to the type of reaction it catalyses. These groups are: 1. oxidoreductases; 2. transferases; 3. hydrolases; 4. lyases; 5. isomerases; 6. ligases. Within each group there are two levels of subclasses, each designated by Arabic numbers. Lastly, each individual enzyme is given an identifying number. Thus, each enzyme has a systematic name, which incorporates its type designation (i.e. group name) and the name of its substrate(s), and a unique four-number numerical designation – the EC number (after Enzyme Commission). For example, the systematic name of glutamate dehydrogenase is L-glutamate:NAD$^+$ oxidoreductase (deaminating) and its numerical designation is EC 1.4.1.2 (stops separating numbers), i.e. 1. (oxidoreductases: major type category no. 1); 4. (acting on the $CH-NH_2$ group of donors: subclass no. 4); 1. (with NAD or NADP as acceptor: subsubclass no. 1); 2. (the systematic enzyme name). Histidine decarboxylase has the systematic name L-histidine carboxy-lyase, and the EC number EC 4.1.1.22. Because of their unwieldy nature, systematic names still tend to be used for clarification rather than as replacements for the trivial names. When an enzyme is the principal subject of a paper, chapter, etc., the EC number and full systematic name (or the reaction equation and source) should be cited initially, with the trivial name used thereafter. In other contexts, trivial names alone suffice, although the EC number and source should, ideally, be given at their first mention.

enzymic Use enzymatic.

eo- Prefix denoting early or primeval (e.g. Eocene).

Eoc Abbrev. for Eocene.

Eocene Abbrev.: Eoc (no stop). **1.** (adjective) Denoting the second epoch of the Tertiary period. **2.** (noun; preceded by 'the') The Eocene epoch.

EOF *Computing* Abbrev. for end of file.

eohippus (no cap., not ital.; pl. eohippuses, not eohippi) The earliest horse, of the genus *Hyracotherium* (originally named *Eohippus*).

eon (not aeon) The largest unit of geological time. *See* geological units.

EOP *Virology* Abbrev. for efficiency of plating.

Eötvös, Baron Roland von (1848–1919) Hungarian physicist.
Eötvös torsion balance

EPEC Abbrev. for enteropathogenic *E. coli*, comprising those strains associated with infantile diarrhoea.

Ephedra (cap. E, ital.) A genus of unusual gymnosperms in the class *Gnetopsida. In some classifications it is separated into its own order, Ephedrales (*see* Gneticae). Individual name: **ephedra** (no cap., not ital., pl. ephedras).

ephemeris (pl. ephemerides) *Astron.*

ephemeris time Abbrev.: ET (no stops).

ephemeris second A unit of time retained for special use in astronomy. From 1957 to 1967 it was the fundamental unit of time, defined as a specified fraction of the time taken by the earth to complete a particular orbit of the sun. It is as nearly equal to the SI second (measured by atomic clock) as the precision of measurement allowed in 1967.

epi- (**ep-** before vowels) Prefix denoting **1.** above or upon (e.g. epidermis, epiphyte, epitaxy). **2.** an isomer that differs only in the configuration of one chiral centre (e.g. epicholesterol is the 3α-hydroxy epimer of cholesterol).

epi- (ital., always hyphenated) *Chem.* Prefix sometimes used to denote 1,6-substitution on the naphthalene ring (e.g. *epi*-dichloronaphthalene).

epicarp Use *exocarp.

epidemic 1. (adjective) Describing a disease that affects many people in a population at the same time. The term should be reserved for human diseases; the equivalent term for animal diseases is epizootic. **2.** (noun) An epidemic disease. *Compare* endemic; pandemic.

epidermal growth factor Abbrev.: EGF (no stops).

epidermis The outer layer of the skin. Adjectival forms: **epidermal**, **epidermoid** (not epidermic).

epididymis (not epidydymis; pl. epididymides) *Anat.* Adjectival form: **epididymal**.

epinephrine US name for *adrenaline.

epiphysis (pl. epiphyses) The rounded end of a long bone. Adjectival form: **epiphyseal** (preferred to epiphysial). *Compare* diaphysis.

epizootic 1. (adjective) Describing a disease of animals that affects many individuals of a particular species or type. **2.** (noun) An epizootic disease. *See also* epidemic. *Compare* enzootic; panzootic.

EPM Abbrev. for electron-probe microanalysis.

eponyms A law, theory, theorem, hypothesis, principle, rule, formula, equation, or disease named after a person is usually preceded by the person's name followed by an apostrophe s, as in Boyle's law. When the name ends in s, an apostrophe s is usually used, as in Charles's law; however, for classical names and for certain other names in which the addition of apostrophe s sounds clumsy, an apostrophe alone is used, as in Archimedes' principle, Fajans' rules.

An effect, process, cycle, synthesis, phenomenon, piece of equipment, reagent, constant, function, number, coefficient angle, or field of study is usually preceded by the name itself or an adjectival form of the name, as in Planck constant, Cartesian coordinates, Newtonian telescope.

Eponymous anatomical parts incorporate either the name followed by apostrophe s (as in Cowper's glands) or an adjectival form of the name (as in Fallopian tubes).

When something is named after two or more people, the names are separated by an en dash, rather than a hyphen or the word 'and', as in Stefan–Boltzmann law, Haber–Bosch process; apostrophe s is not used.

Well-known scientific eponyms, together with some technical ones, are found in the dictionary either listed under the relevant biographical entries or as eponymous entries in their own right.

epoxide Any of a class of organic compounds containing a three-membered

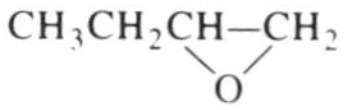

$CH_3CH_2CH—CH_2$ (epoxide ring, O bridging)

(1) 1,2-epoxybutane

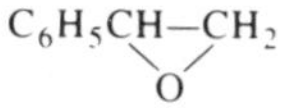

$C_6H_5CH—CH_2$ (epoxide ring, O bridging)

(2) epoxyethylbenzene
(styrene oxide)

heterocyclic ring of two carbon atoms and one oxygen atom. Also called oxirane.

Epoxides are systematically named either by adding the prefix epoxy- to the name of the parent hydrocarbon (numbers are used to indicate the bridged carbon atoms), e.g. 1,2-epoxybutane (1) and epoxyethylbenzene (2).

In nonsystematic nomenclature epoxides are named as oxides of the parent alkene, e.g. styrene oxide (2).

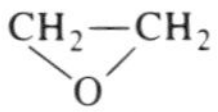

$CH_2—CH_2$ (epoxide ring, O bridging)

epoxyethane
(ethylene oxide)

epoxyethane The recommended name for the compound traditionally known as ethylene oxide.

e.p.p. (or **EPP**; pl. e.p.ps or EPPs) Abbrev. for endplate potential.

EPPS Abbrev. for 4-(2-hydroxyethyl)-1-piperazinepropanesulphonic acid.

EPR Abbrev. for electron paramagnetic resonance.

EPROM Acronym for erasable programmable read-only memory.

epsilon Greek letter, symbol: ϵ (lower case), E (cap.).

ϵ Symbol for **1.** emissivity. **2.** energy imparted. **3.** fifth brightest star in a constellation (*see* stellar nomenclature). **4.** heavy chain of IgE (*see* immunoglobulin). **5.** linear strain. **6.** molar absorption coefficient. **7.** *Astron.* obliquity of ecliptic. **8.** permittivity (ϵ_0 = permittivity of free space, also called the electric constant, ϵ_r = relative permittivity).

ϵ- (always hyphenated) Symbol used in the names of organic compounds to indicate a substituent attached to the fifth carbon atom along from the functional group (e.g. ϵ-methoxystearic acid).

e.p.s.p. (or **EPSP**; pl. e.p.s.ps or EPSPs) Abbrev. for excitatory postsynaptic potential.

Epstein–Barr virus (en dash) Abbrev.: EBV (no stops) or EB virus. Systematic name: human (gamma) herpesvirus 4. Named after M. A. Epstein (1921–) and Y. M. Barr (1932–).

EPT–76 *See* IPTS.

EQ gate *Electronics* Abbrev. for equivalence gate.

eqn. Abbrev. for equation.

Equ (no stop) *Astron.* Abbrev. for Equuleus.

equation of time Abbrev.: ET (no stops).

equi- Prefix denoting equality (e.g. equigranular, equimolecular).

equid (preferred to equine) Any mammal of the horse family, Equidae. Adjectival form: **equid**.

equilibrium constant General symbol: K A physical quantity associated with the equilibrium in a chemical reaction under specified conditions. The **standard equilibrium constant** has been defined so that it is dimensionless and is a function only of temperature; the usual symbol is $K^{\ominus}$ or K°. Several other equilibrium constants exist. These are not dimensionless and their units depend on the equilibrium under consideration.

For gaseous reactions, the equilibrium constant, K_p, is defined in terms of the equilibrium *partial pressures, p, of the reactants and products and their *stoichiometric coefficients, ν. K_p has the unit pascal, Pa, raised to the power $\Delta\nu$, where $\Delta\nu$ is the difference between the sum of ν for the products and the sum of ν for the reactants.

For reactions in solution, the equilibrium constant, K_c, is defined in terms of the equilibrium *concentrations, c, of the components and their stoichiometric coefficients. K_c has the unit mole per cubic metre, mol m^{-3} (or mole per litre, mol L^{-1}), raised to the power $\Delta\nu$.

equine Relating to or resembling horses. The term should not be used as a synonym for *equid.

equinox (pl. equinoxes) Adjectival form: **equinoctial** (not equinoxial).
The **vernal equinox**, symbol: ♈, is now called the **dynamic equinox** in astronomy and is a major point of reference. A **catalogue equinox** is an equinox adopted for the purpose of a stellar catalogue and is close to the dynamic equinox of some particular date.

Equisetum (cap. E, ital.) A genus comprising the horsetails (*see* Sphenopsida). Individual name: **equisetum** (no cap., not ital.; pl. equisetums).

equivalence gate Abbrev.: EQ gate.

Equuleus A constellation. Genitive form: Equulei. Abbrev.: Equ (no stop). *See also* stellar nomenclature.

Er Symbol for erbium.

ER Abbrev. for endoplasmic reticulum.

Erasistratus of Chios (*c.* 304 BC–*c.* 250 BC) Greek anatomist and physician.

Eratosthenes of Cyrene (*c.* 276 BC–*c.* 194 BC) Greek astronomer.
sieve of Eratosthenes

erbium Symbol: Er *See also* periodic table; nuclide.

E region *See* E layer.

erg Symbol: erg The *cgs unit of energy and work, now discouraged. In *SI units, energy and work are measured in joules: 1 erg = 10^{-7} J.

ergo- (**erg-** before vowels) Prefix denoting **1.** work or activity (e.g. ergograph). **2.** ergot (e.g. ergosterol).

ergocalciferol (preferred to calciferol) Permitted synonyms: ercalciol, vitamin D_2.

ergot A disease of cereals caused by the fungus *Claviceps purpurea*, which produces *sclerotia in the flowers of the host; the sclerotia themselves are sometimes called 'ergots'. Consumption of infected cereals causes the disease **ergotism** in humans.

Eri (no stop) *Astron.* Abbrev. for Eridanus.

Erica (cap. E, ital.) A genus of evergreen shrubs (family Ericaceae). Individual name: **erica** (no cap., not ital.; pl. ericas). *See also* heath; heather.

Ericaceae A large family of dicotyledonous plants, commonly known as the heath or heather family. There is no common name for individual members (*see* heath; heather). Adjectival form: **ericaceous**.

Ericsson, John (1803–89) US naval engineer.

Eridanus A constellation. Genitive form: Eridani. Abbrev.: Eri (no stop). *See also* stellar nomenclature.

Erlanger, Joseph (1874–1965) US neurophysiologist.

Erlenmeyer, Richard August Carl Emil (1825–1909) German chemist.
Erlenmeyer flask
Erlenmeyer rule

Ertel potential vorticity *Meteorol.*

Erwinia (cap. E, ital.) A genus of bacteria of the family *Enterobacteriaceae. It includes the species *E. herbicola*, formerly classified as *Corynebacterium beticola*. Individual name: **erwinia** (no cap., not ital.; pl. erwiniae).

erythro- (**erythr-** before vowels) Prefix denoting **1.** redness (e.g. erythrocyte, erythrin). **2.** erythrocytes (e.g. erythropoiesis).

erythro- (ital., always hyphenated) *Chem.* Prefix denoting the diastereomer that has the greatest number of similar substituents on adjacent carbon atoms in the eclipsed configuration (e.g. *erythro*-3-phenyl-1-propanol). *Compare threo-*.

Es Symbol for einsteinium.

ESA Abbrev. or acronym for **1.** European Space Agency. **2.** Environmentally Sensitive Area.

Esaki, Leo (1925–) Japanese physicist.
Esaki diode Also called tunnel diode.

ESCA Acronym for electron spectroscopy for chemical analysis.

Eschenmoser, Albert (1925–) Swiss chemist.

Escherichia coli (ital.; usually referred to in the shortened form, *E. coli*) A species of bacterium widely used in molecular genetics research. The many strains of *E. coli* can be characterized on the basis of biochemical, serological, and pathological attributes. Subdivision of the species relies principally on determining any or all of three categories of cell-surface antigens: the O antigen of the outer membrane lipopolysaccharide; the K antigen of the capsular polysaccharide; and the H or flagellar antigen. A serovar is denoted by an antigenic formula, of the form O8:K25:H9 (classes of antigens separated by colons, with no intervening spaces); the numbers designate the O, K, and H antigens carried by the isolate (i.e. O antigen number 8, K antigen number 25, etc.). When K antigens are absent from a strain the formula has the form O15:H11; strains having no H antigen have formulae of the form O8:K47:H– (an en dash denotes an absent H antigen). There is also a class of fimbrial and fibrillar antigens, the F antigens. *See also* EIEC; EPEC; ETEC.

ESE (no stops) Abbrev. and preferred form for east-south-east or east-south-eastern.

ESO Abbrev. for European Southern Observatory (in Chile).

esophagus US spelling of oesophagus.

ESPRIT Acronym for European strategic programme for research and development in information technology.

ESR Abbrev. for electron spin resonance.

essential fatty acid Abbrev.: EFA (no stops).

EST Abbrev. for Eastern Standard Time (in the USA).

ester Any of a class of organic compounds in which the hydrogen atom belonging to a hydroxyl group in an acid has been replaced by a hydrocarbon group. The acid can be organic or inorganic; esters of carboxylic acids have the general formula RCOOR′ and are known as carboxylic esters.

Esters are systematically named as salts of the acid, e.g. ethyl ethanoate, $CH_3COOCH_2CH_3$ (from ethanoic acid, CH_3COOH), methyl benzoate (or methyl benzenecarboxylate), $C_6H_5COOCH_3$, and butyl sulphate, $(CH_3CH_2CH_2CH_2)_2SO_4$.

In nonsystematic nomenclature esters are named as in systematic nomenclature except that the trivial names of the acids are used, e.g. ethyl acetate, $CH_3COOCH_2CH_3$.

estivation US spelling of aestivation.

estrogen US spelling of oestrogen.

estrus (or **estrum**) US forms of *oestrus.

esu (or **e.s.u.**) Abbrev. for electrostatic unit. *See* cgs units.

Et (no stop) Symbol often used to denote the ethyl group in chemical formulae, e.g. C_2H_5OH can be written as EtOH.

ET Abbrev. for **1.** ephemeris time. **2.** equation of time.

eta Greek letter, symbol: η (lower case), H (cap.).

η Symbol for **1.** eta particle. **2.** overpotential. **3.** viscosity coefficient.

et al. (ital.) Short for *et alii* (Latin: and others) It is used especially in *references, when there are several authors (e.g. A. B. Smith *et al.*).

etalon (no accent) A spectroscopic device to measure wavelength. *Compare* echelon. [from French *étalon*: standard of measurement]

Etard's reaction *Organic chem.*

ETEC Abbrev. for enterotoxigenic *E. coli*, the strains that produce enterotoxins and cause diarrhoea. The thermolabile toxin (LT) and the thermostable toxin (ST) are the best known.

ethanal CH_3CHO The recommended name for the compound traditionally known as acetaldehyde.

ethanal trimer The recommended name for the compound traditionally known as paraldehyde. See formula, p. 122.

ethanamide CH_3CONH_2 The recommended name for the compound traditionally known as acetamide.

ethanedial CHOCHO The recommended name for the compound traditionally known as glyoxal.

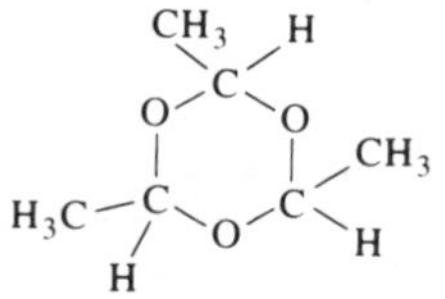

ethanal trimer
(paraldehyde)

ethanediamide $H_2NCOCONH_2$ The recommended name for the compound traditionally known as oxamide.

ethane-1,2-diamine $H_2NCH_2CH_2$-NH_2 The recommended name for the compound traditionally known as ethylenediamine.

ethanedioic acid HOOCCOOH The recommended name for the compound traditionally known as oxalic acid.

ethane-1,2-diol $HOCH_2CH_2OH$ The recommended name for the compound traditionally known as ethylene glycol.

ethanenitrile CH_3CN The recommended name for the compound traditionally known as acetonitrile or methyl cyanide.

ethanoate The recommended name for a salt of ethanoic acid. The traditional name is acetate.

ethanoic acid CH_3COOH The recommended name for the compound traditionally known as acetic acid.

ethanoic anhydride $(CH_3CO)_2O$ The recommended name for the compound traditionally known as acetic anhydride.

ethanol CH_3CH_2OH The recommended name for the compound traditionally known as ethyl alcohol.

ethanoyl- *Prefix denoting the group CH_3CO– derived from ethanoic acid (e.g. ethanoyl chloride). The traditional name is acetyl-.

ethanoylation (preferred to acetylation)

ethenol $CH_2{=}CHOH$ The recommended name for the compound traditionally known as vinyl alcohol.

ethenone CH_2CO The recommended name for the compound traditionally known as keten or ketene.

ethenyl- *Prefix denoting the group $CH_2{=}CH$– (e.g. ethenyl chloride, *N*-ethenylpyrrolidone).

ethenyl ethanoate $CH_2{=}CHOOCCH_3$ The recommended name for the compound traditionally known as vinyl acetate.

ether (not aether) *Physics*

ethers A class of organic compounds containing an oxygen atom joined directly to two alkyl or aryl groups.
Ethers are systematically named as alkoxy derivatives, the larger group being chosen as the root. Examples are methoxyethane, $CH_3OCH_2CH_3$, and ethoxybenzene, $CH_3CH_2OC_6H_5$.
In nonsystematic nomenclature ethers are named according to the two groups attached to the oxygen atom followed by the word ether, e.g. methyl ethyl ether, $CH_3OCH_2CH_3$, and ethyl phenyl ether, $CH_3CH_2OC_6H_5$.

ethoxy- Prefix denoting the ethoxyl group CH_3CH_2O– (e.g. ethoxybenzoic acid, ethoxyethanol).

ethoxybenzene $C_6H_5OC_2H_5$ The recommended name for the compound traditionally known as phenetole.

ethoxyethane $(CH_3CH_2)_2O$ The recommended name for the compound traditionally known as diethyl ether.

ethyl- *Prefix denoting the group CH_3CH_2– (e.g. ethylamine, ethyl ethanoate).

ethyl acetate $CH_3COOC_2H_5$ The traditional name for ethyl ethanoate.

ethyl acetoacetate CH_3COCH_2-$COOC_2H_5$ The traditional name for ethyl 3-oxobutanoate.

ethyl alcohol CH_3CH_2OH The traditional name for ethanol.

ethyl chloride CH_3CH_2Cl The traditional name for chloroethane.

ethyl cyanide CH_3CH_2CN The traditional name for propanenitrile.

ethylene chlorohydrin $ClCH_2CH_2OH$ The traditional name for 2-chloroethanol.

ethylenediamine $H_2NCH_2CH_2NH_2$ The traditional name for ethane-1,2-diamine. *See* en.

ethylenediaminetetraacetic acid *See* edta.

ethylene dibromide $BrCH_2CH_2Br$ The traditional name for 1,2-dibromoethane.

ethylene glycol $HOCH_2CH_2OH$ The traditional name for ethane-1,2-diol.

ethylene oxide The traditional name for *epoxyethane.

ethyl ethanoate $CH_3COOC_2H_5$ The recommended name for the compound traditionally known as ethyl acetate.

ethyl formate $HCOOCH_2CH_3$ The traditional name for ethyl methanoate.

ethylidene- *Prefix denoting the group $CH_3CH=$ (e.g. ethylidene dibromide, 1,1-dibromoethane, CH_3CHBr_2).

ethyl methanoate $HCOOCH_2CH_3$ The recommended name for the compound traditionally known as ethyl formate.

ethyl methyl ether $CH_3CH_2OCH_3$ The traditional name for methoxyethane.

ethyl methyl ketone $CH_3COCH_2CH_3$ The traditional name for butanone.

ethyl 3-oxobutanoate $CH_3COCH_2COOC_2H_5$ The recommended name for the compound traditionally known as acetoacetic ester or ethyl acetoacetate.

ethyne C_2H_2 The recommended name for the compound traditionally known as acetylene. The term acetylene is still used for nonchemical contexts (e.g. oxyacetylene welding).

etiology US spelling of aetiology.

ETSA Abbrev. for ethyl (trimethylsilyl)-acetate.

Ettinghausen effect *Electronics*

Eu Symbol for europium.

Eu (ital.) Symbol for Euler number.

eu- Prefix denoting good, true, or genuine (e.g. euchromatin, eutectic, eutrophic).

euactinomycete (one word) Another name for *sporoactinomycete.

eubacterium (pl. eubacteria) Any of the so-called 'advanced' bacteria, considered to have evolved comparatively recently. It has been proposed that these should constitute a separate kingdom, Eubacteria (cap. E, not ital.). This taxon has not been widely accepted, but the term 'eubacteria' (lower-case e) is used to refer collectively to the advanced groups without implying acceptance of the proposed taxon; it should be avoided as a trivial name for bacteria of the genus **Eubacterium*. *Compare* archaebacterium.

Eubacterium (cap. E, ital.) A genus of obligately anaerobic rod-shaped bacteria. The trivial name 'eubacteria' (*see* eubacterium) is not confined to members of this genus, and its use in this context is best avoided.

Eubrya *See* Bryidae.

eucaryote US spelling of *eukaryote.

Euclid (*c.* 330 BC–*c.* 260 BC) Greek mathematician.
Euclidean geometry
Euclidean space
Euclid's algorithm
Euclid's axioms
Euclid's proof
non-Euclidean geometry

Eudoxus of Cnidus (*c.* 400 BC–*c.* 350 BC) Greek astronomer and mathematician.

Euglenophyta A division of flagellate unicellular algae regarded by zoologists as protozoans of the class *Mastigophora (subclass Phytoflagellata or Phytomastigina). In a recent classification this division is placed in the subkingdom Protistobionta (*see* Protista). Individual name and adjectival form: **euglenoid**.

eukaryote US: **eucaryote**. Adjectival form: **eukaryotic** (US: **eucaryotic**). *Compare* prokaryote.

Euler, Leonhard (1707–83) Swiss mathematician.
Euler angles
***Euler constant**
Euler cycle
***Euler number**
Euler's criterion
Euler's equations
Euler's formula
Euler's identities

Euler's method
Euler's theorem

Euler, Ulf Svante von Usually alphabetized as *von Euler.

Euler-Chelpin, Hans Karl August Simon von (hyphen) (1873–1964) German-born Swedish biochemist.

Euler constant Symbol: γ (Greek gamma) or C A mathematical constant equal to 0.577 215 66 .... It is the limit of

$$\sum_{1}^{n} 1/r - \log_e n, \text{ as } n \rightarrow \infty.$$

Former name: Euler's constant.

Euler number Symbol: *Eu* A dimensionless quantity equal to $\Delta p/\rho v^2$, where Δp is pressure difference, ρ is density, and v is a characteristic speed. *See also* parameter.

Eumycota A division comprising the true fungi, possessing a mycelium, as distinct from the slime moulds, etc. (*see* Myxomycota). Also called Mycobionta. It contains the subdivisions *Mastigomycotina, *Zygomycotina, *Ascomycotina, *Basidiomycotina, and *Deuteromycotina. In a recent classification the true fungi, except for two classes of Mastigomycotina, are regarded as a kingdom, Mycobiota. *See also* fungi.

Euphorbia (cap. E, ital.) A genus of plants (family Euphorbiaceae) including shrubs (e.g. poinsettia) and herbs (spurges). Individual name: **euphorbia** (no cap., not ital.; pl. euphorbias).

Euphorbiaceae A large family of dicotyledonous plants, commonly known as the spurge family. There is no common name for individual members; 'spurge' is restricted to herbaceous plants of the genus **Euphorbia*. Adjectival form: **euphorbiaceous**.

Eureca *Space* Acronym for European retrievable carrier.

European Southern Observatory Abbrev.: ESO (no stops).

European Space Agency Abbrev. or acronym: ESA (no stops).

europium Symbol: Eu *See also* periodic table; nuclide.

eury- (**eur-** before vowels) Prefix denoting broad or wide-ranging (e.g. eurybenthic, euryhaline).

Eustachio, Bartolommeo (*c.* 1520–74) Italian anatomist.
Eustachian (US **eustachian**) **tube**
Eustachian (US **eustachian**) **valve**
The US forms are becoming widely used in Britain.

Eutheria (cap. E) An infraclass (*see* Theria) or subclass of mammals comprising the placental mammals. Also called Placentalia. Individual name and adjectival form: **eutherian** (no cap.; not eutherial).

EUV (or **XUV**) Abbrev. for extreme ultraviolet.

eV Symbol for electronvolt.

EV *Image technol.* Abbrev. for exposure value.

EVA Abbrev. for **1.** ethene and vinyl acetate (copolymers). **2.** extravehicular activity.

evapotranspiration (one word)

Evershed effect *Astron.*

Ewing, James (1866–1943) US pathologist.
Ewing's sarcoma

Ewing, Sir James Alfred (1855–1935) British physicist.

Ewing, William Maurice (1906–74) US oceanographer.
Ewing–Donn hypothesis (en dash)

ex- 1. Prefix denoting out of or away from (e.g. exfoliation, explantation). **2.** *See* exo-.

exa- Symbol: E A prefix to a unit of measurement that indicates 10^{18} times that unit, as in exajoule (EJ). *See also* SI units.

EXAFS Acronym for extended X-ray absorption fine-structure spectroscopy.

excitatory postsynaptic potential Abbrev.: e.p.s.p. (stops; pl. e.p.s.ps) or EPSP (no stops; pl. EPSPs).

excited state *See* electronic states.

Exner function *Meteorol.* Named after E. Exner.

exo- (**ex-** before vowels) Prefix denoting **1.** outer or outside (e.g. exocarp, exoskeleton, exarch). **2.** emitting or passing out of (exothermic, exergonic, exosmosis). *Compare* endo-.

exo- (ital., always hyphenated) *Chem.* Prefix denoting that a substituent is oriented to be on the convex side of a molecule (often a polycyclic compound) (e.g. *exo*-2-chlorobicyclo[2.2.1]heptane). *Compare* *endo-*.

exocarp (preferred to ectocarp and epicarp) The outermost layer of a fruit: forms the skin of fleshy fruits.

exoderm Use *ectoderm.

exoergic (preferred to exothermic in nuclear reactions)

exon (preferred to extron) The coding sequence of a gene. *Compare* intron.

Exopterygota (cap. E) A division of winged insects in which metamorphosis is incomplete. Also called Hemimetabola. Individual name and adjectival form: **exopterygote** (no cap.). *Compare* Endopterygota.

EXOR (or **exor**) *Computing, electronics* Short for exclusive-OR. Also written XOR or xor.

exothermic Denoting a chemical reaction that gives out heat to the surroundings. In the context of a nuclear reaction exoergic is preferred.

exotic 1. (adjective) Describing a species of plant or animal that occurs in a region to which it is not native. **2.** (noun) An exotic species. *Compare* endemic.

exp Symbol for *exponential.

experiment Abbrev.: expt. (stop).

exponential The exponential of x is written $\exp x$ (intervening space or thin space) or e^x. The superscript form is usually used for simple expressions of x; more complex expressions can be put in brackets after the function exp, e.g.

$$\exp (a_1{}^2 + a_1a_2 + a_2{}^3).$$

See also e.

exposure 1. Symbol: H A physical quantity used especially in photography, equal to the product of the *illuminance and the time, Δt (the exposure time), for which the area is exposed. The *SI unit is the lux second (lx s). Also called light exposure. **2.** Symbol: X A physical quantity relating to the electric charge of the ions (of one sign) produced by X-rays or γ-rays when stopped by unit mass of air. The *SI unit is the coulomb per kilogram, which has replaced the roentgen.

exposure value *Image technol.* Abbrev.: EV (no stops).

expt. Abbrev. for experiment.

extent of reaction *See* stoichiometric coefficients.

extero- Prefix denoting outside (e.g. exteroceptor).

extra- Prefix denoting outside or beyond (e.g. extracellular, extraembryonic, extrafloral, extragalactic).

extra-high tension (hyphenated) Abbrev.: EHT (no stops) or e.h.t. (stops).

extravehicular activity *Space* Abbrev.: EVA (no stops).

extremely high frequency Abbrev.: EHF (no stops).

extreme ultraviolet Abbrev.: EUV or XUV (no stops).

extron Use *exon.

Eyde, Samuel (1866–1940) Norwegian engineer and industrialist.
Birkeland–Eyde process (en dash)

eyespot (one word)

Eyring, Henry (1901–) Mexican-born US physical and theoretical chemist.

F

f Symbol for **1.** electron state $l = 3$ (*see* orbital angular momentum quantum number). **2.** femto-. **3.** *f-number.

f Symbol (light ital.) for **1.** *Chem.* activity coefficient, mole fraction basis (f_B = that of substance B). **2.** focal length. **3.** frequency. **4.** *Chem.* fugacity (f_B = that of substance B). **5.** function (mathematical). **6.** furanose (*see* sugars). **7.** *Genetics* repetition frequency.

F Symbol for **1.** false. **2.** farad. **3.** *filial generation. **4.** fluorine. **5.** *f-number. **6.** phenylalanine. **7.** *Astron.* *See* spectral types.
°F Symbol for degree Fahrenheit.

F

F Symbol (light ital.) for **1.** Faraday constant. **2.** force (in nonvector equations; F_m = magnetomotive force); in vector equations it is printed in bold italic (***F***).

f. (stop) Abbrev. for form (in plant taxonomy).

FAA Abbrev. for formalin acetic alcohol.

Fab (cap. F, no stops) Abbrev. and preferred form for antigen-binding fragment (*see* immunoglobulin).
F(ab′)$_2$ An immunoglobulin fragment consisting of two Fabs + a hinge fragment (Fh).

Fabaceae *See* Leguminosae. *Compare* Fagaceae.

Fabre, Jean Henri (1823–1915) French entomologist.

Fabricius ab Aquapendente, Hieronymus (1537–1619) Italian anatomist and embryologist.
bursa of Fabricius

Fabry, Charles (1867–1945) French physicist.
Fabry–Pérot interferometer (en dash) Also named after Alfred Pérot (1863–1925).

fac- (ital., always hyphenated) *Chem.* Prefix denoting facial, indicating geometric isomerization in octahedral inorganic complexes of the type MA_3B_3, where M is a metal atom and A and B are ligands, in which ligands of one type form an equilateral triangle on one of the faces (e.g. *fac*-triaminetrinitrosylcobalt(III)). *Compare mer-*.

face-centred cubic *Crystallog.* Abbrev.: f.c.c. (no caps., stops).

facies (pl. facies) **1.** *Geol.* The characteristics of a rock. **2.** *Biol.* The form of an individual or group of plants or animals. *Compare* fascia. [from Latin: appearance]

facsimile (pl. facsimiles) *See also* fax.

factor *See* coefficient.

factor VIII (Roman numeral) A blood *coagulation factor, deficiency of which results in the most common form of haemophilia. Also called antihaemophilic factor.

factor IX (Roman numeral) A blood *coagulation factor, deficiency of which results in a form of haemophilia. Also called Christmas factor.

factorial Symbol: ! (placed immediately after a number or numerical variable).
$n! = n(n-1)(n-2)(n-3)\ldots 3\cdot 2\cdot 1.$

facula (pl. faculae) *Astron.*

FAD Abbrev. and preferred form for flavin–adenine dinucleotide.

FADH$_2$ Abbrev. and preferred form for the reduced form of the coenzyme flavin–adenine dinucleotide (FAD).

faeces (pl. noun) US: **feces**. Adjectival form: **faecal** (US: **fecal**).

Faenia (cap. F, ital.) A genus of nocardioform actinomycete bacteria. Currently, the sole member is *F. rectivirgula*, the main causative agent of farmer's lung; until 1986, this species was known as *Micropolyspora faeni*.

Fagaceae A family of hardwood trees and shrubs including beech and oak. Adjectival form: **fagaceous**.

Fahrenheit, (Gabriel) Daniel (1686–1736) German physicist.
***degree Fahrenheit**
Fahrenheit temperature scale

fail-safe (adjective and verb; hyphenated)

Fajans, Kasimir (1887–1975) Polish-born US physical chemist.
Fajans' method
Fajans' rules
Fajans–Soddy laws (en dash)

Fallopian tube US: **fallopian tube** (the US form is becoming widely used in Britain). The mammalian oviduct. For nonhuman mammals, use uterine tube. Named after G. *Fallopius.

Fallopius, Gabriel (1523–62) Italian anatomist. Italian name: Gabriele Falloppio.
***Fallopian tube**

Fallot, Étienne Louis Arthur (1850–1911) French physician.
Fallot's tetralogy

fallout (noun; one word)

family A category used in biological classification (*see* taxonomy) consisting of a number of closely related genera (sometimes only one genus). Family names are usually printed in roman (not italic) type with an initial capital letter. Plant family names typically end in -aceae (e.g. Rosaceae) but eight names traditionally end in -ae (*see* Compositae; Cruciferae; Gramineae; Guttiferae; Labiatae; Leguminosae; Palmae; Umbelliferae). Animal family names end in -idae (e.g. Canidae, Pongidae). *See also* subfamily; superfamily; tribe.

Fanno flow *Fluid mech.*

FAO Abbrev. for Food and Agriculture Organization, a UN organization.

farad Symbol: F The *SI unit of electric *capacitance.

$$1\,\mathrm{F} = 1\,\mathrm{C\,V^{-1}}.$$

In practice, the milli-, micro-, nano-, and picofarad are used. Named after Michael *Faraday.

faraday An obsolete unit of charge equal to

$$9.648\,453 \times 10^4\,\mathrm{C}.$$

Compare Faraday constant.

Faraday, Michael (1791–1867) British physicist and chemist.

***farad**

***faraday**

Faraday cage

***Faraday constant**

Faraday dark space

Faraday effect

Faraday rotation

Faraday's law (or **Faraday–Neumann law**) of electromagnetic induction

Faraday's laws of electrolysis

Faraday constant Symbol: F A fundamental constant equal to

$$9.648\,4531 \times 10^4\,\mathrm{C\,mol^{-1}}.$$

It is equal to Le, where L is the Avogadro constant and e the proton charge. It is thus the charge carried by one mole of electrons or singly ionized ions.

FAS Abbrev. for Fellow of the Anthropological Society.

fascia (pl. fasciae) A sheet of connective tissue. Adjectival form: **fascial**. *Compare* facies.

fast breeder reactor (not hyphenated) Abbrev.: FBR (no stops).

fast Fourier transform (not hyphenated) Abbrev.: FFT (no stops).

fathom A unit of depth used in navigation, equal to 6 feet. As Admiralty charts are modernized, the fathom is being replaced by the metre: 1 fathom = 1.8288 m exactly.

fauna (pl. faunas; preferred to faunae) Adjectival form: **faunal**.

fax (no caps.) **1.** A device for sending or receiving facsimile transmissions. **2.** A message sent or received by such a device.

FB element Abbrev. and preferred form for foldback element, a transposable element (*see* transposon) occurring in the genome of *Drosophila melanogaster*.

FBR Abbrev. for fast breeder reactor.

Fc (cap. F, no stops) Abbrev. and preferred form for crystallizable fragment (*see* immunoglobulin).

f.c.c. *Crystallog.* Abbrev. for face-centred cubic.

FCS Abbrev. for Fellow of the Chemical Society, now *FRSC.

fd (no caps.) Type member of the *Inovirus* genus of bacteriophages (filamentous phages). It is also the name of a subgroup of the genus.

FDM Abbrev. for frequency-division multiplexing.

FDS law *Physics* Abbrev. for Fermi–Dirac–Sommerfeld law. *See* Fermi, Enrico.

Fe Symbol for iron. [from Latin *ferrum*]

featherstar (one word) *See* Crinoidea.

feces US spelling of *faeces.

Fechner, Gustav Theodor (1801–87) German physicist and psychologist.

Weber–Fechner law (en dash)

feedback (noun; one word) Verb form: **feed back** (two words).

feedthrough (noun; one word) *Electronics* Verb form: **feed through** (two words).

Fehling, Hermann Christian von (1812–85) German organic chemist.

Fehling's solution

Fehling's test

feldspar (preferred to felspar) Any of a group of rock-forming mineral aluminium silicates.

Felici balance, Felici generator *Elec. eng.*

felid (preferred to feline) Any mammal of the cat family, Felidae. Adjectival form: **felid**.

feline Relating to or resembling cats. The word should not be used as a synonym for *felid.

feline leukaemia virus Abbrev.: FeLV (lower-case e).

Felix, Arthur (1887–1956) Polish bacteriologist.
Weil–Felix reaction (en dash)

felsic *Geol.* Denoting the light-coloured minerals (such as feldspars and quartz) present in a rock; not to be confused with *femic. *Compare* mafic.

felspar Use feldspar.

FeLV (lower-case e) Abbrev. for feline leukaemia virus.

femic *Geol.* Denoting the iron- and magnesium-rich minerals, calculated by the CIPW classification; not to be confused with *felsic. *Compare* salic.

FEMO theory Abbrev. for free-electron molecular orbital theory.

femto- Symbol: f A prefix to a unit of measurement that indicates 10^{-15} times that unit, as in femtometre (fm). *See also* SI units.

fenestra cochleae (preferred to fenestra rotunda) The membrane-covered opening between the middle ear and the scala tympani of the cochlea.

fenestra ovalis Use *fenestra vestibuli.

fenestra rotunda Use *fenestra cochleae.

fenestra vestibuli (preferred to fenestra ovalis) The membrane-covered opening between the middle ear and the vestibule of the inner ear. *Compare* fenestra cochleae.

Fenton's reagent *Chem.*

-fer Noun suffix denoting bearing or containing, esp. in botany (e.g. conifer, crucifer). Adjectival form: **-ferous**.

Fermat, Pierre de (1601–65) French mathematician and physicist.
Fermat numbers
Fermat's last theorem
Fermat's principle

fermi Symbol: fm A unit of length used in nuclear physics, equal to 10^{-15} metre, i.e. one femtometre. The symbols for fermi and femtometre are identical. The femtometre should be used in preference to the fermi. Named after Enrico *Fermi.

Fermi, Enrico (1901–54) Italian–US physicist.
***fermi**
Fermi–Dirac–Sommerfeld law (en dashes)
Fermi–Dirac statistics (en dash)
Fermi level (or **energy**) Symbol: E_F
fermion
Fermi surface
fermium

fermium Symbol: Fm *See also* periodic table; nuclide.

ferns *See* Filicopsida.

-ferous Adjectival suffix denoting bearing or containing (e.g. Carboniferous, fossiliferous, piliferous). *See also* -fer.

Ferranti, Sebastian Ziano de (1864–1930) British electrical engineer.
Ferranti effect
Ferranti–Hawkins system (en dash)
Ferranti meter

Ferrel, William (1817–91) US meteorologist.
Ferrel cell

ferri- Prefix denoting **1.** iron (e.g. ferriferous). *See also* ferro-. **2.** iron in its trivalent form (e.g. ferricyanide). In strict chemical usage 'ferricyanide' would be replaced by hexacyanoferrate(III). *Compare* ferro-.

ferric *Chem.* Denoting compounds in which iron has an oxidation state of +3. The recommended system is to use oxidation numbers, e.g. ferric oxide, Fe_2O_3, has the systematic name iron(III) oxide.

ferricyanide A compound containing the ion $Fe(CN)_6^{3-}$, e.g. potassium ferricyanide, $K_3Fe(CN)_6$. The recommended name is hexacyanoferrate(III).

ferro- Prefix denoting **1.** iron (e.g. ferroelectric, ferromagnetism). *See also* ferri-. **2.** iron in its divalent form (e.g. ferrocyanide). In strict chemical usage 'ferrocyanide' would be replaced by hexacyanoferrate(II). *Compare* ferri-.

ferrocene The trivial name for *bis-(h^5-cyclopentadienyl)iron.

ferrocyanide A compound containing the ion $Fe(CN)_6^{4-}$, e.g. potassium ferrocyanide, $K_4Fe(CN)_6$. The recommended name is hexacyanoferrate(II).

ferrosoferric oxide Fe_3O_4 The traditional name for iron(II) diiron(III) oxide.

ferrous Denoting compounds in which iron has an oxidation state of +2. The recommended system is to use oxidation numbers, e.g. ferrous oxide, FeO, has the systematic name iron(II) oxide.

fertility plasmid *See* F plasmid.

Fessenden, Reginald Aubrey (1866–1932) Canadian electrical engineer.
Fessenden oscillator

FET Abbrev. or acronym (especially in combinations, such as MOSFET) for field-effect transistor.

feticide (not fetocide)

feto- (or **feti-**) Prefix denoting a fetus (e.g. fetoprotein, fetoscopy, feticide). See note at *fetus.

fetus (preferred to foetus; pl. fetuses, not feti) A mammal in the later stages of prenatal development: in humans, any stage after the beginning of the ninth week of pregnancy. *Compare* embryo. Adjectival form: **fetal**. Originally the US spelling, 'fetus' is now replacing 'foetus' in British English and is the preferred spelling in *Nature* and other scientific journals, although 'foetus' is still widely used and acceptable. Note that for words prefixed by *feto-, foeto- is not an acceptable variant.

Feulgen, Robert Joachim (1884–1955) German biochemist.
Feulgen reaction

Feynman, Richard Phillips (1918–88) US theoretical physicist.
Feynman diagram

F factor Use *F plasmid.

FFT *Computing* Abbrev. for fast Fourier transform.

FGS Abbrev. for Fellow of the Geological Society.

Fh (cap. F, no stops) Abbrev. and preferred form for hinge fragment (*see* immunoglobulin).

F horizon A surface *soil horizon consisting of partially decomposed, i.e. f(ermented), litter. Compare L horizon.

fiber US spelling of fibre.

FIBiol (or **F.I.Biol.**) Abbrev. for Fellow of the Institute of Biology. Fellows have the additional designation *CBiol (Chartered Biologist).

Fibonacci, Leonardo (*c.* 1170–*c.* 1250) Italian mathematician. Also called Leonardo of Pisa.
Fibonacci numbers
Fibonacci search

fibre US: **fiber**. Adjectival form: **fibrous**.

fibreglass (one word) US: **fiberglass**.

fibre optics (two words) Adjectival form: **fibre-optic** (hyphenated).

fibre-tracheid (hyphenated) *Bot.*

fibro- Prefix denoting fibres or fibrous tissue (e.g. fibroblast, fibrocartilage, fibroelastic).

fibula (pl. fibulae or fibulas) *Anat.* Adjectival form: **fibular**.

FICE Abbrev. for Fellow of the Institution of Civil Engineers.

FIChemE (or **F.I.Chem.E.**) Abbrev. for Fellow of the Institution of Chemical Engineers.

Fick, Adolf Eugen (1829–1901) German physiologist.
Fick's law *Physics*
Fick's principle *Med.*

-fid Adjectival suffix denoting divided into parts or lobes (e.g. bifid, pinnatifid).

FIEE Abbrev. for Fellow of the Institution of Electrical Engineers.

field-effect transistor (hyphenated) Abbrev. or acronym: FET (no stops).

FIFO (or **fifo**) *Computing* Acronym for first in first out.

FIFST Abbrev. for Fellow of the Institute of Food Science and Technology.

Fijivirus (cap. F, ital.) A genus of viruses. Individual name: **fijivirus** (no cap., not ital.). [from *Fiji* disease *virus*]

filaria (not ital.; pl. filariae) A parasitic nematode worm of any of several genera (*Brugia*, *Wuchereria*, etc.). Note that the word is not a taxon and therefore has no initial capital letter. Adjectival form: **filarial**.

file-transfer protocol (hyphenated) *Computing* Abbrev.: FTP (no stops).

fili- Prefix denoting a thread or threadlike (e.g. filiform). *See also* filo-.

filial generation *Genetics* The first filial generation is denoted F_1 (subscript number); subsequent filial generations are denoted F_2, F_3, etc. Note that F_1 (F_2, F_3, etc.) refer to generations, not individual offspring: to describe such an offspring as 'an F_1 mouse' (for example) is deprecated. *Compare* parental generation.

Filicales The largest order of ferns (*see* Filicopsida). Also called Polypodiales.

Filicopsida A class of pteridophytes comprising the ferns. Also called Pteropsida, Polypodiopsida, or Filicinae. In a recent classification the ferns are regarded as a superclass, Filicatinae, in the infradivision Cormatae (vascular plants).

filo- Prefix denoting a thread or threadlike (e.g. filoplume, filopodium). *See also* fili-.

FIMA Abbrev. for Fellow of the Institute of Mathematics and its Applications.

fimbria (pl. fimbriae) *Anat., biol.* A hairlike process, usually forming part of a fringe; the term is often (though incorrectly) used for the whole fringe. Bacterial fimbriae are usually called pili (*see* pilus). Adjectival forms: **fimbrial, fimbriate**.

FIMechE (or **F.I.Mech.E.**) Abbrev. for Fellow of the Institution of Mechanical Engineers.

FIMinE (or **F.I.Min.E.**) Abbrev. for Fellow of the Institution of Mining Engineers.

Finch, George Ingle (1888–1970) Australian physical chemist.

fine-structure constant Symbol: α (Greek alpha). A dimensionless physical constant equal to

$$7.297\,3531 \times 10^{-3}.$$

Its reciprocal, $1/\alpha$, is equal to

$$137.035\,99.$$

Finsen, Niels Ryberg (1860–1904) Danish physician.

FInstP (or **F.Inst.P.**) Abbrev. for Fellow of the Institute of Physics.

fiord (not fjord) [from Norwegian]

firn (not ital.) *Glaciol.* [from German] Also called névé.

FIS Abbrev. for Fellow of the Institute of Statisticians.

Fischer, Emil Hermann (1852–1919) German organic chemist and biochemist.
Fischer projection formula

Fischer, Ernst Otto (1918–) German inorganic chemist.

Fischer, Franz (1852–1932) German chemist.
Fischer–Tropsch process (en dash) Also named after Hans Tropsch (1839–1935).

Fischer, Hans (1881–1945) German organic chemist.

fish (pl. fishes in zoology, fish in commerce)

Fisher, Sir Ronald Aylmer (1890–1962) British statistician and geneticist.
Fisher's exact probability test
Fisher's z-distribution

fissi- (**fiss-** before vowels) Prefix denoting a splitting or cleft (e.g. fissiparous, fission, fissure).

fission 1. *Physics* The splitting of an atomic nucleus. Adjectival forms: **fissile, fissionable**. In most contexts 'fissile' and 'fissionable' are interchangeable; sometimes 'fissile' is restricted to nuclear fission resulting from the impact of slow neutrons while 'fissionable' refers to fission resulting from any process. **2.** *Biol.* A form of asexual reproduction in single-celled organisms. Adjectival form: **fissiparous**.

FIST Abbrev. for Fellow of the Institute of Science Technology.

Fitch, Val Logsdon (1923–) US physicist.

Fittig, Rudolph (1835–1910) German organic chemist.
Fittig reaction (or **synthesis**)

FitzGerald, George Francis (not Fitzgerald) (1851–1901) Irish physicist.
Lorentz–FitzGerald contraction (en dashes)

Fitzroy, Robert (1805–65) British hydrographer and meteorologist.

Fizeau, Armand Hippolyte Louis (1819–96) French physicist.
Fizeau's experiment

fjord Use fiord.

FK *Astron.* Abbrev. (German) for Fundamental Catalogue: used with a digit, as in FK5, to indicate the particular catalogue.

fl. (ital.) Abbrev. for *floruit* (Latin: flourished). Used if a person's exact dates are not known, to indicate the date or period at which he or she is believed to have been active, e.g. N. Cassegrain (*fl.* 1650–75).

flagellate 1. (adjective) Relating to or describing the protozoan class *Mastigophora (or Flagellata). **2.** (noun) A protozoan of the class Mastigophora. **3.** (adjective) Having flagella. The term 'flagellated' is usually preferred for this sense, which can apply to various cells and tissues, to avoid confusion with the Mastigophora.

flagellum (pl. flagella; not flagellae) A long hairlike locomotory appendage. Adjectival forms: **flagellar, flagellated** (*see also* flagellate).

flammable Readily combustible. Preferred to 'inflammable', which can mistakenly be thought to mean 'not flammable'.

Flamsteed, John (1646–1719) English astronomer.
Flamsteed numbers

flashlight (one word)

flashover (one word)

flashpoint (one word) *Chem.* Abbrev.: f.p. (stops).

flask fungi *See* Pyrenomycetes.

flatworm (one word) *See* Platyhelminthes.

flavin Any of a group of plant pigments, including riboflavin (vitamin B_2); the word is sometimes used as a synonym for riboflavin, especially in the following coenzymes derived from it:
flavin–adenine dinucleotide (en dash) Abbrev. and preferred form: FAD (no stops).
flavin mononucleotide Abbrev. and preferred form: FMN (no stops). Also called riboflavin phosphate.

flavoprotein (one word)

F layer (not hyphenated) *Physics* Also called F region, Appleton layer.

Fleming, Sir Alexander (1881–1955) British bacteriologist.

Fleming, Sir John Ambrose (1849–1945) British physicist and electrical engineer.
Fleming's left-hand rule
Fleming's right-hand rule

Fleming, Williamina (1857–1911) Scottish-born US astronomer.

Flemming, Walther (1843–1905) German cytologist.

Flerov, Georgii Nikolaevich (1913–) Soviet nuclear physicist.

Flexibacter (cap. F, ital.) A genus of gliding nonphotosynthetic bacteria. Individual name: **flexibacter** (no cap., not ital.). *Compare* flexibacterium.

flexibacterium (pl. flexibacteria) Any filamentous gliding nonphotosynthetic bacterium. Not to be confused with 'flexibacter' (*see Flexibacter*).

flexion (not flection) Bending of a limb.

flexor (not flexer) A muscle that bends a limb.

flip-flop (hyphenated) *Electronics*

floe Short for ice floe.

flops (or **FLOPS**) Acronym for floating-point operations per second, a measure of computer power. Hence Mflops (megaflops), Gflops (gigaflops). Note that the s in flops denotes 'per second'.

flora (pl. floras; preferred to florae) Adjectival form: **floral**.

floral formula A graphic representation of the structure of a flower. Flower parts are represented by capital-letter symbols, in the following order: K (calyx), C (corolla), P (perianth; used instead of K and C when the sepals and petals are indistinguishable), A (androecium), G (gynoecium). Each letter is followed by a number indicating the number of parts in each whorl: the symbol ∞ is used when there are more than 12 parts, and the number is bracketed when the parts are fused. If the parts are in distinct groups, or consist of both fused and separate elements, the number is split and linked by a plus sign. If two whorls are united, the respective symbols are linked by a single horizontal bracket. A superior ovary is indicated by a line beneath the G number; an inferior ovary by a line above the G number. The formula starts with the symbol ⊕ (for actinomorphic flowers) or ·|· or ↑ (for zygomorphic flowers). Some examples are as follows:
primrose (*Primula vulgaris*):
$\oplus K(5)\,\overset{\frown}{C(5)A5}\,G\underline{5}$
tulip (*Tulipa* species):
$\oplus P3+3\ \ A3+3\ \ G(\underline{3})$.

Florey, Howard Walter, Lord (1898–1968) Australian pathologist.

flori- (**flor-** before vowels) Prefix denoting flowers, flowering plants, or a flora (e.g. florigen, floristics).

Flory, Paul John (1910–86) US polymer chemist.
Flory temperature

Flourens, Jean Pierre Marie (1794–1867) French physician and anatomist.

fl oz Symbol for *fluid ounce. It should be qualified by the prefix UK or US.

FLS Abbrev. for Fellow of the Linnean Society.

fluence *See* particle fluence.

fluid ounce Symbol: fl oz A UK and US unit of volume, defined differently in the two countries and not used for scientific purposes. *See* gallon.

Fluon (cap. F) A trade name for a form of the 'nonstick' plastic *PTFE.

fluorescence (not flourescence) Adjectival form: **fluorescent**. Verb form: **fluoresce**.

fluorine Symbol: F *See also* periodic table; nuclide.

fluorine monoxide OF_2 The traditional name for oxygen difluoride.

fluoro- (**fluor-** before vowels) Prefix denoting **1.** fluorine (e.g. fluorocarbon, fluoride). **2.** fluorescence (e.g. fluorochrome, fluorescein).

fluorosilicate A compound containing the ion SiF_6^{2-}, e.g. sodium fluorosilicate, Na_2SiF_6. The recommended name is hexafluorosilicate(IV).

fluvio- (**fluv-** before vowels) Prefix denoting a river or rivers (e.g. fluvioglacial, fluvioterrestrial).

Fm Symbol for fermium.

FM (or **f.m.**) Abbrev. for frequency modulation.

FMN Abbrev. and preferred form for flavin mononucleotide.

f-number (hyphenated) *Photog.* The ratio of the focal length of a lens to its aperture. The value is usually denoted by f*n* (or f/*n*, f:*n*), where *n* is the magnitude of the ratio, as in f2.8, f8.

Fo (ital.) Symbol for Fourier number.

focal plane Abbrev.: FP (no stops).

Fock, Vladimir Alexandrovich (1898–1974) Soviet theoretical physicist.
Hartree–Fock approximation (en dash)
Hartree–Fock orbital (en dash)

focus (pl. foci; preferred to focuses) Adjectival form: **focal**. Verb form: **focuses, focusing, focused** (not -uss-).

foetus *See* fetus.

föhn (not foehn) *Meteorol.* [from German]

folacin An imprecise term for any of various compounds derived from folic acid; not recommended in scientific usage.

foldback element *Genetics See* FB element.

folic acid A vitamin of the B complex. Also called pteroylglutamic acid; sometimes referred to as vitamin Bc or vitamin M.

Folkers, Karl August (1906–   ) US organic chemist.

follicle-stimulating hormone (hyphenated) Abbrev.: FSH (no stops).

Food and Agriculture Organization Abbrev.: FAO (no stops).

foot Symbol: ft A unit of length equal to exactly 0.3048 metre. It was the fundamental unit of length in the obsolete system of *fps units but has been superseded in scientific measurements by the metre (*see* SI units).

foot-candle An obsolete unit of illumination (now called *illuminance) equal to one lumen per square foot. The *SI unit of illuminance is the lux: 1 foot-candle = 10.764 lux.

foot poundal Symbol: ft pdl A unit of *energy and *work in the obsolete *fps system of units. The *SI unit, the joule, should now be used:

1 ft pdl = 0.042 14 J.

See also poundal; foot pound-force.

foot pound-force Symbol: ft lbf A unit of *energy and *work in the obsolete *fps system of units. The SI unit, the joule, should now be used:

1 ft lbf = 1.355 82 J.

See also pound-force; foot poundal.

For (no stop) *Astron.* Abbrev. for Fornax.

foramen (pl. foramina) A natural opening in a body part, especially the **foramen magnum** (pl. foramina magna) at the base of the skull or the **foramen ovale** (pl. foramina ovalia) between the two atria in the fetal heart.

Foraminifera (cap. F) US: **Foraminiferida**. An order of mostly marine protozoans. Individual name: **foraminiferan**, often shortened to **foram** (no caps.). Adjectival forms: **foraminiferan, foraminiferal** (consisting of forams).

Forbes, Edward (1815–54) British naturalist.

f orbital *See* orbital.

force Symbol: F A *vector quantity, an external agency that accelerates a body. The *SI unit is the newton. The magnitude of the force is equal to the product ma, where m is the mass of the body and a the acceleration imparted by the force.

Ford, Edmund Brisco (1901–88) British geneticist.

fore- Prefix denoting before or at or near the front (e.g. forebrain, forelimb, foreset, forewing).

forebrain (one word) Anatomical name: prosencephalon.

foregut (one word)

forelimb (one word)

Forest, Lee De Usually alphabetized as *De Forest.

forewing (one word)

form (in plant *taxonomy) Abbrev.: f. (stop).

form- *See* formo-.

-form Adjectival suffix denoting shaped like or resembling (e.g. cruciform, vermiform).

formaldehyde HCHO The traditional name for methanal.

formalin acetic alcohol Abbrev.: FAA (no stops).

formamide $HCONH_2$ The traditional name for methanamide.

format Verb form: **formats**, **formatting**, **formatted**.

formate The traditional name for a salt of methanoic (formic) acid. The recommended name is methanoate.

Formica (cap. F) A trade name for a type of plastic laminate.

formic acid HCOOH The traditional name for methanoic acid.

formo- (often **form-** before vowels) Prefix used in trivial chemical names to denote association with methanoic (formic) acid (e.g. formosul, formamide).

formula (pl. formulae; preferred to formulas) *See also* chemical formulae. Adjectival form: **formulaic**. Verb form: **formulate**.

form. wt (preferred to FW) Abbrev. for formula weight.

formyl- *Prefix denoting the group HCO– derived from methanoic (formic) acid (e.g. formylacetone, formyl chloride).

Methanoyl- is recommended in all contexts.

formylation *Chem.* Use methanoylation.

Fornax A constellation. Genitive form: Fornacis. Abbrev.: For (no stop). *See also* stellar nomenclature.

fornix (pl. fornices) An arched anatomical part, especially of the brain.

Forssmann, Werner Theodor Otto (1904–79) German surgeon and urologist.
Forssmann antibody
Forssmann antigen

Fortin, Nicolas (1750–1831) French instrument maker.
Fortin barometer

Fortran (or **FORTRAN**) *Computing* Acronym for formula translation, a programming language.

fossa (not ital.; pl. fossae) *Anat., astron.* Used in astronomy, with an initial capital, in the approved Latin name of a long narrow shallow depression on the surface of a planet or satellite, as in Alba Fossae and Hephaestus Fossae on Mars.

Foster–Seeley discriminator (en dash) *Electronics*

Foucault, Jean Bernard Léon (1819–68) French physicist.
Foucault knife-edge test
Foucault pendulum

Fourier, (Jean Baptiste) Joseph, Baron (1768–1830) French mathematician.
Fourier analysis
Fourier–Bessel series (en dash)
***Fourier number**
Fourier optics
Fourier series
Fourier synthesis
Fourier transform

Fourier number Symbol: *Fo* A dimensionless quantity equal to $a\Delta t/l^2$, where a is the *thermal diffusivity and Δt and l a characteristic time interval and length, respectively. *See also* parameter.

Fourier-transform spectroscopy *Physics* Abbrev: FTS (no stops). *See also* FT-IR spectroscopy; FT-NMR spectroscopy.

Fourneyron, Benoît (1802–37) French engineer and inventor.

fovea (pl. foveae) *Anat., biol.* A small depression, especially the depression in the retina, surrounded by the *macula, that contains the greatest concentration of cones. The term is not synonymous with macula, though it is sometimes incorrectly used in this way.

Fowler, William Alfred (1911–) US physicist. *See also* B^2FH theory.

Fox, Sidney Walter (1912–) US biochemist.

Fox Talbot, William Henry Usually alphabetized as *Talbot.

FP Abbrev. for focal plane.

f.p. Abbrev. for **1.** freezing point. **2.** flashpoint.

F plasmid (preferred to F factor) A bacterial plasmid determining sex for purposes of conjugation. $\mathbf{F}^+$ and $\mathbf{F}^-$ (superscript plus and minus signs) denote male and female cells, respectively. *See also* Hfr. [from *f*ertility plasmid]
F′ plasmid (prime) A faulty excised F plasmid incorporating some of the host chromosome.

fps (or **f.p.s.**) **units** An obsolete system of units in which the foot, pound, and second were the fundamental units for length, mass, and time. The units of force and energy were the poundal and foot poundal. In the gravitational system of fps units, once popular in engineering, the pound-force and foot pound-force were the units of force and energy and the slug was the unit of mass. The fps system has been largely abandoned in favour of metric systems – originally cgs and mks units and now *SI units.

Fr Symbol for francium.

Fr (ital.) Symbol for Froude number.

fractions Spell out (e.g. two-thirds, three-quarters, nine-fifths, etc., hyphenated) only in nonmathematical text. In mathematical text use decimal fractions where possible (0.667, 0.75, 1.80, etc.). When appropriate, use a 'one-piece' fraction (½, ⅓, ¼, ⅔, ¾, etc.); if this is not available, use a solidus (6/7, 27/64, $^{27}/_{64}$, etc.) rather than a horizontal line between numerator and denominator.

fracto- (**fract-** before vowels) Prefix denoting broken (e.g. fractocumulus).

fraenum Use *frenum.

FRAeS (or **F.R.Ae.S.**) Abbrev. for Fellow of the Royal Aeronautical Society.

FRAgSs (or **F.R.Ag.Ss.**) Abbrev. for Fellow of the Royal Agricultural Societies.

FRAI Abbrev. for Fellow of the Royal Anthropological Institute.

francium Symbol: Fr *See also* periodic table; nuclide.

Franck, James (1882–1964) German-born US physicist.
Franck–Condon principle (en dash) Also named after Edward Condon (1902–74).
Franck–Hertz experiment (en dash)

Frank, Ilya Mikhailovich (1908–) Soviet physicist.

Frankia (cap. F, ital.) A genus of actinomycete bacteria. Individual name: **frankia** (no cap., not ital.; pl. frankiae).
The genus can be divided into two physiologically distinct groups, designated A and B, as well as into two groups on the basis of their spore-forming ability, designated Sp+ and Sp– (plus and minus signs) or P (spore positive) and N (spore negative), respectively. Because of the complexity of host interactions shown by *Frankia* strains, no species names are currently employed. Instead, strains are labelled with a three-letter code denoting the laboratory of origin, plus a two-digit code for the plant host genus, plus up to eight more digits; for example, LLR 01321, DDB 030210 (space after the three-letter code).

Frankland, Sir Edward (1825–99) British organic chemist.

Franklin, Benjamin (1706–90) US scientist, statesman, diplomat, printer, and inventor.

Franklin, Rosalind (1920–58) British X-ray crystallographer.

FRAS Abbrev. for Fellow of the Royal Astronomical Society.

Frasch, Herman (1851–1914) German-born US industrial chemist.
Frasch process

Fraunhofer, Josef von (1787–1826) German physicist and optician.
Fraunhofer diffraction
Fraunhofer lines

FRBS Abbrev. for Fellow of the Royal Botanic Society.

FRCP Abbrev. for Fellow of the Royal College of Physicians.

FRCS Abbrev. for Fellow of the Royal College of Surgeons.

FRCVS Abbrev. for Fellow of the Royal College of Veterinary Surgeons.

Fredholm, Erik Ivar (1866–1927) Swedish mathematician.
Fredholm integral equations
Fredholm operator

free-electron molecular orbital theory Abbrev.: FEMO theory.

free energy *See* Gibbs function; Helmholtz function.

freemartin (one word) A sterile female animal, usually a calf, born as a twin with a normal male.

free radical *Chem.* Use *radical.

free space *Physics* A space free from particles and fields of force. *Compare* vacuum.

freeze-dry (hyphenated; **freeze-dries, freeze-drying, freeze-dried**) Noun form: **freeze-drying**.

freezing point (two words) Abbrev.: f.p. (stops). Do not use as a synonym for *melting point.
freezing-point constant Suggested symbol (IUPAC): K_f; recommended unit: kelvin kilogram per mole (K kg mol^{-1}). Also called cryoscopic constant.

Frege, (Friedrich Ludwig) Gottlob (1848–1925) German philosopher and mathematician.

F region *See* F layer.

Frenet formulae *Maths.* Named after F. Frenet (1816–1900).

Frenkel defect *Crystallog.*

frenum (preferred to fraenum; pl. frena) A fold of tissue, especially that beneath the tongue.

Freon (cap. F) A trade name for any of a series of fluorocarbons or chlorofluorocarbons (CFCs).

frequency Symbol: *f* or ν (Greek nu). A physical quantity, the number of complete oscillations or cycles of a periodic phenomenon in unit time. It is the reciprocal of the period. The *SI unit is the hertz.

frequency-division multiplexing (hyphenated) Abbrev.: FDM (no stops).

frequency modulation Abbrev.: FM (no stops) or f.m. (stops).

FRES Abbrev. for Fellow of the Royal Entomological Society.

freshwater (adjective; one word)

fresnel A unit of frequency equal to 10^{12} hertz, i.e. one terahertz. Its use is discouraged. Named after A. J. *Fresnel.

Fresnel, Augustin Jean (1788–1827) French physicist.
***fresnel**
Fresnel–Arago laws (en dash)
Fresnel biprism
Fresnel diffraction
Fresnel double mirror
Fresnel equations
Fresnel integrals
Fresnel lens
Fresnel rhomb
Fresnel zones
Huygens–Fresnel principle (en dash)

Freud, Sigmund (1856–1939) Austrian psychoanalyst. Adjectival form: **Freudian.**

Freund, Jules Thomas (1891–1960) Hungarian-born US bacteriologist.
Freund's adjuvant

Freundlich, Herbert Max Finlay (1880–1941) German-born US physical chemist.
Freundlich isotherm

Frey-Wyssling, Albert Friedrich (hyphen) (1900–) Swiss botanist.

FRGS Abbrev. for Fellow of the Royal Geographical Society.

FRHS Abbrev. for Fellow of the Royal Horticultural Society.

FRIC Abbrev. for Fellow of the Royal Institute of Chemistry, now *FRSC.

Fricke dosimeter *Radiation chem.*

Friedel, Charles (1832–99) French chemist.
Friedel–Crafts reaction (en dash)

Friedel, Georges (1865–1933) French crystallographer.
Friedel's law

Friedman, Herbert (1916–) US physicist and astronomer.

Friedmann, Aleksandr Alexandrovich (1888–1925) Soviet astronomer.
Friedmann universe

Friedreich, Nikolaus (1825–82) German neurologist.
Friedreich's ataxia

Fries, Elias Magnus (1794–1878) Swedish mycologist.

Frisch, Karl von (1886–1982) Austrian zoologist, entomologist, and ethologist.

Frisch, Otto Robert (1904–79) Austrian-born British physicist.

Fritsch, Felix Eugen (1879–1954) British algologist.

FRMetS (or **F.R.Met.S.**) Abbrev. for Fellow of the Royal Meteorological Society.

FRMS Abbrev. for Fellow of the Royal Microscopical Society.

Frobenius, (Ferdinand) Georg (1849–1917) German mathematician.
Frobenius norm
Frobenius's theorem

Fröhlich, Alfred (1871–1953) Austrian neurologist.
Fröhlich's syndrome

Froude, William (1810–79) British engineer and naval architect.
***Froude number**
Froude's law

Froude number Symbol: *Fr* A dimensionless quantity equal to $v/\sqrt{(lg)}$, where v is a characteristic speed, l a characteristic length, and g the *acceleration of free fall. *See also* parameter.

FRPS Abbrev. for Fellow of the Royal Photographic Society.

FRS Abbrev. for Fellow of the Royal Society.

FRSC Abbrev. for Fellow of the Royal Society of Chemistry. Fellows have the additional designation *CChem (Chartered Chemist).

FRSE Abbrev. for Fellow of the Royal Society of Edinburgh.

Fru (no stop) Abbrev. for fructose.

fructi- (**fruct-** before vowels) Prefix denoting fruit (e.g. fructiferous, fructose).

fructose Abbrev.: Fru (no stop). *See* sugars.

frugivorous (not fructivorous) Fruit-eating.

fruit fly (not hyphenated) *See also Drosophila.*

frustum (not frustrum; pl. frusta) *Geom.*

FSD Abbrev. for full-scale deflection.

FSE Abbrev. for Fellow of the Society of Engineers.

FSH Abbrev. for follicle-stimulating hormone.

FSS Abbrev. for Fellow of the Royal Statistical Society.

ft Symbol for foot.

FT-IR spectroscopy Abbrev. for Fourier-transform infrared spectroscopy.

FT-NMR spectroscopy Abbrev. for Fourier-transform nuclear magnetic resonance spectroscopy.

FTP *Computing* Abbrev. for file-transfer protocol.

FTS Abbrev. for Fourier-transform spectroscopy.

Fuc (no stop) Abbrev. for fucose.

Fuchs, Leonhard (1501–66) German botanist and physician.
fuchsia or, as generic name (*see* genus), ***Fuchsia***

Fuchs, Sir Vivien Ernest (1908–) British explorer and geologist.

fuco- (**fuc-** before vowels) Prefix denoting seaweeds (e.g. fucosterol, fucoxanthin, fucin).

fucose Abbrev.: Fuc (no stop). *See* sugars.

fugacity Symbol: f_B for a substance B or, when a complicated formula is to be written, $f(B)$, as in $f(N_2O_4)$. A physical quantity that can be used in place of *partial pressure in chemical reactions involving real gases and mixtures. It is defined, for a component B, by $d\mu = RT d(\ln f_B)$, where μ is the *chemical potential of component B. The *SI unit is the pascal.

-fuge Noun suffix denoting disliking or repelling (e.g. calcifuge, centrifuge). Adjectival form: **-fugous** (e.g. calcifugous) or **-fugal** (e.g. centrifugal).

Fukui, Kenichi (1918–) Japanese theoretical and physical chemist.

fulcrum (pl. fulcrums or fulcra)

full-scale deflection (hyphenated) Abbrev.: FSD (no stops).

fumaric acid HOOCCH=CHCOOH The traditional name for *trans*-butenedioic acid.

function *Maths.* Symbol: f Function of x is written $f(x)$. *See also* mathematical symbols.

Fundamental Catalogue *Astron. See* FK.

fundamental particles *Physics* Particles that cannot be described as compound (in the present state of knowledge) and are thus regarded as fundamental constituents of matter. They are also called elementary particles. There are believed to be three kinds: *quarks, *leptons, and *gauge bosons. Each particle has a corresponding *antiparticle. The group of particles known as the *hadrons (i.e. mesons and baryons) cannot be considered fundamental and should be referred to as subatomic particles.

fungi (pl. noun, no cap.; sing. fungus) A group of eukaryotic organisms formerly regarded as plants and classified as a subdivision, Fungi (cap. F), of the division Thallophyta, but now usually placed in a separate kingdom, Fungi (or Mycota), comprising the divisions *Myxomycota (including the slime moulds) and *Eumycota (true fungi). In a recent classification the fungi are classified as the kingdom Mycobiota, but this excludes the slime moulds and two classes of Eumycota, which are placed in separate subdivisions of plants (Protistobionta and Chromobionta, respectively). The term 'higher fungi' is used for the ascomycetes and basidiomycetes; 'lower fungi' include the Mastigomycotina and Zygomycotina. For genetic nomenclature of fungi, *see* gene. Adjectival form: **fungal** (not fungous).

fungi- Prefix denoting a fungus (e.g. fungicide, fungiform).

Fungi Imperfecti *See* Deuteromycotina.

Funk, Casimir (1884–1967) Polish-born US biochemist.

furan C_4H_4O A heterocyclic oxygen-containing compound.

furlong *See* yard.

furyl- *Prefix denoting a group derived from *furan by removal of a hydrogen atom (e.g. furylacrylic acid, furyl chloride).

fuse *See also* fuze. Adjectival form: **fusible** (not fusable). Derived noun: **fusion**.

fuze US spelling of fuse, only in the sense of an igniter of an explosive charge.

FW Abbrev. for formula weight. Use form. wt.

FZS Abbrev. for Fellow of the Zoological Society.

G

g Symbol for **1.** gaseous state (*see* gas; state symbols). **2.** gerade (often as a subscript to a term symbol). **3.** gluon. **4.** (as superscript) grade. **5.** gram.

g Symbol (light ital.) for **1.** acceleration of free fall (g_n = standard value). **2.** Landé factor. **3.** statistical weight.

G Symbol for **1.** gauss. **2.** giga-. **3.** glycine. **4.** guanine. **5.** guanosine. **6.** gynoecium (in a *floral formula). **7.** *Astron. See* spectral types.

G Symbol (light ital.) for **1.** conductance. **2.** Gibbs function (G_m = molar Gibbs function). **3.** gravitational constant. **4.** shear modulus.

G_0 (subscript zero) Denoting a phase of the *cell cycle. *See* restriction point.

G_1 (subscript number) Denoting a phase of the *cell cycle.

G_2 (subscript number) Denoting a phase of the *cell cycle.

G^3 (superscript number) *Electronics* Abbrev. for gadolinium gallium garnet.

Ga Symbol for gallium.

GA Symbol for gibberellin, followed by a subscript number to denote individual gibberellins; for example, gibberellic acid is GA_3.

GABA Acronym and preferred form for γ-aminobutyric acid: a neurotransmitter. Neurones releasing GABA are described as **GABAergic**.

gabbro (pl. gabbros) *Geol.* Adjectival forms: **gabbroic**, **gabbroitic**.

Gabor, Dennis (1900–79) Hungarian-born British physicist.

Gabriel, Siegmund (1851–1924) German chemist.
Gabriel synthesis

Gadolin, Johan (1760–1852) Finnish chemist.
gadolinium

gadolinium Symbol: Gd *See also* periodic table; nuclide.

Gaede molecular pump *Vacuum tech.* Named after W. Gaede.

Gaertner, August (not Gärtner) (1848–1934) German bacteriologist.
Gaertner's bacillus Old name for *Salmonella enteritidis*.

Gaertner, Gustav (not Gärtner) (1855–1937) Austrian physician.
Gaertner's phenomenon

gage US spelling of gauge, especially in technical senses.

Gahn, Johan Gottlieb (1745–1818) Swedish chemist and mineralogist.

Gajdusek, Daniel Carleton (1923–) US virologist.

gal 1. Symbol for gallon. It should be qualified by the prefix UK or US. **2.** Symbol: Gal A *cgs unit of acceleration of free fall used in geophysics to express the earth's gravitational field. It is equal to one centimetre per second per second (cm s^{-2}). Named after *Galileo.

Gal (no stop) Abbrev. for galactose.

galacto- (**galact-** before vowels) Prefix denoting milk (e.g. galactocele, galactagogue, galactose).

galactose Abbrev.: Gal (no stop). *See* sugars.

galaxy (no cap.) Any large assembly of stars, gas, etc. Adjectival form: **galactic**.
According to the Hubble classification, elliptical galaxies are denoted by the letter E followed by a digit, 0–7, that indicates its apparent degree of flattening; E7 has the most elliptical outline.
Spiral galaxies are denoted by S and barred spirals by SB. Each has subgroups a, b, and c (denoted by Sa, SBa, Sb, etc.) following a sequence of progressively increased openness and prominence of the spiral arms and decreased importance of the nucleus; intermediate characteristics are denoted by ab and bc. Galaxies intermediate in outline between ellipticals and spirals are denoted S0 (zero).
Irregular galaxies were formerly denoted Irr I and Irr II, depending on the apparent age of the member stars. Irregulars are now denoted Irr I. The Irr IIs are examples of perturbed galaxies and have been reclassified as starburst galaxies, interacting galaxies, etc.

Galaxy (cap. G, preceded by 'the') The galaxy to which the sun belongs. Also called Milky Way System. Adjectival form: **Galactic**.

Galen (*c.* 130–*c.* 200) Greek physician. Adjectival form: **Galenic** (*compare* galenical).
Galenism

galenical (noun) A pharmaceutical preparation of plant or animal origin.

Galileo Galilei (1564–1642) Italian astronomer and physicist.
***gal**
Galilean satellites
Galilean telescope
Galilean transformation
Galileo mission
Galileo's law

Gall, Franz Joseph (1758–1828) German physician and anatomist.

gall bladder (two words) US: **gallbladder** (one word).

Galle, Johann Gottfried (1812–1910) German astronomer.

gallic acid $C_6H_3(OH)_3$ The traditional name for 3,4,5-trihydroxybenzoic acid.

gallinaceous Describing any bird of the order Galliformes, which includes domestic poultry, grouse, partridges, etc.

gallium Symbol: Ga *See also* periodic table; nuclide.

	Metric equiv. (dm^3)	
	UK	US (liq. measure)
gallon	4.546 09	3.785 41
quart	1.136 52	0.946 35
pint	0.568 26	0.473 18
fluid ounce	0.028 41	0.029 57

gallon The traditional UK and US unit of volume, defined differently in the two countries and thus not used for scientific purposes. The UK gallon, symbol gal or UKgal, has been defined since 1976 as exactly 4.546 09 cubic decimetres (dm^3). The US gallon, symbol gal or USgal, is used for liquid measure only and is equal to 231 cubic inches, 3.785 41 dm^3. (The US bushel can be used for dry measure.) Thus
1 UKgal = 1.200 95 USgal.
Submultiples of the gallon also have different values in the two countries (see table). In both countries
1 gallon = 4 quarts = 8 pints.
In the UK, 1 gallon = 160 fluid ounces; in the US, 1 gallon = 128 fluid ounces.

gallstone (one word)

Galois, Evariste (1811–32) French mathematician.
Galois field
Galois group
Galois theory

GALP Abbrev. for glyceraldehyde 3-phosphate. Use *phosphoglyceraldehyde.

Galton, Sir Francis (1822–1911) British anthropologist and explorer.
Galton's law
Galton whistle

Galvani, Luigi (1737–98) Italian anatomist and physiologist.
galvanic cell
galvanize
galvanometer

gam- *See* gamo-.

gameto- (**gamet-** before vowels) Prefix denoting gametes (e.g. gametogenesis, gametophyte, gametangium).

gamma Greek letter, symbol: γ (lower case), Γ (cap.).
γ Symbol for **1.** activity coefficient, concentration basis (γ_B = that of substance B). **2.** chemical shift. **3.** electrical conductivity. **4.** growth rate. **5.** gyromagnetic ratio. **6.** heavy chain of IgG (*see* immunoglobulin). **7.** photon. **8.** a plane angle. **9.** ratio of *specific heat capacities. **10.** shear strain. **11.** *Photog.* slope of the Hurter–Driffield curve. **12.** surface tension. **13.** third brightest star in a constellation (*see* stellar nomenclature).
γ- (always hyphenated) Symbol used in the names of organic compounds to indicate a substituent attached to the third carbon atom along from the functional group (e.g. γ-aminobutyric acid).

Gamma (cap. G, followed by the genitive form of a constellation name) Usually the third brightest star in that constellation. *See* stellar nomenclature.

gamma-aminobutyric acid Usually written **γ-aminobutyric acid**. Acronym and preferred form: *GABA (no stops).

gamma function *Maths.* Symbol: $\Gamma(x)$

gamma globulin Often written **γ globulin**. The two fractions are designated γ_1 globulin and γ_2 globulin (note subscript numbers).

gammaherpesvirus (one word) Any member of the subfamily Gammaherpesvirinae (vernacular name: lymphoproliferative virus group).

gamma iron Also written **γ-iron**.

gamma ray (two words) Also written **γ-ray** (hyphenated). Hyphenated when used adjectivally (e.g. gamma-ray astronomy, gamma-ray source).

gamo- (**gam-** before vowels) Prefix denoting union or sexual reproduction (e.g. gamophyllous, gamosepalous).

-gamous *See* -gamy.

Gamow, George (1904–68) Soviet-born US physicist.
Alpher–Bethe–Gamow theory (en dashes) Also called alpha–beta–gamma theory or $\alpha\beta\gamma$ theory.
Gamow barrier

-gamy Noun suffix denoting sexual union (e.g. allogamy, homogamy). Adjectival form: **-gamous**.

ganglion (pl. ganglia) Adjectival forms: **gangliar**, **gangliate**, **gangliform**, **ganglioid**, **ganglionic**.

GARP Acronym for global atomospheric research programme.

Garrod, Sir Archibald Edward (1857–1936) British physician.

Gartner, Herman Treschow (not Gärtner or Gaertner) (1785–1827) Danish anatomist.
Gartner's duct

gas (pl. gases) **1.** Adjectival forms: **gas**, **gaseous**. Verb form: **gasify**, **gasifying**, **gasified**. Derived noun: **gassing**.
It is recommended that the gaseous state of a substance X be denoted X(g), where X may be a chemical name or formula. The term gas is sometimes reserved for a gas at a temperature above its critical temperature. *Compare* vapour.
2. Short for gasoline, US name for petrol.

gas chromatography Abbrev.: GC (no stops). *Chem.*

gas chromatography–mass spectroscopy (en dash) *Chem.* Abbrev.: GCMS (no stops).

gas constant *See* molar gas constant.

gas-cooled (hyphenated)

gas-filled (hyphenated)

gas law *See* molar gas constant.

gas–liquid chromatography (en dash) Abbrev.: GLC (no stops). *Chem.*

gas–solid chromatography (en dash) Abbrev.: GSC (no stops). *Chem.*

Gassendi, Pierre (1592–1655) French physicist and philosopher.

Gasteromycetes (cap. G) A class of basidiomycete fungi (*see* Basidiomycotina) that includes the puffballs, earthstars, stinkhorns, bird's-nest fungi, etc. Individual name and adjectival form: **gasteromycete** (no cap.).

gasteropod, Gasteropoda Use gastropod, *Gastropoda.

gastro- (sometimes **gastr-** before vowels) Prefix denoting a stomach (e.g. gastroenteritis, gastrointestinal, gastrovascular, gastrectomy, gastritis).

Gastropoda (cap. G; not Gasteropoda) A class of molluscs containing the snails, slugs, etc. Individual name and adjectival form: **gastropod** (no cap.; not gastropodous).

Gastrotricha (cap. G) A phylum of minute aquatic invertebrates formerly regarded as a class of *aschelminths. Individual name: **gastrotrich** (no cap.). Adjectival form: **gastrotrichan**.

gastrula (pl. gastrulas or gastrulae) *Embryol.* Adjectival form: **gastrular**.

Gattermann, Ludwig (1860–1920) German organic chemist.
Gattermann reactions

gauge (not guage) *See also* gage.

gauge bosons *Physics* *Fundamental particles that mediate the interactions between particles. The *photon, symbol γ, mediates the electromagnetic interaction; the gluon, symbol g, mediates the strong interaction; and the W^+, W^-, and Z^0 particles mediate the weak interaction. The graviton is predicted as mediating the gravitational interaction. The *antiparticle of a gauge boson is the same as the particle, except for the W^+ particle, whose antiparticle is the W^-.

Gause's principle *Genetics* Named after G. F. Gause.

gauss (no cap.) Symbol: G or Gs The *cgs (electromagnetic) unit of magnetic flux density, now discouraged. In *SI units, magnetic flux density is measured in tesla: $1\ \mathrm{G} = 10^{-4}\ \mathrm{T}$. Named after K. F. *Gauss.

Gauss, Karl Friedrich (1777–1855) German mathematician.
degaussing
***gauss**
Gaussian distribution Also called normal distribution.
Gaussian eyepiece
Gaussian noise
Gaussian points
Gaussian units
Gauss's formulae
Gauss's proof
Gauss's theorem

Gay-Lussac, Joseph-Louis (hyphens) (1778–1850) French chemist and physicist.
Gay-Lussac's law Use *Charles's law.
Gay-Lussac tower

Gb Symbol for gilbert.

G-banding (cap. G, hyphenated) Abbrev. and preferred form for Giemsa banding, a technique in which characteristic banding patterns in chromosomes are revealed using Giemsa stain.

GC Abbrev. for gas chromatography.

GCD (or **g.c.d.**) Abbrev. for greatest common divisor.

GCMS Abbrev. for gas chromatography–mass spectroscopy.

Gd Symbol for gadolinium.

GDP Abbrev. and preferred form for guanosine 5′-diphosphate.

Ge Symbol for germanium.

Geer, Gerard Jacob de (1858–1943) Swedish geologist.

gegenions *Chem.* [from German: counter ions]

gegenschein *Astron.* [from German: opposing light]

Geiger, Hans Wilhelm (1882–1945) German physicist.
Geiger counter (preferred to Geiger–Müller counter)
Geiger–Marsden experiment (en dash) Also named after E. Marsden (1889–1970).

Geiger–Nuttal rule (en dash) Also named after J. M. Nuttal (1890–1958).

Geikie, Sir Archibald (1835–1924) British geologist.

Geissler, Heinrich (1814–79) German mechanic and glassblower.
Geissler pump

Gelfand, Israil Moiseyevich (1913–) Soviet mathematician and biologist.

Gellibrand, Henry (1597–1636) English mathematician and astronomer.

Gell-Mann, Murray (hyphen) (1929–) US theoretical physicist.

Gem (no stop) *Astron.* Abbrev. for Gemini.

gem- (ital., always hyphenated) *Chem.* Prefix denoting *geminal*, indicating that similar substituents are attached to the same atom (e.g. *gem*-dibromoethane, *gem*-dichlorohexafluorotrisilane). *Compare vic-*.

Gemini A constellation. Genitive form: Geminorum. Abbrev.: Gem (no stop). *See also* stellar nomenclature.
Geminids (meteor shower)

Geminivirus (cap. G, ital.) Approved name for the group of plant viruses of which maize streak virus is the type member. Individual name: **geminivirus** (no cap., not ital.).

gemma (pl. gemmae) Adjectival forms: **gemmate**, **gemmiferous**, **gemmiparous**.

gen. (stop) Abbrev. for genus.

gen- *See* geno-.

-gen Noun suffix denoting something producing or produced (e.g. androgen, antigen, carcinogen, collagen, hydrogen). Adjectival form: **-genic** (e.g. androgenic, antigenic, carcinogenic) or **-genous** (e.g. collagenous, hydrogenous).

gene For most species the full names of genes are printed in normal roman type, all lower-case, irrespective of whether the gene is dominant or recessive. However, in the fruit fly, *Drosophila melanogaster*, the names of dominant mutants are given an initial capital letter, e.g. Bar, Segregation Distorter. The same convention is applied to certain plants, such as the tomato, which has the dominant genes Curl and Lax, for example.
The symbols for genes generally comprise one or more italic letters (or other characters), usually derived from the gene name. In most species, dominant genes have an initial capital letter, while recessive genes are all lower-case. For example, in maize the symbol for the dominant gene 'hairy sheath' is *Hs* (cap. H, ital.), while that for the recessive gene 'shrunken endosperm' is *sh* (lower case, ital.); in *D. melanogaster*, the symbol for the dominant Bar-eye gene is *B*, and that for the recessive eye colour gene 'carnation' is *car*. However, certain groups of organisms, notably yeasts and humans (see below), follow slightly different rules.
When a gene has only a single locus, the same symbol applies to both gene and locus. However, when a characteristic is governed by more than one locus, italic Arabic numbers are used to distinguish the different loci. For example, in maize the symbols denoting the various loci determining the character 'glossy leaf' (symbol: *gl*) take the form *gl3*, *gl8*, *gl10*, etc. (ital., no space). In mice, rats, rabbits, hamsters, and guinea pigs the locus numbers are hyphenated; e.g. glucose phosphate isomerase locus 1 in mice is denoted *Gpi-1*.
Conventions vary for distinguishing alleles at the same locus. In *D. melanogaster* and mice such alleles are denoted by italic superscript characters suffixed to the symbol for the particular locus. For example, in *D. melanogaster* mutant alleles at the white-eye locus (*w*) are designated w^e (the eosin allele; superscript italic e), w^i (the ivory allele), etc. Similarly, mutant alleles of the albino gene (*c*) in mice are denoted c^{ch} (chinchilla), c^h (himalayan), and c^p (platinum). Normal alleles are designated by a plus sign (*see* wild type); for example, the normal allele at the white-eye locus is designated w^+. Any defective mutant may be denoted by a superscript minus sign (in this example, w^-).
Humans. Gene symbols are printed in all capital italic letters. Loci are designated by italic Arabic numbers, and alleles by italic superscript Arabic numbers. For example,

the symbol for the gene encoding proline-rich protein type H is *PRH*. There are two loci, *PRH1* (no space) and *PRH2*, with various alleles at each locus, including $PRH1^{1}$, $PRH1^{4}$, $PRH2^{1}$, etc. Exceptions to this convention include the human *oncogenes.

Yeasts and other fungi. Gene symbols for yeasts (*Saccharomyces* species) comprise three italic letters, all lower-case for recessive genes and all capitals for dominant genes; for example, *gal* is the symbol for a recessive gene involved in galactose fermentation, whereas *MAL* symbolizes a dominant gene controlling maltose fermentation. Loci are identified by an Arabic number (not ital.) immediately following the gene symbol: e.g. *ser*1, *met*2, etc. Alleles at a particular locus are also designated by an Arabic number, as a hyphenated suffix; e.g. *arg*2-6. Complementation groups are designated by a suffixed capital letter, e.g. *aro*1A, *aro*1B, etc. (no spaces). Resistance or sensitivity to a drug may be denoted by a superscript capital R or S, respectively, suffixed to the gene symbol and preceding any other characters; e.g. the dominant gene conferring resistance to copper sulphate is designated $CUP^{\mathrm{R}}1$; the corresponding sensitive recessive allele is denoted $cup^{\mathrm{S}}1$. Linkage groups are designated by Roman numerals.

In other fungi the three-letter gene notation does not apply and dominant alleles are designated by an initial capital only. In *Aspergillus nidulans* locus designations typically take the form *lacA*, *methA*, *pantoB*, *rA*, etc. (ital., including the capital letter suffix). In *Podospora anserina* locus designations include *su*-4 (Arabic numeral suffix hyphenated, not ital.), *m*, and *49*.

Bacteria. Genes are designated by italic lower-case three-letter abbreviations derived from the full name of the product of the gene (or operon) or its function. There may be an additional letter or letters (ital., usually caps.) to distinguish the various member genes of an operon or functionally related group. For example, the genes responsible for the ability of *E. coli* to utilize lactose (or other β-galactosides) include *lacZ*, *lacY*, and *lacA*. A cluster of functionally related genes may be written in the form *lacZYA* (no spaces); note that *ZYA* indicates the order of the genes on the chromosome. Defective mutants are designated by a superscript minus sign, e.g. $lacI^{-}$. Dominance of such a defective mutant is shown by an additional roman superscript d, e.g. $lacI^{-\mathrm{d}}$ (not by an initial capital, as in eukaryotes). Wild-type alleles are denoted by a superscript plus, e.g. $lacI^{+}$. When a gene has numerous mutant forms, these are simply numbered, e.g. *leuA58*, *leuB79* (all ital., no intervening space).

Viruses. Genes are given various designations (always italicized), including three-letter abbreviations derived from the full name of the product or function, single capital letters (*A*,*B*,*C*, etc.), lower-case letters suffixed by Roman numerals (e.g. *rII*, *cI*: all ital., no space), and Arabic numbers. *See also* gene products; genotype; oncogene; phenotype.

gene products These are often denoted by the abbreviations given to the corresponding *genes printed in roman (not italic) type with an initial capital letter. For example, the *lexA* gene encodes the protein LexA, and the *recA* gene encodes the protein RecA. Note, however, it is permissible to refer to 'the *lexA* protein', meaning the product of the *lexA* gene, especially when the product is not well characterized. To distinguish products of various molecular weights encoded by different but functionally and/or spatially related DNA sequences, the protein product is designated by a lower-case p followed by a number corresponding to the product's molecular weight in kilodaltons. The gene label may be added as a suffix (often superscript). Hence, $\mathrm{p}62^{\mathrm{c}-myc}$ (no spaces; superscript gene label) denotes a 62 kDa product of the c-*myc* oncogene.

generalized coordinates Symbol: $\boldsymbol{q}$ or q_i. Used in general dynamics.

generalized momenta Symbol: $\boldsymbol{p}$ or p_i. Used in general dynamics. They are related to *generalized coordinates q_i by:

$$p_i = \partial L/\partial q_i,$$

where L is the *Lagrangian function.

general relativity Abbrev.: GR (no stops).

-genesis Noun suffix denoting origin or development (e.g. gametogenesis, nucleogenesis, parthenogenesis). Adjectival form: **-genetic**.

genetics Adjectival form: **genetic** (not genetical). For genetic nomenclature and symbols, *see* cytogenetics nomenclature; gene; gene products; genotype.

-genic *See* -gen; -geny.

genitalia (pl. noun) The external reproductive organs of a male or female.

genito- Prefix denoting the reproductive organs (e.g. *genitourinary).

genitourinary (one word) Relating to the reproductive and excretory systems. The term is synonymous with *urinogenital, but the two words tend to be used in different contexts, e.g. genitourinary medicine, urinogenital system.

geno- (**gen-** before vowels) Prefix denoting heredity or genes (e.g. genotype, genome).

genome (not genom) The basic haploid set of chromosomes. In plants, in which polyploidy is common, the basic number of chromosomes in a genome is designated as *x* (italic type). Thus in diploid plants $2n = 2x$, in tetraploid plants $2n = 4x$, and in hexaploid plants $2n = 6x$.

genotype The genetic constitution of an organism, i.e. its complement of alleles (*see* gene). A genotype consisting of a small number of genes is designated by the symbols for the relevant alleles in the form of a genotypic formula. For example, in the case of a single gene existing in the allelic forms *A* (dominant) and *a* (recessive), the genotype of a diploid organism may be written *A*/*A* (solidus) for a homozygous dominant, *a*/*a* for a homozygous recessive, and *A*/*a* for a heterozygote. A triploid organism homozygous for *A* is designated *A*/*A*/*A*, and so on for other polyploid states.

In the case of two genes on the same chromosome in a diploid organism, the double homozygote for the dominant alleles *A* and *B* is expressed as *A A*/*B B* (spaces separating symbols denote alleles on the same chromosome) and the double heterozygote as *A B*/*a b* or *A b*/*a B*. The double dominant haploid gamete is represented as *A B* (the solidus form is not used for haploids). Alleles on nonhomologous chromosomes are separated by semicolons and spaces; e.g. *A b*/*a B*; *Y z*/*y Z* indicates that *Y* and *Z* are on a different chromosome from *A* and *B*. When multiple loci are involved, and especially when the symbols for the alleles consist of several characters, the allele labels must be spaced to avoid confusion.

In prokaryotes the genes are usually in the haploid state and are situated on a single chromosome (i.e. are linked). Genotypes are written in the form $lacY^+$ $lacZ^-$ (space). Where particular genes are in the diploid state (e.g. due to the presence in the cell of a plasmid), the genotype is given in the form F′ $lacY^-$ $lacZ^+$/$lacY^+$ $lacZ^-$ (spaces, solidus), indicating that the $lacY^-$ and $lacZ^+$ alleles are carried by an *F′ plasmid. *See also* gene products; phenotype; transposon.

-genous *See* -gen.

genu (pl. genua) The knee or any kneelike anatomical part. Adjectival form: **genual**.

genus (pl. genera) Abbrev.: gen. (stop). A category used in biological classification (*see* taxonomy) consisting of a number of closely related species (occasionally only one species). Names of genera are printed in italics with an initial capital letter (e.g. *Rosa*, *Corvus*, *Corynebacterium*); when used as a common name for individuals of the genus (which is usual practice for cultivated plants and bacteria), the name is printed in roman type (especially for the plural form) usually with no initial capital letter (e.g. acer, fuchsia, salmonella). Note that species are referred to as belonging to (for example) 'the genus *Rosa*', not 'the *Rosa* genus'. *See also* binomial nomenclature. Adjectival form: **generic**.

-geny Noun suffix denoting origin or way of development (e.g. ontogeny, orogeny, phylogeny). Adjectival form: **-genic**.

geo- Prefix denoting **1.** the planet earth (e.g. geo-isotherm (hyphenated), geomagnetism, geomorphology, geothermal). **2.** ground or gravity (e.g. geophyte, geotropism).

Geodermatophilus (cap. G, ital.) A genus of actinomycete bacteria. Individual name: **geodermatophilus** (no cap., not ital.; pl. geodermatophili).

Geoffroy Saint-Hilaire, Étienne (hyphen) (1772–1844) French naturalist.
Geoffroyism
Geoffroy's cat (*Felis geffroyi*)

geological Preferred form in British English; **geologic** is the US preferred form.

geological units Any of the divisions used in geological time scales. There are two sets of units, listed below in descending order of size. The time-stratigraphic units are each defined by a specific rock formation formed during a particular time interval in a type area. The time units correspond exactly to the equivalent time-stratigraphic units. For example, the Devonian system is a time-stratigraphic division named after a particular rock formation in Devon, England. It gives its name to the equivalent division of geological time, the Devonian period. Named units of the higher orders (e.g. Cambrian, Ordovician, etc.) are entries in the dictionary, and the entire geological time scale is shown in Appendix 4.

Time-stratigraphic units	Time units
eonothem	eon
erathem	era
system	period
series	epoch
stage	age

geometry Abbrev.: geom. (stop). Adjectival forms: **geometric, geometrical**.

gerade Abbrev.: g (no stop). The parity of an orbital is defined by a symmetry operation: inversion through a centre. An orbital is described as gerade (German: even) if after inversion the sign of the wave function remains the same. *See also* ungerade.

geranium (pl. geraniums) Any plant of the genus *Pelargonium* (not **Geranium*) that is cultivated for ornament. *See also* pelargonium.

Geranium (cap. G, ital.) A genus of plants that includes cranesbill and herb Robert; not to be confused with horticultural *geraniums of the related genus *Pelargonium*.

Gerard, John (1545–1612) British herbalist.

Gerhardt, Charles Frédéric (1816–56) French chemist.

germanium Symbol: Ge *See also* periodic table; nuclide.

German measles (cap. G) Medical name: rubella.

Germer, Lester Halbert (1896–1971) US physicist.
Davisson–Germer experiment (en dash)

Gesner, Conrad (1516–65) Swiss naturalist, encyclopedist, and physician.

GH Abbrev. for *growth hormone, prefixed by a lower-case letter (or letters) to denote the source; e.g. hGH (human growth hormone).

GHA Abbrev. for Greenwich hour angle.

G horizon A predominantly g(leyed) *soil horizon. 'G' may also be used as a descriptive suffix to mean strongly gleyed.

Giaever, Ivar (1929–) Norwegian-born US physicist.

giant *Astron.* A large star in luminosity classes II (bright giant) or III; a class IV star is a subgiant.

giant molecular cloud *Astron.* Abbrev.: GMC (no stops).

Giauque, William Francis (1895–1982) Canadian-born US physical chemist.

gibberellic acid Symbol: GA_3. The best known *gibberellin.

gibberellin Any of a class of plant growth substances. Individual gibberellins are designated GA followed by a subscript number, i.e. GA_1, GA_2, etc. *See also* gibberellic acid.

Gibbs, Josiah Willard (1839–1903) US mathematician and theoretical physicist.
Gibbs–Duhem equation (en dash)
***Gibbs function**
Gibbs–Helmholtz equation (en dash)
Gibbsian ensembles
Gibbs phase rule (or simply **phase rule**)

Gibbs function (preferred to Gibbs free energy) Symbol: G A thermodynamic property of a system defined as:

$$H - TS,$$

where H is the *enthalpy, T the *thermodynamic temperature, and S the *entropy. The *SI unit is the joule.

Molar Gibbs function, symbol G_m, is Gibbs function divided by *amount of substance; it is measured in joules per mole. *Compare* Helmholtz function.

Giemsa, Gustav (1867–1948) German chemist and bacteriologist.

Giemsa banding *Genetics* Abbrev. and preferred form: *G-banding (hyphenated).

Giemsa stain

giga- Symbol: G **1.** A prefix to a unit of measurement that indicates 10^9 times that unit, as in gigahertz (GHz) or giga-electronvolt (GeV). *See also* SI units. **2.** A prefix used in computing to indicate a multiple of 2^{30} (i.e. 1 073 741 824), as in gigabyte.

GIGO (or **gigo**) *Computing* Acronym for garbage in garbage out.

gilbert (no cap.) Symbol: Gb An obsolete *cgs (electromagnetic) unit of *magnetomotive force, equal to $10/4\pi$ ampere-turns. It is no longer used. Named after William *Gilbert.

Gilbert, Grove Karl (1843–1918) US geologist.

Gilbert, Walter (1932–) US molecular biologist.

Gilbert, William (1544–1603) English physicist and physician.

***gilbert**

Gilchrist, Percy (1851–1935) British chemist.

Thomas–Gilchrist process (en dash) Use Thomas process.

Gill, Sir David (1843–1914) British astronomer.

gingiva (pl. gingivae) The epithelium in the oral cavity that covers the jawbones and surrounds the teeth; the gum. Adjectival form: **gingival**.

ginkgo (not gingko; pl. ginkgos) A gymnosperm tree, commonly known as maidenhair tree, the sole species of the genus *Ginkgo* (cap. G, ital.) and order Ginkgoales in the class *Coniferopsida. In a recent classification it is separated into its own class, Ginkgoopsida (double o; *see* Gymnospermae).

Ginzburg, Vitaly Lazarevich (1916–) Soviet physicist and astrophysicist.

Ginzburg–Landau theory (en dash)

Giorgi, Giovanni (1871–1950) Italian physicist.

Giorgi system *See* mks system.

Giraldès, Joaquim Pedro Casado (1808–75) Portuguese-French surgeon.

Giraldès' organ Now called paradidymis.

GKS *Computing* Abbrev. for graphics kernel system.

glacier Adjectival forms: **glacial, glaciated.** Derived noun: **glaciation**.

gladiolus (pl. gladioli; not gladioluses) Generic name: *Gladiolus* (cap. G, ital.).

Gladstone–Dale law (en dash) *Physics*

Glaser, Donald Arthur (1926–) US physicist.

Glashow, Sheldon Lee (1932–) US physicist.

Glauber, Johann Rudolf (1604–68) German chemist.

***Glauber's salt**

Glauber's salt $Na_2SO_4{\cdot}10H_2O$ The traditional name for hydrated sodium sulphate.

Glazebrook, Sir Richard Tetley (1854–1935) British physicist.

Glc (no stop) Abbrev. for glucose.

GLC Abbrev. for gas–liquid chromatography.

Gleditsch, J. G. (1714–86) German botanist.

gleditsia or, as generic name (*see* genus), ***Gleditsia*** (not gleditschia, *Gleditschia*)

gley A type of soil. [from Ukrainian *glei*: clay]

glia (preferred to neuroglia) Specialized connective tissue in the central nervous system. Adjectival form: **glial**.

Glisson, Francis (1597–1677) English physician.

Glisson's capsule

Gln (or **gln**; no stop) Abbrev. for glutamine. *See* amino acid.

***Gloeocapsa* group** (cap. G, ital.) A provisional assemblage of strains of *cyanobacteria previously assigned to the botanical genera *Gloeocapsa* and *Chroococcus*.

glomerulus (pl. glomeruli) A knot of capillaries within a Bowman's capsule in the vertebrate kidney. Adjectival form: **glomerular**.

glomus (pl. glomera, not glomi) A small body providing a direct communication between a tiny artery and a vein.

glosso- (**gloss-** before vowels) Prefix denoting a tongue (e.g. glossopharyngeal, glossectomy).

Glu (or **glu**; no stop) Abbrev. for glutamic acid. *See* amino acid.

gluco- (**gluc-** before vowels) Prefix denoting glucose (e.g. glucocorticoid, gluconeogenesis, glucose). *Compare* glyco-.

glucose Abbrev.: Glc (no stop; preferred to G). (+)-glucose is the recommended name for the compound traditionally known as dextrose. *See* sugars.

glucuronic acid (preferred to glycuronic acid)

gluon *See* gauge bosons.

glutamic acid Abbrev.: Glu or glu (no stop). Symbol: E Recommended name: 2-aminopentanedioic acid. *See* amino acid.

glutamine Abbrev.: Gln or gln (no stop). Symbol: Q *See* amino acid.

gluteus (not glutaeus; pl. glutei) *Anat.* Adjectival form: **gluteal**.

Gly (or **gly**; no stop) Abbrev. for glycine. *See* amino acid.

glyceraldehyde 3-phosphate Abbrev.: GALP (no stops). Use *phosphoglyceraldehyde.

glycerine US: **glycerin.** Use *glycerol.

glycerol (preferred to glycerine) $CH_2OHCH(OH)CH_2OH$ Recommended name: propane-1,2,3-triol.

glyceryl- *Prefix denoting a group derived from glycerol by removing one or more hydrogen atoms from hydroxyl groups, especially the group $CH_2OCHOCH_2O$ formed by loss of all three hydrogen atoms.

glycine Abbrev.: Gly or gly (no stop). Symbol: G Recommended name: aminoethanoic acid. *See* amino acid.

glyco- Prefix denoting carbohydrate (e.g. glycogen, glycolipid, glycolysis, glycoprotein). *Compare* gluco-.

glycol The traditional name for a diol, in particular ethylene glycol (ethane-1,2-diol).

glycollic acid $CH_2OHCOOH$ The traditional name for hydroxyethanoic acid.

glycoside A compound consisting of a pyranose sugar linked to a nonsugar molecule. Many glycosides are **glucosides,** i.e. the sugar is glucose, but the terms should not be used synonymously. Adjectival form: **glycosidic.**

glycuronic acid Use glucuronic acid.

glyoxal CHOCHO The traditional name for ethanedial.

glyoxalic acid CHOCOOH The traditional name for oxoethanoic acid.

glyoxylate (preferred to glyoxalate)

gm Do not use as symbol for gram; the correct symbol is g. The symbol g m (intervening space) is the symbol for gram metre.

gm^2 (properly g/m^2 or g m^{-2}) Abbrev. for grams per square metre, used in measuring the thickness ('weight') of paper and card. Preferred form is *gsm.

GMC *Astron.* Abbrev. for giant molecular cloud.

Gmelin, Leopold (1788–1853) German chemist.

GMP Abbrev. and preferred form for guanosine 5′-phosphate (guanosine monophosphate).

GMST Abbrev. for Greenwich Mean Sidereal Time.

GMT Abbrev. for Greenwich Mean Time.

gnatho- Prefix denoting the jaw (e.g. gnathobase, gnathostome).

Gnathostomata (cap. G) A superclass comprising all vertebrates with jaws, i.e. fishes, amphibians, reptiles, birds, and mammals. Individual name: **gnathostome**

(no cap.). Adjectival form: **gnathostomatous.** *Compare* Agnatha.

-gnathous Adjectival suffix denoting the jaw (e.g. prognathous). Noun form: **-gnathism.**

gneiss *Geol.* Adjectival forms: **gneissic, gneissoid, gneissose.**

Gnetales An order of gymnosperms. In some classifications it includes the three genera *Ephedra*, *Gnetum*, and *Welwitschia* (*see* Gnetopsida); in others it includes only *Gnetum* (*see* Gneticae).

Gneticae A subdivision of the *Pinophytae containing the orders Ephedrales, Welwitschiales, and Gnetales. Individual name: **gnetophyte.**

Gnetopsida A class of gymnosperms containing the genera *Ephedra*, *Gnetum*, and *Welwitschia*. In some classifications it is reduced to the status of an order, Gnetales; in others it is regarded as a subdivision, Gneticae, of the division *Pinophyta.

gnotobiotics The study of organisms (**gnotobiotes**) or conditions that are either germ-free or are associated with a known microorganism. Adjectival form: **gnotobiotic.**

goat pox (two words) *See also* poxvirus.

Goddard, Robert Hutchings (1882–1945) US physicist.

Gödel, Kurt (1906–78) Austrian-born US mathematician.
Gödel numbering
Gödel's incompleteness theorems
Gödel's proof

Godwin, Sir Harry (1901–85) British botanist.

Goeppert-Mayer, Maria (hyphen) (1906–72) German-born US physicist.

Golay cell *Physics*

gold Symbol: Au *See also* periodic table; nuclide.

Gold, Thomas (1920–) Austrian-born US astronomer.

Goldbach, Christian (1690–1764) German mathematician.
Goldbach's conjecture

golden-brown algae *See* Chrysophyta.

Goldhaber, Maurice (1911–) Austrian-born US physicist.
Goldhaber triangle

Goldschmidt, Johann (Hans) Wilhelm (1861–1923) German chemist.
Goldschmidt radius
Goldschmidt reaction

Goldschmidt, Richard Benedict (1878–1958) US geneticist.

Goldschmidt, Victor Moritz (1888–1947) Swiss-born Norwegian geochemist.

Goldstein, Joseph Leonard (1940–) US physician and geneticist.

Golgi, Camillo (1843–1926) Italian cytologist and histologist.
Golgi apparatus (or **body**)
Golgi cell
Golgi tendon organ Use tendon organ.

gon *See* grade.

gon- *See* goni-; gono-.

-gon Noun suffix denoting angles of a geometric figure (e.g. hexagon, polygon). Adjectival form: **-gonal.**

gonadotrophin US: **gonadotropin.** Also called gonadotrophic hormone (US: gonadotropic hormone).
chorionic gonadotrophin Abbrev.: CG, prefixed by a lower-case letter to indicate the source; e.g. hCG (human chorionic gonadotrophin).

Gondwanaland (or **Gondwana**) *Geol.*

goni- (**gon-** before vowels) Prefix denoting an angle (e.g. goniometer, gonion).

-gonium Noun suffix denoting a reproductive cell (e.g. archegonium, oogonium).

gono- (**gon-** before vowels) Prefix denoting reproduction (e.g. gonocyte, gonoduct, gonophore, gonad).

gonococcus (not ital.; pl. gonococci) Common name for the bacterium (*Neisseria gonorrhoea*) responsible for gonorrhoea. Adjectival form: **gonococcal.**

gonorrhoea US: **gonorrhea.** Adjectival form: **gonorrhoeal** (US: **gonorrheal**).

Goode's interrupted homolosine projection *Cartog.* Named after Paul Goode.

Goodpasture, Ernest William (1886–1960) US pathologist.
Goodpasture syndrome

Goodrich, Edwin Stephen (1868–1946) British zoologist.

Goodricke, John (1764–86) Dutch-born British astronomer.

Gorer, Peter Alfred (1907–61) British immunologist.

Gossage, William (1799–1877) British chemist.

Goudsmit, Samuel Abraham (1902–78) Dutch-born US physicist.

Gould, Benjamin Apthorp (1824–96) US astronomer.
Gould Belt

Gouy balance, Gouy layer *Physical chem.* Named after G. Gouy.

gr UK symbol for grain. It should not be used as a symbol for gram.

Gr (ital.) Symbol for Grashof number.

GR Abbrev. for general relativity.

Graaf, Regnier de (1641–73) Dutch anatomist.
Graafian (US graafian) follicle

Graaff, Robert Jemison Van de Usually alphabetized as *Van de Graaff.

grad Symbol for gradient. *See* vector.

grade Symbol: ^{g} (superscript) A unit of angle used in some European countries, equal to 1/100 of a right angle, i.e. $\pi/200$ radian. Also called gon.

-grade Adjectival suffix denoting type of movement (e.g. plantigrade, retrograde, unguligrade).

graft hybrid A plant *chimera. In *binomial nomenclature, a plus sign (+) preceding the binomial denotes that the plant is a graft hybrid, e.g. +*Laburnocytisus adami*.

Graham, Thomas (1805–69) Scottish chemist.
Graham's law
Graham's salt

grain UK symbol: gr A unit of mass that is a submultiple of the pound (1 lb = 7000 grains) and is equal to exactly 0.064 798 91 gram. The grain has the same value in avoirdupois, troy, and apothecaries' units. 5760 grains = 1 pound troy.

gram (not gramme) Symbol: g (not g., gm, or gr). The fundamental unit of mass in the system of *cgs units. In *SI units, the kilogram, equal to 1000 grams, is the base unit of mass but the names of multiples and submultiples are formed using the gram rather than the kilogram (e.g. milligram not microkilogram). One gram = 0.035 27 oz, $2.204\ 62 \times 10^{-3}$ lb. *See also* gsm.

-gram Noun suffix denoting something recorded (e.g. hologram, spectrogram).

Gram, Hans Christian Joachim (1853–1938) Danish bacteriologist.
Gram-negative (US: **gram-negative**)
Gram-positive (US: **gram-positive**)
Gram's stain

gram-atom A *cgs unit equal to the *relative atomic mass of a chemical element in grams. The gram-atom has been superseded by the mole.

Gramineae (not Graminae) A family of monocotyledonous plants, commonly known as the grass family. Alternative name: Poaceae. Individual members are known as *grasses. The adjectival form, **graminaceous**, is rarely used.

gramme Use *gram.

gram-molecule A *cgs unit equal to the *relative molecular mass of a chemical compound in grams. The gram-molecule has been superseded by the mole.

Gram-negative, Gram-positive (cap. G, hyphenated) US: **gram-negative, gram-positive**. *Bacteriol.* Named after H. C. J. *Gram.

gram-weight Symbol: gm wt An obsolete unit of force in the *cgs system of units, sometimes used instead of the dyne: 1 gm wt = g dynes; g is the *acceleration of free fall (in cm s^{-2}) and, like the gram-weight, varies with locality. *Compare* kilogram-force.

grana pl. of *granum.

grand unified theory *Physics, astron.* Acronym: GUT (no stops).

granite Adjectival form: **granitic**. Derived noun: **granitization**.

grano- Prefix denoting granite (e.g. granoblastic, granodiorite).

granum (pl. grana; usually referred to in the pl.) *Bot.* A stack of thylakoids (fluid-filled sacs) within a chloroplast. Adjectival form: **granal**.

-graph Noun suffix denoting **1.** a recording instrument (e.g. barograph, spectrograph). **2.** something recorded (e.g. micrograph, photograph). Adjectival form: **-graphic**.

graphic 1. (or **graphical**) Denoting written or printed characters or marks (e.g. graphical symbols). **2.** Denoting, using, or determined by a graph or other drawing (e.g. graphic representation, graphic formula). In some contexts **graphical** is more common (e.g. graphical information, graphical display unit, graphical method). **3.** *Geol.* Denoting a texture of rocks resembling writing on their exposed surfaces (e.g. graphic granite). *See also* graphics.

graphical symbols *See* electronics, graphical symbols.

graphics Pictorial information produced electronically, photographically, etc. (e.g. computer graphics, raster graphics). Also used adjectivally (e.g. graphics adaptor). *See also* graphic.

graphics kernel system Abbrev.: GKS (no stops).

graphite *See* carbon.

graphs The axes of graphs should always be labelled. To avoid ambiguity the label should be in the form:

physical quantity/unit

Examples of such labels, which may be written out in full or in symbols, are

current/microamps ($I/\mu A$)
energy/joules (E/J).

The same applies to headings in *tables.

-graphy Noun suffix denoting **1.** a descriptive or representational science or field (e.g. cartography, geography, oceanography). **2.** a recording technique (e.g. holography, lithography). Adjectival form: **-graphic**.

Grashof number Symbol: *Gr* A dimensionless quantity equal to $l^3 g\gamma\Delta T/\nu^2$, where l is a characteristic length, g is the *acceleration of free fall, γ the cubic expansion coefficient (*see* linear expansion coefficient), ΔT a characteristic temperature difference, and ν the kinematic *viscosity. *See also* parameter. Named after Franz Grashof (1826–93).

grass Any plant of the family *Gramineae, which includes cereal plants. Note that 'grass of Parnassus' (*Parnassia palustris*), belonging to the family Parnassiaceae, is not a true grass.

Grassi, Giovanni Battista (1854–1925) Italian zoologist.

Grassmann, Herman Günther (1809–77) German mathematician.

gravi- Prefix denoting gravity or gravitation (e.g. gravimetric, gravipause).

gravitational constant Symbol: G A fundamental constant equal to

$$6.672\,59 \times 10^{-11}\,\mathrm{N\,m^2\,kg^{-2}}.$$

Also called constant of gravitation.

gray (no cap.) **1.** Symbol: Gy The *SI unit of *absorbed dose of ionizing radiation and of specific *energy imparted by radiation.

$$1\,\mathrm{Gy} = 1\,\mathrm{J\,kg^{-1}}.$$

The gray has recently replaced the rad: one rad is equal to 10^{-2} Gy. Named after L. H. Gray (1905–65). *Compare* sievert.
2. US spelling of grey.

Gray, Asa (1810–88) US botanist.

greater than Symbol: > The symbol ⩾ denotes 'greater than or equal to'. The symbol ≫ denotes 'much greater than'. These symbols usually have a space or thin space on either side, as in $x > y$.

greatest common divisor Abbrev.: GCD (no stops) or g.c.d. (stops).

Great Red Spot *Astron.* Abbrev.: GRS (no stops).

Greek alphabet Most letters of the Greek alphabet are used as scientific or mathematical symbols. There is an entry in this dictionary for each letter of the Greek alphabet (*see* alpha; beta; gamma; etc.) giving its symbol (α, β, γ, etc.) and indicating its use in scientific and mathematical contexts. The entire Greek alphabet is listed in Appendix 8.

Green, George (1793–1841) British mathematician.

Green's theorem

green algae *See* Chlorophyta.

Greenstein, Jesse Leonard (1909–) US astronomer.

Greenwich hour angle Abbrev.: GHA (no stops).

Greenwich Mean Time Abbrev.: GMT (no stops).

Greenwich Mean Sidereal Time Abbrev.: GMST (no stops).

Gregorian calendar Instituted by Pope Gregory XIII (1502–85). *Compare* Julian calendar.

Gregory, James (1638–75) Scottish mathematician and astronomer.

Gregorian telescope

Gregory formula

grenz rays (not Grenz) *Physics, radiol.* [from German *Grenze*: boundary]

Grévy's zebra (*Equus grevyi*)

grey US: **gray**. The unit is spelt *gray.

greywacke US: **graywacke**. *Geol.* [from German *Grauwacke*: grey rock]

Griffith, Frederick (1881–1941) British microbiologist.

Griffith's typing

Grignard, François Auguste Victor (1871–1935) French chemist.

Grignard reaction

***Grignard reagent**

Grignard reagent Any of a class of organometallic compounds of general formula RMgX, where R is any hydrocarbon group and X is any halogen atom. Named after F. A. V. *Grignard.

Grignard reagents are systematically named by prefixing the name of the magnesium halide by the name of the hydrocarbon group, e.g. dimethylethylmagnesium bromide, $(CH_3)_3CMgBr$, and 2-methylpropylmagnesium chloride, $(CH_3)_2$-$CHCH_2MgCl$.

In nonsystematic nomenclature Grignard reagents are named as in systematic nomenclature but the trivial name of the hydrocarbon group is used, e.g. *t*-butylmagnesium bromide, $(CH_3)_3$-$CMgBr$.

Grimaldi, Francesco Maria (1618–63) Italian physicist.

Gros Clark, Sir Wilfrid Edward Le Usually alphabetized as *Le Gros Clark.

ground *Elec. eng., etc.* US name for earth.

ground state *See* electronic states.

Grove, Sir William Robert (1811–96) British physicist.

Grove cell

growth hormone Abbrev.: *GH (no stops). The term is synonymous with *somatotrophin, but the two terms tend to be used in different contexts: growth hormone in human medicine and biochemistry and somatrotrophin in veterinary and agricultural sciences.

growth rate Symbol: γ (Greek gamma). A parameter relating to an increasing periodic function $f(t)$, defined by the equation:

$$f(t) = \exp(\gamma t) \sin \omega t,$$

where ω is *angular frequency. The *SI unit is the reciprocal of the second (s^{-1}). *Compare* damping coefficient.

GRS *Astron.* Abbrev. for Great Red Spot (on Jupiter).

GRSC Abbrev. for Graduate of the Royal Society of Chemistry.

Gru (no stop) *Astron.* Abbrev. for Grus.

Grubb, Sir Howard (1844–1931) British engineer and instrument maker.

Grüneisen parameter *Physics*

Grüneisen's law *Heat*

Grus A constellation. Genitive form: Gruis. Abbrev.: Gru (no stop). *See also* stellar nomenclature.

Gs Symbol for gauss.

GSC Abbrev. for gas–solid chromatography.

gsm Abbrev. for gram per square metre, a unit for expressing the thickness ('weight') of paper and card.

GTP Abbrev. and preferred form for guanosine 5′-triphosphate.

guanidine $(H_2N)_2CNH$ The traditional name for iminourea. *Compare* guanosine.

guanine Symbol: G A purine base. *See also* base pair. *Compare* guanosine.

guanosine Symbol: G A *nucleoside consisting of *guanine combined with D-ribose. *Compare* guanidine.

guanosine 5′-diphosphate Abbrev. and preferred form: GDP (no stops).

guanosine 5′-phosphate Abbrev. and preferred form: GMP (no stops).

guanosine 5′-triphosphate Abbrev. and preferred form: GTP (no stops).

Gudden–Pohl effect (en dash) *Physics*

Guericke, Otto von (1602–86) German physicist and engineer.

Guillaume, Charles Edouard (1861–1938) Swiss metrologist.

Guillemin effect *Magnetism* Named after E. A. Guillemin (1898–).

Guldberg, Cato Maximilian (1836–1902) Norwegian chemist.

Guldberg–Waage theory (en dash)

Gulf Stream (initial caps.) The ocean current flowing north-eastwards from the Gulf of Mexico to the Newfoundland Banks, where it branches: commonly (but incorrectly) used as a synonym for *North Atlantic Drift.

Gum nebula *Astron.* Named after Colin Gum (1924–60).

Gunn, John Battiscombe (1928–) British physicist.

Gunn diode

Gunn effect

Gurney–Mott theory (en dash) *Photog.* Named after R. W. Gurney and W. F. *Mott.

GUT *Physics, astron.* Acronym for grand unified theory.

Gutenberg, Johann (*c.* 1400–*c.* 68) German printer.

Gutenberg discontinuity *Geol.* Named after Bens Gutenberg (1889–1960).

Guthrie test (for phenylketonuria) Named after R. Guthrie (1916–).

Guttiferae A family of dicotyledonous plants including hypericum. Proposed alternative name: Hypericaceae.

Guttman, L. (1916–87) US psychologist.

Guttman scale

Guyot, Arnold Henry (1807–84) Swiss-born US geologist and geographer.

guyot

Gy Symbol for gray. *See also* SI units.

gymno- Prefix denoting naked or uncovered (e.g. gymnosperm).

Gymnospermae (cap. G) A subdivision of seed plants whose seeds are not enclosed in carpels; it comprises the classes *Cycadopsida (cycads), *Coniferopsida, and *Gnetopsida. Also called Coniferophytina. Individual name: **gymnosperm** (no cap.). Adjectival form: **gymnosperm** or **gymnospermous**.

Alternative classifications regard the gymnosperms as a division, *Pinophyta, or a superclass, Spermatophytatinae; in the latter scheme the class Coniferopsida is split into two new classes: Ginkgoopsida (including only *Ginkgo*) and Pinopsida.

gynaecium, gynaeceum Use *gynoecium.

gynecium US spelling of *gynoecium.

gyno- (**gyn-** before vowels) Prefix denoting female (e.g. gynodioecious, gynophore, gynandrous).

gynoecium (not gynaecium or gynaeceum; pl. gynoecia) US: **gynecium**. The carpels of a flower, collectively.

-gynous Adjectival suffix denoting a plant ovary (e.g. epigynous, perigynous). Noun form: **-gyny**.

gyro- (**gyr-** before vowels) Prefix denoting **1.** circular, part-circular, or sinusoidal (e.g. gyrator, gyrose). **2.** a spinning motion (e.g. gyrocompass, gyromagnetic, gyroscope).

gyromagnetic ratio Symbol: γ (Greek gamma). A physical quantity, the *magnetic moment (μ) of a particle divided by its angular momentum. The proton gyromagnetic ratio, symbol γ_p, is a constant equal to

$$2.675\,221\,28 \times 10^8\,\mathrm{s}^{-1}\,\mathrm{T}^{-1}.$$

It is given by $4\pi\mu_p/h$, where h is the Planck constant.

gyrus (pl. gyri) *Anat.*

H

h Symbol for **1.** hecto-. **2.** helion. **3.** hour.

h Symbol (light ital.) for **1.** heat transfer coefficient. **2.** height. **3.** *Electronics* hybrid fourpole matrix parameter (*see* electronics, letter symbols). **4.** Planck constant; $h/2\pi$ (sometimes called the Dirac constant) is often written $\hbar$. **5.** specific enthalpy.

H Symbol for **1.** adenosine, cytidine, or thymidine (or uridine) (unspecified). **2.** heavy chain (of an *immunoglobulin molecule). **3.** henry. **4.** histamine receptor. **5.** histidine. **6.** hydrogen.

Hα (or H_α; no space, Greek alpha) Designation of the first line of the Balmer series in the hydrogen spectrum. Lines of decreasing wavelength are designated Hβ (or H_β), Hγ (or H_γ), etc. *See also* Ly α.

H Symbol for **1.** (bold ital.) angular impulse. **2.** (light ital.) dose equivalent. **3.** (light ital.) enthalpy (H_m = molar enthalpy, ΔH_m = molar enthalpy change). **4.** (light ital.) Hamiltonian. **5.** (light ital.) light exposure. **6.** (light ital.) magnetic field strength (in nonvector equations; *H*); in vector equations it is printed in bold italic (***H***).

H_0 (light ital. H) Symbol for Hubble constant.

***H*-** (ital., always hyphenated) *Chem.* Prefix denoting hydrogenation of an unsaturated ring compound (e.g. 3*H*-pyrazole, in which the H is in the 3-position).

H–2 (en dash) The major histocompatibility system (*see* MHC) in mice. The two principal classes of antigens are designated I and II (Roman numerals). Class I antigens are encoded by at least four genetic loci (class I genes), designated *H–2K*, *H–2D*, *H–2L*, and *H2–R* (en dashes; ital. for genes, roman for the antigens encoded by them). Class II antigens, also called Ia (cap. I) antigens, are encoded by the *I* (ital. cap. I) locus, which is divided into subregions (class II genes) designated *I–A*, *I–B*, *I–J*, *I–E*, and *I–C* (en dashes; italic for genes, roman for antigens). *See also* S region.

ha Symbol for hectare.

Ha Symbol for hahnium. *See* element 105.

Ha (ital.) Symbol for Hartmann number.

HA Abbrev. for **1.** hyaluronic acid. **2.** *histamine.

HABA Abbrev. for 2-(4-hydroxyphenylazo)benzoic acid.

Haber, Fritz (1868–1934) German physical chemist.
Born–Haber cycle
Haber–Bosch process (en dash)
Haber process

Hadamard, Jacques-Salomon (hyphen) (1865–1963) French mathematician.
Cauchy–Hadamard formula (en dash)
Hadamard codes
Hadamard matrices

Hadfield, Sir Robert Abbott (1858–1940) British metallurgist.

Hadley, George (1685–1768) English meteorologist.
Hadley cell

Hadley, John (1682–1744) English mathematician and inventor.

hadro- (**hadr-** before vowels) Prefix denoting thick, heavy, fat (e.g. hadrosaur, hadron).

hadrons Strongly interacting subatomic particles composed of *quarks and gluons (*see* gauge bosons). There are two kinds: mesons, composed of a quark–antiquark pair, $q\bar{q}$, and baryons, composed of three quarks, qqq, bound together by the exchange of gluons.

The hadrons are grouped into isotopic spin multiplets on the basis of isotopic spin, *I*. Each multiplet contains $2I + 1$ hadrons with similar masses but different isospin components. Each multiplet has a name and symbol usually involving a Roman or Greek letter; an exception is the nucleon multiplet of the proton (p) and neutron (n), for which the single symbol N is sometimes used. The members of a multiplet are distinguished by writing charge as a superscript. Examples of mesons are:

$S = 0$	$\pi^+, \pi^0, \pi^-, \eta, D^+, D^0$
$S = 1$	K^+, K^0, F^+

Examples of baryons are:

$S = 0$ p, n, Λ_c^+
$S = -1$ $\Lambda, \Sigma^+, \Sigma^0, \Sigma^-$
$S = -2$ Ξ^0, Ξ^-
$S = -3$ Ω^-

Λ_c^+ is the Λ particle in which the s quark has been replaced by a c quark.

With the discovery of large numbers of unstable hadrons (more correctly called particle resonances), it has become impractical to give each one an individual name. The custom is to call it by the name of a similar hadron of lower mass, the symbol (sometimes asterisked) being followed by its approximate mass in MeV. Baryons are named after the lowest-mass baryon with the same *I* and *S*; an example is $\Lambda(1520)$, Λ having a mass 1116 MeV. The same system applies to some mesons. A complete set of nomenclature rules in high-energy physics is in the process of formulation.

Haeckel, Ernst Heinrich (1834–1919) German biologist.

Haeckel's law of recapitulation

haem US: **heme**.

haemato- (**haemat-** before vowels) US: **hemato-** (**hemat-**). Prefix denoting blood or blood-coloured (e.g. haematochrome, haematology, haematite, haematoma. *See also* haemo-.

haematoxylin US: **hematoxylin**. A blue dye. Not to be confused with *Haematoxylon*, the genus of the tree from which it is obtained.

haemo- (**haem-** before vowels) US: **hemo-** (**hem-**). Prefix denoting blood; use in preference to haema- (e.g. haemocoel, haemocyanin, haemagglutination, haemerythrin). *See also* haemato-.

haemoglobin US: **hemoglobin**. Abbrev.: Hb (no stops). The types of normal human haemoglobin are designated Hb A, Hb A_2, and Hb F (found only in the fetus). Each is composed of two pairs of globin chains with attached haem groups, each chain being designated by a Greek letter (α, β, γ, or δ); subscript numerals indicate the number of chains of the same type occurring in the molecule. Thus the chains of Hb A are designated $\alpha_2\beta_2$ (no space), those of Hb A_2 are $\alpha_2\delta_2$, and those of Hb F are $\alpha_2\gamma_2$. Abnormal haemoglobins are designated either by letters (C–Q, S) or by the name of the place where they were first identified (e.g. Hb Bart's, Hb Chad).

haemolysis (pl. haemolyses) US: **hemolysis**. Adjectival form: **haemolytic** (US: **hemolytic**).

α-haemolysis (Greek alpha, hyphenated; preferred to alpha-haemolysis) A form of haemolysis characterizing certain bacterial strains (described as **α-haemolytic**), especially of the genus **Streptococcus*.

α′-haemolysis (prime) A variant type of α-haemolysis.

β-haemolysis (Greek beta, hyphenated; preferred to beta-haemolysis) A form of haemolysis characterizing certain bacterial strains (described as **β-haemolytic**), especially of the genus **Streptococcus*.

Haemophilus (cap. H, ital.) A genus of bacteria of the family Pasteurellaceae. Individual name: **haemophilus** (no cap., not ital.; pl. haemophili). The species *H. influenzae* and *H. parainfluenzae* comprise several biovars, each designated by a Roman numeral.

haemopoiesis US: **hemopoiesis**. Blood-cell formation. Adjectival form: **haemopoietic** (US: **hemopoietic**).

hafnium Symbol: Hf *See also* periodic table; nuclide.

Hahn, Otto (1879–1968) German chemist.

hahnium *See* element 105.

Haidinger fringes *Interferometry* Named after W. K. Haidinger.

hailstone (one word)

hairline (one word)

Haldane, John Burdon Sanderson (1892–1964) British geneticist, son of J. S. Haldane.

Haldane's rule

Haldane, John Scott (1860–1936) British physiologist, father of J. B. S. Haldane.

Hale, George Ellery (1868–1938) US astrophysicist.

Hales, Stephen (1677–1761) English plant physiologist and chemist.

half (pl. halves) Verb form: **halve**.
Two-word terms in which the first word is 'half' are usually hyphenated (e.g. half-life, half-width). In the case of the *fraction, spell out (one-half, three-halves, etc., hyphenated) only in nonmathematical text. In mathematical text use decimal fractions where possible (0.5, 1.5, etc.) or, when appropriate (e.g. in statistics, particle physics), a 'one-piece' fraction (½); if this is not available use a solidus (3/2, 7/2, $^7/_2$, etc.).

half duplex (two words) *Telecom.* Hyphenated when used adjectivally (e.g. half-duplex operation).

half frame (two words) *Photog.* Hyphenated when used adjectivally (e.g. half-frame camera).

half-life (hyphenated) *See* decay constant.

halftone (one word)

Hall, Asaph (1829–1907) US astronomer.

Hall, Charles Martin (1863–1914) US chemist.
Hall–Héroult process (en dash)

Hall, Edwin Herbert (1855–1938) US physicist.
Hall coefficient Symbol: R_H or A_H
Hall effect
Hall mobility Symbol: μ_H (Greek mu)
Hall probe

Hall, James (1811–98) US geologist.

Hall, Sir James (1761–1832) British geologist.

Halley, Edmund (1656–1742) British astronomer and physicist.
Halley's comet

hallux (pl. halluces) The first (innermost) digit of a hindlimb; the big toe in humans. *Compare* pollex.

halo (pl. haloes; preferred to halos)

halo- (sometimes **hal-** before vowels) Prefix denoting **1.** salt or the sea (e.g. halophyte, halosere, halite). **2.** a halogen (e.g. haloamine, haloform, halothane, halide).

haloalkane Any of a class of organic compounds in which one or more hydrogen atoms of an alkane have been replaced by halogen atoms. Traditional name: alkyl halide.
Haloalkanes are systematically named as halogen derivatives of the parent alkane, e.g. bromoethane, CH_3CH_2Br (from ethane, CH_3CH_3), and 1-bromo-2,3-dichlorobutane, $CH_2BrCHClCHClCH_3$. In nonsystematic nomenclature monohaloalkanes are named as alkyl halides, e.g. ethyl iodide, CH_3CH_2I, and *t*-butyl chloride, $(CH_3)_3CCl$. Dihaloalkanes are named according to which carbon atoms the hydrogen atoms are attached to. If both halogen atoms are attached to the same carbon atom the dihaloalkanes are named as alkylidene dihalides, e.g. ethylidene dibromide, CH_3CHBr_2. If the halogen atoms are attached to adjacent carbon atoms the dihaloalkanes are named as dihalides of the parent alkene, e.g. ethylene dibromide, CH_2BrCH_2Br (from ethylene, $CH_2{:}CH_2$). If the halogen atoms are attached to the terminal carbon atoms of the chain the dihaloalkanes are named as polymethylene dihalides, e.g. trimethylene dibromide, $CH_2BrCH_2CH_2Br$.

halobacteria (pl. noun; sing. halobacterium) Bacteria of the genus *Halobacterium*. Some authors use 'halobacteria' as the trivial name for members of any genus in the order Halobacteriales, but this should be avoided; 'halophile' is a less confusing alternative.

Halobacterium (cap. H, ital.) A genus of salt-requiring bacteria. Individual name: **halobacterium** (no cap., not ital.; pl. *halobacteria).

Halococcus (cap. H, ital.) A genus of salt-requiring bacteria. Individual name: **halococcus** (no cap., not ital.; pl. halococci).

haloform The traditional name for a trihalomethane.

halogens The elements in group VIIB of the periodic table.

halophile Any organism that thrives in or requires salty conditions, especially any bacterium of the order Halobacteriales (*see* halobacteria). Adjectival form: **halophilic**.

haltere (preferred to halter) A balancing organ in the Diptera (two-winged flies).

Hamilton, William Donald (1936–) British theoretical biologist.

Hamilton, Sir William Rowan (1805–65) Irish mathematician.
Cayley–Hamilton theory (en dash)
***Hamiltonian**
Hamilton–Jacobi theory (en dash)
Hamilton's equations
Hamilton's principle
Lagrange–Hamilton theory (en dash)

Hamiltonian Symbol: H A function used in general dynamics:

$$H = \sum_i p_i \dot{q}_i - L,$$

where p_i are the *generalized momenta, q_i are the *generalized coordinates, and L is the *Lagrangian function. Also called Hamiltonian function.

Hammet acidity function *Chem.* Named after L. P. Hammet (1894–).

Hamming codes *Information theory*

hamster polyoma virus Abbrev.: HaPV (lower-case a).

hamulus (pl. hamuli) *Anat.* A hooklike process. Adjectival forms: **hamular, hamulate.**

Hand, Alfred (1868–1949) US paediatrician.
Hand–Schüller–Christian disease or **Schüller–Christian disease** (en dashes) Another name for reticuloendotheliosis.

Handley Page, Sir Frederick Usually alphabetized as *Page.

handset (one word) *Telecom.*

handshake (one word) *Computing*

Hankel functions *Maths., physics* Named after Hermann Hankel (1838–73).

Hansen, Armauer Gerhard Henrik (1841–1912) Norwegian bacteriologist.
Hansen's bacillus Old name for *Mycobacterium leprae.*
Hansen's disease Old name for leprosy.

Hantzsch, Arthur Rudolf (1857–1935) German chemist.
Hantzsch synthesis

haplo- (**hapl-** before vowels) Prefix denoting single or simple (e.g. haplobiontic, haplosere, haplostele, haplotype).

haploid Denoting a nucleus, cell, or organism having a single set of unpaired chromosomes. The haploid number is designated as n (ital.); for example in humans $n = 23$. *See also* genome; genotype. Noun form: **haploidy.** *Compare* haplont.

haplont Any organism whose somatic cells are *haploid. Adjectival form: **haplontic.**

hapteron (pl. haptera) An organ that attaches a plant, especially an aquatic plant, to a substrate. Also called holdfast.

hapto- (**hapt-** before vowels) Prefix denoting **1.** attachment (e.g. haptonema, hapten). **2.** touch (e.g. haptonasty, haptotropism).

Haptophyta A division of algae that have a haptonema (attachment organ) between the two flagella. It contains a single class, Haptophyceae, which contains genera formerly included in *Chrysophyta. In a recent classification this division is placed in the subkingdom *Chlorobionta.

HaPV (lower-case a) Abbrev. for hamster polyomavirus.

Ha-*ras* (or **H-*ras***; hyphenated) *See* oncogene.

Harcourt, Sir William Venables Vernon (1789–1871) British chemist.

Harden, Sir Arthur (1865–1940) British biochemist.

hardware (one word)

hardwired (one word) *Computing*

Hardy, Godfrey Harold (1877–1947) British mathematician.
Hardy classes
Hardy–Weinberg equation (en dash)
Hardy–Weinberg equilibrium (en dash)
Hardy–Weinberg law (en dash)

Hardy, Sir William Bate (1864–1934) British biologist and chemist.

Hare, Robert (1781–1858) US chemist.

Hargreaves, James (died 1778) British inventor.
Hargreaves' spinning-jenny (hyphen)

Hariot (or Harriot), Thomas (1560–1621) English mathematician, astronomer, and physicist.

Harkins, William Draper (1873–1951) US physical chemist.

Haro galaxies *Astron.* Named after G. Haro.

Harris, Geoffrey Wingfield (1913–71) British endocrinologist.

Hartig net *Bot.* Named after H. Hartig.

Hartley oscillator *Elec. eng.*

Hartmann number Symbol: *Ha* A dimensionless quantity equal to $Bl\sqrt{(\sigma/\rho\nu)}$, where B is the *magnetic flux density, l a characteristic length, σ the electric *conductivity, ρ the density, and ν the kinematic *viscosity. *See also* parameter.

Hartree, Douglas Rayner (1897–1958) British mathematician.
hartree
Hartree diagram
Hartree–Fock approximation (en dash)
Hartree–Fock orbital (en dash)

Harvey, William (1578–1657) English physician.

Hashimoto, Hakaru (1881–1934) Japanese surgeon.
Hashimoto's disease (or **thyroiditis**)

Hassall, Arthur Hill (1817–94) British chemist and physician.
Hassall's corpuscles *Anat.*

Hassel, Odd (1897–1981) Norwegian chemist.

Hatch–Slack pathway (en dash) *Bot.* Named after M. D. Hatch (1932–) and R. Slack.

Hauksbee, Francis (*c.* 1670–*c.* 1713) English physicist.

Hauptman, Herbert Aaron (1917–) US mathematician and physicist.

Hausdorff, Felix (1868–1942) German mathematician.
Hausdorff space

haustorium (pl. haustoria) An absorptive organ produced by a parasitic plant. Adjectival form: **haustorial** (this term is also applied to similar organs in nonparasitic plants).

Haüy, René Just (1743–1822) French mineralogist and crystallographer.

Havers, Clopton (1657–1702) English anatomist.
Haversian (US **haversian**) **canal**
Haversian (US **haversian**) **system**

Havilland, Sir Geoffrey De Usually alphabetized as *De Havilland.

Hawking, Stephen William (1942–) British theoretical physicist.
Hawking radiation

Haworth, Sir (Walter) Norman (1883–1950) British chemist.
Haworth formula

Hb (no stop) Abbrev. or symbol for *haemoglobin.

HBLV Abbrev. for human B-lymphotropic virus.

HbO_2 Symbol for oxyhaemoglobin.

HBT (or **HJBT**) *Electronics* Abbrev. for heterojunction bipolar transistor.

HBV Abbrev. for hepatitis B virus.

HCF (or **h.c.f.**) *Maths.* Abbrev. for highest common factor.

hCG (lower-case h; preferred to HCG) Abbrev. for human chorionic gonadotrophin.

HCI Abbrev. for human–computer interface (or interaction).

h.c.p. *Crystallog.* Abbrev. for hexagonal close-packed.

HD *Astron.* Prefix used to designate a star listed in the Henry Draper Catalogue.

HDAL *Chem.* Abbrev. for hexadecenal.

H–D curve (en dash) *Photog.* Abbrev. for Hurter–Driffield curve. *See* Hurter, Ferdinand.

(*Z,Z*)-7,11-HDDA Abbrev. for *cis*-7,*cis*-11-hexadecadien-1-yl acetate.

HDL Abbrev. for high-density lipoprotein.

HDODA Abbrev. for 1,6-hexanediol diacrylate.

(*Z*)-11-HDOL Abbrev. for *cis*-11-hexadecen-1-ol.

He Symbol for helium.

heartbeat (one word)

heartwood (one word)

heat Symbol: *Q* A physical quantity, the *energy in the course of being transferred from a hotter body or region to a cooler one as a result of the *temperature difference between them. The *SI unit is the joule. *See also* British thermal unit; calorie. Also called quantity of heat.

The energy in a body or region before or after transfer is also sometimes called heat, but this is now deprecated; the name *internal energy should be used for this quantity. When a body (internal energy, U) suffers a change there is a change of internal energy, ΔU, given according to the first law of thermodynamics by:

$$\Delta U = Q + W,$$

where Q is the heat absorbed by the body from the surroundings and W is the work done simultaneously by the body on the surroundings. To call both U and Q 'heat' clearly leads to confusion.

Two-word terms in which the first word is 'heat' (e.g. heat transfer) should not be hyphenated unless they are used adjectivally (e.g. heat-transfer coefficient).

heat capacity Symbol: C_p (at constant pressure, ital. subscript) or C_V (at constant volume, ital. subscript). A physical quantity, dQ/dT, where dT is the temperature increase resulting from the addition of a small quantity of heat dQ. The *SI unit is usually the joule per kelvin.

The **molar heat capacity**, symbol C_m, is heat capacity divided by *amount of substance. The SI unit is normally J/(mol K).

The **specific heat capacity**, symbol c_p, c_V, or C_{sat} (at saturation), is heat capacity divided by mass. The SI unit is normally J/(kg K). The ratio c_p/c_V is dimensionless and usually has the symbol γ (Greek gamma). The former term 'specific heat', as an alternative to specific heat capacity, should be avoided.

heat flow rate Symbol: Φ (Greek cap. phi). A physical quantity, the rate of heat flow through a surface. The *SI unit is the watt.

The density of heat flow rate, symbol q, is the heat flow rate divided by area; it is measured in watts per square metre.

heath An unspecified shrub of the genus **Erica*. The name is qualified for individual species, but note that the common names of many *Erica* species include the word *heather, e.g. bell heather (*E. cinerea*), bog heather (*E. tetralix*), and the name 'heathers' is often used interchangeably with 'heaths' for all members of the genus. In addition, several plants known as heaths belong to different genera, e.g. sea heath (*Frankenia laevis*). Use the binomial names for *Erica* species to avoid confusion.

heather Used without qualification, this usually refers to the ericaceous shrub *Calluna vulgaris*, also called ling. Qualified, or in the plural, it refers to many *Erica* species (*see* heath).

Heaviside, Oliver (1850–1925) British electronic engineer and physicist.

Heaviside layer Also called E layer.

Heaviside–Lorentz units (en dash)

heavy chain *Immunol. See* immunoglobulin.

heavy-water reactor (heavy-water hyphenated) Abbrev.: HWR (no stops).

Heberden, William (not Heberdon) (1710–1801) British physician.

Heberden's nodes

hectare Symbol: ha A unit of area equal to 100 ares, i.e. 10^4 m^2. One hectare = 2.471 05 acres.

hecto- Symbol: h A prefix to a unit of measurement that indicates one hundred times that unit, as in hectometre (hm). *See also* SI units; hectare.

-hedron Noun suffix denoting surfaces of a geometric solid (e.g. tetrahedron). Adjectival form: **-hedral**.

HEED Acronym for high-energy electron diffraction.

Heezen, Bruce Charles (1924–77) US oceanographer.

Heidelberg man *See Homo.*

Heidenhain, Rudolf Peter Heinrich (1834–97) German physiologist.

Heidenhain's cell

Heidenhain's law

Heidenhain's stain

height Abbrev.: ht (no stop). *See* length.

height equivalent to a theoretical plate Abbrev: HETP (no stops). *Chem.* A measure of the contacting efficiency of the packing in a packed column.

Heilbron, Sir Ian (1886–1959) British organic chemist.

Heisenberg, Werner Karl (1901–76) German physicist.
Heisenberg uncertainty principle Use uncertainty principle.

HeLa cell (cap. H and L) A carcinomatous cell type, maintained in culture and used in research since 1951. Named after *He*nrietta *La*cks, the patient from whom it originated.

helicoid 1. (noun) A surface resembling a screw thread. Adjectival form: **helicoidal**. 2. (adjective) *Biol.* Spiral-shaped (e.g. a helicoid shell).

helio- (**heli-** before vowels) Prefix denoting the sun (e.g. heliocentric, heliotropism).

helion *See* alpha particle.

heliophyte Any plant that can only thrive in conditions of strong light; commonly known as a sun plant. *Compare* helophyte.

Heliozoa (cap. H) US: **Heliozoia**. An order of mainly freshwater amoeboid protozoans. Individual name and adjectival form: **heliozoan** (no cap.; not heliozoic).

helium Symbol: He *See also* periodic table; nuclide.

helix (pl. helices) Adjectival forms: **helical**, ***helicoid** (the latter is also used as a noun).
α helix (no hyphen) *Biochem.*

Helmert's formula *Geophysics* Named after F. R. Helmert (1843–1917).

Helmholtz, Hermann Ludwig Ferdinand von (1821–94) German physiologist and theoretical physicist.
Gibbs–Helmholtz equation (en dash)
Helmholtz coils
***Helmholtz function**
Helmholtz resonator

Helmholtz function (preferred to Helmholtz free energy) Symbol: A A thermodynamic property of a system defined as:

$$U - TS,$$

where U is the *internal energy, T the *thermodynamic temperature, and S the *entropy. The *SI unit is the joule.
Molar Helmholtz function, symbol A_m, is Helmholtz function divided by *amount of substance; it is measured in joules per mole. *Compare* Gibbs function.

Helmont, Jan Baptista van (1579–1644) Flemish chemist and physician.

helophyte A plant whose perennating organs are situated in mud at the bottom of a pond or lake. *Compare* heliophyte.

Helotiales An order of ascomycete fungi (*see* Ascomycotina) containing some of the cup fungi, formerly included in the class Discomycetes.

hemato- US spelling of *haemato-.

heme US spelling of haem.

hemelytron (preferred to hemielytron; pl. hemelytra; usually used in the pl.) Either of the forewings of heteropteran bugs, being leathery at the base and membranous at the apex. *Compare* elytron.

hemi- Prefix denoting half (e.g. hemicellulose, hemicryptophyte, hemicyclic, hemizygous).

Hemiascomycetes A former class of ascomycete fungi (*see* Ascomycotina). Most of its members are now included in the order *Endomycetales. In some classifications these fungi are regarded as a subclass, Hemiascomycetidae, of the class Ascomycetes.

Hemibasidiomycetes *See* Teliomycetes.

Hemichordata (cap. H) A phylum of marine invertebrates comprising the acorn worms. Individual name and adjectival form: **hemichordate** (no cap.).

Hemiptera (cap. H) An order of insects comprising the *bugs, divided into the suborders *Heteroptera and *Homoptera. Individual name and adjectival form: **hemipteran** (no cap.; not hemipterous or hemipteral).

hemo- US spelling of *haemo-.

hemoglobin US spelling of *haemoglobin.

hemolysis (pl. hemolyses) US spelling of *haemolysis.

Hench, Philip Showalter (1896–1965) US biochemist.

Henderson–Hasselbalch equation (en dash) *Biochem.* Named after L. J. Henderson (1878–1942) and K. A. Hasselbalch.

Henle, Friedrich Gustav Jacob (1809–85) German physician, anatomist, and pathologist.
Henle's loop (or **loop of Henle)**

Henoch, Eduard Heinrich (1820–1910) German paediatrician.
Henoch–Schönlein purpura (en dash)
Henoch's purpura

henry (no cap.; pl. henrys or henries) Symbol: H The *SI unit of *self-inductance and *mutual inductance.

$$1\,\mathrm{H} = 1\,\mathrm{Wb}\,\mathrm{A}^{-1}.$$

In practice the milli-, micro-, nano-, and picohenry are used. Named after Joseph *Henry.

Henry, Joseph (1797–1878) US physicist.
***henry**

Henry, William (1774–1836) British physician and chemist.
Henry's law

Hensen, Viktor (1835–1924) German physiologist and oceanographer.
Hensen net
Hensen's node Also called primitive knot.

Hepaticae *See* Hepaticopsida.

Hepaticatinae *See* Bryophyta.

Hepaticopsida A class of bryophytes comprising the liverworts. Also called Hepaticae or Marchantiopsida. In a recent classification liverworts are classified as a superclass, Hepaticatinae (*see* Bryophyta).

hepatitis Inflammation of the liver. The different types of viral hepatitis are designated by capital letters: hepatitis A (formerly called infectious hepatitis), hepatitis B (formerly called serum hepatitis), hepatitis C, D, and E.

hepato- (hepat- before vowels) Prefix denoting the liver (e.g. hepatocyte, hepatitis).

HEPES Abbrev. for 4-(2-hydroxyethyl)-1-piperazineethanesulphonic acid.

hepta- (hept- before vowels) Prefix denoting seven (e.g. heptagon, heptamerous, heptane).

heptanedioic acid $HOOC(CH_2)_5$-$COOH$ The recommended name for the compound traditionally known as pimelic acid.

heptaoxodiphosphoric(V) acid H_4P_2-O_7 The recommended name for the compound traditionally known as pyrophosphoric acid.

heptyl- *Prefix denoting the group C_7H_{15}– (e.g. heptylbenzoic acid, heptyl chloride).

Her (no stop) *Astron.* Abbrev. for Hercules.

herb Any nonwoody flowering plant, i.e. a plant whose aerial parts die back at the end of the growing season. The word is also used loosely to denote any aromatic culinary plant, but this should be avoided in technical usage. Adjectival form: **herbaceous.**

Herbaceae A taxon of dicotyledonous plants in John Hutchinson's classification (1948) containing all the plant families with predominantly herbaceous members. This is now regarded as an artificial grouping and the taxon is no longer used. *Compare* Lignosae.

herbarium (pl. herbaria)

Herbig, George Howard (1920–) US astronomer.
Herbig–Haro objects (en dash) Also named after G. Haro.

Hercules A constellation. Genitive form: Herculis. Abbrev.: Her (no stop). *See also* stellar nomenclature.

Herelle, Félix d' Usually alphabetized as *d'Herelle.

Hering, Karl Ewald Konstantin (not Herring) (1834–1918) German psychologist and physiologist.

Hermite, Charles (1822–1901) French mathematician.
Hermite interpolation
Hermite polynomial
Hermitian conjugate
Hermitian matrix

Hero of Alexandria (not Heron) (*fl.* mid-1st century AD) Greek mathematician and inventor.
Hero's formula

Herophilus of Chalcedon (*fl.* 335–280 BC) Greek anatomist and physician.

Héroult, Paul Louis Toussaint (1863–1914) French chemist.
Hall–Héroult process (en dash)

herpes simplex virus An *alphaherpesvirus, strains of which can be differentiated as either type 1 or type 2 (Arabic numerals). These are also known as human (alpha) herpesvirus 1 and human (alpha) herpesvirus 2, respectively. **Herpes simplex** (cold sore) is usually caused by type 1 strains.

herpesvirus (one word) Any member of the family Herpesviridae (vernacular name: herpesvirus group). Names of the species and genera often take the form human herpesvirus 1, equid herpesvirus 3, etc., with a distinguishing Arabic numeral. *See also* alphaherpesvirus; betaherpesvirus; gammaherpesvirus.

herpes zoster The medical name for shingles.

Herschbach, Dudley Robert (1932–) US chemist.

Herschel, Caroline Lucretia (1750–1848) German-born British astronomer, sister of William Herschel.

Herschel, Sir John Frederick William (1792–1871) British astronomer, son of William Herschel.

Herschel, Sir (Frederick) William (1738–1822) German-born British astronomer, father of John Herschel.
Herschelian–Cassegrain telescope (en dash)
Herschel's condition
Herschel effect
Herschel's fringes

Hershey, Alfred Day (1908–) US biologist.

hertz (no cap.; pl. hertz) Symbol: Hz The *SI unit of frequency, the same as cycles per second:

$$1\ \text{Hz} = 1\ \text{s}^{-1}.$$

Named after H. R. *Hertz.

Hertz, Gustav (1887–1975) German physicist.
Franck–Hertz experiment (en dash)

Hertz, Heinrich Rudolf (1857–94) German physicist.
***hertz**
Hertzian oscillator
Hertzian waves (radio waves)

Hertzsprung, Ejnar (1873–1967) Danish astronomer.
Hertzsprung–Russell diagram (en dash)
Hertzsprung gap

Herzberg, Gerhard (1904–) German-born Canadian spectroscopist.

Hess, Germain Henri (1802–50) Swiss-born Russian chemist.
Hess's law

Hess, Harry Hammond (1906–69) US geologist.

Hess, Victor Francis (1883–1964) Austrian-born US physicist.

Hess, Walter Rudolf (1881–1973) Swiss neurophysiologist.

hetero- (**heter-** before vowels) Prefix denoting difference or dissimilarity (e.g. heterocercal, heterochromatin, heterocyclic, heterolytic, heteroecious). *Compare* homo-.

Heterobasidiomycetidae *See* Basidiomycotina.

heterocyclic Designating an organic compound containing closed rings in which one or more of the ring atoms are elements other than carbon.

In systematic nomenclature monocyclic compounds are named so that the number, kind, and positions of the hetero atoms present and the degree of unsaturation are specified. The ring size is indicated by the stem -ir-, -et-, -ol-, -in-, -ep-, -oc-, -on-, or -ec- for 3-, 4-, 5-, 6-, 7-, 8-, 9-, or 10-membered rings, respectively. The nature of the hetero atom present is indicated by the prefix oxa-, thia-, aza-, sila-, or phospha- for oxygen, sulphur, nitrogen, silicon, or phosphorus, omitting the final 'a' where necessary. When two or more of the

	Stem + suffix			
	Ring containing nitrogen		Ring containing no nitrogen	
Ring size	Unsaturated[a]	Saturated	Unsaturated[a]	Saturated
3	-irine	-iridine	-irene	-irane
4	-ete	-etidine	-ete	-etane
5	-ole	-olidine	-ole	-olane
6	-ine	[b]	-in	-ane
7	-epine	[b]	-epin	-epane
8	-ocine	[b]	-ocin	-ocane
9	-onine	[b]	-onin	-onane
10	-ecine	[b]	-ecin	-ecane

[a] Corresponding to the maximum number of noncumulative double bonds.
[b] The prefix perhydro- is attached to the name of the parent unsaturated compound.

same hetero atoms are present the prefixes di-, tri-, etc., are added, e.g. dioxa-, triaza-. When two or more different hetero atoms are present the prefixes are cited starting with the hetero atom belonging to the highest group in the periodic table and of lowest atomic number in that group, thus (in descending order) oxygen, sulphur, nitrogen, phosphorus, silicon; e.g. oxathia- and thiaza-. The degree of unsaturation is usually specified by a suffix, which varies according to ring size and whether or not the ring contains nitrogen, as shown in the table.

The degree of hydrogenation is indicated by the prefixes dihydro-, tetrahydro-, etc., or by prefixing the name of the parent unsaturated compound with *H* (denoting hydrogen) preceded by a number denoting the position of saturation. Finally, the numbering of the ring starts with the hetero atom of highest priority and proceeds around the ring so as to give substituents or other hetero atoms the lowest numbers possible.

Polycyclic compounds are named in the following manner: when one heterocyclic ring is present this is chosen as the parent compound, and the name of the fused ring system is attached as a prefix (e.g. benzo-, naphtho-, etc.). When two or more heterocyclic rings are present an oxygen-containing ring is given precedence over a sulphur-containing ring (and sulphur over phosphorus) in accord with the rule for precedence for monocyclic compounds, but with the exception that a nitrogen-containing ring is given overall precedence. The positions of the ring junctions are indicated by lettering the sides of the parent compound *a*, *b*, *c*, *d*, etc., starting with the 1,2-bond; the positions of the ring junctions of any fused rings are specified by prefixing these letters by numbers denoting the positions of attachment on the fused rings. The appropriate letters (in italic type) and numbers (in roman type) are enclosed in brackets and prefix the name of the parent compound. Finally, the numbering of the periphery of a polycyclic compound is done by orienting the compound such that the greatest number of rings lie along a horizontal axis and, of any other rings present, a maximum lies furthest to the right above the horizontal axis; numbering then starts with the uppermost ring furthest to the right and proceeds in a clockwise direction, omitting the ring junctions. Some examples are shown in the illustration.

Many heterocyclic organic compounds have widely used trivial names, e.g. pyridine, furan, and azepine. These trivial names are often used in less systematic nomenclature, e.g. thieno[2,3-*b*]furan. To name a polycyclic compound using trivial names the largest heterocyclic ring system that has a simple name is chosen as the parent system, e.g. 4-methylbenzo[*h*]isoquinoline.

Compare homocyclic.

heterogeneous nuclear ribonucleoprotein Abbrev.: hnRNP (no stops).

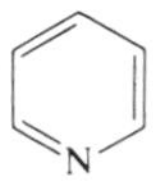

azine 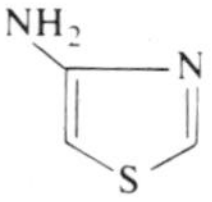 | 4-amino-1,3-thiazole  | 2,5-dihydroxole (2*H*,5*H*-oxole)

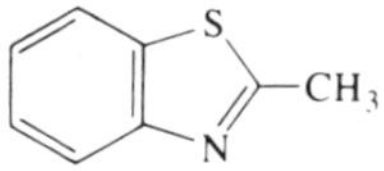

2-methylbenzothiazole 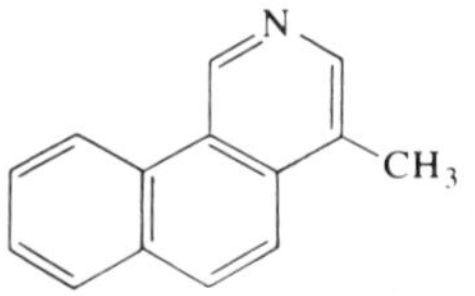 | 4-methylbenzo[*h*]isoquinoline

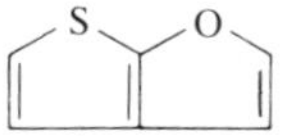

thieno[2,3-*b*]furan 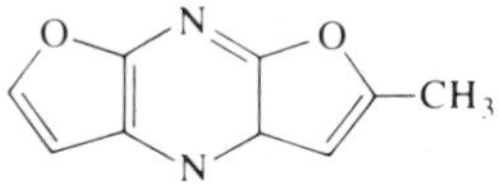 | 2-methyldifuro[2,3-*b*:3′,2′-*e*]pyrazine

heterogeneous nuclear RNA Abbrev.: hnRNA (no stops).

heterojunction bipolar transistor Abbrev.: HBT or HJBT (no stops).

heterokont Describing an organism or cell having two different types of flagella, as occurs in the motile stages of algae of the division *Xanthophyta (hence the former name of this taxon, Heterokontae). The term 'heterokont' has no taxonomic significance and is used purely as a descriptive term.

Heteroptera (cap. H) A suborder of bugs (*Hemiptera) including the bedbug and water boatman. Individual name and adjectival form: **heteropteran** (no cap.; not heteropterous).

heterozygous Having different alleles of any one gene. Heterozygous individuals are called **heterozygotes**. Noun form: **heterozygosity**. *See also* genotype.

HETP Abbrev. for *height equivalent to a theoretical plate.

Heusler alloys *Metallurgy* Named after Fritz Heusler.

Hevelius, Johannes (1611–87) German astronomer.

Hevesy, George Charles von (1885–1966) Hungarian-born Swedish chemist.

Hewish, Antony (1924–) British radio astronomer.

hex (no stop) Abbrev. for hexadecimal notation.

hexa- (**hex-** before vowels) Prefix denoting six (e.g. hexacanth, hexadecimal, hexahedron, hexane, hexose).

1,2,3,4,5,6-hexachlorocyclohexane $C_6H_6Cl_6$ The recommended name for the compound traditionally known as benzene hexachloride.

hexacyanocobaltate(III) A compound containing the ion $Co(CN)_6^{3-}$, e.g. potassium hexacyanocobaltate(III). The traditional name is cobalticyanide.

hexacyanoferrate(II) A compound containing the ion $Fe(CN)_6^{4-}$, e.g. potassium hexacyanoferrate(II). The traditional name is ferrocyanide.

hexacyanoferrate(III) A compound containing the ion $Fe(CN)_6^{3-}$, e.g. potassium hexacyanoferrate(III), $K_3Fe(CN)_6$. The traditional name is ferricyanide.

hexadecane $CH_3(CH_2)_{14}CH_3$ The recommended name for the compound traditionally known as cetane.

hexadecanoic acid $CH_3(CH_2)_{14}COOH$ The recommended name for the compound traditionally known as palmitic acid.

hexadecan-1-ol $CH_3(CH_2)_{14}CH_2O$ The recommended name for the compound traditionally known as cetyl alcohol.

hexadecenal Abbrev.: HDAL (no stops).

hexadecimal notation Abbrev.: hex (no stop).

2,4-hexadienoic acid CH_3CH:CHCH:CHCOOH The recommended name for the compound traditionally known as sorbic acid.

hexafluorosilicate(IV) Denoting a compound containing the ion $SiF_6{}^{2-}$, e.g. sodium hexafluorosilicate(IV). The traditional name is fluorosilicate.

hexagonal close-packed *Crystallog.* Abbrev.: h.c.p. (no caps.; stops).

hexahydrophenol The traditional name for *cyclohexanol.

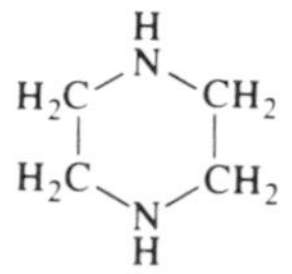

hexahydropyrazine
(piperazine)

hexahydropyrazine The recommended name for the compound traditionally known as piperazine.

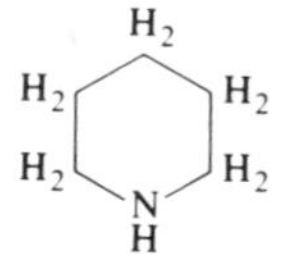

hexahydropyridine
(piperidine)

hexahydropyridine The recommended name for the compound traditionally known as piperidine.

hexamine $(CH_2)_6N_4$ The traditional name for 1,3,5,7-tetraazaadamantane.

hexane-1,6-dicarboxylic acid $HOOC(CH_2)_6COOH$ The recommended name for the compound traditionally known as suberic acid.

hexanedioic acid $HOOC(CH_2)_4COOH$ The recommended name for the compound traditionally known as adipic acid.

hexanoic acid $CH_3(CH_2)_4COOH$ The recommended name for the compound traditionally known as caproic acid.

Hexapoda Use *Insecta.

hexokinase (not hexakinase) An enzyme acting on glucose (a hexose sugar).

hexyl- *Prefix denoting the group $C_6H_{13}-$ (e.g. hexylamine, hexyl chloride).

Heymans, Corneille Jean François (1892–1968) Belgian physiologist and pharmacologist.

Heyrovský, Jaroslav (1890–1967) Czech physical chemist.

Heyrovský–Ilkovic equation (en dash) Use Ilkovic equation.

Hf Symbol for hafnium.

HF Abbrev. for high frequency.

Hfr (lower-case f and r) Abbrev. and preferred form for high frequency of recombination: used to denote a cell with an *F plasmid integrated into its chromosome. Genes of an Hfr cell are transferred to F^- cells with much higher frequency than those of a non-Hfr cell.

Hg Symbol for mercury. [from Latin *hydrargyrum*]

hGH (lower-case h; preferred to HGH) Abbrev. for human *growth hormone.

H horizon A surface *soil horizon consisting of well-decomposed litter, i.e. h(umus).

HHV Abbrev. for human herpesvirus.

hiatus (pl. hiatuses; preferred to hiatus)

hibernation *Zool.* Dormancy during the winter months. *Compare* aestivation.

hi-fi (pl. hi-fis) Acronym for high fidelity.

Higgins, William (1763–1825) Irish chemist.

Higgs boson *Physics* Named after Peter Higgs (1929–).

high Two- and three-word terms in which the first word is 'high' are hyphenated when used adjectivally (e.g. high-carbon steel, high-energy particle, high-level language, high-pass filter; high-resolution graphics, high-electron-mobility transistor, high-mobility-group protein).

high-density lipoprotein (hyphenated) Abbrev.: HDL (no stops).

high-energy electron diffraction (hyphenated) *Physics* Abbrev.: HEED (no stops).

highest common factor Abbrev.: HCF (no stops) or h.c.f. (stops).

high frequency Abbrev.: HF (no stops). Hyphenated when used adjectivally (e.g. high-frequency transformer).

highlight (one word)

high-performance liquid chromatography Abbrev.: HPLC (no stops).

high pressure Abbrev.: HP (no stops). Hyphenated when used adjectivally (e.g. high-pressure turbine).

high-pressure liquid chromatography Abbrev.: HPLC (no stops).

high tension Abbrev.: HT (no stops). *Elec. eng.* Hyphenated when used adjectivally (e.g. high-tension cable). Also called high voltage.

high water Abbrev.: HW (no stops).

Hilbert, David (1862–1943) German mathematician.
Hilbert's axioms
Hilbert space

Hildebrand, Joel Henry (1881–1983) US chemist.

Hilditch, Thomas Percy (1886–1965) British chemist.

Hill, Archibald Vivian (1886–1977) British physiologist and biochemist.

Hill, James Peter (1873–1954) British embryologist.

Hillier, James (1915–) Canadian-born US physicist.

Hill reaction *Bot.* Named after Robin Hill (1899–1991).

hilum (pl. hila) **1.** *Bot.* A scar on a seed coat marking the point at which the seed was attached to the fruit wall. **2.** (preferred to hilus) *Anat.* A hollow on the surface of an organ where blood vessels, ducts, etc., enter or leave it.

hilus (pl. hili) *Anat.* Use *hilum.

hindbrain (one word)

hindgut (one word)

hindlimb (one word)

hindwing (one word)

Hinshelwood, Sir Cyril Norman (1897–1967) British chemist.

Hipparchus (*c.* 170 BC–*c.* 120 BC) Greek astronomer and geographer.

hippocampus (pl. hippocampi) *Anat.* Adjectival form: **hippocampal**.

Hippocrates of Cos (*c.* 460 BC–*c.* 370 BC) Greek physician.
Hippocratic oath

Hirayama families *Astron.* Named after K. Hirayama.

hi-res Acronym for high resolution.

Hirsch, Sir Peter Bernhard (1925–) British metallurgist.

Hirst, Sir Edmund Langley (1898–1975) British chemist.

Hirudinea (cap. H) A class of annelid worms comprising the leeches. Individual name and adjectival form: **hirudinean** (no cap.).

His (or **his**; no stop) Abbrev. for histidine. *See* amino acid.

His, Wilhelm (1831–1904) Swiss anatomist and physiologist.

His, Wilhelm (1863–1934) Swiss anatomist, son of Wilhelm His (1831–1904).
***bundle of His**

histamine Abbrev.: HA (no stops). A neurotransmitter with numerous physiological actions. Histamine receptors are designated H_1, H_2, and H_3 (subscript numbers).

histidine Abbrev.: His or his (no stop). Symbol: H *See* amino acid.

histiocyte (not histocyte)

histo- (**hist-** before vowels) Prefix denoting tissue (e.g. histochemistry, histocompatibility, histogenesis, histamine).

histocompatibility antigens *See* H-2; HLA system; MHC.

histone One of a class of proteins associated with DNA in eukaryotic cell nuclei. The five classes of histones are designated H1, H2A, H2B, H3, and H4 (H5 is a variant of H1 found in avian erythrocytes). Histone aggregates may be represented by, for example, $H3_2H4_2$, $H2A_2H2B_2$, etc.

Hitchings, George Herbert (1905–) US biochemist and pharmacologist.

Hittorf, Johann Wilhelm (1824–1914) German chemist and physicist.
Hittorf dark space Use Crookes dark space.
Hittorf's principle

HIV Abbrev. and preferred form for human immunodeficiency virus – the human AIDS virus. Two serovars are recognized at present, designated HIV-1 and HIV-2 (hyphenated, Arabic numerals). Former names: HTLV-III, LAV.

HJBT (or **HBT**) *Electronics* Abbrev. for heterojunction bipolar transistor.

HLA system (not HL-A system) Abbrev. and preferred form for human leucocyte antigen system, the major histocompatibility system (*see* MHC) in humans. The two principal classes of antigens are designated I and II (Roman numerals). Class I antigens are encoded by three genetic loci, designated *HLA-A*, *HLA-B*, and *HLA-C* (hyphens; ital. for genes, roman for the antigens encoded by them); class II antigens are encoded by the *HLA-D* locus, which itself has three subregions denoted *HLA-DP*, *HLA-DR*, and *HLA-DQ* (italic for genes, roman for antigens).

HMI Abbrev. for human–machine interface.

HMPA Abbrev. for hexamethylphosphoramide.

HMPT Abbrev. for hexamethylphosphorous triamide.

hnRNA (lower-case h, n) Abbrev. for heterogeneous nuclear RNA.

hnRNP (lower-case h, n) Abbrev. for heterogeneous nuclear ribonucleoprotein.

Ho Symbol for holmium.

Hoagland, Mahlon Bush (1921–) US biochemist.

Hodge, Sir William Vallance Douglas (1903–75) British mathematician.

Hodgkin, Alan Lloyd (1914–) British physiologist.

Hodgkin, Dorothy Crowfoot (1910–) British chemist.

Hodgkin, Thomas (1798–1866) British physician.
Hodgkin's disease

Hoff, Jacobus van't Usually alphabetized as *van't Hoff.

Hoffmann, Roald (1937–) Polish-born US chemist.
Woodward–Hoffmann rules (en dash)

Hofmann, August Wilhelm von (1818–92) German chemist.
Hofmann reaction
Hofmann rule

Hofmeister, Wilhelm Friedrich Benedict (1824–77) German botanist.

Hofstadter, Robert (1915–90) US physicist.

Hogness box (cap. H) Another name for *TATA box.

Hohmann orbit *Astronautics* Named after Walter Hohmann.

Hol (no stop) Abbrev. for Holocene.

hol- *See* holo-.

holdfast (one word) *See* hapteron.

hole number density *See* number density.

Hollerith, Herman (1860–1929) US inventor.
Hollerith code

Holley, Robert William (1922–) US biochemist.

Holmes, Arthur (1890–1965) British geologist.

holmium Symbol: Ho *See also* periodic table; nuclide.

holo- (sometimes **hol-** before vowels) Prefix denoting complete or entire (e.g. holoblastic, holocrine, holoenzyme, hologram, holozoic, Holarctic).

Holocene (preferred to Recent) Abbrev.: Hol (no stop). **1.** (adjective) Denoting the second epoch of the Quaternary period, i.e. the most recent of all geological epochs. **2.** (noun; preceded by 'the') The Holocene epoch.

Holostei (cap. H) An infraclass of ray-finned fishes containing the garpike and bowfin. Individual name and adjectival form: **holostean** (no cap.); not to be confused with 'holosteous' (having a bony skeleton).

Holothuroidea (cap. H) A class of echinoderms containing the sea cucumbers. Individual name and adjectival form: **holothurian** (no cap.; not holothuroid).

homeo- (now usually preferred to homoeo- in scientific contexts) Prefix denoting similar or like (e.g. homeogenetic, homeomorphic, homeopathy, homeostasis). *See also* homo-; homoio-.

homeo box (two words; preferred to homoeo box) *Genetics* A recurring sequence of bases found in the homeotic genes of insects and vertebrates. It encodes a so-called **homeo domain** (two words) in the protein product.

homeotic (preferred to homoeotic) *Genetics* Denoting or effecting a major mutational change.

hominid Any member of the primate family Hominidae, the only extant species of which is *Homo sapiens* (modern man). The family also includes extinct species of **Homo* and **Australopithecus*. *Compare* hominoid.

hominoid Any member of the primate superfamily Hominoidea, which includes apes and man. *Compare* hominid.

Homo (cap. H, ital.) The genus of hominids that contains modern man and a number of representatives known only from (often scanty) fossil remains. The latter are often designated by common names derived from the site where these fossils were first discovered. The genus contains the following species:

H. erectus: descended from *H. habilis*; includes Java man (*H. erectus erectus*; originally named *Pithecanthropus erectus*) and Peking man (*H. erectus pekinensis*; originally named *Sinanthropus pekinensis*, then *Pithecanthropus pekinensis*).

H. habilis: the earliest species; includes the so-called '1470 skull' (named after its catalogue number).

H. sapiens: descended from *H. erectus*; includes modern man and Cro-Magnon man (*H. sapiens sapiens*), Heidelberg man (originally named *H. heidelbergensis* and regarded by some authorities as a representative of *H. erectus*), Neanderthal man (*H. sapiens neanderthalensis*; originally named *H. neanderthalensis*), Rhodesian man (*H. sapiens rhodesiensis*), Solo man (*H. sapiens soloensis*), Steinheim man, and Swanscombe man.

HOMO Acronym for highest occupied molecular orbital.

homo- Prefix denoting **1.** the same or common (e.g. homocyclic, homoecology, homogametic, homologous, homospory, homotaxis, homozygous). **2.** *Chem.* a homologous relationship between compounds, e.g. homovanillic acid, $HOC_6H_3(OCH_3)CH_2COOH$, and vanillic acid, $HOC_6H_3(OCH_3)COOH$. *See also* homeo-; homoio-. *Compare* hetero-.

Homobasidiomycetidae *See* Basidiomycotina.

homocyclic Designating a chemical compound containing a ring of atoms of the same type. *Compare* heterocyclic.

homoeo- Use *homeo-.

homogeneous Preferred to homogenous for all senses implying similarity of constituent parts or uniformity of nature. However, homogenous is used in reference to *homogeny. Noun form: **homogeneity**.

homogeneously staining region Abbrev.: HSR (no stops). A region occurring at the site of gene amplification on a chromosome, visualized upon staining by a lack of the usual banding pattern.

homogeny *Biol.* Similarity in structure as a result of common ancestry; the term homology is usually preferred. Adjectival form: **homogenous** (*compare* homogeneous).

homoio- Prefix denoting the same; use *homeo- or *homo- except in certain biological senses (e.g. homoiomerous, *homoiothermic).

homoiothermic (preferred to homoiothermal) US: **homeothermic**. Describing animals (**homoiotherms**; US: **homeotherms**) whose body temperature is maintained at a relatively constant level. The term is often incorrectly used as a synonym for *endothermic. Noun form: **homoiothermy** (US: **homeothermy**). *Compare* poikilothermic.

homologue US: **homolog**.

homonym A taxonomic name (e.g. of a species or genus) that has been applied to more than one group of organisms. Only one can be valid, usually the name published first (the senior homonym); the later name (junior homonym) is usually suppressed. Note that this does not apply across kingdom boundaries; e.g. the wasp *Ammophila* and the grass *Ammophila* are not regarded as homonyms. *Compare* synonym.

Homoptera (cap. H) A suborder of bugs (*Hemiptera) including the aphids and cicadas. Individual name and adjectival form: **homopteran** (no cap.; not homopterous).

homovanillic acid Abbrev.: HVA (no stops).

homozygous Having identical alleles of any one gene. Homozygous individuals are called **homozygotes**. Noun form: **homozygosity**. *See also* genotype.

Honda, Kotaro (1870–1954) Japanese metallurgist.

Hooke, Robert (1635–1703) English physicist.
Hooke's joint
Hooke's law

Hooker, Sir Joseph Dalton (1817–1911) British plant taxonomist and explorer, son of William Hooker.

Hooker, Sir William Jackson (1785–1865) British botanist, father of Joseph Hooker.

hook-up (noun; hyphenated) *Telecom., etc.* Verb form: **hook up** (two words).

Hope, Thomas Charles (1766–1844) British chemist.
Hope's apparatus

Hopkins, Sir Frederick Gowland (1861–1947) British biochemist.
Hopkins–Cole reaction (en dash) Also named after Sidney William Cole.

Hopkinson, John (1849–98) British physicist and electrical engineer.

Hoppe-Seyler, (Ernst) Felix Immanuel (hyphen) (1825–95) German biochemist.

Hor (no stop) *Astron.* Abbrev. for Horologium.

Hordeivirus (cap. H, ital.) Approved name for the *barley stripe mosaic virus group. Individual name: **hordeivirus** (no cap., not ital.).

hormogonium (not hormonogonium; pl. hormogonia) A form of bacterial trichome (chain of cells) produced during multiplication.

Horner, Johann Friedrich (1831–86) Swiss ophthalmologist.
Horner's syndrome

hornworts (or **horned liverworts**) *See* Anthocerotopsida.

Horologium A constellation. Genitive form: Horologii. Abbrev.: Hor (no stop). *See also* stellar nomenclature.

Horrocks, Jeremiah (*c.* 1617–41) English astronomer and clergyman.

horsepower Symbol: hp A unit of *power in the obsolete *fps system of units, defined as 550 foot pound-force per second, which is equivalent to 745.700 watts. The metric horsepower is equal to 735.499 watts, i.e. 0.986 horsepower; its use is now discouraged.

horsetail (one word) *Bot. See* Sphenopsida.

horst *Geol.* [from German: thicket]

Hounsfield, Sir Godfrey Newbold (1919–) British engineer.

hour Symbol: h A unit of time equal to 60 minutes or 3600 seconds. Although not an *SI unit, the hour may be used with the SI units and can form compound units with them, for example km/h.

Hoyle, Sir Fred (1915–) British astronomer. *See also* B^2FH theory.

hp Symbol for horsepower.

HP Abbrev. for high pressure.

HPLC Abbrev. for **1.** high-pressure liquid chromatography. **2.** high-performance liquid chromatography.

H–R diagram (en dash) *Astron.* Abbrev. for Hertzsprung–Russell diagram. *See* Hertzsprung, Ejnar.

HSAB Abbrev. for hard and soft acids and bases.

HSI Abbrev. for human–system interface. *See* human–computer interface.

HSR Abbrev. for *homogeneously staining region.

HST Abbrev. for Hubble Space Telescope.

H strand (cap. H) Abbrev. and preferred form for heavy strand (of mitochondrial DNA). *Compare* L strand.

ht (no stop) Abbrev. for height. *See also* length.

HT *Elec. eng.* Abbrev. for high tension.

5-HT Abbrev. for 5-hydroxytryptamine, another name for serotonin. Serotonin receptors are designated 5-HT_1 (subdivided into 5-HT_{1A} and 5-HT_{1B}) and 5-HT_2 (note subscript characters).

HTLV Abbrev. and preferred form for human T-lymphotropic virus (also known as human T-cell leukaemia virus). There are two types, designated HTLV-I and HTLV-II (hyphenated, Roman numerals). HTLV-III was an early name for the human AIDS virus, now renamed *HIV.

Hubble, Edwin Powell (1889–1953) US astronomer and cosmologist.
Hubble classification
***Hubble constant**
Hubble diagram
Hubble effect
Hubble's law
Hubble Space Telescope Abbrev.: HST (no stops)

Hubble constant Symbol: H_0 A constant used in cosmology, the rate at which the expansion velocity of the universe changes with distance. Its value is currently estimated as between 50 and 100 kilometres per second per megaparsec ($1.7–3.4 \times 10^{-18}\,s^{-1}$).

Huggins, Sir William (1824–1910) British astronomer and astrophysicist.

Hulst, Hendrik Christoffell van de Usually alphabetized as *van de Hulst.

human B-lymphotropic virus Abbrev.: HBLV (no stops).

human chorionic gonadotrophin Abbrev.: hCG (lower-case h; this is preferred to HCG).

human–computer interface (en dash) Abbrev.: HCI (no stops). Also called human–system interface (abbrev.: HSI), human–machine interface (abbrev.: HMI), man–machine interface (abbrev.: MMI).

human growth hormone Abbrev.: hGH (no stops). *See* growth hormone.

human herpesvirus Abbrev.: HHV (no stops). *See* herpesvirus.

human immunodeficiency virus Abbrev. and preferred form: *HIV (no stops).

human leucocyte antigen system Abbrev. and preferred form: *HLA system (no stops).

human T-lymphotropic virus Abbrev. and preferred form: *HTLV (no stops).

Humason, Milton La Salle (1891–1972) US astronomer.

Humboldt, (Friedrich Wilhelm Heinrich) Alexander, Baron von (1769–1859) German explorer and scientist.
Humboldt current Use Peru current.
Humboldt glacier

Hume-Rothery, William (hyphen) (1899–1968) British metallurgist.
Hume-Rothery rule

humerus (pl. humeri) The long bone of the upper arm (or upper forelimb of animals). Adjectival form: **humeral** (*compare* humoral).

humidity A measure of the water-vapour content of air. In meteorology this is now usually given as **specific humidity**, symbol q, which is the (dimensionless) ratio of the mass of water vapour to the mass of air containing it (also expressed in g/kg) or in terms of the **mixing ratio**, $q/(1 - q)$.

humoral (not humoural) Literally, relating to a body fluid (**humour**; US: **humor**); used especially to designate circulating (as opposed to cell-bound) antibodies. Not to be confused with humeral (*see* humerus).

hundredweight Symbol: cwt *See* pound.

Hund rule *Chem.* Named after Friedrich Hund (1896–).

Hunsaker, Jerome Clarke (1886–1984) US aeronautical engineer.

Hunter, John (1728–93) British surgeon and anatomist.

Huntington, George (not Huntingdon) (1851–1916) US neurologist.
Huntington's chorea

Hurter, Ferdinand (1844–98) Swiss chemist.
Hurter–Driffield curve (en dash) Also named after V. C. Driffield (1848–1915). Also called characteristic curve.

Hutchinson, John (1884–1972) British botanist.

Hutchinson, Sir Jonathan (1828–1913) British surgeon.
Hutchinson's teeth
Hutchinson's triad

Hutton, James (1726–97) British geologist.

Huxley, Sir Andrew Fielding (1917–) British physiologist, half-brother of Julian Huxley.

Huxley, Hugh Esmor (1924–) British molecular biologist.

Huxley, Sir Julian Sorell (1887–1975) British biologist, grandson of T. H. Huxley and half-brother of Andrew Huxley.

Huxley, Thomas Henry (1825–95) British biologist, grandfather of Julian and Andrew Huxley.
Huxley's layer

Huygens, Christiaan (not Huyghens) (1629–95) Dutch physicist and astronomer.
Huygens construction
Huygens eyepiece
Huygens–Fresnel principle (en dash)
Huygens' principle

HVA Abbrev. for homovanillic acid.

HW Abbrev. for high water.

HWR Abbrev. for heavy-water reactor.

Hya (no stop) *Astron.* Abbrev. for Hydra.

hyaline Translucent: applied to types of cartilage, rock, etc. Not to be confused with **hyalin**, a glassy substance resulting from tissue degeneration.

hyalo- (**hyal-** before vowels) Prefix denoting glassy or transparent (e.g. hyaloplasm, hyaluronic).

hyaluronic acid Abbrev.: HA (no stops).

hybrid The offspring of individuals of different varieties or species. In *binomial nomenclature a hybrid is indicated either by a multiplication sign between the generic name and specific epithet, e.g. *Platanus* × *acerifolia* (London plane), or by a hybrid formula, e.g. *Digitalis grandifolia* × *D. purpurea*.

hybrid four-pole matrix parameter *See* electronics, letter symbols.

hybrid orbital *See* orbital.

hyda- Prefix denoting water (e.g. hydathode, hydatid).

hydr- *See* hydro-.

Hydra A constellation. Genitive form: Hydrae. Abbrev.: Hya (no stop). *See also* stellar nomenclature. *Compare* Hydrus.

Hydra (cap. H, ital.) A genus of hydrozoan coelenterates. Individual name: **hydra** (no cap.; not ital.; pl. hydras).

hydracrylic acid CH_2OHCH_2COOH The traditional name for 2-hydroxypropionic acid.

hydrate *See* inorganic chemical nomenclature.

hydrazobenzene $C_6H_5NHNHC_6H_5$ The traditional name for *N,N'*-diphenylhydrazine.

hydrazoic acid HN_3 The traditional name for hydrogen azide.

hydrazone Any of a class of organic compounds containing the group $=C=NNH_2$, where the carbon atom is joined to two hydrocarbon groups, or to one hydrocarbon group and a hydrogen atom.
Hydrazones are systematically named by adding the word hydrazone after the name of the corresponding aldehyde or ketone, e.g. ethanal hydrazone, CH_3CHNNH_2 (from ethanal, CH_3CHO), and acetone hydrazone, $(CH_3)_2CNNH_2$.
In nonsystematic nomenclature hydrazones are named as in systematic nomenclature but the trivial names of the corresponding aldehydes or ketones are used, e.g. acetaldehyde hydrazone, CH_3CHN-NH_2.

hydro- (sometimes **hydr-** before vowels) Prefix denoting **1.** water, a liquid, or fluids (e.g. hydroacoustics, hydrodynamics, hydroelectric, hydrolysis, hydrophyte, hydrosere, hydrotropism). **2.** hydrogen in a chemical compound (e.g. hydrocarbon, hydrochloric, hydride, hydriodic, hydroxide).

hydrocyanic acid HCN The traditional name for hydrogen cyanide.

hydrogen Symbol: H Isotopes are usually represented as ^1H (protium), ^2H (deuterium), ^3H (tritium). The molecule H_2 is called dihydrogen when it is necessary to distinguish it from the atomic form H (sometimes called monohydrogen). *See also* periodic table; nuclide.

hydrogen azide HN_3 The recommended name for the compound traditionally known as hydrazoic acid.

hydrogencarbonate (one word) A compound containing the ion HCO_3^-, e.g. sodium hydrogencarbonate, $NaHCO_3$. The traditional name is bicarbonate.

hydrogen cyanide HCN The recommended name for the compound traditionally known as hydrocyanic acid or prussic acid.

hydrogenethanedioate A compound containing the ion $HC_2O_4^-$, e.g. potassium hydrogenethanedioate, HC_2O_4K. The traditional name is binoxalate.

hydrogenfluoride (one word) A compound containing the ion HF_2^-, e.g. ammonium hydrogenfluoride, NH_4HF_2.

hydrogensulphate (one word) US: **hydrogensulfate.** A compound containing the ion HSO_4^-, e.g. sodium hydrogensulphate, $NaHSO_4$. The traditional name is bisulphate.

hydrogensulphite (one word) US: **hydrogensulfite.** A compound containing the ion HSO_3^-, e.g. sodium hydrogensulphite, $NaHSO_3$. The traditional name is bisulphite.

hydrophilic Having an affinity for water. *Compare* hydrophobic.

hydrophobic Lacking an affinity for water. *Compare* hydrophilic.

hydroquinone $C_6H_4(OH)_2$ The traditional name for benzene-1,4-diol.

2-hydroxybenzaldehyde HOC_6H_4-CHO The recommended name for the compound traditionally known as salicylaldehyde.

2-hydroxybenzamide HOC_6H_4-$CONH_2$ The recommended name for the compound traditionally known as salicylamide.

2-hydroxybenzoic acid HOC_6H_4-COOH The recommended name for the compound traditionally known as salicylic acid.

3-hydroxybutanal $CH_3CH(OH)CH_2$-CHO The recommended name for the compound traditionally known as aldol.

2-hydroxybutanedioic acid CH_2-(COOH)CH(OH)COOH The recommended name for the compound traditionally known as malic acid.

3-hydroxy-2-butanone CH_3CH-$(OH)COCH_3$ The recommended name for the compound traditionally known as acetoin.

2-hydroxy-1,2-diphenylethanone $C_6H_5COCH(OH)C_6H_5$ The recommended name for the compound traditionally known as benzoin.

hydroxyethanoic acid $CH_2OHCOOH$ The recommended name for the compound traditionally known as glycollic acid.

2-hydroxyethylamine $HOCH_2CH_2$-NH_2 The recommended name for the compound traditionally known as 2-aminoethyl alcohol.

hydroxyl- (**hydroxy-**) Prefix denoting the hydroxyl group, –OH (e.g. hydroxylamine, hydroxyacetone).

4-hydroxy-4-methylpentan-2-one $HOC(CH_3)_2CH_2COCH_3$ The recommended name for the compound traditionally known as diacetone alcohol.

2-hydroxyphenylmethanol C_6H_5-CH_2OH The systematic name for the compound traditionally known as salicyl alcohol.

4-hydroxyproline Abbrev.: Hyp or hyp (no stop). *See* amino acid.

$$\begin{array}{c} OH \\ | \\ HOOC\text{—}C\text{—}COOH \\ | \\ CH_2COOH \end{array}$$

2-hydroxypropane-1,2,3-tricarboxylic acid (citric acid)

2-hydroxypropane-1,2,3-tricarboxylic acid The systematic name for the compound traditionally known as citric acid. Citric acid is preferred in nonchemical contexts.

2-hydroxypropanoic acid CH_3-$CHOHCOOH$ The systematic name for the compound traditionally known as lactic acid.

hydroxypropanone CH_3COCH_2OH The systematic name for the compound traditionally known as acetol.

2-hydroxypropionic acid CH_2-$OHCH_2COOH$ The systematic name for the compound traditionally known as hydracrylic acid.

5-hydroxytryptamine Abbrev.: 5-HT (no stops). Use *serotonin, but note that the abbreviation 5-HT is retained for designating serotonin receptors.

Hydrozoa (cap. H) A class of coelenterates (phylum Cnidaria) containing the jellyfishes and *Hydra*. Individual name and adjectival form: **hydrozoan** (no cap.).

Hydrus A constellation. Genitive form: Hydri. Abbrev.: Hyi (no stop). *See also* stellar nomenclature. *Compare* Hydra.

hyeto- Prefix denoting rain (e.g. hyetograph).

hygro- (**hygr-** before vowels) Prefix denoting moisture (e.g. hygrophyte, hygroscopic, hygristor).

Hyi (no stop) *Astron.* Abbrev. for Hydrus.

hylo- (**hyl-** before vowels) Prefix denoting wood (e.g. hylophagous).

Hyman, Libbie Henrietta (1888–1969) US zoologist.

hymeno- Prefix denoting a membrane (e.g. hymenophore, *Hymenoptera).

Hymenomycetes (cap. H) A class of basidiomycete fungi (*see* Basidiomycotina) that includes the agarics. Individual name: **hymenomycete** (no cap.).

Hymenoptera (cap. H) An order of insects including the ants, bees, and wasps. Individual name: **hymenopteran** (no cap.). Adjectival form: **hymenopterous** or **hymenopteran**.

hyo- Prefix denoting the hyoid bone (e.g. hyomandibular, hyostylic).

hyp- *See* hypo-.

hyper- Prefix denoting over, greater than, excessive, or beyond (e.g. hyperbaric, hyperplasia, hypersonic, hypertonic, hypertrophy).

hyperbola (pl. hyperbolas or hyperbolae) Adjectival form: **hyperbolic**. Derived noun: **hyperboloid**.

A hyperbola is a plane curve, a hyperboloid is a curved surface. The adjective hyperbolic is often used in optics, astronomy, etc., to describe a hyperboloid surface (as in hyperbolic mirror), since it is the hyperbolic sections of the surface that are relevant.

hyperbolic functions *Maths.* The symbols for these functions are formed by adding the letter h to the end of the corresponding trigonometric function, e.g.

cosh *x*, sinh *x*, tanh *x*, sech *x*, etc.
(pronounced cosh, shine, thann, sech).

The symbols for the inverse hyperbolic functions are formed by adding the letters ar to the beginning of corresponding hyperbolic function, e.g.

arcosh *x*, arsinh *x*, artanh *x*, etc.
(pronounced arc-cosh, arc-shine, etc.).

hypercharge *See* isospin.

hyperchromic Denoting an increase in absorption intensity as a result of a chromophore. *Compare* hypochromic.

Hypericaceae *See* Guttiferae.

hypha (pl. hyphae) *Bot.* Adjectival form: **hyphal**.

Hyphochytriomycetes A class of fungi of the subdivision *Mastigomycotina. In a recent classification these organisms are regarded as a division, Hypochytriomycota, in the subkingdom *Chromobionta.

Hyphomicrobium (cap. H, ital.) A genus of prosthecate bacteria. Individual name: **hyphomicrobium** (no cap., not ital.; pl. hyphomicrobia).

Hyphomonas (cap. H, ital.) A genus of prosthecate bacteria. Individual name and adjectival form: **hyphomonad** (no cap., not ital.).

Hyphomycetes (cap. H) A class of imperfect fungi (*see* Deuteromycotina) that do not possess fruiting bodies. It includes many moulds (e.g. *Aspergillus*, *Penicillium*) and the imperfect yeasts, formerly contained in the class Blastomycetes. Individual name and adjectival form: **hyphomycete** (no cap.).

hypo- (usually **hyp-** before vowels) Prefix denoting **1.** beneath, lower down (e.g. hypodermis, hypogeal, hyponasty, hypabyssal). **2.** less than, lacking, or deficient (e.g. hypo-osmotic (hyphenated), hypotonic). **3.** *Chem.* a lower oxidation state in comparison to another compound (e.g. hypophosphorous acid, H_3PO_2, in relation to phosphorous acid, H_3PO_3). In modern chemical nomenclature, the prefix is not used; for examples, see individual entries.

hypobromite A compound containing the ion BrO^-, e.g. sodium hypobromite, NaBrO. The recommended name is bromate(I).

hypobromous acid HOBr The traditional name for bromic(I) acid.

hypochlorite A compound containing the ion ClO^-, e.g. sodium hypochlorite, NaClO. The recommended name is chlorate(I).

hypochlorous acid HOCl The traditional name for chloric(I) acid.

hypochromic Denoting a decrease in absorption intensity as a result of a chromophore. Not to be confused with *hypsochromic. *Compare* hyperchromic.

hypophosphite A compound containing the ion $H_2PO_2^-$, e.g. potassium hypophosphite, KH_2PO_2. The recommended name is phosphinate.

hypothalamus (pl. hypothalami) *Anat.* Adjectival form: **hypothalamic**.

hypothesis (pl. hypotheses) Adjectival form: **hypothetical**. Verb form: **hypothesize**.

hypso- (**hyps-** before vowels) Prefix denoting **1.** height (e.g. hypsodont, hypsographic, hypsometer). **2.** shorter wavelengths (e.g. hypsochrome).

hypsochromic Denoting the shift of an absorption maximum to shorter wavelength. Not to be confused with *hypochromic. *Compare* bathochromic.

Hyracoidea (cap. H) An order of mammals comprising the hyraxes. Individual name and adjectival form: **hyracoidean** (no cap.; not in common usage).

hysteresis *Physics* Adjectival form: **hysteresis** (not hysteretic), as in hysteresis loss.

hystero- (**hyster-** before vowels) Prefix denoting **1.** lagging or later (e.g. hysteranthous, hysteresis). **2.** the uterus (e.g. hysterectomy).

Hz Symbol for hertz. *See also* SI units.

I

i Symbol for the imaginary number $\sqrt{-1}$.

i Symbol for **1.** (light ital.) instantaneous current (*see also* electronics, letter symbols). **2.** (light ital.) van't Hoff factor. **3.** (bold ital.) a unit coordinate vector.

I Symbol for **1.** inductive (or inductomeric) effect (*see* electron displacement). **2.** inosine. **3.** iodine. **4.** isoleucine.

I Symbol (light ital.) for **1.** electric current (*see also* electronics, letter symbols). **2.** intensity. **3.** ionic strength. **4.** ionization energy or ionization potential. **5.** isotopic spin (or isospin) quantum number. **6.** (or I_v) luminous intensity. **7.** moment of inertia. **8.** nuclear spin quantum number. **9.** (or I_e) radiant intensity. **10.** unit matrix.

IA Abbrev. for intra-arterial.

-ia Noun suffix denoting certain *classes in animal taxonomy (e.g. Mammalia, Reptilia). Adjectival form (no initial cap.): **-ian**.

IAA Abbrev. and preferred form for indoleacetic acid.

Ia antigens (cap. I) Abbrev. and preferred form for I-region-associated antigens. *See* H–2.

IAEA Abbrev. for International Atomic Energy Agency.

IAU Abbrev. for International Astronomical Union.

IBA Abbrev. for indole 3-butyric acid.

Ibn Musa *See* al-Khwarizmi.

Ibn Rushd *See* Averroës.

ibn-Sina *See* Avicenna.

iBu, i-Bu Use Bu^i (*see* Bu).

IC Abbrev. for **1.** *Electronics* integrated circuit. **2.** *Astron.* Index Catalogue, either of two supplements to the *NGC.

-ic Adjectival suffix denoting **1.** characteristic of, resembling, involving (e.g. algebraic, cyclic, electronic, seismic). *See also* -ical. **2. a.** the presence of a metal in its highest oxidation state (e.g. ferric, stannic). In systematic chemical nomenclature, the oxidation state is used (e.g. iron(III) chloride for ferric chloride). **b.** a higher oxidation state in an acid (e.g. hydrochloric, sulphuric). In systematic chemical names, the suffix is used for all oxo acids of the element, with the oxidation number attached. For example sulphuric acid, H_2SO_4, is strictly sulphuric(VI) acid while sulphurous acid, H_2SO_3, is sulphuric(IV) acid.

-ical Adjectival suffix denoting characteristic of, resembling, or involving (e.g. chemical, mathematical, zoological). *See also* -ic.

ICBN Abbrev. for International Code of Botanical Nomenclature. *See also* binomial nomenclature; taxonomy.

ICE Abbrev. for **1.** Institution of Civil Engineers. **2.** internal-combustion engine. *See also* IC engine.

ice age (two words) Also called glacial period. The 'Ice Age' (initial caps.) is an informal name for the Pleistocene epoch.

iceberg (one word)

iceblink (one word)

icecap (one word)

ice crystal (two words) Hyphenated when used adjectivally (e.g. ice-crystal structure).

icefall (one word)

ice field (two words)

ice fish (two words)

ice floe (two words)

Iceland spar (cap. I)

IC engine Abbrev. for internal-combustion engine. *See also* ICE.

ice point (two words)

ice sheet (two words)

ice shelf (two words)

I.Chem.E. (or **IChemE**) Abbrev. for Institution of Chemical Engineers.

ichthyo- (**ichthy-** before vowels) Prefix denoting fish (e.g. ichthyology, ichthyosis).

Ichthyopterygia (cap. I) A subclass of extinct marine reptiles comprising the ichthyosaurs. Individual name and adjectival form: **ichthyopterygian** (no cap.). *Compare* ichthyopterygium.

ichthyopterygium (pl. ichthyopterygia) A limb adapted for swimming. *Compare* Ichthyopterygia.

ichthyosaur Any extinct marine reptile of the order Ichthyosauria (subclass *Ichthyopterygia), which includes the genus *Ich-*

thyosaurus. Adjectival form: **ichthyosaurian**.

ichthyosaurus (pl. ichthyosauruses or ichthyosauri) Any *ichthyosaur of the genus *Ichthyosaurus* (cap. I, ital.). The word should not be used for ichthyosaurs of other genera.

ICI *See* CIE.

ICNB Abbrev. for International Code of Nomenclature of Bacteria. *See also* taxonomy.

ICNCP Abbrev. for International Code of Nomenclature of Cultivated Plants. *See also* binomial nomenclature; taxonomy.

ICNV Abbrev. for International Code of Nomenclature of Viruses. *See also* taxonomy.

icon (not ikon) *Computing*

-ics Noun suffix denoting a body of knowledge or study of (e.g. aeronautics, cladistics, electronics, mathematics). Words ending in -ics take a sing. form of verb. Adjectival form: ***-ic** or ***-ical**.

ICSH Abbrev. for interstitial-cell-stimulating hormone.

ICTV Abbrev. for International Committee on Taxonomy of Viruses. *See* virus.

ICZN Abbrev. for International Code of Zoological Nomenclature. *See also* binomial nomenclature; taxonomy.

-id Noun suffix denoting **1.** (usually pl.) a meteor shower from the direction of a specified constellation (e.g. Geminids, Perseids). **2.** a member of a zoological family (e.g. bovid, hominid). *See also* -idae.

-idae Noun suffix denoting **1.** a *family in animal taxonomy (e.g. Bovidae, Canidae). Adjectival form (no initial cap.): **-id** (e.g. bovid). **2.** a *subclass in plant taxonomy (e.g. Hemiascomycetidae, Magnoliidae).

-ide Noun suffix characteristic of the more electronegative element in simple compounds (e.g. sodium chloride, hydrogen sulphide).

idio- Prefix denoting separate or peculiar to (e.g. idioblast, idiomorphic).

IDP Abbrev. and preferred form for inosine 5′-diphosphate.

IE Abbrev. for immunoelectrophoresis.

i.e. (stops) Abbrev. for *id est* (Latin: that is).

IEA Abbrev. for International Energy Agency.

IEC Abbrev. for International Electrotechnical Commission.

IED Abbrev. for Information Engineering Directorate.

IEE Abbrev. for Institution of Electrical Engineers, a UK organization. *Compare* IEEE.

IEEE (no stops; usually pronounced I triple E) Abbrev. for Institute of Electrical and Electronics Engineers, a US organization. *Compare* IEE.

IERE Abbrev. for Institution of Electronic and Radio Engineers.

IF Abbrev. for **1.** intermediate frequency. **2.** *Genetics* initiation factor. In prokaryotes the different initiation factors are denoted by Arabic numerals: for instance in *E. coli* they are designated IF1, IF2, and IF3 (IF3 was originally termed the 'dissociation factor'). In eukaryotes, initiation factors are denoted by an Arabic numeral followed, in some cases, by capital letters; the designations are prefixed by a lower-case e (for eukaryote), e.g. eIF3, eIF4A, eIF4D, etc.

iff *Logic* Short for if and only if. Symbol: $\Leftrightarrow$

IFIP Abbrev. for International Federation for Information Processing.

-iformes Noun suffix denoting an *order of birds (e.g. Gruiformes, Passeriformes).

IFRB Abbrev. for International Frequency Registration Board.

Ig (cap. I, no stops) Abbrev. for *immunoglobulin used, always in combination with a capital letter, to designate classes of immunoglobulin:

- **IgA**, divided into two subclasses: IgA1 and IgA2
- **IgD**
- **IgE**
- **IgG**, divided into four subclasses: IgG1, IgG2, IgG3, and IgG4
- **IgM.**

IGFET *Electronics* Acronym for insulated-gate field-effect transistor.

igneous (not ignious) *Geol.*

IGY Abbrev. for International Geophysical Year (1.7.57 to 31.12.58).

I-Hsing (*c.* 681 – *c.* 727) Chinese mathematician and astronomer.

ijolite A coarse-grained igneous rock. *Compare* iolite.

IJsselmeer (not Ijsselmeer) Former name: Zuider Zee.

IKBS *Computing* Abbrev. for intelligent knowledge-based system.

ikon *Computing* Use icon.

IL Abbrev. for interleukin used, always prefixed to an Arabic numeral, to designate individual interleukins; for example, IL-1, IL-2, etc.

IL-2R Abbrev. for interleukin-2 receptor.

I²L (or **IIL**) *Electronics* Symbol for integrated injection logic.

Ilarvirus (cap. I, ital.) Approved name for the *tobacco streak virus group. Individual name: **ilarvirus** (no cap., not ital.). [from *i*sometric *la*bile *r*ingspot *virus*]

Ile (or **ile**; no stop) Abbrev. for isoleucine. *See* amino acid.

ileum (pl. ilea) The terminal portion of the small intestine in mammals. Not to be confused with *ilium or with **ileus** (intestinal obstruction). Adjectival form: **ileal** (not ileac).

ilium (pl. ilia) A bone of the pelvis; not to be confused with *ileum. Adjectival form: **iliac** (not ilial).

Ilkovic equation (preferred to Heyrovský – Ilkovic equation) *Polarography*

illuminance Symbol: E or E_v (v stands for visible). A physical quantity that, at a point of a surface, is the *luminous flux falling on an element of the surface divided by the area of that element. The *SI unit is the lux (lx). *Compare* irradiance.

Ilyushin, Sergei Vladimirovich (1894 – 1977) Soviet aircraft designer.

IM Abbrev. for intramuscular.

im- (before b, m, and p) Prefix denoting **1.** not (e.g. imparipinnate). **2.** into or within (e.g. imbibition, implantation, implosion, imprinting). *See also* in-; ir-.

image photon counting system Abbrev.: IPCS (no stops).

imago (pl. imagines in entomology, imagos in psychology)

imbrication Adjectival form: **imbricate** or **imbricated**.

Imbrie, John (1925–) US geologist.

IMechE (or **I.Mech.E.**) Abbrev. for Institution of Mechanical Engineers.

imine Any of a class of organic compounds containing the group =C=NH.

Imines are systematically named by adding the word imine after the name of the corresponding aldehyde or ketone, e.g. ethanal imine, CH_3CHNH (from ethanal, CH_3CHO), propanone imine, $(CH_3)_2$-CNH, and ethanal methylimine, CH_3-$CHNCH_3$.

In nonsystematic nomenclature imines are named as in systematic nomenclature but the trivial names of the corresponding aldehydes or ketones are used, e.g. acetaldehyde imine, CH_3CHNH.

iminourea $(H_2N)_2C{=}NH$ The recommended name for the compound traditionally known as guanidine.

immune response gene Abbrev. and preferred form: Ir gene (cap. I, no stops).

immuno- Prefix denoting immunity or immune (e.g. immunoassay, immunodeficiency, immunofluorescence, immunosorbent, immunosuppression).

immunoelectrophoresis (one word) Abbrev.: IE (no stops).

immunoglobulin A type of protein with antibody activity, made up of two types of polypeptide chain, light and heavy. Immunoglobulins are divided into five classes (designated Ig followed by a capital letter), according to the type of heavy chain (designated by a lower-case Greek letter) they possess: IgA (α chain), IgD (δ chain), IgE (ϵ chain), IgG (γ chain), and IgM (μ chain). Some classes are divided into subclasses designated by Arabic numerals (e.g. IgG1 (with heavy chain $\gamma 1$), IgG2 (with heavy chain $\gamma 2$), etc.). Subcategories of a particular chain are denoted by a subscript character (e.g. μ_m is a membrane-bound μ

chain). Light chains, which are designated κ and λ, do not determine immunoglobulin class. Note that Ig is only used in combination to designate immunoglobulin classes; it is not used alone as an abbreviation of immunoglobulin.

Immunoglobulins can be degraded by enzymes into various fragments, denoted by a capital F followed by an appropriate qualifying lower-case letter (or letters). The best known are Fab (antigen-binding fragment), Fc (crystallizable fragment), Fh (hinge fragment), and $F(ab')_2$ (two Fabs + Fh).

Different regions of heavy (H) and light (L) chains can be recognized: these are denoted by the capital letters C (constant), V (variable), D (diversity), and J (joining). These letters may be followed by a subscript H or L (according to the chain) and by Arabic numerals if there is more than one such region; for example, J_H1 and J_H2 are two joining regions of a heavy chain. These regions may be further specified according to the class of immunoglobulin molecule; for example, $C_\gamma1$, $C_\gamma2$, and $C_\gamma3$ (note subscript gammas) are constant regions of the γ chain of IgG. The genes coding for these regions are designated in the same way as the regions, but using italic type. Thus the variable region of a lambda chain (V_λ) is encoded by the gene V_λ (italic V).

IMP Abbrev. and preferred form for inosine 5′-phosphate.

IMPATT *Electronics* Acronym for impact ionization avalanche transit time.

impedance Symbol: Z A physical quantity associated with an alternating-current circuit. It is the complex representation of potential difference divided by the complex representation of electric current.

Impedance may be written as a complex number:

$$Z = R + iX,$$

where R, the real part, is the *resistance and X, the imaginary part, is the reactance; $i = \sqrt{-1}$. The **modulus of impedance** is then:

$$|Z| = \sqrt{(R^2 + X^2)}.$$

The *SI unit of impedance, resistance, and reactance is the ohm.

imperfect fungi Fungi that do not undergo sexual reproduction. *See* Deuteromycotina. *Compare* perfect fungi.

impulse Symbol: $\boldsymbol{I}$ A *vector quantity equal to the time integral $\int \boldsymbol{F}\,dt$, where $\boldsymbol{F}$ is the *force acting over time t. The *SI unit is the newton second (N s).

in Symbol for inch.

In Symbol for indium.

in- Prefix denoting **1.** not (e.g. incompatibility, indehiscent, indeterminacy, inelastic, innominate, inorganic). **2.** into or within (e.g. inbreeding, innervation, intine, intron, invagination). *See also* im-; ir-.

-in Noun suffix characteristic of neutral compounds (e.g. albumin, casein, gelatin). Former spellings of certain of these compounds used the suffix -ine and are still so spelt in popular use (e.g. gelatine).

-inae Noun suffix denoting **1.** a *subfamily in animal taxonomy (e.g. Melinae, Sciurinae). **2.** a subtribe (*see* tribe) in plant taxonomy (e.g. Scandicinae).

inch Symbol: in A unit of length now equal to exactly 0.0254 metre.

incompatibility group *Genetics See* plasmid.

incus (pl. incudes) The anvil-shaped (middle) ear ossicle of mammals.

Ind *Astron.* Abbrev. for Indus.

index 1. (pl. indexes) A systematic list. *See also* indexing. **2.** (pl. indices; preferred to indexes) *Maths., physics See also* coefficient. **3.** (pl. indexes) An indicator.

Index Catalogue *Astron.* Abbrev.: IC (no stops). *See* NGC.

indexing Indexing of scientific books, journals, etc., follows usual indexing rules. A few specifically scientific problems arise: 1. Greek letters. In chemical compounds, star names, etc., a Greek letter appearing as the first character of an index title must be written out (and any hyphen dropped), e.g.
α-Centauri is indexed as Alpha Centauri
α chain is indexed as alpha (α) chain
α-iron is indexed as alpha iron
λ (a bacteriophage) is indexed as lambda (λ) phage.

However, for names of chemical compounds that start with a Greek letter, the Greek is ignored in alphabetization. For example, γ-aminobutyric acid is written γ-aminobutyric acid but alphabetized under A.

2. Numerals and prefixes. Chemical compounds in which the first character or characters are numerals or other prefixes (e.g. 0-, *s*-, *cis*-) are ignored for alphabetization but taken into account in ordering the entries, e.g.

2,3-dihydroxybenzene
2,4-dihydroxybenzene
cis-1,2-dimethylcyclohexane.

3. Capitalization. The initial letter of an index title should not be capitalized unless it is customarily written with a capital letter, e.g.

Euglenophyta
eukaryote
europium
Eustachian tube.

4. Eponymous entries. An eponymous index entry with a lower-case initial letter usually precedes a biographical entry of the same name, e.g.

newton
Newton, Sir Isaac.

5. Italicization. Index titles should be italicized if the word is usually printed in italic type, e.g.

Eugenia caryophyllata
Euglena
fons et origo.

indium Symbol: In *See also* periodic table; nuclide.

indoleacetic acid Abbrev.: IAA (no stops).

indole 3-butyric acid Abbrev.: IBA (no stops).

inductance *See* self-inductance; mutual inductance.

induction Adjectival form: **induced**.

inductive effect *See* electron displacement.

inductomeric effect *See* electron displacement.

Indus A constellation. Genitive form: Indi. Abbrev.: Ind (no stop). *See also* stellar nomenclature.

-ine 1. Noun suffix characteristic of basic nitrogen compounds (e.g. cocaine, epinephrine). **2.** Adjectival suffix denoting characteristic of, consisting of, or resembling (e.g. crystalline, feline).

-ineae Noun suffix denoting a *suborder in plant taxonomy.

inert gases Use noble gases.

inertia Adjectival forms: **inertia, inertial.** In most mathematical contexts 'inertial' is the commoner form (e.g. inertial force, inertial mass) but in some technological contexts 'inertia' is more common (e.g. inertia-reel seat belt, inertia switch).

inf (no stop) *Maths.* Abbrev. for infimum.

infimum (pl. infima) *Maths.* Abbrev.: inf (no stop).

infinity Symbol: ∞ Adjectival forms: **infinite** (extremely great, boundless, immeasurable, etc.), **infinitesimal** (extremely or immeasurably small). The smallest infinite number is denoted $\aleph_0$ (Hebrew aleph, subscript zero).

inflammable Use *flammable.

inflexion *Maths.* Use inflection.

influenza virus (two words) Either of two species of virus, designated A and B, constituting a genus in the family Orthomyxoviridae. A third type, formerly called influenza virus C, is now placed in the separate genus *mesoinfluenza virus. The name is sometimes printed as one word (influenzavirus); although this form is consistent with most other viral names, it has yet to achieve widespread acceptance.

Information Engineering Directorate Abbrev.: IED (no stops).

information retrieval Abbrev.: IR (no stops).

information storage and retrieval Abbrev.: ISR (no stops).

information technology Abbrev.: IT (no stops).

infra- Prefix denoting **1.** below in position, range, or extent (e.g. infraorbital, infrared, infrasound, infraspecific). **2.** an intermediate rank in animal *taxonomy (e.g. infraclass).

infrared (one word) Abbrev.: IR (no stops).

Infrared Astronomical Satellite Abbrev.: IRAS (no stops).

Infrared Telescope Facility Abbrev.: *IRTF (no stops).

infundibulum (pl. infundibula) *Anat.* Adjectival form: **infundibular**.

Ingenhousz, Jan (1730–99) Dutch plant physiologist and physician.

Ingold, Sir Christopher Kelk (1893–1970) British chemist.

Ingram, Vernon Martin (1924–) German-born US biochemist.

inHg *See* mmHg.

inH$_2$O *See* mmHg.

inhibitory postsynaptic potential Abbrev.: i.p.s.p. (stops; pl. i.p.s.ps) or IPSP (no stops; pl. IPSPs).

-ini Noun suffix denoting a *tribe in animal taxonomy (e.g. Bovini, Caprini).

initiation factor *Genetics* Abbrev.: *IF (no stops).

innate releasing mechanism *Ethology* Abbrev.: IRM (no stops).

innominate (not inominate) Denoting various anatomical parts more correctly known by their specific names.
innominate artery, vein Use brachiocephalic artery, vein.
innominate bone Use hip bone.

inoculate (not innoculate) To introduce microorganisms into a suitable culture medium or pathogens into a plant or animal. The word should not be used as a synonym for *vaccinate. Derived noun: **inoculation**.

inoculum (not innoculum; pl. inocula) Inoculated material. *See* inoculate.

inorganic chemical nomenclature Binary inorganic compounds are named using the name of the more electropositive element followed by a shortened form of the name of the more electronegative element with the suffix -ide. The oxidation number, in Roman numerals (small caps.) enclosed in parentheses, follows the name of the least electronegative constituent with no intervening space, e.g. antimony(III) chloride, $SbCl_3$, and copper(II) oxide, CuO. If the compound is a common one or the structure is unambiguous, the oxidation number can be omitted, e.g. carbon dioxide, CO_2, and calcium chloride, $CaCl_2$. Certain hydrides have acceptable common names, e.g. water, H_2O, ammonia, NH_3, hydrazine, N_2H_4, diphosphane, P_2H_4, trisilane, Si_3H_8, and plumbane, PbH_4.
The names of heteropolyatomic anions generally terminate in the suffix -ate, e.g. the phosphate(V) ion, PO_4^{3-}, and the peroxosulphate(VI) ion, SO_5^{2-}; however, the suffix -ide can be used in some cases, e.g. the hydroxide ion, OH^-, the imide ion, NH^{2-}, the amide ion, NH_2^-, and the cyanide ion, CN^-.
Many heteropolyatomic cations have names ending in the suffix -yl, e.g. the nitrosyl ion, NO^+, and the uranyl(VI) ion, UO_2^{2+}; those derived by addition of protons to a monatomic anion end in the suffix -onium, e.g. the phosphonium ion, PH_4^+. The NH_4^+ ion retains its common name of ammonium. The names of cations derived from nitrogen bases with names ending in -amine are formed by replacing this suffix by -ammonium, e.g. the hydroxylammonium ion, $HONH_3^+$, while names of cations derived from nitrogen bases having names with endings other than -amine are formed by adding the suffix -ium to the name of the base, omitting any terminal vowel, e.g. the pyridinium ion, $C_5H_5N^+H$, from pyridine.
For general use IUPAC recommends short forms of certain systematic names, e.g. the disulphate(IV) ion, $S_2O_5^{2-}$, instead of the pentaoxodisulphate(IV) ion and the phosphate(V) ion, PO_4^{3-}, instead of the tetraoxophosphate(V) ion) and certain common names, e.g. the nitrate ion, NO_3^-, the nitrite ion, NO_2^-, the sulphite ion, SO_3^{2-}, the sulphate ion, SO_4^{2-}, the chromyl group, CrO_2, and the sulphuryl group, SO_2.
It is often possible to specify a characteristic central atom in a heteropolyatomic compound or ion, in which case the compound

can be regarded as a coordination compound for the purpose of nomenclature.

Hydrates are named in two ways. The number of water molecules can be indicated, as in copper(II) sulphate-5-water, $CuSO_4{\cdot}5H_2O$. Alternatively, the compound can be named as a coordination compound, with water as a ligand, as in tetraaquacopper(II) tetraoxosulphate(VI)-1-water, $[Cu(H_2O)_4]SO_4{\cdot}H_2O$. If the extent of hydration is uncertain the word 'hydrated' is placed before the name of the compound, e.g. hydrated aluminium nitrate, $Al(NO_3)_3{\cdot}nH_2O$.

Two series of prefixes are used in inorganic nomenclature: mono-, di-, tri-, tetra-, penta-, etc., and bis-, tris-, tetrakis-, pentakis-, etc. The first series is used to indicate: (i) the relative proportions of different elements, e.g. diiodine hexachloride, I_2Cl_6; (ii) the number of substituted atoms, e.g. monochlorosilane, $SiClH_3$; (iii) the number of identical coordinated ligands, e.g. potassium hexacyanoferrate(III), $K_3[Fe(CN)_6]$; (iv) the number of identical central atoms in condensed acids and their salts, e.g. potassium dichromate(VI), $K_2Cr_2O_7$; and (v) the number of identical linked atoms, e.g. trisilane, Si_3H_8. The second series is used to indicate the number of identical substituents in a compound and to avoid ambiguity, e.g. tris(decyl)phosphine, $(P(C_{10}H_{21})_3$, avoids confusion with the tridecyl group $(C_{13}H_{27})$ and dicyanotetrakis(methyl isocyanide)iron(II), $[Fe(CN)_2(CH_3NC)_4]$, indicates clearly that the ligand is methyl isocyanide (CH_3NC).

It is recommended that names of minerals are written in parentheses after the names of the compounds (or elements) constituting them, e.g. calcium sulphate-2-water (gypsum), carbon (diamond), and iron(II) disulphide (pyrites).

inosine Symbol: I *See* nucleoside.

inosine 5′-diphosphate Abbrev. and preferred form: IDP (no stops).

insosine 5′-phosphate Abbrev. and preferred form: IMP (no stops).

inosine 5′-triphosphate Abbrev. and preferred form: ITP (no stops).

inositol 1,4,5-triphosphate Abbrev.: IP_3 (subscript 3).

input (noun and adjective) Verb form: **inputs, inputting, input** (not inputted).

input/output (or **input-output**) Abbrev.: I/O (no stops).

Insecta (cap. I) The class of arthropods comprising the insects. Also called Hexapoda.

Insectivora (cap. I) An order of mammals including the shrews, hedgehogs, moles, etc. (*See also* Macroscelidea.) Individual name: **insectivore** (no cap.). The adjectival form **insectivorous** is applied to the insect-eating habit rather than to the taxonomic group, as many insectivores are omnivorous.

insectivore 1. A member of the *Insectivora. **2.** Any animal that feeds on insects. Adjectival form (for second sense only): **insectivorous**.

insectivorous plant Use *carnivorous plant.

insertion sequence *Genetics* Abbrev.: *IS (no stops).

insol. Abbrev. for insoluble.

insolation Exposure to solar radiation (not to be confused with insulation).

insoluble Abbrev.: insol. (stop).

install (not instal) Noun form: **installation** (not instalation).

Institute of Electrical and Electronics Engineers Abbrev.: *IEEE (no stops).

Institute of Physics Abbrev.: InstP (no stops) or Inst.P. (stops).

Institution of Chemical Engineers Abbrev.: I.Chem.E. (stops) or IChemE (no stops).

Institution of Civil Engineers Abbrev.: ICE (no stops).

Institution of Electrical Engineers Abbrev.: *IEE (no stops).

Institution of Electronic and Radio Engineers Abbrev.: IERE (no stops).

Institution of Mechanical Engineers Abbrev.: IMechE (no stops) or I.Mech.E. (stops).

insulator (not insulater)

INT Abbrev. for Isaac Newton Telescope (La Palma).

integer Any member of the set of positive and negative whole numbers and zero. Adjectival forms: **integer, integral**; use integer when there may be ambiguity (e.g. integer equation). *See also* sets (for denotation). *Compare* natural numbers.

integral (not intergral) *Maths.* **1.** (noun) Denoted $\int$ or for a definite integral $\int_a^b$, where *a* and *b* are the lower and upper *limits. The line integral around a closed curve is denoted $\oint$ (integral sign plus circle). Adjectival form: **integral**. Derived nouns: **integration, integrand. 2.** (adjective) Relating to or containing integrals (e.g. integral sign). **3.** (adjective) *See* integer.

integrated circuit Abbrev.: IC (no stops). Informal name: chip.

integrated injection logic Abbrev.: I^2L or IIL (no stops).

Integrated Services Digital Network Abbrev.: ISDN (no stops).

intelligence quotient Abbrev. and preferred form: IQ (no stops).

intelligent knowledge-based system *Computing* Abbrev.: IKBS (no stops).

Intelsat Acronym for International Telecommunications Satellite Consortium.

intensity *See* luminous intensity; radiant intensity; sound intensity.

inter- Prefix denoting between or mutual (e.g. intercalary, intercellular, intercostal, interface, interglacial, internode, interstellar).

interleukin Any of a family of serum proteins important in the immune response. Members are designated IL-1 to IL-6 (note hyphens); IL-2 is the most important.

intermediate frequency Abbrev.: IF (no stops).

internal-combustion engine (hyphenated) Abbrev.: IC engine or ICE (no stops).

internal energy Symbol: U A thermodynamic property equal to the sum of the kinetic and potential *energies of all the atoms and/or molecules, ions, and electrons in a system. The *SI unit is the joule. The **zero-point energy**, symbol U_0, is the residual internal energy at absolute zero.
Only a change, ΔU, in the internal energy of a system can be determined. It is recommended by IUPAC that the equation for ΔU be written:

$$\Delta U = Q + W,$$

where Q is the heat abstracted by the system from its surroundings and W is the work done by the system on the surroundings. (This equation was previously written with a minus rather than a plus sign.) *See also* heat.
Specific internal energy, symbol u, is internal energy divided by mass; it is measured in J kg^{-1}. **Molar internal energy**, symbol U_m, is internal energy divided by *amount of substance; it is measured in J mol^{-1}.

internal transmission density Symbol: D_i A physical quantity given by:

$$\log_{10}(\Phi_o/\Phi_{tr}),$$

where Φ_o is the incident radiant (or luminous) flux and Φ_{tr} is the transmitted flux. Also called absorbance. Former name: optical density.

International Atomic Energy Agency Abbrev.: IAEA (no stops).

International Astronomical Union Abbrev.: IAU (no stops).

International Atomic Time *See* TAI.

International Codes of Nomenclature *Biol. See* binomial nomenclature; taxonomy.

International Commission on Illumination Abbrev.: ICI (no stops). *See* CIE.

International Committee on Taxonomy of Viruses Abbrev.: ICTV (no stops). *See* virus.

International Committee on Weights and Measures *See* CIPM.

International Electrotechnical Commission Abbrev.: IEC (no stops).

International Energy Agency Abbrev.: IEA (no stops).

International Federation for Information Processing Abbrev.: IFIP (no stops).

International Frequency Registration Board Abbrev.: IFRB (no stops).

International Geophysical Year Abbrev.: IGY (no stops). The year 1.7.57 to 31.12.58.

international nautical mile *See* nautical mile, international.

International Organization for Standardization Abbrev.: *ISO (no stops).

International Practical Temperature Scale Abbrev.: *IPTS (no stops).

International Standard Book Number Abbrev.: ISBN (no stops).

International Statistical Institute Abbrev.: ISI (no stops).

International Table calorie *See* calorie.

International Telecommunication Union Abbrev.: ITU (no stops).

International Ultraviolet Explorer Abbrev.: IUE (no stops).

International Union for the Conservation of Nature and Natural Resources Abbrev.: IUCN (no stops).

International Union of Pure and Applied Chemistry Abbrev. or acronym: IUPAC (no stops).

International Union of Pure and Applied Physics Abbrev.: IUPAP (no stops).

international unit Abbrev.: IU (no stops). A unit of measurement of the vitamin content of foods.

intero- Prefix denoting inside, within (e.g. interoceptor).

interplanetary scintillations Abbrev.: IPS (no stops).

interrupt Used as a noun and adjective as well as a verb in computing.

intersecting storage ring *Physics* Abbrev.: ISR (no stops).

intersection *Maths.* Symbol: ∩ For *sets A and B:

$$A \cap B = \{x \mid x \in A \text{ and } x \in B\}.$$

Compare union.

interstellar medium Abbrev.: ISM (no stops).

interstice (usually used in the pl.) Adjectival form: **interstitial**.

interstitial-cell-stimulating hormone Abbrev.: ICSH (no stops). A hormone that stimulates androgen production by the interstitial cells of the testis in male mammals. The same hormone acts on the ovary of female mammals, in which it is called *luteinizing hormone.

intertropical convergence zone (not hyphenated) *Meteorol.* Abbrev.: ITCZ (no stops).

intra- Prefix denoting within (e.g. intra-atomic (hyphenated), intracellular, intrafascicular, intramolecular, intrauterine).

intra-arterial (hyphenated) Abbrev.: IA (no stops).

intramuscular Abbrev.: IM (no stops).

intrauterine device Abbrev.: IUD (no stops). More correctly termed intrauterine contraceptive device (abbrev.: IUCD).

intravenous Abbrev.: IV (no stops).

intrinsic magnetic flux density Another name for magnetic polarization.

intrinsic number density *See* number density.

intro- (**intr-** before vowels) Prefix denoting in or into (e.g. introgression, intromittent, introrse).

intron The noncoding sequence of a gene. *Compare* exon.

intussusception **1.** Invagination of one part of the bowel into another. **2.** The deposition of cellulose in spaces in a plant cell wall. [from Latin *intus*, within; not *inter*, between]

in vacuo (ital.) Latin: in a vacuum.

Invar (cap. I) A trade name for an iron–nickel–carbon alloy with low coefficient of expansion.

inverse The inverse of a mathematical function f is often denoted f^{-1}. Verb form: **invert**. Derived noun: **inversion**.

inverse hyperbolic functions *See* hyperbolic functions.

invertebrate Any animal that does not possess a vertebral column. Such animals were formerly classified in the now obsolete group Invertebrata, but the term invertebrate has been retained for descriptive purposes, to distinguish these animals from the *vertebrates.

in vitro (ital.) Performed or occurring outside the natural environment of the organism's body. [Latin: in glass]

***in vitro* fertilization** Abbrev.: IVF (no stops).

in vivo (ital.) Performed or occurring within the body of an organism. [Latin: in a living thing]

I/O *Computing* Abbrev. for input/output.

iodate A compound containing the ion IO_3^-, e.g. potassium iodate, KIO_3. The recommended name is iodate(V).

iodate(V) A compound containing the ion IO_3^-, e.g. potassium iodate(V), KIO_3. The traditional name is iodate.

iodic(V) acid HIO_3 The recommended name for the compound traditionally known as iodic acid.

iodine Symbol: I *See also* periodic table; nuclide.

iodine pentoxide I_2O_5 The traditional name for iodine(V) oxide.

iodine trichloride I_2Cl_6 The traditional name for diiodine hexachloride.

iodo- Prefix denoting iodine (e.g. iodoethane, iodoform, iodomethane).

iodomethane CH_3I The recommended name for the compound traditionally known as methyl iodide.

iodoso- Prefix denoting the group –IO (e.g. iodosobenzene).

iolite A violet-coloured mineral of metamorphic rocks. Also called cordierite. *Compare* ijolite.

ion Adjectival forms: **ion, ionic**. Verb form: **ionize**. Derived noun: **ionization**, also used adjectivally (e.g. ionization potential).

ion exchange Abbrev.: IX (no stops).

ionization potential Symbol: *I* Abbrev.: IP (no stops). The minimum energy required to remove an electron from an atom or molecule, usually expressed in electronvolts. The energy required to remove the least tightly bound electron is the first ionization potential. The second ionization potential is the energy required to remove only the second most tightly bound electron, producing a singly charged ion in an excited state. Higher ionization potentials are similarly defined.

This use of the term is preferred to one sometimes used in inorganic chemistry, in which the second ionization potential is taken to be the energy required to remove two electrons, producing a doubly charged ion. If this definition is used, it should be made clear.

ion-pair (hyphenated) *Chem.*

iota Greek letter, symbol: ι (lower case), Ι (cap.).

Ip Symbol sometimes used to denote the isopropylidene group in chemical formulae, e.g. $(CH_3)_2C=C(CN)COOC_2H_5$ can be written as $IpC(CN)COOC_2H_5$.

IP Abbrev. for ionization potential.

IP_3 (subscript 3) Abbrev. for inositol 1,4,5-triphosphate.

IPA Abbrev. and preferred form for isopentenyl adenosine, a cytokinin and minor base in some tRNA molecules.

Ipatieff, Vladimir Nikolayevich (1867–1952) Russian-born US chemist.

Ipatieff bomb

IPCS *Astron., etc.* Abbrev. for image photon counting system.

iPr, i-Pr Use Pr^i (*see* Pr).

IPS *Astron.* Abbrev. for interplanetary scintillations.

IPSE *Computing* Acronym for integrated project support environment.

i.p.s.p. (or **IPSP**; pl. i.p.s.ps or IPSPs) Abbrev. for inhibitory postsynaptic potential.

IPTS Abbrev. for International Practical Temperature Scale, an easily and accurately reproducible temperature scale, introduced in 1927, that is the practical realization of the *thermodynamic temperature scale. It is based on a number of fixed points and interpolation procedures. The 1968 version (IPTS–68) is expressed in both thermodynamic and *Celsius temperatures, which are given the symbols T_{68} and t_{68} respectively. It defines the temperature down to 13.81 kelvin.

EPT–76 is a provisional scale for lower temperatures, down to 0.519 K; temperatures on this scale have the symbol T_{76}.

IQ Abbrev. and preferred form for intelligence quotient.

Ir Symbol for iridium.

IR Abbrev. for **1.** infrared. **2.** information retrieval.

ir- (before r) Prefix denoting **1.** not (e.g. irrational, irreversible). **2.** into or within (e.g. irradiance). *See also* im-; in-.

IRAS Abbrev. for Infrared Astronomical Satellite.

Ir gene (cap. I) Abbrev. and preferred form for immune response gene.

iridescence (not irridescence)

iridium Symbol: Ir *See also* periodic table; nuclide.

irido- (**irid-** before vowels) Prefix denoting **1.** the iris of the eye (e.g. iridomotor, iridectomy). **2.** the plant genus *Iris* (e.g. Iridaceae). **3.** a rainbow or spectrum (e.g. iridescent).

iris 1. (pl. irides) Part of the eye. Adjectival form: **iridic. 2.** (pl. irises) Any plant of the genus *Iris* (cap. I, ital.).

I^2R loss (I and R not ital.) The rate of dissipation of electrical energy when a current, I, passes through a circuit of resistance R.

IRM *Ethology* Abbrev. for innate releasing mechanism.

iron Symbol: Fe *See also* periodic table; nuclide.

Iron and Steel Institute Abbrev.: ISI (no stops).

iron(II) diiron(III) oxide Fe_3O_4 The recommended name for the compound traditionally known as ferrosoferric oxide.

iron(II) oxide FeO The recommended name for the compound traditionally known as ferrous oxide.

iron(III) oxide Fe_2O_3 The recommended name for the compound traditionally known as ferric oxide.

ironstone (one word) *Geol.*

irradiance Symbol: E or E_e (e stands for energetic). A physical quantity that, at a point on a surface, is the *radiant flux falling on an element of the surface divided by the area of that element. The *SI unit is the watt per square metre. *Compare* illuminance.

IRTF Abbrev. for Infrared Telescope Facility (Mauna Kea, Hawaii).

IS *Genetics* Abbrev. for insertion sequence, a type of *transposon. Different insertion sequences are denoted by Arabic numerals, e.g. IS1, IS10 (no space). IS elements incorporated as modules in *transposons may be differentiated by an additional capital R or L, to denote left or right orientation, e.g. IS10R, IS10L.

Isaac Newton Telescope Abbrev.: INT (no stops).

Isaacs, Alick (1921–67) British virologist and biologist.

ISBN Abbrev. for International Standard Book Number.

ischium (pl. ischia) A bone of the pelvis. Adjectival forms: **ischiac, ischial**.

ISDN *Computing, telecom.* Abbrev. for Integrated Services Digital Network.

-ise *See* -ize.

Ishihara, Shinobu (1879–1963) Japanese ophthalmologist.
Ishihara test

ISI Abbrev. for **1.** International Statistical Institute. **2.** Iron and Steel Institute.

Ising model *Physics* Named after Ernst Ising.

islets of *Langerhans *Anat.* Also called pancreatic islets.

ISM Abbrev. for interstellar medium.

ISO Abbrev. or acronym for International Organization for Standardization (not International Standards Organization), the body by which international standards are established for units of measurement.

iso- 1. (sometimes **is-** before vowels) Prefix denoting equal or identical (e.g. isoagglutination, isoantigen, isoelectric, isogamy, isomorphism, isallobar, isentropic). **2.** *Chem.* Prefix traditionally denoting that a compound is an isomer of a specified compound (e.g. *isobutane, *isoprene).

isobar *Chem., physics, meterol.* Adjectival form: **isobaric**. *See also* nuclide.

isoBu Use Bu[i] (*see* Bu).

isobutane $CH_3CH(CH_3)CH_3$ The traditional name for 2-methylpropane.

isobutanoic acid $(CH_3)_2CHCOOH$ The traditional name for 2-methylpropanoic acid.

isobutyl alcohol $(CH_3)_2CHCH_2OH$ The traditional name for 2-methylpropan-1-ol.

isobutylene, isobutene $(CH_3)_2CCH_2$ Traditional names for 2-methylpropene.

isobutyric acid $(CH_3)_2CHCOOH$ The traditional name for 2-methylpropanoic acid.

isocline *Physics, geol., aeronautics* Adjectival forms: **isoclinal, isoclinic.**

isocyano compound Any of a class of organic compounds containing the group –NC joined directly to a carbon atom.

Isocyano compounds are systematically named by adding the prefix isocyano- to the name of the parent hydrocarbon, e.g. isocyanodimethylethane, $(CH_3)_3CNC$ (from dimethylethane, $(CH_3)_3CH$), and isocyanobenzene, C_6H_5NC.

In nonsystematic nomenclature isocyano compounds are named by adding either of the words isonitrile or isocyanide after the name of hydrocarbon group attached to the isocyano group, e.g. *t*-butyl isonitrile, $(CH_3)_3CNC$.

isoenzyme (preferred to isozyme)

Isoetales An order of the class *Lycopsida containing the genera *Isoetes* (quillworts) and *Stylites*. Some authors recognize only the former genus.

isogonal (or **isogonic**) *Maths., physics* Noun form: **isogon.**

isokont Describing an organism having flagella similar in form and length, as occurs in motile algae of the division *Chlorophyta (hence the former name of this taxon, Isokontae). The term 'isokont' has no taxonomic significance and is used purely as a descriptive term.

isoleucine Abbrev.: Ile or ile (no stop). Symbol: I *See* amino acid.

isomer *Chem., physics* Adjectival form: **isomeric**. Derived noun: **isomerism.**

isometric (preferred to isometrical)

isomorphism *Maths., crystallog., biol., etc.* Adjectival form: **isomorphic** (preferred to isomorphous, especially in biology).

iso-osmotic (preferred to isosmotic and isotonic) *Biochem.* Having the same osmotic pressure.

isopentenyl adenosine Abbrev. and preferred form: *IPA (no stops).

Isopoda (cap. I) An order of crustaceans comprising the woodlice. Individual name and adjectival form: **isopod** (no cap.: not isopodan or isopodous).

isoPr Use Pr[i] (*see* Pr).

isoprene $H_2CC(CH_3)CH=CH_2$ The traditional name for methylbuta-1,3-diene.

isopropylbenzene $C_6H_5CH(CH_3)_2$ The traditional name for (1-methylethyl)benzene.

isospin (or **isotopic spin**) Symbol: I A quantum number associated with a group of elementary particles of identical spin but different charge and/or strangeness. I can take values

$$0, 1/2, 1, 3/2, \ldots$$

the value being the same for each member of the group. Individual members have different values of **isospin quantum number**, symbol I_3, which can have values

$$-I, -I+1, \ldots, I-1, I.$$

The hypercharge, symbol Y, is linked to charge Q by the equation:

$$I_3 = Q - Y/2.$$

isospin quantum number *See* isospin.

isostasy *Geol.* Adjectival form: **isostatic.**

isotaxy *Chem.* Adjectival form: **isotactic.**

isothermal *Physics, eng.* Occurring at constant temperature. *Compare* adiabatic. Associated noun: **isotherm.**

isotones *Physics* *Nuclides with the same neutron number but different proton numbers. *Compare* isotopes. Adjectival form: **isotonic**.

isotonic **1.** Denoting muscles that have equal tonicity. **2.** Relating to *isotones. **3.** *Biochem.* Use iso-osmotic.

isotopes *Nuclides with the same proton number but different neutron numbers, i.e. different nucleon numbers. *Compare* isotones. Adjectival form: **isotopic**. *Compare* isotropic.

isotropic Not varying with direction; not to be confused with isotopic (*see* isotopes). Noun form: **isotropy**.

isovaleric acid $(CH_3)_2CHCH_2COOH$ The traditional name for 3-methylbutanoic acid.

isozyme Use isoenzyme.

ISR Abbrev. for **1.** information storage and retrieval. **2.** *Physics* intersecting storage ring.

ISSN Abbrev. for International Standard Serial Number (US Library of Congress).

isthmus (pl. isthmuses in geography, isthmi in anatomy)

IT Abbrev. for information technology.

ITCZ *Meteorol.* Abbrev. for intertropical convergence zone.

-ite Noun suffix denoting **1.** a mineral or rock (e.g. bauxite, dolomite, granite). It is recommended that the systematic chemical name should precede the mineral name, as in hydrated aluminium oxide (bauxite). **2.** *Biochem.* a product (e.g. catabolite, metabolite). **3.** *Biol.* a division of a body or organ (e.g. somite). **4.** *Chem.* traditionally, a salt or ester of an acid with a name ending in *-ous (e.g. chlorite, nitrite, sulphite).

iteration Adjectival forms: **iterative, iterated**.

ITP Abbrev. and preferred form for inosine 5′-triphosphate.

ITU Abbrev. for International Telecommunication Union, a UN organization.

i-type *Electronics* Denoting an (almost) intrinsic semiconductor. Abbrev. in combinations: i-, as in p-i-n diode.

IU Abbrev. for *international unit or units.

IUCD Abbrev. for intrauterine contraceptive device.

IUCN Abbrev. for International Union for the Conservation of Nature and Natural Resources.

IUD Abbrev. for intrauterine device. In medical terminology the term intrauterine contraceptive device (abbrev.: IUCD) is preferred.

IUE *Astron.* Abbrev. for International Ultraviolet Explorer.

-ium Noun suffix denoting **1.** an element (e.g. calcium, lithium, titanium). **2.** a chemical group forming a positive ion (e.g. ammonium, oxonium).

IUPAC Acronym (pronounced **yoo**-pak) or abbrev. for International Union of Pure and Applied Chemistry.

IUPAP Abbrev. for International Union of Pure and Applied Physics.

IV Abbrev. for intravenous.

Ivanovsky, Dmitri Iosifovich (1864–1920) Russian botanist and microbiologist.

IVF Abbrev. for *in vitro* fertilization.

IX Abbrev. for ion exchange.

-ize (or **-ise**) In British English, the -ize form of this suffix is the recommended ending for most verbs, with -ise as an acceptable variant (e.g. ionize, metabolize, photosynthesize); it is the form used throughout this book. Exceptions include advertise, advise, comprise, devise, excise, improvise, incise, supervise, surmise, and televise; note that these verbs have the ending -ise in both British and American English, as they do not derive from Greek verbs ending in *-izō*. *See also* -yse.

J

j Symbol for the imaginary number $\sqrt{-1}$ (used, instead of i, especially in electrical engineering).

j Symbol for **1.** (light ital.) current density. **2.** (bold ital.) a unit coordinate vector.

J 1. Symbol for i. joining region (of an *immunoglobulin chain). ii. joule. 2. Abbrev. for Jurassic.

J Symbol for 1. (bold ital.) angular momentum. 2. (light ital.) coupling constant. 3. (light ital.) current density. 4. (bold ital.) magnetic polarization. 5. (light ital.) *Obsolete* mechanical equivalent of heat. 6. (light ital.) nuclear spin quantum number. 7. (light ital.) rotational quantum number. 8. (light ital.) sound intensity. 9. (light ital.) total angular momentum quantum number.

jaagsiekte (not ital.; from Afrikaans) Another name for pulmonary adenomatosis of sheep.

Jackson, John Hughlings (1835–1911) British physician and neurologist.
Jacksonian epilepsy

Jackson method *Data processing* Named after Michael Jackson.

Jacob, François (1920–) French biologist.
Jacob–Monod hypothesis (or **model**) (en dash) Also called operon model.

Jacobi, Karl Gustav Jacob (1804–51) German mathematician.
Hamilton–Jacobi theory (en dash)
Jacobian elliptic functions

Jacobson, Ludvig Levin (1783–1843) Danish anatomist and physician.
Jacobson's nerve Also called tympanic nerve.
Jacobson's organ Also called vomeronasal organ.

Jacquard, Joseph-Marie (1752–1834) French inventor.
Jacquard loom

Jahn–Teller effect (en dash) *Chem.* Named after H. A. Jahn and E. *Teller.

Jamin interferometer *Physics*

JANET *Computing* Acronym for Joint Academic Network.

jansky Symbol: Jy A unit used in astronomy to measure spectral flux density, i.e. radiant flux density per unit bandwidth:

$$1\,\text{Jy} = 10^{-26}\,\text{W m}^{-2}\,\text{Hz}^{-1}.$$

Named after Karl *Jansky.

Jansky, Karl Guthe (1905–50) US radio engineer.
***jansky**
Jansky noise

Janssen, Pierre Jules César (1824–1907) French astronomer.

Janssen, Zacharias (1580–*c.* 1638) Dutch instrument maker.

Java man *See Homo.*

JCL *Computing* Abbrev. for job control language.

JD *Astron.* Abbrev. for Julian date.

Jeans, Sir James Hopwood (1877–1946) British mathematician, physicist, and astronomer.
Rayleigh–Jeans formula (en dash)

Jeffreys, Sir Harold (1891–) British astronomer and geophysicist.

jellyfish (one word) *See* Scyphozoa.

Jenner, Edward (1749–1823) British physician.

Jensen, Johannes Hans Daniel (1907–73) German physicist.

Jerne, Niels Kaj (1911–) Danish immunologist and microbiologist.

JET *Physics* Acronym for Joint European Torus, built by the European Commission at Culham, Oxfordshire.

Jet Propulsion Laboratory (in California) Abbrev.: JPL (no stops).

JFET (pronounced **jay**-fet) *Electronics* Acronym for junction field-effect transistor.

job control language *Computing* Abbrev.: JCL (no stops).

Johannsen, Wilhelm Ludwig (1857–1927) Danish botanist and geneticist.

Johnson Space Center (in Texas) Abbrev.: JSC (no stops).

Joliot-Curie, Frédéric (hyphen; originally Frédéric Joliot) (1900–58) French physicist, husband of Irène Joliot-Curie.

Joliot-Curie, Irène (hyphen; originally Irène Curie) (1897–1956) French physicist, daughter of Pierre and Marie Curie.

joliotium The Soviet name for nobelium (element 102).

Joly, John (1857–1933) Irish geologist and physicist.

Jones, Sir Ewart Ray Herbert (1911–) British chemist.

Jones, Sir Harold Spencer Usually alphabetized as *Spencer Jones.

Jordan, (Marie-Ennemond) Camille (1838–1922) French mathematician.
Jordan curve
Jordan matrix

Jordan, Ernst Pascual (1902–80) German theoretical physicist and mathematician.
Brans–Dicke–Jordan theory (en dashes; preferred to Brans–Dicke theory)

Josephson, Brian David (1940–) British physicist.
Josephson effects
Josephson junction

joule (no cap.) Symbol: J The *SI unit of energy, work, and quantity of heat.

$$1\,\mathrm{J} = 1\,\mathrm{N\,m} = 1\,\mathrm{W\,s}.$$

Electrical work is usually measured in watt-hours: 1 W h = 3600 J. In atomic and nuclear physics and accelerator technology, energy is normally measured in electronvolts: 1 eV = 1.602×10^{-19} J (approx.). Named after James *Joule.

Joule, James Prescott (1818–89) British physicist.
***joule**
Joule effect
Joule heating
Joule's equivalent *Obsolete* Also called mechanical equivalent of heat.
Joule's laws
Joule–Thomson coefficient (en dash; preferred to Joule–Kelvin coefficient) Symbol: μ (Greek mu).
Joule–Thomson effect (en dash; preferred to Joule–Kelvin effect)

Jovian *See* Jupiter.

JPL Abbrev. for Jet Propulsion Laboratory (California).

JSC Abbrev. for Johnson Space Center (Texas).

Julian calendar Named after Julius Caesar (100 BC–44 BC). *Compare* Gregorian calendar.

Julian century An astronomical constant equal to a period of 36 525 days. The **Julian year**, equal to 365.25 days, is denoted by the prefix J, as in J2000.0.

Julian date A date expressed as the number of days since noon GMT on 1 Jan 4713 BC. Abbrev.: JD (no stops). Named after Julius Caesar Scaliger by his son, Joseph Scaliger (1540–1609), who devised the system.

Jung, Carl Gustav (1875–1961) Swiss psychologist and psychiatrist. Adjectival form: **Jungian**.

Jupiter A planet. Adjectival form: **Jovian**.

Jurassic Abbrev.: J (no stop). **1.** (adjective) Denoting the middle period of the Mesozoic era. **2.** (noun; preceded by 'the') The Jurassic period. The period is divided into the Lower Jurassic, Middle Jurassic, and Upper Jurassic epochs (initial caps.); abbrevs. (respectively): J_1, J_2, and J_3 (subscript numerals). These epochs correspond to the Lias, Dogger, and Malm series, respectively.

Jussieu, Antoine-Laurent de (hyphen) (1748–1836) French plant taxonomist.

Jy Symbol for jansky.

K

k Symbol for kilo-.

k Symbol for **1.** (light ital.) Boltzmann constant. **2.** (light ital.) circular wavenumber. **3.** (light ital.) Gaussian gravitational constant. **4.** (light ital.) radius of gyration. **5.** (light ital.) rate coefficient (or constant). **6.** (light ital.) thermal conductivity. **7.** (bold ital.) a unit coordinate vector.

K Symbol for **1.** calyx (in a *floral formula). **2.** Cretaceous. **3.** guanosine or thymidine (or uridine) (unspecified). **4.** kaon (K^+, K^-, K^0). **5.** kelvin. **6.** *Computing* *kilo- (discouraged: use k). **7.** lysine. **8.** potassium. **9.** *Astron. See* spectral types.

K Symbol (light ital.) for **1.** bulk modulus. **2.** equilibrium constant ($K^{\oplus}$ (or $K^{\ominus}$) = standard equilibrium constant, K_p refers to a pressure basis, K_c to a concentration basis). **3.** kerma. **4.** kinetic energy. **5.** luminous efficacy. **6.** thermal conductivity.

K_a (light ital. K) Symbol for acid dissociation constant.

K_m (light ital.) Symbol for Michaelis constant.

K_s (light ital. K) Symbol for solubility product.

Ka band *See* K band.

Kahn, Reuben Leon (1887–) Lithuanian-born US bacteriologist.
Kahn test (or **reaction**)

Kaluza–Klein theory (en dash) *Physics* Named after Theodor F. E. Kaluza and Oskar Klein.

kame (noun) *Geol.*

Kamen, Martin David (1913–) Canadian-born US biochemist.

Kamerlingh Onnes (not hyphenated), **Heike** (1853–1926) Dutch physicist.

Kammerer, Paul (1880–1926) Austrian zoologist.

Kamp, Peter van de Usually alphabetized as *van de Kamp.

kampfzone (no cap.) The zone of elfin forest (*see* krummholz) between the timber and tree lines. [from German *Kampfzone*, literally 'struggle zone']

KAO *Astron.* Abbrev. for Kuiper Airborne Observatory.

kaon Another name for K meson. *See* hadrons.

Kapitza, Pyotr (or **Peter**) **Leonidovich** (1894–1984) Soviet physicist.
Kapitza resistance
Kapitza's law

kappa Greek letter, symbol: κ (lower case), K (cap.).
κ Symbol for **1.** (or K_T) compressibility. **2.** conductivity (thermal and electrolytic). **3.** a type of light chain of an *immunoglobulin molecule.

Kapteyn, Jacobus Cornelius (1851–1922) Dutch astronomer.
Kapteyn selected areas
Kapteyn's star

Karle, Jerome (1918–) US crystallographer.

Karl Fischer reagent *Chem.*

Kármán, Theodor von (1881–1963) Hungarian physicist and aeronautical engineer.
Kármán (vortex) streak

Karnaugh, Maurice (1924–) US physicist.
Karnaugh map

Karrer, Paul (1889–1971) Swiss chemist.

karyo- (**kary-** before vowels) Prefix denoting a cell nucleus (e.g. karyogamy, karyokinesis, karyotype). *See also* caryo-.

-karyon Noun suffix denoting a cell nucleus (e.g. heterokaryon, homokaryon). Derived noun form: **-karyosis**. Adjectival form: **-karyotic**.

Kastler, Alfred (1902–84) French physicist.

kata- (**kat-** or **kath-** before vowels) Prefix denoting down or against; used in preference to *cata- in some words (e.g. katabatic, katophorite, katharometer).

kata- (ital., always hyphenated) *Chem.* Prefix sometimes used to denote 1,7-substitution of a naphthalene ring (e.g. *kata*-dichloronaphthalene). *Compare amphi-*; *ana-*; *epi-*; *peri-*; *pros-*.

katabolism Use catabolism.

Kater's pendulum *Physics* Named after Henry Kater (1777–1835).

Katz, Bernard (1911–) German-born British neurophysiologist.

Kay, John (1704–*c.* 1764) British inventor.

Kayser–Fleischer ring (en dash) *Med.* Named after B. Kayser (1869–1954) and B. Fleischer (1848–1904).

kb Symbol for *kilobase or kilobases.

K band (not hyphenated) *Telecom., radar* A microwave band with the frequency range 12–40 GHz, which includes the subdivisions **Ku band** (K-under), 12–19 GHz, and **Ka band** (K-above), 26–40 GHz.

K capture (not hyphenated) *Physics*

K cell Abbrev. for killer cell.

KE Abbrev. for kinetic energy.

Kekulé von Stradonitz, Friedrich August (1829–96) German chemist.
Kekulé structure

Kellner, Karl (1851–1905) Austrian chemical engineer.
Castner–Kellner process (en dash)

kelvin (no cap.; pl. kelvin, preferred to kelvins) Symbol: K The *SI unit of *thermodynamic temperature, superseding degree Kelvin (symbol °K). It is one of the seven SI base units, defined as the fraction 1/273.16 of the triple point of water. The unit and its symbol are also used to express an interval or a difference of temperature. The units kelvin and degree Celsius are identical, the latter being used to express Celsius temperature or temperature difference on the Celsius scale. Named after Lord *Kelvin.

Kelvin, Sir William Thomson, Lord (1824–1907) British theoretical and experimental physicist. The name Kelvin is applied to instruments and concepts devised by Lord Kelvin; the name Thomson is usually preferred in connection with certain effects.
Joule–Kelvin coefficient, effect Use Joule–Thomson coefficient, effect (en dash).
***kelvin**
Kelvin balance
Kelvin bridge (or **double bridge)**
Kelvin contacts
Kelvin effect *Elec. eng.*
Kelvin effect *Thermoelectricity* Use Thomson effect.
Kelvin temperature scale Now called thermodynamic temperature scale.
Kelvin–Varley slide (en dash)

Kendall, Edward Calvin (1886–1972) US biochemist.

Kendrew, Sir John Cowdery (1917–) British biochemist.

Kennelly, Arthur Edwin (1861–1939) British-born US electrical engineer.
Kennelly–Heaviside layer (en dash) Use Heaviside layer or E layer.

Kenyapithecus (cap. K, ital.) A genus of fossil apes from E Africa, similar to **Dryopithecus*.

kephalin Use phosphatidylethanolamine.

Kepler, Johannes (1571–1630) German astronomer.
Keplerian orbit
Keplerian telescope
Kepler's laws
Kepler's rule
Kepler's star

kerato- (**kerat-** before vowels) Prefix denoting **1.** horn or horny tissue (e.g. keratogenous, keratin, keratosis). *See also* cerato-. **2.** the cornea (e.g. keratoplasty).

kerma Symbol: K Acronym for kinetic energy released in matter, a physical quantity associated with indirectly ionizing (i.e. uncharged) particles. It is equal to the sum of the initial kinetic energies of all charged particles liberated in an element of matter divided by the mass of the element. The *SI unit is the gray (Gy), which has replaced the rad.

Kernig, Vladimir Michalovich (1840–1917) Russian physician.
Kernig's sign

kerosene (not kerosine) Another name for paraffin, especially in the USA.

Kerr, John (1824–1907) British physicist.
Kerr cell
Kerr effects

Kerr, Roy Patrick (1934–) New Zealand mathematician.
Kerr black hole

ketene (or **keten**) $CH_2=CO$ The traditional name for ethenone.

keto- (**ket-** before vowels) Prefix denoting **1.** a ketone or ketone derivative (e.g. ketogenic, ketene, ketoxime). **2.** a ketose (e.g. ketohexose).

keto–enol tautomerism (en dash, not hyphen)

ketone Any of a class of organic compounds containing the carbonyl group $=CO$ directly joined to two carbon atoms.
Ketones are systematically named by adding the suffix -one to the name of the parent hydrocarbon, e.g. propanone, CH_3COCH_3 (from propane, $CH_3CH_2CH_3$), 1-phenylpropanone, $C_6H_5CH_2COCH_3$, and diphenylmethanone, $(C_6H_5)_2CO$.

In nonsystematic nomenclature ketones are named according to the two groups attached to the carbonyl group followed by the word ketone, e.g. benzyl methyl ketone, $C_6H_5CH_2COCH_3$. A ketone in which the carbonyl group is directly joined to a benzene ring has the suffix -phenone, e.g. acetophenone, $C_6H_5COCH_3$, and benzophenone, $(C_6H_5)_2CO$.

Kettlewell, Henry Bernard Davis (1907–79) British geneticist and lepidopterist.

keyword (one word)

kg (no stop) Symbol for kilogram. *See also* SI units.

kgf Symbol for kilogram-force.

Khorana, Har Gobind (1922–) Indian-born US chemist.

kidney Adjectival form: **renal** (e.g. renal artery, renal dialysis, renal tubule).

Kikuchi, Seishi (1902–) Japanese physicist.
Kikuchi lines

Kilburn, Tom (1921–) British computer scientist.

killer cell Abbrev.: K cell.

kilo (pl. kilos) Short for kilogram; use the name in full or its symbol (kg) in scientific contexts.

kilo- 1. Symbol: k (not K) A prefix to a unit of measurement that indicates 10^3 times that unit, as in kilometre (km), kilowatt (kW), or kiloamp (kA). *See also* SI units; kilogram. **2.** Symbol: k (K is discouraged) A prefix used in computing to indicate a multiple of 2^{10} (i.e. 1024), as in kilobyte or kilobit. The symbol is often used to mean kilobyte, as in 512 k memory, or kilobit, as in 32 k RAM; this usage is discouraged, the forms kbyte and kbit being preferred.

kilobase Symbol: kb A unit of length equal to 1000 bases or base pairs. It is used in molecular biology to express the length of a DNA or RNA segment. **Kilobase pair** (symbol kbp) is used synonymously.

kilogram (not kilogramme) Symbol: kg (no stop). The *SI unit of mass. It is one of the seven SI base units, defined as the mass of the international prototype of the kilogram. The decimal multiples and submultiples are formed by adding the SI prefixes to the word gram, for example milligram (mg) rather than microkilogram. One kilogram = 2.204 62 pounds. A mass of 1000 kg is called a tonne. *See also* atomic mass unit.

kilogram-force Symbol: kgf A metric unit of *force. The *SI unit, the newton, should now be used:
$1\ \text{kgf} = 9.806\ 65\ \text{kg m s}^{-2} = 9.806\ 65\ \text{N}$,
where $9.806\ 65\ \text{m s}^{-2}$ is the value of the standard *acceleration of free fall. *Compare* pound-force.

kilopond Symbol: kp The name in some European countries for the *kilogram-force.

kilowatt hour Symbol: kWh or kW h A unit widely used to express electrical work:

$$1\ \text{kW h} = 1000\ \text{W} \times 3600\ \text{s} = 3.6 \times 10^6\ \text{J}.$$

kinaesthesia (preferred to kinaesthesis) US: **kinesthesia**. Adjectival form: **kinaesthetic** (US: **kinesthetic**).

kine- (**kin-** before vowels) Prefix denoting movement or action (e.g. kinematics, kinase).

kinematic viscosity *See* viscosity.

-kinesis Noun suffix denoting movement (e.g. karyokinesis). Adjectival form: **-kinetic**.

kinesthesia US spelling of *kinaesthesia.

kinetic energy *See* energy.

kineto- (**kinet-** before vowels) *Biol.* Prefix denoting movement (e.g. kinetochore, kinetosome, kinetin). *See also* kine-.

kingdom The highest category used in biological classification (*see* taxonomy). Names of kingdoms are printed in roman (not italic) with an initial capital letter. There are two traditional kingdoms, the Animalia (animals) and Plantae (plants), and unicellular organisms are sometimes placed in the kingdom *Protista. Modern classifications place fungi in a separate kingdom (Mycota, Fungi, or Mycobiota, depending on the authority) and assign the bacteria and blue-green algae to a fourth kingdom, Procaryotae (*see* prokaryote; sometimes divided into two separate kingdoms). Kingdoms are divided into subkingdoms, then into phyla (for animals) or divisions (for plants).

Kinsey, Alfred Charles (1894–1956) US zoologist and sociologist.

kip In the USA, a unit of force equal to 1000 *pound-force. *See also* ksi.

Kipp, Petrus Jacobus (1808–64) German chemist.
Kipp's apparatus

Kipping, Frederic Stanley (1863–1949) British chemist.

Ki-*ras* (or **K-*ras***; hyphenated) *See* oncogene.

Kirchhoff, Gustav Robert (1824–87) German physicist.
Kirchhoff's laws
Kirchhoff's radiation law

Kirkwood, Daniel (1814–95) US astronomer.
Kirkwood gaps

Kirschner value *Food technol.*

Kirwan, Richard (1733–1812) Irish chemist and mineralogist.

Kitasato, Baron Shibasaburo (1852–1931) Japanese bacteriologist.

Kittel, Charles (1916–) US physicist.

Kitt Peat National Observatory (in Arizona) Abbrev.: KPNO (no stops).

Kjeldahl, Johan Gustav Christoffer Thorsager (1849–1900) Danish chemist.
Kjeldahl flask
Kjeldahl method

Klaproth, Martin Heinrich (1743–1817) German chemist.

Klebs, Theodor Albrecht Edwin Swiss-born US bacteriologist.
Klebsiella
Klebs–Löffler bacillus (en dash) Old name for *Corynebacterium diphtheriae* (cause of diphtheria).

Klebsiella (cap. K, ital.) A genus of bacteria of the family *Enterobacteriaceae. Individual name: **klebsiella** (no cap., not ital.; pl. klebsiellae).

Kleene, Stephen Cole (1909–) US mathematician.
Kleene star
Kleene's theorems

Klein, (Christian) Felix (1849–1925) German mathematician.
Cayley–Klein parameters (en dash)
Klein bottle

Klein–Gordon equation (en dash) *Physics* Named after Oskar Klein and W. Gordon.

Kleimann–Low nebula (en dash) Abbrev.: *KL nebula.

Kline, Benjamin S. (1886–) US pathologist.
Kline test

Klinefelter, Harry Fitch, Jr (1912–) US physician.
Klinefelter's syndrome

klinostat Use clinostat.

Klitzing, Klaus von Usually alphabetized as *von Klitzing.

KL nebula *Astron.* Abbrev. for Kleinmann–Low nebula (en dash). Named after D. E. Kleinmann and J. J. Low.

Klug, Sir Aaron (1926–) South African-born British biophysicist.

km (no stop) Symbol for kilometre (or kilometres). *See* metre.

K meson (not hyphenated) Also called kaon. *See* hadrons.

kn Symbol for knot.

Kn (ital.) Symbol for Knudsen number.

knockdown (noun; one word) A measure of biocide effect.

knot Symbol: kn A unit of speed equal to one nautical mile per hour.

Knudsen, Martin (1871–1949) Danish physicist.
Knudsen flow Also called molecular flow.
***Knudsen number**

Knudsen number Symbol: *Kn* A dimensionless quantity equal to λ/l, where λ is the mean free path, and l a characteristic length. *See also* parameter.

Koch, (Heinrich Hermann) Robert (1843–1910) German bacteriologist.
Koch's bacillus Old name for *Mycobacterium tuberculosis* (cause of tuberculosis).
Koch's postulates

Koch–Weeks bacillus (en dash) Old name for *Haemophilus aegyptius* (cause of conjunctivitis).

Kodak (cap. K) A trade name for a series of photographic products and cameras.

Koeppen, Wladimir Peter Usually alphabetized as *Köppen.

Kohlrausch, Friedrich Wilhelm Georg (1840–1910) German physicist.
Kohlrausch's law

Kolbe, Adolph Wilhelm Hermann (1818–84) German chemist.
Kolbe method
Kolbe synthesis

Kölliker, Rudolph Albert von (1817–1905) Swiss histologist and embryologist.

Kolmogorov, Andrei Nikolaievich (1903–87) Soviet mathematician.

Kondo effect *Physics* Named after Jun Kondo.

Köppen (or **Koeppen**), **Wladimir Peter** (1846–1940) Russian-born German climatologist.
Köppen classification

Kornberg, Arthur (1918–) US biochemist.

Korolev, Sergei Pavlovich (1907–66) Soviet rocket engineer.

Korsakoff, Sergei Sergeevich (1854–1900) Russian physician.
Korsakoff's syndrome

Kossel, Albrecht (1853–1927) German biochemist.

Kovalevskaya, Sofya Vasilyevna (1850–91) Russian mathematician. Also called Sonya Koyalevski.

Kovar (cap. K) A trade name for an alloy of iron, cobalt, and nickel.

kp Symbol for kilopond.

KPNO Abbrev. for Kitt Peak National Observatory (Arizona).

Kr Symbol for krypton.

Kramer, Paul Jackson (1904–) US plant physiologist.

Kranz structure *Bot.*

Krebs, Sir Hans Adolf (1900–81) German-born British biochemist.

Krebs cycle (no apostrophe) Also called TCA (tricarboxylic acid) cycle. Former name: citric acid cycle.

Kronecker, Leopold (1823–91) German mathematician.
Kronecker delta Symbol: δ_{ij} (Greek delta)

Kronig–Penney model (en dash) *Physics* Named after R. de L. Kronig and W. G. *Penney.

krummholz (no cap.) The dwarfed trees of elfin forest, between the timber and tree lines (*see* kampfzone). [from German *Krummholz*, literally 'bent wood']

krypton Symbol: Kr *See also* periodic table; nuclide.

K shell (not hyphenated) *Physics*

ksi (or **k.s.i.**) Abbrev. for kip per square inch.

***K*-strategist** (hyphenated) An organism adapted to living in a stable community (with a population close to its carrying capacity, *K*), expending little of its resources on reproduction. *Compare r*-strategist.

Ku Symbol for kurchatovium. *See* element 104.

Ku band *See* K band.

Kuhn, Richard (1900–67) Austrian-born German chemist.

Kuiper, Gerard Peter (1905–73) Dutch-born US astronomer.
Kuiper Airborne Observatory Abbrev.: KAO (no stops).

Kundt, August Eduard Eberhard Adolph (1839–94) German physicist.
Kundt's rule
Kundt tube

Kupffer, Karl Wilhelm von (1829–1902) Bavarian anatomist and embryologist.
Kupffer cell

Kurchatov, Igor Vasilievich (1903–60) Soviet physicist.
kurchatovium Also called rutherfordium. *See* element 104.

Kurti, Nicholas (1908–) Hungarian-born British physicist.

Kusch, Polykarp (1911–) US physicist.

Kutta, Martin Wilhelm (1867–1944) German mathematician.
Runge–Kutta method (en dash)

kV Symbol for kilovolt (or kilovolts). *See* volt.

kW Symbol for kilowatt (or kilowatts). *See* watt.

kWh Symbol for kilowatt hour.

L

l Symbol for **1.** *liquid state. *See also* state symbols. **2.** *litre.

l Symbol (light ital.) for **1.** length. **2.** *orbital angular momentum quantum number. **3.** specific latent heat.

L 1. Symbol for **i.** leucine **ii.** light chain (of an *immunoglobulin molecule). **iii.** *litre. **2.** Abbrev. for live (on plugs).

L Symbol (light ital.) for **1.** angular momentum (in nonvector equations, L); in vector equations it is printed in bold italic ($\boldsymbol{L}$). **2.** Avogadro constant. **3.** Lagrangian function. **4.** latent heat. **5.** length. **6.** *Biochem.* linking number. **7.** longitude (geographical). **8.** (or L_v) luminance. **9.** luminosity. **10.** *orbital angular momentum quantum number. **11.** (or L_e) radiance. **12.** self-inductance (L_{12} = mutual inductance). **13.** sound intensity (L_N = loudness level, L_p = sound pressure level, L_W = sound power level, power-level difference).

***l*-** (ital., always hyphenated) *See* laevorotatory.

L- (small cap., always hyphenated) Prefix denoting a structural relationship to laevorotatory glyceraldehyde (L-(−)-glyceraldehyde) (e.g. L-lactic acid, L-threonine). An L-isomer is not necessarily *laevorotatory. *Compare* D-.

La Symbol for lanthanum.

label Verb form: **labels, labelling** (US: **labeling**), **labelled** (US: **labeled**).

Labiatae (cap. L) A large family of dicotyledonous plants, commonly known as the mint family. Alternative name: Lamiaceae. Individual name and adjectival form: **labiate** (no cap.).

-labile Adjectival suffix denoting change or displacement (e.g. thermolabile).

labium (pl. labia) A lip-shaped part, especially the fused second maxillae forming the lower lip of insects. Adjectival form: **labial**. *Compare* labrum.
labia majora (pl.; sing. labium majus) and **labia minora** (pl.; sing. labium minus) Two pairs of skin folds forming the vulva of mammals.

Labrador Current (initial caps.)

labrum (pl. labra) A cuticular plate forming the upper lip of insects. Adjectival form: **labral**. *Compare* labium.

Labyrinthodontia (cap. L) A subclass of extinct amphibians. Individual name and adjectival form: **labyrinthodont** (no cap.).

Lac *Astron.* Abbrev. for Lacerta.

Lacaille, Nicolas Louis de (1713–62) French astronomer.

laccolith (not lacolith or lakkolith) *Geol.* [from Greek *lakko*: cistern]

Lacerta A constellation. Genitive form: Lacertae. Abbrev.: Lac (no stop). *See also* stellar nomenclature.

Lack, David Lamber (1910–73) British ornithologist.

La Condamine, Charles-Marie de (hyphen) (1701–74) French geographer and explorer.

***lac* operon** (ital. *lac*) Abbrev. and preferred form for lactose operon. *See* gene.

lacrimal (preferred to lachrymal in anatomy and zoology) Derived nouns: **lacrimation, lacrimator**.

lactic acid $CH_3CH(OH)COOH$ The traditional name for 2-hydroxypropanoic acid.

lacto- (**lact-** before vowels, **lacti-**) Prefix denoting milk (e.g. lactogenic, lactose, lactiferous).

Lactobacillus (cap. L, ital.) A genus of fermentative lactate-producing bacteria. Individual name: **lactobacillus** (no cap., not ital.; pl. lactobacilli). Many authors still recognize the traditional division of the genus into three subgenera: *Thermobacterium*, *Streptobacterium*, and *Betabacterium*. However, some authors now prefer the terms 'group I', 'group II', and 'group III' (Roman numerals) instead of these names.

lacuna (pl. lacunae; preferred to lacunas) *Anat., biol.* Adjectival forms: **lacunar, lacunate**.

Lacus *See* mare.

laevo- (**laev-** before vowels) US: **levo-** (**lev-**). Prefix denoting on or towards the left (e.g. laevorotatory).

laevo- (ital., always hyphenated) *See* laevorotatory.

laevodopa Use levodopa (*see* L-dopa).

laevorotatory US: **levorotary**. Laevorotatory compounds are denoted by the prefix (−)- (in parentheses, always hyphenated), e.g. (−)-tartaric acid. The alternatives *l*- and *laevo*- are not recommended. *See also* L-. *Compare* dextrorotatory.

Lagomorpha (cap. L) An order of mammals that includes the rabbits and hares. Individual name and adjectival form: **lagomorph** (no cap.; not lagomorphic or lagomorphous).

Lagrange, Comte Joseph Louis (1736–1813) Italian-born French mathematician and theoretical physicist.
Lagrange–Hamilton theory (en dash)
Lagrange's equations
Lagrange's interpolation formula
Lagrange's theorem
***Lagrangian function** (or **Lagrangian**)
Lagrangian points

Lagrangian function Symbol: L A function used in general dynamics:

$$L = T(q_i, \dot{q}_i) - V(q_i, \dot{q}_i),$$

where q_i are the *generalized coordinates and T and V are kinetic and potential energy, respectively. Also called Lagrangian.

LAI Abbrev. for leaf area index.

Lalande, Joseph Jerome Le François de (1732–1807) French astronomer.

Lamarck, Jean Baptiste Pierre Antoine de Monet, Chevalier de (not Lamark) (1744–1829) French biologist. Adjectival form: **Lamarckian.**
Lamarckism
neo-Lamarckism

Lamb, Sir Horace (1849–1934) British mathematician and geophysicist.

Lamb, Hubert Horace (1913–) British climatologist.

Lamb, Willis Eugene, Jr (1913–) US physicist.
Lamb shift

lambda Greek letter, symbol: λ (lower case), Λ (cap.).
λ Symbol for **1.** *Chem.* absolute activity (λ_B = that of substance B). **2.** celestial longitude. **3.** damping coefficient. **4.** decay constant. **5.** a type of light chain of an *immunoglobulin molecule. **6.** thermal conductivity. **7.** wavelength (λ_C = Compton wavelength, λ_D = Debye length). **8.** *See* phage λ.
Λ Symbol for **1.** lambda particle. **2.** logarithmic decrement. **3.** molar conductivity of an electrolyte or ion (Λ^∞ (preferred to Λ_0) = molar conductivity at infinite dilution). **4.** permeance.

lambda phage Usually written **λ phage** (Greek lambda). *See* phage λ.

lambdoid (or **lambdoidal**; not lamdoid, lamdoidal) Denoting a suture of the skull, meeting the sagittal suture at a point called the **lambda**.

Lambert, Johann Heinrich (1728–77) German mathematician, physicist, astronomer, and philosopher.
Lambert–Beer law (en dash; preferred to Bouguer–Lambert–Beer law)
Lambert's law

lamella (pl. lamellae) *Anat., biol.* Adjectival forms: **lamellar, lamellate, lamelliform**.

Lamellibranchia Use Bivalvia (*see* bivalve).

Lamiaceae *See* Labiatae.

lamina (pl. laminae) *Anat., bot., geol.* Adjectival forms: **laminar, laminate, laminiform**.

LAN *Computing* Acronym for local-area network.

Lancefield, Rebecca Craighill (1895–1981) US bacteriologist.
Lancefield classification *See Streptococcus*.

land Related adjective: **terrestrial**. Terms incorporating the word 'land' are usually written as one word (e.g. landform, land-locked, landmass, landrace, landscape, landslide (or landslip). Two-word terms (e.g. land breeze, land bridge, land line) are hyphenated when used adjectivally (e.g. land-line connection).

Land, Edwin Herbert (1909–91) US physicist.
Land camera (trade name)

Landau, Lev Davidovich (1908–68) Soviet theoretical physicist.
Ginzburg–Landau theory (en dash)
Landau levels

Landé, Alfred (1888–1975) German physicist.
Landé factor Symbol: g Use g factor.

Landsteiner, Karl (1868–1943) Austrian-born US pathologist.

Langerhans, Paul (1847–1888) German pathological anatomist.
islets of Langerhans Also called pancreatic islets.

Langevin, Paul (1872–1946) French physicist.
Langevin ion
Langevin theory

Langhaus, Theodor (1839–1915) German pathologist.
Langhaus cells

Langmuir, Irving (1881–1957) US chemist.
Langmuir–Child law (en dash; preferred to Child's law)
Langmuir effect
Langmuir isotherm
Lewis–Langmuir theory (en dash)

Lankester, Sir Edwin Ray (1847–1929) British zoologist.

lanthanides The traditional name for lanthanoids.

lanthanoids The recommended name for the group of elements from atomic number 57 (lanthanum) to 71 (lutetium). The traditional names are lanthanides, lanthanons, and rare-earth elements.

lanthanons Use lanthanoids.

lanthanum Symbol: La *See also* periodic table; nuclide; lanthanoids.

La Palma Observatory Official name: Roque de los Muchachos Observatory. An observatory on La Palma, Canary Islands.

lapillus (pl. lapilli; usually referred to in the pl.) *Geol.*

lapis lazuli (two words)

Laplace, Marquis Pierre Simon de (1749–1827) French mathematician, astronomer, and physicist.
Ampère–Laplace law (en dash)
De Moivre–Laplace theorem (en dash)
Laplace's (differential) equation
Laplace's nebular hypothesis
Laplace transform
***Laplacian**

Laplacian (not Laplacean) Symbol: ∇^2 The differential operator

$$(\partial^2/\partial x^2 + \partial^2/\partial y^2 + \partial^2/\partial z^2).$$

Also called Laplacian operator. *See also* del.

Lapworth, Charles (1842–1920) British geologist.

LAR Abbrev. for leaf area ratio.

large calorie *See* calorie.

large-scale integration *Electronics* Abbrev.: LSI (no stops).

Larmor, Sir Joseph (1857–1942) Irish physicist.
Larmor circular frequency Symbol: ω_L (Greek omega)
Larmor precession
Larmor radius

Lartet, Édouard Armand Isidore Hippolyte (1801–71) French palaeontologist.

larva (pl. larvae) *Zool.* Adjectival form: **larval**. *Compare* lava.

larynx (pl. larynges) Adjectival form: **laryngeal**.

Las Campanas Observatory (not Cerro Las Campanas) An observatory near La Serena, Chile, on Cerro Las Campanas.

laser Acronym for light amplification by stimulated emission of radiation.

Lassa fever (cap. L) Named after the Nigerian village where the disease was first described.

Lassell, William (1799–1880) British astronomer.

lat (no stop) Abbrev. for latitude.

lat- (ital., always hyphenated) *Chem.* Prefix denoting a lateral configuration of ligands in a geometrical isomer of a square-pyramidal inorganic complex (e.g. *lat*-dibromodicarbonylcyclopentadienylrhenium(III)). *Compare diag-*.

latent heat Symbol: L A physical quantity, the total heat absorbed or released in a phase transformation (e.g. liquid to gas) at constant temperature. The *SI unit is the joule.

Latent heat is now often expressed as the change in the relevant thermodynamic functions, e.g. $T{\cdot}\Delta S$ or ΔH, where T is thermodynamic temperature, S is entropy, and H is enthalpy.

The **specific latent heat**, symbol l, is the latent heat absorbed or released by unit mass of a substance during an isothermal change of phase. It is measured in joules per kilogram in SI units.

The **molar latent heat**, symbol l_m, is the latent heat absorbed or released by unit amount of substance during an isothermal change of phase. It is measured in joules per mole in SI units.

latex (pl. latexes, not latices)

lati- Prefix denoting broad in extent, range, or scope (e.g. latifoliate, latirostral, latitude).

laticifer (not lacticifer) A latex-containing structure in a plant. Adjectival form: **laticiferous**.

latitude Abbrev.: lat (no stop).

latus rectum (not ital.; pl. latera recta) *Geom.* [from Latin: straight side]

Laue, Max Theodor Felix von (1879–1960) German physicist.
Laue pattern

Laurent, Auguste (1807–53) French chemist.

lauric acid $CH_3(CH_2)_{10}COOH$ The traditional name for dodecanoic acid.

lauryl alcohol $CH_3(CH_2)_{10}CH_2OH$ The traditional name for dodecan-1-ol.

LAV Abbrev. for lymphadenopathy-associated virus, an early name for the human AIDS virus, now renamed *HIV. *See also* HTLV.

lava (pl. lavas) *Geol. Compare* larva.

Lavoisier, Antoine Laurent (1743–94) French chemist.

Lawes, Sir John Bennet (1814–1900) British agricultural chemist.

Lawrence, Ernest Orlando (1901–58) US physicist.
lawrencium

lawrencium Symbol: Lr *See also* periodic table; nuclide.

lb Symbol for **1.** pound. **2.** binary *logarithm.

lbf Symbol for pound-force.

LCAO Abbrev. for linear combination of atomic orbitals.

L capture (not hyphenated) *Physics*

LCC *Electronics* Abbrev. for leadless chip carrier.

LCD Abbrev. for **1.** liquid-crystal display. **2.** (or **l.c.d.**) lowest common denominator.

LCM (or **l.c.m.**) Abbrev. for lowest common multiple.

LD Abbrev. for lethal dose.
LD_{50} Abbrev. for median lethal dose.

LDA Abbrev. for lithium diisopropylamide.

LDL Abbrev. for low-density lipoprotein.

L-dopa (small cap. L) The laevorotatory and pharmacologically active form of *dopa, used to treat parkinsonism. Also called levodopa.

LDQ Abbrev. for Linnett double quartet.

Le (ital.) Symbol for Lewis number.

lead *Chem.* Symbol: Pb *See also* periodic table; nuclide.

lead(II) carbonate hydroxide $Pb_3(OH)_2(CO_3)_2$ The recommended name for the compound traditionally known as white lead.

lead dioxide PbO_2 The traditional name for lead(IV) oxide.

lead(IV) ethanoate $Pb(OOCCH_3)_4$ The recommended name for the compound traditionally known as lead tetraacetate.

lead(IV) hydride PbH_4 The recommended name for the compound traditionally known as plumbane.

leadless chip carrier (not hyphenated) *Electronics* Abbrev.: LCC (no stops).

lead monoxide PbO The traditional name for lead(II) oxide.

lead(II) oxide PbO The recommended name for the compound traditionally known as lead monoxide or plumbous oxide.

lead(IV) oxide PbO_2 The recommended name for the compound traditionally known as lead dioxide or plumbic oxide.

lead tetraacetate $Pb(OOCCH_3)_4$ The traditional name for lead(IV) ethanoate.

lead tetraethyl $(C_2H_5)_4Pb$ The traditional name for tetraethyl-lead(IV).

leaf area index Abbrev.: LAI (no stops). The total surface area of a plant's leaves divided by the area of ground available to the plant: used in assessing the number of plants that can be cultivated in a given area. *Compare* leaf area ratio.

leaf area ratio Abbrev.: LAR (no stops). The total surface area of a plant's leaves divided by the dry weight of the plant: used in assessing the plant's available energy balance. *Compare* leaf area index.

Leakey, Louis Seymour Bazett (1903–72) British anthropologist and archaeologist, husband of Mary Leakey, father of Richard Leakey.

Leakey, Mary (1913–) British palaeoanthropologist, wife of Louis Leakey, mother of Richard Leakey.

Leakey, Richard Erskine Frere (1944–) Kenyan palaeontologist, son of Louis and Mary Leakey.

leap year *See* year.

least significant bit *Computing* Abbrev.: LSB (no stops) or l.s.b. (stops).

least significant digit Abbrev.: LSD (no stops).

Leavitt, Henrietta Swan (1868–1921) US astronomer.

Lebedev, Pyotr Nicolayevich (1866–1912) Russian physicist.

Lebedev, Sergey Vasilyevich (1874–1934) Russian chemist.

Le Bel, Joseph Achille (1847–1930) French chemist.

Lebesgue, Henri Léon (1875–1941) French mathematician.

Lebesgue integral

Leblanc, Maurice (1857–1923) French electrical engineer.

Leblanc connection

Leblanc exciter

Leblanc, Nicolas (1742–1806) French chemist.

Leblanc process

Le Chatelier, Henri Louis (1850–1936) French chemist.

Le Chatelier's principle

Lecher, Ernst (1856–1926) Austrian physicist.

Lecher wires

lecithin Use phosphatidylcholine.

Leclanché, Georges (1839–82) French engineer and inventor.

Leclanché cell

Leclerc, Georges Louis *See* Buffon, Comte de.

Lecoq de Boisbaudran, Paul-Émile (hyphen) (1838–1912) French chemist.

LED Abbrev. for light-emitting diode.

Lederberg, Joshua (1925–) US geneticist.

Lederman, Leon M. (1922–) US physicist.

Lee, Tsung-Dao (hyphen) (1926–) Chinese-born US physicist.

Lee, Yuan Tseh (1936–) Chinese-born US chemist.

LEED Acronym for low-energy electron diffraction.

Leeuwenhoek, Antoni van (1632–1723) Dutch microscopist.

Legendre, Adrien-Marie (hyphen) (1752–1833) French mathematician.

Legendre functions

Legendre polynomials

Legendre's differential equation

Legg–Calvé–Perthes disease (en dashes) Also called Perthes' disease. Named after A. T. Legg (1874–1939), J. Calvé (1875–1954), and G. C. *Perthes.

Legionella (cap. L, ital.) A genus of bacteria that cause pneumonia in humans (including legionnaires' disease). Individual name: **legionella** (no cap., not ital.; pl. legionellae).

legionnaires' disease (no initial caps.) Named after an outbreak at an American Legion convention in 1976.

Le Gros Clark, Sir Wilfrid Edward (1895–1971) British anatomist and anthropologist.

legume 1. A type of dry fruit typical of the family *Leguminosae: a pod. **2.** Any plant of the family Leguminosae.

Leguminosae (cap. L) A large family of dicotyledonous plants, commonly known as the pea family. Alternative name: Fabaceae. It includes three subfamilies: Papilionoideae, Caesalpinioideae, and Mimosoideae, sometimes regarded as separate families: Papilionaceae, Caesalpiniaceae, and Mimosaceae, respectively. Individual name: **legume** (no cap.). Adjectival form: **leguminous**.

Lehn, Jean-Marie Pierre (hyphen) (1939–) French chemist.

Leibniz, Gottfried Wilhelm (not Leibnitz) (1646–1716) German mathematician, philosopher, historian, and physicist.

Leibniz's theorem

Leishman, Sir William Boog (1865–1926) British bacteriologist.

Leishman body

leishmania or, as generic name (*see* genus), ***Leishmania***

Leishman's stain

lel- (ital., always hyphenated) *Chem.* Prefix denoting a geometrical isomer in octahedral tris(chelate) structures in which the carbon-to-carbon bonds of, for example, ethylenediamine are parallel to the three-fold axis of the complex (e.g. *lel*-tris (ethylenediamine)cobalt(III) chloride). *Compare ob-*.

Leloir, Luis Frederico (1906–87) Argentinian biochemist.

Lemaître, Abbé Georges Édouard (1894–1966) Belgian astronomer and cosmologist.

lemma (pl. lemmata or lemmas) A bract subtending a grass flower.

Lenard, Philipp Eduard Anton (1862–1947) German physicist.

length A scalar physical quantity indicating linear extent. The *SI unit is the metre; the angstrom, astronomical unit, parsec, and international nautical mile may also be used. The symbols l, L, and a are common; s is often used for length of a path (especially if the corresponding vector quantity, *displacement, is involved), d or δ for thickness, h for height, b for breadth, r or R for radius, and d or D for diameter.

Lennard-Jones, Sir John Edward (hyphen) (1894–1954) British theoretical chemist.

Lennard-Jones potential

lens (pl. lenses) Adjectival forms: **lens, lenticular**.

Lenz, Heinrich Friedrich Emil (1804–65) Russian physicist.

Lenz's law

Leo A constellation. Genitive form: Leonis. Abbrev.: Leo (no stop). *See also* stellar nomenclature.

Leonids (meteor shower)

Leo Minor A constellation. Genitive form: Leonis Minoris. Abbrev.: LMi (no stops). *See also* stellar nomenclature.

Leonardo da Vinci (1452–1519) Italian artist, engineer, and inventor.

Leonardo of Pisa *See* Fibonacci, Leonardo.

Lep (no stop) *Astron.* Abbrev. for Lepus.

LEP Acronym for Large Electron-Positron (collider), at CERN.

lepido- (lepid- before vowels) Prefix denoting a scale or scaly (e.g. lepidolite, lepidopteran, lepidote).

Lepidodendrales *See* Lycopsida.

Lepidoptera (cap. L) An order of insects comprising the butterflies and moths. Individual name and adjectival form: **lepidopteran** (no cap.; not lepidopterous).

Leporipoxvirus (cap. L, ital.) Approved name for a genus of *poxviruses. Vernacular name: myxoma subgroup. Individual name: **leporipoxvirus** (no cap., not ital.).

lepto- (**lept-** before vowels) Prefix denoting slender, fine, or small (e.g. leptocephalus, leptosporangium, leptotene, lepton).

leptons A family of *fundamental particles that do not take part in the strong interaction: the *electron, *muon, and *tauon, the three *neutrinos associated with them, and all their antiparticles.

Leptospira (cap. L, ital.) A genus of bacteria of the family *Leptospiraceae. Individual name: **leptospire** (no cap., not ital.). Adjectival form: **leptospiral**.

Leptospiraceae (cap. L) A family of *spirochaete bacteria. It contains the single genus **Leptospira*. A second genus, *Leptonema*, has been proposed but not accepted.

Lepus A constellation. Genitive form: Leporis. Abbrev.: Lep (no stop). *See also* stellar nomenclature.

less than Symbol: $<$ The symbol $\leqslant$ denotes 'less than or equal to'. The symbol $\ll$ denotes 'much less than'. These symbols usually have a space or thin space on either side, e.g. $x < y$.

LET Abbrev. for linear energy transfer.

lethal dose Abbrev.: LD (no stops).
median lethal dose Abbrev.: LD_{50} (subscript number).

letter quality Abbrev.: LQ (no stops).

Leu (or **leu**; no stop) Abbrev. for leucine. *See* amino acid.

leucine Abbrev.: Leu or leu (no stop). Symbol: L *See* amino acid.

Leuckart, Karl Georg Friedrich Rudolf (1822–98) German zoologist.

leuco- (**leuc-** before vowels) US: **leuko-** (**leuk-**). Prefix denoting **1.** white or lack of colour (e.g. leucocratic, leucocyte, leucite). **2.** leucocytes (e.g. leucopoiesis; but note *leukaemia).

leucocyte US: **leukocyte**; the US spelling is now widely used in British English. Not a synonym for *lymphocyte, which is one of the various types of leucocyte. Also called white blood cell (abbrev.: WBC).

Leuconostoc (cap. L, ital.) A genus of lactic acid-producing coccoid bacteria. Individual name: **leuconostoc** (no cap., not ital.).

Leu-enkephalin (hyphenated) *See* enkephalin.

leukaemia US: **leukemia**. Adjectival form: **leukaemic** (US: **leukemic**).

leukemia US spelling of leukaemia.

leukocyte US spelling of *leucocyte.

level Verb form: **levels, levelling** (US: **leveling**), **levelled** (US: **leveled**).

Levene, Phoebus Aaron Theodor (1869–1940) Russian-born US biochemist.

Leverrier, Urbain Jean Joseph (1811–77) French astronomer.

Levi-Montalcini, Rita (hyphen) (1909–) Italian cell biologist.

Lévi-Strauss, Claude (hyphen) (1908–) French anthropologist.

levo- US spelling of *laevo-, but note levodopa (not laevodopa; *see* L-dopa).

levorotary US term for *laevorotatory.

Lewis, Gilbert Newton (1875–1946) US physical chemist.
Lewis acid
Lewis base
Lewis–Langmuir theory (en dash)

Lewis number Symbol: *Le* A dimensionless quantity equal to a/D, where a is the *thermal diffusivity and D the *diffusion coefficient. *See also* parameter.

Leydig, Franz von (1821–1908) German anatomist.
Leydig cells Use interstitial cells (of the testis).

LF Abbrev. for low frequency.

lg Symbol for common *logarithm.

LGO Abbrev. for ligand group orbital.

LH Abbrev. for *luteinizing hormone.

L horizon A surface *soil horizon consisting mainly of undecomposed l(itter). Compare F horizon.

L'Hospital (or **L'Hôpital), Guillaume François Antoine, Marquis de** (1661–1704) French mathematician.
L'Hospital's rule or **Bernoulli–L'Hospital rule** (en dash)

LH-RH (hyphenated) Abbrev. for luteinizing-hormone-releasing hormone.

Lhwyd, Edward (1660–1709) Welsh geologist and botanist.

Li Symbol for lithium.

Li, Choh Hao (1913–) Chinese-born US biochemist.

Lib (no stop) *Astron.* Abbrev. for Libra.

Libby, Willard Frank (1908–80) US chemist.

Libra A constellation. Genitive form: Librae. Abbrev.: Lib (no stop). *See also* stellar nomenclature.

LIC *Electronics* Acronym for linear integrated circuit.

lichee, lichi Use litchi.

lichen An organism consisting of both fungal and algal (or cyanobacterial) cells in a symbiotic relationship. As the thalli and fruiting structures of the lichens are fungal in structure, no taxonomic significance is attached to the algal component and the classification is based on the fungal characteristics. Formerly classified together in the taxon Lichenes, lichens are now variously classified as *Ascomycotina, *Basidiomycotina, or *Deuteromycotina, depending on the fungal component.

lidar Acronym for light detection and ranging.

Lie, (Marius) Sophus (1842–99) Norwegian mathematician.
Lie algebra
Lie group

Lieberkühn, Johann Nathaniel (1711–56) German anatomist.
Lieberkühn's glands or **crypts of Lieberkühn**. Also called intestinal glands.

Liebermann–Burchard reaction (en dash) *Chem.* Also called Liebermann's reaction. Named after K. T. Liebermann (1842–1914) and H. Burchard.

Liebig, Justus von (1803–73) German chemist.
Liebig condenser

Liesegang, Raphael Eduard (1869–1947) German chemist.
Liesegang rings

life Two-word terms in scientific contexts in which the first word is 'life' should not be hyphenated unless used adjectivally (e.g. life cycle, life form, life history, life science, life span). Adjectives such as life-sized are always hyphenated.

lifetime (one word)

LIFO (or **lifo**) *Computing* Acronym for last in first out.

ligand A molecule or ion that donates an electron pair to an atom or ion forming a complex.

Ligands are systematically named as follows: anionic ligands are generally named by replacing the final -e by -o, e.g. the ethanoate anion, CH_3COO^-, is named ethanoato-. Exceptions to this rule are fluoro- (F^-), chloro- (Cl^-), bromo- (Br^-), iodo- (I^-), oxo- (O^{2-}), hydrido- (or hydro-) (H^-), hydroxo- (OH^-), peroxo- ($O_2{}^{2-}$), and cyano- (CN^-). Water and ammonia are named aqua- and ammine-, respectively. The groups NO and CO are named nitrosyl- and carbonyl-, respectively, and are considered as neutral ligands.

Nonsystematic nomenclature is generally the same as systematic nomenclature except that the trivial names of organic ligands are used, e.g. acetato-, CH_3COO^-, and water is named aquo-.

See also coordination compound.

ligand group orbital Abbrev.: LGO (no stops).

light Two-word terms in which the first word is 'light' (e.g. light cone, light meter, light pen, light reaction, *light year) should not be hyphenated unless used adjectivally (e.g. light-pen position, light-reaction stage). Adjectives such as light-emitting and light-sensitive are always hyphenated.

light chain *Immunol. See* immunoglobulin.

light-emitting diode Abbrev.: LED (no stops).

light exposure Another name for exposure (def. 1).

lighthouse (one word)

lightning (not lightening) *Meteorol.*

light-water reactor (hyphenated) Abbrev.: LWR (no stops).

lightweight (one word)

light-year (or **light year**) Abbrev.: l.y. A unit of distance used in astronomy (mainly in nonspecialist texts), equal to 9.4605×10^{15} metres, 0.3066 parsec. Distances given in light-years express the time that light or other electromagnetic radiation would take to travel that distance. Analogous but smaller units, such as light-month and light-day, are also used. The parsec is preferred in professional texts.

ligno- (**lign-** (before vowels), **ligni-**) Prefix denoting wood (e.g. lignocellulose, lignose, lignicolous).

Lignosae A taxon of dicotyledonous plants in John Hutchinson's classification (1948) containing all the plant families with predominantly woody members. This is now regarded as an artificial grouping and the taxon is no longer used. *Compare* Herbaceae.

-like In British English, monosyllabic words should not be hyphenated when combined with '-like' (e.g. plantlike) unless they end in l (e.g. oil-like). Words of two or more syllables are always hyphenated (e.g. carbon-like, albumin-like). In American English, words of one or two syllables are not hyphenated, irrespective of their endings, when forming '-like' words (e.g. oillike, carbonlike, petallike); words of three or more syllables are always hyphenated (e.g. albumin-like, haemoglobin-like).

Liliaceae A family of monocotyledonous plants, commonly known as the lily family. In some classifications it includes plants regarded by other authorities as assigned to separate families (e.g. Alliaceae, Alstroemericaceae, Amaryllidaceae, etc.). There is no common name for individual members; *lily without qualification is restricted to members of the genus *Lilium*. Adjectival form: **liliaceous**.

Lilienthal, Otto (1848–96) German aeronautical engineer.

Liliidae A subclass of monocotyledonous plants containing between two and five orders, including Liliales and Orchidales, depending on the classification used. In a recent classification these plants are regarded as a superorder, Lilianae, and the name 'Liliidae' is given to the subclass comprising all the monocotyledons.

Liliopsida In some plant classification schemes, a class of the *Magnoliophyta comprising the monocotyledons.

lily An unspecified plant of the genus *Lilium* (family *Liliaceae); the word is qualified for individual species, e.g. tiger lily (*L. tigrinum*), Madonna lily (*L. candida*). Note that the word is also used in the common names of plants of other genera and families, e.g. arum lily (*Zantedeschia aethiopica*), lily of the valley (*Convallaria majalis*), and water lilies (family Nymphaeaceae).

lime CaO The traditional name for calcium oxide.

limewater aqueous $Ca(OH)_2$ The traditional name for a solution of calcium hydroxide in water.

limits For an integral, summation, or product, the limits are written in a subscript position (lower limit) and superscript position (upper limit) either immediately below and above the symbol or to its right, e.g.

$$\int_a^b x^2 dx \qquad \sum_a^b n^2 \qquad \prod_a^b (n+1)^3.$$

limno- (**limn-** before vowels) Prefix denoting a lake or marsh (e.g. limnology, limnoplankton).

Linacre, Thomas (*c.* 1460–1524) English physician and humanist.

Lindblad, Bertil (1895–1965) Swedish astronomer.

Linde, Karl von (1842–1934) German engineer.

Linde process

Lindemann, Carl Louis Ferdinand von (1875–1939) German mathematician.

Lindemann, Frederick Alexander, Viscount Cherwell (1886–1957) German-born British physicist.

line Adjectival forms: **linear** (relating to a line, lines, length, or one dimension), **lineal** (relating to direct line or descent). Derived nouns: **linearity, lineation, lineage**.

linea (not ital.; pl. lineae) *Anat., astron.* Used in astronomy, with an initial capital, in the approved Latin name of a curved or straight linear feature on the surface of a planet or satellite, as in Asterius Linea on Europa.

linear absorption coefficient Symbol: α (Greek alpha) or *a*. The part of the *linear attenuation coefficient attributable to absorption. The *SI unit is the reciprocal of the metre (m^{-1}).

linear attenuation coefficient Symbol: μ (Greek mu). A physical quantity relating to the attenuation of a wave or beam of particles along a path in a particular medium. The attenuation may be due to absorption, scattering, etc. The *SI unit is the reciprocal of the metre (m^{-1}).

linear combination of atomic orbitals *Chem.* Abbrev.: LCAO (no stops).

linear density *See* density.

linear energy transfer Abbrev.: LET (no stops).

linear expansion coefficient Symbol: α_l (Greek alpha, subscript letter l). A physical quantity defined as:

$$(1/l)(\mathrm{d}l/\mathrm{d}T),$$

where l is length and T is thermodynamic temperature. The *SI unit is the reciprocal of the kelvin (K^{-1}) or the reciprocal of the degree Celsius ($°C^{-1}$).

The cubic expansion coefficient, symbol α_V, is defined as:

$$(1/V)(\mathrm{d}V/\mathrm{d}T),$$

where V is volume. The SI unit is K^{-1} or $°C^{-1}$.

linear momentum Another name for momentum.

linear strain *See* strain.

lines per minute Abbrev.: lpm (no stops).

linking number Symbol: *L* (ital.) The number of times one strand of a DNA helix coils around the other strand in the right-handed direction. It must be an integer.

lin-log *Elec. eng.* Short for linear-logarithmic (response).

Linnaeus, Carolus (1707–78) Swedish botanist. Swedish name: Carl von Linné.

Linnaean system *See* binomial nomenclature.

Linnean Society (not Linnaean)

Linnett double quartet Abbrev.: LDQ (no stops).

Liouville, Joseph (1809–82) French mathematician.

Liouville number

Liouville's theorem

Lipmann, Fritz Albert (1899–1986) German-born US biochemist.

lipo- (lip- before vowels) Prefix denoting fat or fatty (e.g. lipolysis, lipopolysaccharide, lipoprotein, lipase).

lipoprotein (one word)

high-density lipoprotein Abbrev.: HDL (no stops).

low-density lipoprotein Abbrev.: LDL (no stops).

very low-density lipoprotein Abbrev.: VLDL (no stops).

Lippershey, Hans (*c.* 1570–*c.* 1619) Dutch spectacle-maker.

Lippes loop *Med.*

Lippmann, Gabriel (1845–1921) French physicist.

Lipscomb, William Nunn (1919–) US inorganic chemist.

liq. Abbrev. for liquid.

liquefied natural gas Abbrev.: LNG (no stops).

liquefied petroleum gas Abbrev.: LPG (no stops).

liquefy (not liquify) Adjectival form: **liquefiable**. Noun forms: **liquefier, liquefaction**.

liquid Abbrev.: liq. (stop). Verb forms: **liquidize, *liquefy**.

It is recommended that the liquid state of a substance X be denoted X(l), where X may be a chemical name or formula.

Two-word terms in which the first word is 'liquid' (e.g. liquid crystal, liquid helium) are not hyphenated unless used adjectivally (e.g. liquid-drop model, liquid-helium temperatures).

liquid-crystal display (hyphenated) Abbrev.: LCD (no stops).

liquid-metal reactor (hyphenated) Abbrev.: LMR (no stops).

liquified natural gas Abbrev.: LNG (no stops).

Lissajous, Jules Antoine (1822–80) French physicist.
Lissajous figures

Lister, Joseph, Lord (1827–1912) British physician.
Listeria
listeriosis

Listeria (cap. L, ital.) A genus of Gram-positive rod-shaped bacteria. According to the Seeliger–Donker-Voet scheme (en dash; second name hyphenated), strains can be assigned to one of 16 serovars. These are designated by an Arabic numeral (1 to 7) plus a lower-case letter, e.g. 3a, 3b, 4a, etc. Serovars 1 and 2 are combined and designated 1/2, hence 1/2a, 1/2b, and 1/2c (solidus, no space). Adjectival form: **listerial**.

litchi (not lichee, lichi, or lychee; pl. litchis) Generic name: *Litchi* (cap. L, ital.).

-lite Noun suffix denoting a mineral or crystal (e.g. chrysolite, microlite, zeolite).

liter US spelling of *litre.

-lith Noun suffix denoting rock, stone, or a stone (e.g. batholith, megalith, otolith, statolith). Adjectival form: **-lithic**.

lithia Li_2O The traditional name for lithium oxide.

lithium Symbol: Li *See also* periodic table; nuclide.

lithium aluminium hydride $LiAlH_4$ The traditional name for lithium tetrahydridoaluminate(III).

lithium oxide Li_2O The recommended name for the compound traditionally known as lithia.

lithium tetrahydridoaluminate(III) $LiAlH_4$ The recommended name for the compound traditionally known as lithium aluminium hydride.

litho- Prefix denoting **1.** stone or rock (e.g. lithography, lithophyte, lithosol, lithosphere). **2.** *Med.* a calculus (e.g. lithotomy, lithotrite).

litre US: **liter**. Symbol: L or l A unit of volume now identical to the cubic decimetre (dm^3) but not to be used for high-precision measurements. The symbol L is now preferred because of possible confusion between the lower case letter l and the digit one. SI prefixes may be attached to the unit, as in millilitre (mL or ml).

$1\ mL = 10^{-3}\ L = 10^{-3} \times (10^{-1}\ m)^3 = 10^{-6}\ m^3 = 1\ cm^3$.

One litre is equal to 1.759 75 UK pints. The litre as defined in 1901 (in terms of the volume of 1 kg of water) and used up to 1964 is now described as the litre (1901):

1 litre (1901) = 1.000 028 litre.

See also SI units.

Littrow mounting *Spectroscopy*

litzendraht wire (or **litz wire**) *Elec. eng.* [from German]

live *Elec. eng.* Abbrev. (on plugs): L (no stop).

liver Adjectival form: **hepatic** (e.g. hepatic portal system, hepatic vein).

liver fluke (two words)

liverwort (one word) *See* Hepaticopsida.

liveware (one word) *Computing*

Lloyd's mirror *Interferometry*

lm Symbol for lumen. *See also* SI units.

LMC *Astron.* Abbrev. for Large Magellanic Cloud.

LMi *Astron.* Abbrev. for Leo Minor.

LMR Abbrev. for liquid-metal reactor.

LMT *Astron.* Abbrev. for local mean time.

ln Symbol for natural *logarithm.

LNG Abbrev. for liquefied natural gas.

loadstone Use lodestone.

Lobachevsky, Nikolai Ivanovich (1793–1856) Russian mathematician.
Lobachevskian geometry
Lobachevskian space
Lobachevsky's method

local-area network (hyphenated) Abbrev.: LAN (no stops).

local mean time (not hyphenated) Abbrev.: LMT (no stops).

local sidereal time (not hyphenated) *Astron.* Abbrev.: LST (no stops).

local standard of rest *Astron.* Abbrev.: LSR (no stops).

Lockyer, Sir Joseph Norman (1836–1920) British astronomer.

locus (not ital.; pl. loci) *Genetics, maths.*

lodestone (one word; not loadstone)

Lodge, Sir Oliver Joseph (1851–1940) British physicist.

loess *Geol.* Adjectival form: **loessial.** [from German]

Loewi, Otto (1873–1961) German-born US physiologist.

Löffler, Friedrich August Johannes (1852–1915) German bacteriologist.
***Klebs–Löffler bacillus** (en dash)

log Abbrev. and symbol for *logarithm.

logarithmic decrement *See* damping coefficient; relaxation time.

logarithms The logarithm to the base a of x is written:
$\log_a x$ (subscript a, followed by a thin space or a space).
Common logarithms, $a = 10$, are represented by:
$\log_{10} x$ or $\lg x$.
Binary logarithms, $a = 2$, are denoted:
$\log_2 x$ or $\mathrm{lb}\, x$.
Natural logarithms, $a = \mathrm{e}$ (2.718 . . .), are denoted:
$\log_\mathrm{e} x$ or $\ln x$.
Adjectival form: **logarithmic.**

logic *Computing, electronics* Used adjectivally, in preference to *logical, for hardware that uses digital logic (e.g. logic circuit, logic array, logic gate).

logical 1. Of or relating to logic. In computing and electronics, use *logic as an adjective when it relates to hardware. **2.** Conceptual or virtual or involving conceptual entities, as opposed to physical or actual (e.g. logical connection, logical device).

-logical *See* -logy.

logic symbols Symbolic logic symbols are shown in the left-hand side of the table; $p \wedge q$ means 'p and q' while $p \vee q$ means 'p or q or both'. Symbols for the logic operations used in computing are shown on the right-hand side, together with some alternative denotations. A OR B means 'A or B or both' while A XOR B means 'A or B but not both'. NAND and NOR operations are negations of AND and OR operations. The operators AND, OR, etc. are used to name the equivalent logic gates used in electronics. *See also* Appendix 2.

-logous *See* -logy.

-logy Noun suffix denoting **1.** the study or science of (e.g. biology, cytology, ecology, geology). Adjectival form: **-logical** (US (sometimes): **-logic**). **2.** words, ratio (e.g. analogy, homology). Adjectival form: **-logous.** Derived noun form: **-logue** (US (sometimes): **-log**).

Logic symbols

conjunction	$\wedge$	AND	$\wedge$	
disjunction	$\vee$	OR	$\vee$	
negation	$\neg$	NOT	$\neg$	
implication	$\Rightarrow$	NAND	$\triangle$	$\mid$
equivalence	$\Leftrightarrow$			
universal quantifier	$\forall$	NOR	∇	$\downarrow$
existential quantifier	$\exists$	XOR	xor	

Lomonosov, Mikhail Vasilievich (1711–65) Russian scientist and scholar.

London, Fritz (1900–54) German-born US physicist, brother of Heinz London.
London equation
London forces

London, Heinz (1907–70) German-born British physicist, brother of Fritz London.
London equation

long 1. (no stop) Abbrev. for longitude. **2.** (adjective) Two-word terms in which the first word is 'long' are hyphenated when used adjectivally (e.g. long-persistence screen, long-range navigation, long-term memory, long-wave radio). Adjectives such as long-lived and long-tailed are always hyphenated.

long-day plant (hyphenated)

long hundredweight *See* pound.

longi- Prefix denoting long or length (e.g. longicorn, longitudinal).

longitude Abbrev.: long (no stop).

longshore (one word) *Oceanog.*

long-sighted (hyphenated) US: **farsighted** (one word). Noun form: **long-sightedness** (US: **farsightedness**). Medical name: hypermetropia (US: hyperopia).

long terminal repeat *Genetics* Abbrev.: LTR (no stops).

long ton *See* pound.

Longuett-Higgins, Hugh Christopher (hyphen) (1923–) British theoretical chemist.

long wave (hyphenated when used adjectivally) Abbrev.: LW (no stops).

Lonsdale, Dame Kathleen (1903–71) British crystallographer.

Loomis, Elias (1811–89) US meteorologist.

lopho- Prefix denoting an anatomical ridge or crest (e.g. lophophore, lophotrichous).

Lorentz, Hendrik Antoon (1853–1928) Dutch theoretical physicist.
Lorentz–FitzGerald contraction (en dash; preferred to FitzGerald–Lorentz contraction)
Lorentz force
Lorentz transformation

Lorenz, Konrad Zacharias (1903–89) Austrian zoologist and ethologist.

Loschmidt, Johann Joseph (1821–95) Austrian chemist.
***Loschmidt constant**

Loschmidt constant Symbol: n_0 A fundamental constant equal to

$$2.686\,763 \times 10^{25}\,\mathrm{m}^{-3}.$$

It is the ratio L/V_{m}, where L is the Avogadro constant and V_{m} is the molar volume of an ideal gas.

Lossev effect *Electronics*

Lotka, Alfred James (1880–1949) US biologist.
Lotka–Volterra equations (en dash)

loudness level Symbol: L_{N} The physical quantity whose unit is the phon.

loudspeaker (one word)

Love, Augustus Edward Hough (1863–1940) British mathematician and geophysicist.
Love numbers
Love waves

Lovell, Sir (Alfred Charles) Bernard (1913–) British radio astronomer.

Lovibond comparator *Chem.*

low Two- and three-word terms in which the first word is 'low' are hyphenated when used adjectivally (e.g. low-frequency amplifier, low-hysteresis steel, low-level waste, low-pass filter, low-melting-point alloy).

Lowell, Percival (1855–1916) US astronomer.

low-energy electron diffraction *Physics* Abbrev.: LEED (no stops).

lowest common denominator (not hyphenated) Abbrev.: LCD (no stops) or l.c.d. (stops).

lowest common multiple (not hyphenated) Abbrev.: LCM (no stops) or l.c.m. (stops).

low frequency Abbrev.: LF (no stops).

Lowry, Thomas Martin (1874–1936) British chemist.

Brønsted–Lowry theory (en dash)

low-temperature carbonization *Chem.* Abbrev.: LTC (no stops).

low water (hyphenated when used adjectivally) Abbrev.: LW (no stops).

LOX (or **lox**) Acronym for liquid oxygen.

LPG Abbrev. for liquefied petroleum gas.

lpm (no stops) Abbrev. for lines per minute.

LQ Abbrev. for letter quality.

Lr Symbol for lawrencium.

LRSC Abbrev. for Licentiate of the Royal Society of Chemistry.

LSB (or **l.s.b.**) *Computing* Abbrev. for least significant bit.

LSD Abbrev. for **1.** lysergic acid diethylamide. **2.** least significant digit.

L shell (not hyphenated) *Physics*

LSI *Electronics* Abbrev. and preferred form for large-scale integration.

LSR *Astron.* Abbrev. for local standard of rest.

LST *Astron.* Abbrev. for local sidereal time.

L strand Abbrev. and preferred form for light strand (of mitochondrial DNA). *Compare* H strand.

LT Abbrev. for thermolabile toxin. *See* ETEC.

LTA Abbrev. for lead tetraacetate.

LTC Abbrev. for low-temperature carbonization.

LTH Abbrev. for luteotrophic hormone.

LTR *Genetics* Abbrev. for long terminal repeat.

Lu Symbol for lutetium.

Lubbock, John, Lord Avebury (1834–1913) British biologist, politician, and banker.

Luc, Jean André de Usually alphabetized as *de Luc.

luci- Prefix denoting light (e.g. luciferin, lucifugous).

'Lucy' *See Australopithecus.*

lugworm (one word)

lumen 1. Symbol: lm The *SI unit of *luminous flux.

$$1 \text{ lm} = 1 \text{ cd sr}.$$

2. (pl. lumina) The cavity in a tubular organ or a plant cell. Adjectival form: **luminal**.

Lumière, Auguste (1862–1954) French cinematographer, brother of Louis Lumière.

Lumière, Louis (1864–1948) French cinematographer, brother of Auguste Lumière.

luminance Symbol: L or L_v (v stands for visible). A physical quantity, the product of the *luminous intensity per unit area of a surface viewed from a particular direction, and the secant of the angle Θ between the surface and that direction, i.e.

$$(dI/dA) \sec \Theta,$$

where I is luminous intensity and A is area. It can also be expressed as the *luminous flux emitted by unit area into unit solid angle. The *SI unit is the candela per square metre.

luminosity 1. Symbol: L A physical quantity, the total energy radiated per second from a celestial body. The *SI unit is the watt or the joule per second. The luminosity of a body can be expressed as a multiple or submultiple of the solar luminosity, symbol $L_\odot$, which is equal to 3.9×10^{26} W. **2.** The effectiveness of light of any particular wavelength in producing the sensation of brightness.

luminous efficacy Symbol: K A physical quantity, the ratio of *luminous flux to total *radiant flux, Φ_v/Φ_e. The *SI unit is the lumen per watt.

luminous emittance Former name for luminous exitance.

luminous exitance Symbol: M or M_v (v stands for visible). A physical quantity that, at a point of a surface, is the *luminous flux leaving an element of the surface divided by the area of that element. The *SI unit is the lumen per square metre. *Compare* radiant exitance.

luminous flux Symbol: Φ (Greek cap. phi) or Φ_v (v stands for visible). A physical quantity, the *luminous intensity, I, of a light source multiplied by or integrated over the solid angle into which the light is emitted, i.e. $\int I \, d\Omega$. The *SI unit is the lumen.

luminous intensity Symbol: I or I_v (v stands for visible). A physical quantity, the quantity of light emitted per second by a point source in a given direction in unit solid angle. The *SI unit is the candela. *Compare* radiant intensity.

Lummer, Otto (1860–1925) German physicist.

Lummer–Brodhun photometer (en dash)

Lummer–Gehrcke interferometer (en dash)

LUMO Acronym for lowest unoccupied molecular orbital.

lunar mass *See* moon.

lunar month *See* month.

lunar radius *See* moon.

lungfish (one word) *See* Dipnoi.

luni- Prefix denoting the moon (e.g. lunisolar, lunitidal).

Lup (no stop) *Astron.* Abbrev. for Lupus.

Lupus A constellation. Genitive form: Lupi. Abbrev.: Lup (no stop). *See also* stellar nomenclature.

Luria, Salvador Edward (1912–) Italian-born US biologist.

lustre US: **luster**. Adjectival form: **lustrous**.

luteinizing hormone Abbrev.: LH (no stops). A hormone that stimulates corpus luteum formation (among other effects) in the mammalian ovary. The same hormone acts on the testis of male mammals, in which it is called *interstitial-cell-stimulating hormone.

luteinizing-hormone-releasing hormone Abbrev.: LH-releasing hormone or LH-RH (hyphenated, no stops).

luteotrophic hormone (or **luteotrophin**) US: **luteotropic hormone** (or **luteotropin**). Abbrev.: LTH (no stops). Other names for prolactin.

lutetium (not lutecium) Symbol: Lu *See also* periodic table; nuclide.

lux (pl. lux) Symbol: lx The *SI unit of *illuminance.

$$1\ \mathrm{lx} = 1\ \mathrm{lm\ m^{-2}}.$$

LW Abbrev. for **1.** low water. **2.** long waves or long-wave.

Lwoff, André Michael (1902–) French biologist.

LWR Abbrev. for light-water reactor.

lx Symbol for lux. *See also* SI units.

l.y. Abbrev. for light-year.

Ly α (intervening space, Greek alpha) Designation of the first line in the Lyman series in the hydrogen spectrum. Lines of decreasing wavelength are designated Ly β, Ly γ, etc. *See also* Hα.

lychee Use litchi.

Lycoperdales An order of gasteromycete fungi that contains the puffballs and earthstars. Not to be confused with Lycopodiales (*see* Lycopsida).

lycopod Any pteridophyte plant of the order Lycopodiales (*see* Lycopsida), especially any member of the genus *Lycopodium* (cap. L, ital.).

Lycopsida A class of pteridophytes comprising the clubmosses and their allies, containing the extant orders Lycopodiales (lycopods), Selaginellales (including *Selaginella*), and Isoetales (including the quillworts) and the extinct orders Lepidodendrales and Protolepidodendrales. In a recent classification these plants are classified as a superclass, Lycopodiatinae, in the infradivision Cormatae (vascular plants).

Lyell, Sir Charles (1797–1875) British geologist.

Lyman, Theodore (1874–1954) US physicist.

Lyman series

Lyme disease (cap. L) Named after Lyme, Connecticut, where it was first diagnosed (1975).

lymphadenopathy-associated virus Abbrev.: *LAV (no stops).

lymph node (not lymph gland)

lymphocyte A type of leucocyte (white blood cell). The two subdivisions, based on immunological properties, are *B cells (or B lymphocytes) and *T cells (or T lymphocytes).

Lyn *Astron.* Abbrev. for Lynx.

Lynen, Feodor (1911–79) German chemist.

Lynx A constellation. Genitive form: Lyncis. Abbrev.: Lyn (no stop). *See also* stellar nomenclature.

lyo- Prefix denoting a solvent, dissolving, or dispersion (e.g. lyophilic, lyophobic, lyosorption, lyotropic).

Lyot, Bernard Ferdinand (1897–1952) French astronomer.
Lyot filter

lyophilic Denoting a colloid in which the particles have an affinity for the solvent. *Compare* lyophobic.

lyophobic Denoting a colloid in which the particles do not have an affinity for the solvent. *Compare* lyophilic.

Lyr (no stop) *Astron.* Abbrev. for Lyra.

Lyra A constellation. Genitive form: Lyrae. Abbrev.: Lyr (no stop). *See also* stellar nomenclature.
April Lyrids (meteor shower)
June Lyrids (meteor shower)
RR Lyrae stars

Lys (or **lys**; no stop) Abbrev. for lysine. *See* amino acid.

Lysenko, Trofim Denisovich (1898–1976) Soviet biologist.
Lysenkoism

lysergic acid diethylamide Abbrev.: LSD (no stops).

lysi- *See* lyso-.

lysigeny The formation of a space, especially in a plant structure, by the destruction of cells. Adjectival forms: **lysigenic**, **lysigenous**. *Compare* lysogeny.

lysine Abbrev.: Lys or lys (no stop). Symbol: K *See* amino acid.

lysis (pl. lyses) *Biol.* Destruction of a cell. The word is often used in combination (*see* -lysis). Adjectival form: **lytic**.

-lysis Noun suffix denoting breakdown, dissolution, or decomposition (e.g. analysis, autolysis, electrolysis, glycolysis, plasmolysis). Adjectival form: **-lytic**.

lyso- (or **lysi-**) Prefix denoting dissolution or decomposition (e.g. lysosome, lysozyme, lysimeter).

Lysobacter (cap. L, ital.) A genus of non-fruiting gliding bacteria. Individual name: **lysobacter** (no cap., not ital.).

lysogeny The incorporation of viral bacteriophage nucleic acid into the bacterial host's DNA without lysis of the host cell. Adjectival form: **lysogenic**. *Compare* lysigeny.

-lyte Noun suffix denoting a substance that can be broken down (e.g. electrolyte, hydrolyte). Adjectival form: **-lytic**.

M

m Symbol for **1.** medium absorption, used in infrared spectroscopy. **2.** metre. **3.** milli-. **4.** (as a subscript) *molar.

m Symbol for **1.** (light ital.) *Astron.* apparent magnitude. **2.** (bold ital.) magnetic moment. **3.** (light ital.) *magnetic quantum number. **4.** (light ital.) mass (m_e, m_n, m_p = electron, neutron, proton mass, respectively, m_u = atomic mass constant). **5.** (light ital.) molality (m_B = molality of substance B).

M Symbol for **1.** adenosine or cytidine (undefined). **2.** mega-. **3.** mesomeric effect (*see* electron displacement). **4.** *Astron.* a Messier Catalogue object (followed by a number). **5.** a metal in chemical formulae, e.g. MOH. **6.** methionine. **7.** sea mile. **8.** *Astron. See* spectral types.

M Symbol for **1.** (light ital.) *Astron.* absolute magnitude. **2.** (or M_v; light ital. M) luminous exitance. **3.** (light ital.) *magnetic quantum number. **4.** (bold ital.) magnetization. **5.** (light ital.) mass, especially molar mass (M_r = relative molecular mass, $M_\odot$, M_M, $M_\oplus$ = solar, lunar, terrestrial mass, respectively). **6.** (bold ital.) moment of a force. **7.** (light ital.) mutual inductance. **8.** (or M_e; light ital. M) radiant exitance.

***m*-** (ital., always hyphenated; preferred to *meta-*) *Chem.* Prefix denoting *meta*, indicating substitution in the 1,3 positions in a benzene ring (e.g. *m*-dichlorobenzene). The use of numbered substitution positions is recommended (e.g. 1,3-dichlorobenzene). *Compare o-*; *p-*.

Ma (ital.) Symbol for Mach number.

McAdam, John Loudon (1756–1836) British engineer.
macadam or, as trade name, **Tarmacadam**

MacArthur, Robert Helmer (1930–72) US ecologist.

McBurney, Charles (1845–1913) US surgeon.
McBurney's point

McCarty, Maclyn (1911–) US microbiologist.

McClintock, Barbara (1902–) US biologist.

McCollum, Elmer Verner (1879–1967) US biochemist.

McConnell, Harden Marsden (1927–) US theoretical chemist and biochemist.

Mach, Ernst (1838–1916) Austrian physicist.
Mach angle
machmeter (no initial capital letter)
***Mach number**
Mach's principle

Mach number Symbol: *Ma* A dimensionless quantity equal to v/c_a, where v is a characteristic speed and c_a is the speed of sound. *See also* parameter.

Macintosh, Charles (not Mackintosh) (1766–1834) Scottish industrial chemist.
mackintosh (not macintosh)

Maclaurin, Colin (1698–1746) British mathematician.
Maclaurin series

MacLeod, Colin Munro (1909–72) Canadian microbiologist.

McLeod, Herbert (1841–1923) British scientist.
McLeod gauge

MacLeod, John James Rickard (1876–1935) British physiologist.

MacMahon, Percy Alexander (1854–1919) British mathematician.

McMillan, Edwin Mattison (1907–) US physicist.

macro *Computing* Short for macroinstruction.

macro- Prefix denoting large, relatively large, or long (e.g. macroaxis, macroevolution, macrofauna, macroinstruction, macromolecule, macronutrient, macrophage).

macrophyll Use megaphyll.

Macroscelidea (cap. M) An order of mammals comprising the elephant shrews, formerly classified as a family (Macroscelididae) of the order Insectivora. Individual name and adjectival form: **macroscelidean** (no cap.; not in common usage).

macrospore Use megaspore.

macrosporophyll Use megasporophyll.

macula (not ital.; pl. maculae) **1.** *Anat., zool.* A small area distinguishable from surrounding tissue. The macula in the retina (called the **macula lutea**, or 'yellow spot', in the human eye) surrounds the fovea, the area of greatest visual acuity. The terms 'macula' and 'fovea' are sometimes used, incorrectly, as synonyms. Adjectival form: **macular**. **2.** *Astron.* A dark spot on the surface of a planet or satellite. The word is used, with an initial capital, in the approved Latin name of such features, as in Thera Macula and Tyre Macula on Europa.

macula lutea (pl. maculae luteae) *See* macula.

Madelung constant *Chem.* Named after E. Madelung.

maduromycete Any actinomycete bacterium of a group that includes **Actinomadura*, *Microbispora*, *Streptosporangium*, and several other genera. The name has no taxonomic significance and is used for descriptive purposes only. Adjectival form: **maduromycete**.

MAFF Abbrev. for Ministry of Agriculture, Fisheries and Food.

mafic *Geol.* Denoting the dark-coloured (usually ferromagnesian) minerals present in a rock. *Compare* felsic.

Magellan, Ferdinand (1480–1521) Portuguese explorer.
Magellanic Clouds (initial caps.) *Astron.*
Large Magellanic Cloud Abbrev.: LMC (no stops).
Small Magellanic Cloud Abbrev.: SMC (no stops).

Magendie, François (1783–1855) French physiologist.
foramen of Magendie

magma (pl. magmas; not magmata) Adjectival form: **magmatic**.

Magnadur (cap. M) A trade name for a ceramic material used to make permanent magnets.

Magnalium (cap. M) A trade name for an aluminium-based alloy of high reflectivity for light and UV.

magnesia MgO The traditional name for magnesium oxide.

magnesium Symbol: Mg *See also* periodic table; nuclide.

magnesium oxide MgO The recommended name for the compound traditionally known as magnesia.

magnet Adjectival form: **magnetic** (not magnetical). Verb form: **magnetize**. Derived nouns: **magnetism**, ***magnetization**. *See also* magneto-.

magnetic circular dichroism Abbrev.: MCD (no stops). *Chem.*

magnetic constant Another name for permeability of vacuum. *See* permeability.

magnetic dipole moment Symbol: $\boldsymbol{m}$ or μ (Greek mu, bold if available). A physical quantity, a *vector, that indicates the strength of a magnet: the product of a magnet's dipole moment and the ambient *magnetic flux density gives the torque on the magnet. The *SI unit is the ampere metre squared or joule per tesla.

magnetic field strength Symbol: $\boldsymbol{H}$ A *vector quantity, the ratio of *magnetic flux density $\boldsymbol{B}$ to the *permeability, μ, of the medium. The direction at a given point is that of the flux density at that point. The *SI unit is the ampere per metre. Also called magnetic field.

magnetic flux Symbol: Φ (Greek cap. phi). A physical quantity, the *scalar product $\boldsymbol{B}\cdot dA$, where $\boldsymbol{B}$ is the *magnetic flux density across an element of area dA. The *SI unit is the weber.

magnetic flux density Symbol $\boldsymbol{B}$ A physical quantity that indicates the strength of a magnetic field, usually in terms of its effects on, for example, a current-carrying wire. It is a *pseudovector: its magnitude is not invariably proportional to the *magnetic field strength since this would require the permeability to be constant. The *SI unit is the tesla. Also called magnetic induction.

magnetic induction Another name for magnetic flux density.

magnetic-ink character recognition Abbrev.: MICR (no stops).

magnetic moment 1. Symbol: μ (Greek mu). A property of a particle that arises from its spin. The *SI unit is the ampere square metre or joule per tesla. The electron magnetic moment, symbol μ_e, is equal to

$$9.284\,7701 \times 10^{-24}\,\mathrm{J\,T^{-1}}.$$

The proton magnetic moment, symbol μ_p, is equal to

$$1.410\,607\,61 \times 10^{-26}\,\mathrm{J\,T^{-1}}.$$

2. Another name for magnetic dipole moment.

magnetic polarization Symbol: $\boldsymbol{J}$ or $\boldsymbol{B}_i$ A physical quantity, $\mu_0\boldsymbol{M}$, where $\boldsymbol{M}$ is the *magnetization and μ_0 is the *permeability of vacuum. The *SI unit is the *tesla. Also called intrinsic magnetic flux density.

magnetic potential Former name for magnetomotive force.

magnetic potential difference Symbol: U_m A physical quantity, a *scalar, equal to the line integral of the *magnetic field strength between two points. The *SI unit is the ampere.

magnetic quantum number A number, either integral or half-integral, used in specifying the effect of a strong magnetic field on an electron, atom, etc. The symbol is m for a single entity or M for a whole system. The symbols m_l, m_s, and m_j (or M_l, etc.) are the quantum numbers of the components l, s, and j (or L, etc.) in a direction defined by a magnetic field, where l is the *orbital angular momentum quantum, s is the *spin quantum number, and j is the *total angular momentum quantum number.

magnetic resonance imaging *Med.* Abbrev.: MRI (no stops).

magnetic Reynolds number *See* Reynolds number.

magnetic susceptibility *See* permeability.

magnetic tape unit Abbrev.: MTU (no stops).

magnetic vector potential Symbol: $\boldsymbol{A}$ A *vector physical quantity given by:

$$\text{curl}\,\boldsymbol{A} = \boldsymbol{B},\ \text{div}\,\boldsymbol{A} = 0,$$

where $\boldsymbol{B}$ is the *magnetic flux density.

magnetization Symbol: $\boldsymbol{M}$ A physical quantity,

$$(\boldsymbol{B}/\mu_0) - \boldsymbol{H},$$

where $\boldsymbol{B}$ is the *magnetic flux density, $\boldsymbol{H}$ the *magnetic field strength, and μ_0 the *permeability of vacuum. The *SI unit is the ampere per metre. It resolves as a *vector but does not add as a vector because of saturation effects. *See also* magnetic polarization.

magneto- Prefix denoting magnetism or a magnetic property (e.g. magnetoelectric, magnetohydrodynamics, magnetometer, magneto-optics).

magnetohydrodynamics (one word) Abbrev.: MHD (no stops).

magnetomotive force Symbol: $F_{\rm m}$ Abbrev.: mmf. A *scalar physical quantity, the line integral of the *magnetic field strength, $\boldsymbol{H}$, around a closed path. The *SI unit is the ampere. Former name: magnetic potential.

magnification *Optics, etc.* Symbol: M Indicated by the numerical value followed after a space by × (multiplication sign), as in 100 ×.

magnify (magnifies, magnifying, magnified) Derived noun: ***magnification**.

magnitude A dimensionless physical quantity, the logarithm of the reciprocal of the brightness of a celestial body. The **apparent magnitude**, symbol m, is the brightness (intensity) observed from earth. The apparent magnitudes, m_1 and m_2 of two bodies are related by the equation:

$$m_1 - m_2 = 2.5 \log_{10}(I_2/I_1),$$

where I is intensity. Magnitude may have a positive, zero, or negative value: the brighter the object, the lower its magnitude. The **absolute magnitude**, symbol M, is the apparent magnitude that a celestial body would have at a distance of 10 parsecs.

Magnoliidae A subclass of dicotyledonous plants containing a variable number of orders depending on the classification used; Cronquist's widely used classification recognizes eight: Magnoliales, Laurales, Piperales, Aristolochiales, Illiciales, Nymphaeales, Ranunculales, and Papaverales. It is equivalent to the former taxon Ranales. In a recent classification these plants are regarded as a superorder, Magnolianae, and the name 'Magnoliidae' is given to the subclass comprising all the dicotyledons.

Magnoliophyta In some plant classification schemes, a division comprising the flowering plants. It is divided into two classes: Magnoliopsida (dicotyledons) and Liliopsida (monocotyledons). *See* Angiospermae.

Magnoliopsida In some plant classification schemes, a class of the *Magnoliophyta comprising the dicotyledons. In a recent classification the name 'Magnoliopsida' is given to a class comprising all the flowering plants (*see* Angiospermae).

Magnus, Heinrich G. (1802–70) German scientist.

Magnus effect

MAI *Forestry* Abbrev. for mean annual increment.

Maiman, Theodore Harold (1927–) US physicist.

mainframe (one word) *Computing*

mains (usually takes a pl. form of verb) Adjectival form: **mains** (e.g. mains transformer, mains voltage).

main-sequence stars (hyphenated) *Astron.* Denoted by V (Roman numeral). Former name: dwarfs.

major histocompatability complex Abbrev. and preferred form: *MHC (no stops).

Maksutov telescope *Astron.* Named after D. D. Maksutov (1896–1964).

malaco- Prefix denoting **1.** soft (e.g. Malacostraca). **2.** molluscs (e.g. malacology).

Malacostraca (cap. M) A subclass of crustaceans including crabs, lobsters, shrimps, etc. Individual name and adjectival form: **malacostracan** (no cap.; not malacostracous).

maleic acid HOOCCH=CHCOOH The traditional name for *cis*-butenedioic acid.

maleic anhydride The traditional name for *butenedioic anhydride.

malic acid $CH_2(COOH)CH(OH)COOH$ The traditional name for 2-hydroxybutanedioic acid.

malleable (not maleable) Noun form: **malleability**.

malleolus (pl. malleoli) A bony protuberance on either side of the ankle. Adjectival form: **malleolar**. *Compare* malleus.

Mallet, Robert (1810–81) Irish industrialist and seismologist.

malleus (pl. mallei) The hammer-shaped outermost ear ossicle of mammals. *Compare* malleolus.

Mallophaga (cap. M) An order (or suborder: *see* Phthiraptera) of insects comprising the biting lice. Individual name and adjectival form: **mallophagan** (no cap.).

malonic acid $CH_2(COOH)_2$ The traditional name for propanedioic acid.

Malpighi, Marcello (1628–94) Italian histologist.
Malpighiaceae *Bot.*
Malpighian (US malpighian) body (or **corpuscle)**
Malpighian (US malpighian) layer
Malpighian (US malpighian) tubule

Malthus, Thomas Robert (1766–1834) British economist. Adjectival form: **Malthusian**.

Malus, Étienne Louis (1775–1812) French military engineer and physicist.
Malus's law

mamilla (also US **mammilla**; pl. mamillae) A nipple or nipple-like part. Adjectival form: **mamillary** (also US **mammillary**). *Compare* mamma.

mamma (pl. mammae) A breast. Adjectival form: **mammary**. *Compare* mamilla.

Mammalia (cap. M) The class of vertebrates comprising the mammals, subdivided into the *Prototheria, *Metatheria, and *Eutheria. Adjectival form: **mammalian**.

mammalogy (not mammology)

mammilla A variant US spelling of *mamilla.

Man (no stop) Abbrev. for mannose.

manganate A compound containing the ion MnO_4^{2-}, e.g. potassium manganate. The recommended name is manganate(VI).

manganate(VI) A compound containing the ion MnO_4^{2-}, e.g. potassium manganate(VI). The traditional name is manganate.

manganate(VII) A compound containing the ion MnO_4^-, e.g. potassium manganate(VII). The traditional name is permanganate.

manganese Symbol: Mn *See also* periodic table; nuclide.

manganese dioxide MnO_2 The traditional name for manganese(IV) oxide.

manganese heptoxide Mn_2O_7 The traditional name for manganese(VII) oxide.

manganese(II) oxide MnO The recommended name for the compound traditionally known as manganous oxide.

manganese(III) oxide Mn_2O_3 The recommended name for the compound traditionally known as manganic oxide or manganese sesquioxide.

manganese(IV) oxide MnO_2 The recommended name for the compound traditionally known as manganese dioxide.

manganese(VII) oxide Mn_2O_7 The recommended name for the compound traditionally known as manganese heptoxide.

manganese sesquioxide Mn_2O_3 The traditional name for manganese(III) oxide.

manganic Denoting compounds in which manganese has an oxidation state of +3. The recommended system is to use oxidation numbers, e.g. manganic oxide, Mn_2O_3, has the systematic name manganese(III) oxide.

manganous Denoting compounds in which manganese has an oxidation state of +2. The recommended system is to use oxidation numbers, e.g. manganous oxide, MnO, has the systematic name manganese(II) oxide.

man–machine interface (en dash) *Computing* Abbrev.: MMI (no stops).

mannose Abbrev.: Man (no stop). *See* sugars.

mano- Prefix denoting pressure (e.g. manometer, manostat).

manoeuvre US: **maneuver**. Adjectival form: **manoeuvrable** (US: **maneuverable**).

Manson, Sir Patrick (1844–1922) British physician.

mantis Any insect of the genus *Mantis* (cap. M, ital.); also used, as an alternative to **mantid**, for any other insect of the family Mantidae (suborder Mantodea). The plural, 'mantises', is rarely used: 'mantids' is preferred, especially for members of the family. *See* Dictyoptera.

Mantoux, Charles (1877–1947) French physician.
Mantoux test

MAO Abbrev. for monoamine oxidase.
MAO inhibitor or **MAOI** Abbrev. for monoamine oxidase inhibitor.

map unit *Genetics* A unit of length used in chromosome mapping, equal to the length of chromosome over which recombination occurs with 1% frequency. For short distances (<10 map units) recombination frequency is virtually proportional to distance. For greater distances, because of multiple crossovers, a correction factor must be employed to obtain accurate map distances from recombination frequencies.

Marchantiopsida *See* Hepaticopsida.

Marconi, Guglielmo (1874–1937) Italian electrical engineer.

mare (not ital.; pl. maria) A large relatively smooth dark-coloured area on the surface of a planet or satellite. For a particular feature there is an approved Latin name, as in Mare Imbrium (Sea of Rains), on the moon, and Mare Australe (Southern Sea), on Mars.

A very large mare is given the name Oceanus (Latin: ocean), as in the lunar Oceanus Procellarum (Ocean of Storms). Small maria are given the name Sinus (Latin: bay), Palus (Latin: marsh), or Lacus (Latin: lake), in order of decreasing size, examples being Sinus Iridum (Bay of Rainbows), Palus Putredinus (Marsh of Decay), and Lacus Veris (Lake of Spring), all on the moon.

Marey, Étienne-Jules (hyphen) (1830–1904) French physiologist.

Marggraf, Andreas Sigismund (1709–82) German chemist.

margo (pl. margones) *Bot.* Part of the structure of a pit in a plant cell wall.

Margulis, Lynn (1938–   ) US biologist.

maria *See* mare.

Marignac, Jean Charles Galissard de (1817–94) Swiss chemist.

Mariotte, Edmé (*c.* 1620–84) French physicist.
Mariotte's law Use *Boyle's law.

Markarian galaxies *Astron.* Named after B. E. Markarian.

Markov, Andrei Andreevich (1856–1922) Russian mathematician. Adjectival form: **Markovian**.
Markov chain
Markov process

Markovnikov, Vladimir Vasilyevich (1837–1904) Russian chemist.
Markovnikov rule

Mars A planet. Adjectival form: **Martian**.

Marsh, James (1794–1846) British chemist.
Marsh's test

Marsh, Othniel Charles (1831–99) US palaeontologist.

Marsupialia (cap. M) The sole order of the *Metatheria, comprising mammals possessing a *marsupium in which the young develop. Individual name and adjectival form: **marsupial** (no cap.).

marsupium (pl. marsupia) The abdominal pouch of the Marsupialia. Adjectival form: **marsupial.**

Martin, Archer John Porter (1910–) British chemist.

Martin, Pierre-Émile (hyphen) (1824–1915) French engineer.
Siemens–Martin process (en dash)

Marvel, Carl Shipp (1894–1988) US chemist.

mascon *Astron.* Short for mass concentration.

maser Acronym for microwave amplification by stimulated emission of radiation.

Maskelyne, Nevil (1732–1811) British astronomer.

mass Symbol: m (or M) A fundamental physical quantity that both measures a body's inertia and determines the mutual gravitational attraction between it and another body. The *SI unit is the kilogram or sometimes the tonne. In general language mass and *weight are often used synonymously, although in a scientific context this is incorrect, as weight is a force.

Massachusetts Institute of Technology Abbrev.: MIT (no stops).

mass concentration *See* concentration.

mass density *See* density.

mass–energy relation (en dash)

mass excess Symbol: Δ (Greek cap. delta). A physical quantity given by $(m_a - Am_u)$, where m_a is the atomic mass of a nuclide ^{A}X, A is the nuclide's mass number, and m_u is the unified *atomic mass constant. The *SI unit is the kilogram or (when energy equivalence is being considered) the electronvolt.

mass fraction Symbol: w_B (for a substance B) or, when a complicated formula is to be written, w(B), e.g. $w(NaClO_3)$. A dimensionless physical quantity, the ratio of the mass of substance B to the mass of the mixture.

Massieu function *Thermodynamics* Symbol: J

mass–luminosity relation (en dash) Abbrev.: M–L relation (en dash). *Astron.*

mass number Another name for nucleon number.

mass spectrometry Abbrev.: MS (no stops).

mastigo- Prefix denoting flagella or fine hairs (e.g. mastigoneme, Mastigophora).

Mastigomycotina A subdivision of true fungi (Eumycota) containing the classes *Chytridiomycetes, *Hyphochytriomycetes, and *Oomycetes. In a recent classification this subdivision is not used: the first two classes are regarded as separate divisions and the Oomycetes are placed in the division *Chromophyta.

Mastigophora (cap. M) A class of protozoans possessing one or more flagella. Also called Flagellata (*see* flagellate). Individual name and adjectival form: **mastigophoran** (no cap.).

materials science (not material) Similarly **materials handling, materials testing.**

mathematical symbols Recommended general symbols and letter symbols for common mathematical functions are given in Appendix 5. Both sets of symbols are printed in roman (upright) type; in contrast *physical quantities and numerical variables are generally set in italic type. An en dash is used for a minus sign.
A space or thin space is usually set on one or both sides of a symbol, e.g.

$$a^2 + b^2 - 2ab, \quad x > y,$$
$$5.3 \times 10^{-3}, \quad \sin(-x^2).$$

Greek letters are used as symbols for summation, product, finite increase of x, variations of x, and variation of f. The symbol for partial derivative, ∂, is often called curly dee. The total differential of a function f, symbol df, is given by:

$$\mathrm{d}f(x,y) = (\partial f/\partial x)_y \mathrm{d}x + (\partial f/\partial y)_x \mathrm{d}y.$$

There are also symbols used in connection with *sets, *vectors, and *matrices. *See also* arithmetic operations; brackets; complex numbers; hyperbolic functions; logic symbols.

mathematics Abbrev.: maths (no stop) or maths. (US: math or math.). **1.** (takes a sing. form of verb) The field of study itself. **2.** (takes a pl. form of verb) The operations, etc., involved in a particular study or solution. Adjectival form: **mathematical** (not mathematic). Derived noun: **mathematician.**

Mather, Kenneth (1911–90) British geneticist.

matric potential Symbol: Ψ_m *See* water potential.

Matrix symbols

matrix	A, $\{a_{ij}\}$
product of A and B	AB
inverse of A	A^{-1}
unit matrix	E, I
transpose of matrix A	A^{T}
complex conjugate of A	A^{*}
Hermitian conjugate of A	$A^{\dagger}$
determinant of A	det A
trace of A	Tr A

matrix (pl. matrices) Symbols used in mathematics and physics in connection with matrices are shown in the table. An italic capital letter conventionally denotes a matrix in its entirety and the corresponding lower-case letter, indexed by a pair of subscripts, denotes an element in that matrix; the first subscript is row number, the second is column number. In computing, the notation for matrices is determined by the programming language.

matrix parameter *See* electronics, letter symbols.

Matthews, Drummond Hoyle (1931–) British geologist.

Matthias, Bernd Teo (1918–80) German-born US physicist.

Matthiessen's rule *Physics* Named after A. Matthiessen.

Matuyama, Motonori (1884–1958) Japanese geologist.
Matuyama reversed epoch

Maudslay, Henry (1771–1831) British engineer and inventor.

Mauna Kea Observatory An observatory on the dormant volcano Mauna Kea on Big Island, Hawaii.

Maunder, Edward Walter (1851–1928) British astronomer.
Maunder minimum

Maupertuis, Pierre-Louis Moreau de (hyphen) (1698–1759) French mathematician, physicist, and astronomer.

Maury, Antonia Caetana de Paiva Pereira (1866–1952) US astronomer.

Maury, Matthew Fontaine (1806–73) US oceanographer.

max (no stop) Abbrev. for maximum.

maxilla (pl. maxillae) *Anat., zool.* Adjectival form: **maxillary.**

maximum (pl. maxima) Abbrev.: max (no stop). Adjectival forms: **maximum, maximal.** Verb form: **maximize.**

maximum and minimum thermometer (not hyphenated)

maximum permissible dose (not hyphenated)

maxwell Symbol: Mx The *cgs (electromagnetic) unit of magnetic flux, now discouraged. In *SI units, magnetic flux is measured in webers: 1 Mx = 10^{-8} Wb. Named after J. C. *Maxwell.

Maxwell, James Clerk (1831–79) British physicist. Adjectival form: **Maxwellian.**
***maxwell**
Maxwell–Boltzmann distribution (en dash)
Maxwell's demon
Maxwell's equations
Maxwell's thermodynamic relations

May, Robert McCredie (1936–) Australian-born US theoretical ecologist.

Mayer, Julius Robert von (1814–78) German physician and physicist.

Maynard Smith, John (1920–) British biologist.

Mayow, John (1640–79) English physiologist and chemist.

Mayr, Ernst Walter (1904–) German-born US zoologist.

mb Symbol for millibar, often used in meteorology instead of mbar.

MBBA Abbrev. for *N*-(4-methoxybenzylidene)-4-butylaniline.

MBE Abbrev. for molecular-beam epitaxy.

Mc- Names starting with Mc- (e.g. McLeod) are alphabetized as Mac- in this dictionary.

MCD Abbrev. for magnetic circular dichroism.

m chromosome (lower-case m) A tiny chromosome found in moss nuclei.

MCPA Abbrev. and preferred form for 2-methyl-4-chlorophenoxyacetic acid, used as a weedkiller.

MCPB Abbrev. for 4-(2-methyl-4-chlorophenoxy)butyric acid, used as a weedkiller.

MCPBA Abbrev. for *m*-chloroperoxybenzoic acid.

Md Symbol for mendelevium.

MDEA Abbrev. for *N*-methyldiethanolamine.

MDI Abbrev. for methylene diisocyanate.

Me (no stop) **1.** Symbol often used to denote the methyl group in chemical formulae, e.g. CH_3OH can be written as MeOH. **2.** *Astron. See* spectral types.

mean *Stats.* Denoted by $\bar{x}$ for a sample of observations, x_i, or a group of numbers, x_i. Usually denoted by $\langle x \rangle$ for a time-varying quantity, $x(t)$.

mean annual increment Abbrev.: MAI (no stops). *Forestry*

mean energy imparted *See* energy imparted.

mean free path (not hyphenated) Symbol: λ (Greek lambda). The SI unit is the metre.

mean life *See* decay constant.

mean sea level (not hyphenated) Abbrev.: msl (no stops).

mean time between failures Abbrev.: MTBF (no stops).

measles Medical name: morbilli.

meatus (pl. meatuses or meatus) *Anat., zool.*

mechanoreceptor (one word) *Physiol.*

Meckel, Johan Friedrich (1724–74) German anatomist and botanist, grandfather of Johann (double n) Meckel.

Meckel, Johann Friedrich (1781–1833) German anatomist and surgeon.
Meckel's cartilage
Meckel's diverticulum

Medawar, Sir Peter Brian (1915–87) British immunologist.

media 1. (pl. mediae) *Anat.* **a.** The middle layer of a blood vessel. **b.** The central region of an organ or the body. Adjectival form: **medial. 2.** (pl. noun) *See* medium.

median 1. (noun) *Stats., geom.* Adjectival form: **median** (not medial). **2.** (adjective) *Anat.* Situated in or towards the plane that divides the body into right and left halves. Not to be confused with medial (*see* media).

median effective dose Abbrev. and preferred form: ED_{50}.

median lethal dose Abbrev. and preferred form: LD_{50}.

Medical Research Council Abbrev.: MRC (no stops).

medium (pl. media) *Computing, biol., etc.* The plural, media, should always take a plural form of verb, despite its widespread use as a singular noun in the communications sense.

medium frequency Abbrev.: MF (no stops).

medium-scale integration (hyphenated) Abbrev.: MSI (no stops).

medium wave (hyphenated when used adjectivally) Abbrev.: MW (no stops).

medulla (pl. medullas or medullae) The central region of many organs and parts. Adjectival forms: **medullary, medullated.**
medulla oblongata (pl. medullae oblongatae)

medusa (pl. medusae) *Zool.* Adjectival form: **medusan.**

mega- 1. Symbol: M A prefix to a unit of measurement that indicates 10^6 times that unit, as in megavolt (MV) or megahertz (MHz). *See also* SI units. **2.** A prefix used in computing to indicate a multiple of 2^{20} (i.e. 1 048 576), as in megabyte. **3.** A general prefix denoting large size (e.g. meganucleus, megasporangium).

megaphyll (preferred to macrophyll) *Bot.*

megaspore (preferred to macrospore) *Bot.*

megasporophyll (preferred to macrosporophyll) *Bot.*

megawatt Symbol: MW *See* watt.

meio- Prefix denoting reduction, less, or small (e.g. meiocyte, meiofauna, *meiosis). *See also* mio-.

meiosis (pl. meioses) Cell division resulting in daughter nuclei with half the chromosome number of the parent nucleus. Not to be confused with 'miosis' (constriction of the pupil of the eye). Adjectival form: **meiotic**. *Compare* mitosis.

Meissner, Fritz Walther (1882–1974) German physicist.
Meissner effect

Meissner, Georg (1829–1905) German anatomist.
Meissner's corpuscle
Meissner's plexus

Meitner, Lise (1878–1968) Austrian-born Swedish physicist.

MEK Abbrev. for methyl ethyl ketone (butanone).

melamine (no cap., not a trade name)

melano- Prefix denoting black or dark coloration (e.g. melanocratic, melanocyte).

melanocyte-stimulating hormone (hyphen) Abbrev.: MSH (no stops).

Melinex (cap. M) US: **Mylar**. A trade name for a type of strong transparent polyester.

melting point (two words) Abbrev.: m.p. (stops). Do not use freezing point as a synonym. A pure substance under standard conditions of pressure has a single reproducible melting point; this may or may not be the case for its freezing point so that the two are not necessarily equal.

melting temperature *Cell biol.* Symbol T_m The midpoint of the temperature range over which the strands of a given sample of DNA separate.

meltwater (one word) *Glaciol.*

Men (no stop) *Astron.* Abbrev. for Mensa.

Mendel, Gregor Johann (1822–84) Austrian monk and botanist. Adjectival form: **Mendelian.**
Mendelism
Mendel's laws

Mendeleev, Dmitri Ivanovich (1834–1907) Russian chemist.
Mendeleev's law of octaves
***mendelevium**

mendelevium Symbol: Md *See also* periodic table; nuclide.

MEng (or **M.Eng.**) Abbrev. for Master of Engineering.

Ménière, Prosper (acute and grave accents) (1799–1862) French physician.
Ménière's disease

meninges (pl. noun; sing. meninx) The three membranes enclosing the central nervous system of vertebrates, comprising the dura mater (pachymeninx) and the arachnoid mater and pia mater (the leptomeninges). The singular form, meninx, is rarely used: individual meninges are referred to by their respective names. Adjectival form: **meningeal**.

meningococcus (no cap., not ital.; pl. meningococci) Common name for one of the bacteria (*Neisseria meningitidis*) responsible for meningitis. Adjectival form: **meningococcal**.

meninx (pl. meninges) **1.** The mesodermal layer surrounding the embryonic vertebrate brain. **2.** *See* meninges.

meniscus (pl. menisci; preferred to meniscuses) Adjectival forms: **meniscus, meniscoid**.

mensa (not ital.; pl. mensae) *Astron.* A flat-topped steep-sided elevation. The word, with an initial capital, is used in the approved Latin name of such elevations on the surface of a planet or satellite, as in Protonilus Mensae on Mars.

Mensa A constellation. Genitive form: Mensae. Abbrev.: Men (no stop). *See also* stellar nomenclature; mensa.

-mer *Chem.* Noun suffix denoting a particular form of a compound (e.g. isomer, polymer). Adjectival form: **-meric**.

mer- (ital., always hyphenated) *Chem.* Prefix denoting *meridional*, indicating geometric isomerization in octahedral inorganic complexes of the type MA_3B_3, where M is a metal and A and B are ligands, in which ligands of one type span three positions such that two are opposite each other (e.g. *mer*-triaminetrinitrosylcobalt(III)). *Compare fac-*.

Mercalli, Giuseppe (1850–1914) Italian geologist.
Mercalli scale

mercaptan The traditional name for a thiol.

mercapto- Prefix denoting the group –SH attached to a carbon atom (e.g. mercaptoethanol).

Mercator, Gerhardus (1512–94) Dutch cartographer and geographer.
Mercator projection

Mercer, John (1791–1866) British chemist.
mercerizing

mercuric Denoting compounds in which mercury has an oxidation state of +2. The recommended system is to use oxidation numbers, e.g. mercuric chloride, $HgCl_2$, has the systematic name mercury(II) chloride.

mercurous Denoting compounds in which mercury has an oxidation state of +1. The recommended system to to use oxidation numbers, e.g. mercurous chloride, HgCl, has the systematic name mercury(I) chloride.

mercury Symbol: Hg *See also* periodic table; nuclide.

Mercury A planet. Adjectival form: **Mercurian.**

-mere Noun suffix denoting a division or part (e.g. actinomere, blastomere, metamere). Adjectival form: **-meric**.

meri- Prefix denoting a part or division (e.g. mericarp, meristele, meristem). *See also* mero-.

-meric *See* -mer; -mere.

meridian (not meridion) *Astron., cartog.* Adjectival forms: **meridian, meridional**.

MERLIN *Astron.* Acronym for Multi-Element Radio-linked Interferometer Network (operated from Jodrell Bank, Cheshire).

mero- Prefix denoting a part or partial (e.g. meroblastic, merocrine, merozoite). *See also* meri-.

-merous *Biol.* Adjectival suffix denoting a number or type of parts (e.g. heptamerous).

Merrifield, (Robert) Bruce (1921–) US biochemist.

Mersenne, Marin (1588–1648) French mathematician, philosopher, and theologian.
Mersenne numbers

mes- *See* meso-.

mesa (pl. mesas) *Geol., electronics*

Meselson, Matthew Stanley (1930–) US molecular biologist.
Meselson–Stahl experiment (en dash)

MESFET *Electronics* Acronym for metal-semiconductor field-effect transistor.

mesitylene $C_6H_3(CH_3)_3$ The traditional name for 1,3,5-trimethylbenzene.

meso- (**mes-** before vowels) Prefix denoting the middle or intermediate (e.g. mesoderm, mesophyte, mesosphere, Mesozoic, mesarch, mesencephalon).

meso- (ital., always hyphenated) *Chem.* Prefix denoting an optically inactive optical isomer of a compound that can exist in other optically active forms (e.g. *meso*-tartaric acid).

mesoglea (preferred to mesogloea) The gelatinous layer between the ectoderm and endoderm of coelenterates.

mesoinfluenza virus (two words) Either of two species of virus, designated C and D, similar to the *influenza viruses but now placed in a separate genus in the family Orthomyxoviridae.

mesomeric effect *See* electron displacement.

mesons *See* hadrons.

mesophilic Denoting microorganisms (known as **mesophiles**) whose optimum growth is at moderate temperatures (20–

45 °C). Not to be confused with mesophyllic (*see* mesophyll).

mesophyll Photosynthetic leaf tissue. Adjectival form **mesophyllic** (*compare* mesophilic).

Mesozoic Abbrev.: Mz (no stop). **1.** (adjective) Denoting an era in the geological time scale. **2.** (noun; preceded by 'the') The Mesozoic era.

messenger RNA Abbrev.: mRNA (no stops). *See* RNA.

Messier, Charles (1730–1817) French astronomer.
Messier catalogue
Messier numbers

mesyl- *Prefix denoting the group CH_3SO_2-. Methanesulphonyl- is recommended in all contexts.

Met (or **met**; no stop) Abbrev. for methionine. *See* amino acid.

meta- (**met-** before vowels) Prefix denoting **1.** after or behind (e.g. metacarpus, metachronal, metameric, metanephros, metaxylem, metencephalon). **2.** change or transformation (e.g metabolism, metamorphosis, metastable). **3.** analysing, describing, or going beyond (e.g. metalanguage, metamathematics). **4.** an isomeric or polymeric form of a compound (e.g. metaphosphate, metaldehyde).

meta- (ital., always hyphenated) *Chem. See m-*.

metabisulphite Denoting a compound containing the ion $S_2O_5^{2-}$, e.g. sodium metabisulphite, $Na_2S_2O_5$. The recommended name is disulphate(IV).

metal Adjectival form: **metallic**. Verb form: **metallize**. Derived noun: **metallization**.

metallo- (**metall-** before vowels, **metalli-**) Prefix denoting metal (e.g. metallocene, metallography, metallurgy, metalliferous).

metallocene Any of a class of organometallic compounds in which a metal atom is coordinated to two cyclopentadienyl rings.
A common method of nomenclature for metallocenes is to indicate the number of atoms of each cyclopentadienyl group attached to the metal atom by adding the prefix monohapto-, trihapto-, or pentahapto- (often abbreviated to h^1, h^3, or h^5) to the systematic name of the complex. Another system is to indicate σ or π bonding between each cyclopentadienyl group and the metal by adding the prefix σ- or π-, respectively, to the systematic name of the complex; this system is less informative than the former as the bonding is not described as fully. Examples are bis(pentahaptocyclopentadienyl)iron or bis(h^5-cyclopentadienyl)iron or bis(π-cyclopentadienyl)iron.

metallography The microscopic study of the structure of metals (*compare* metallurgy). Adjectival form: **metallographic**.

metalloid (noun; preferred to semimetal)

metallurgy The science concerned with producing metals from their ores (*compare* metallography). Adjectival form: **metallurgical**.

metal-oxide semiconductor (hyphenated) Acronym and preferred form: *MOS.

metamorphism The process, involving the action of heat and pressure, by which **metamorphic** rocks are formed. Derived verb: **metamorphose**. *Compare* metamorphosis.

metamorphosis (pl. metamorphoses) The transformation of the larval form of an animal, especially an amphibian or insect, to the adult form. Adjectival form: **metamorphic**. Derived verb: **metamorphose**. *Compare* metamorphism.

Metatheria (cap. M) An infraclass (*see* Theria) or subclass of mammals containing a single order, *Marsupialia. Individual name and adjectival form: **metatherian** (no cap.).

Metazoa (cap. M) A subkingdom comprising all multicellular animals except the sponges (*see* Parazoa). Individual name and adjectival form: **metazoan** (no cap.; not metazoic).

Metchnikoff, Elie (1845–1916) Russian-born French zoologist and bacteriologist.

Met-enkephalin (hyphenated) *See* enkephalin.

meteor A streak of light seen when a small particle of interplanetary dust (a **meteoroid**) enters and burns up in the earth's atmosphere. A **meteorite** is a large mineral aggregate from interplanetary space that reaches the earth's surface.

meter 1. A measuring instrument. *See also* -meter. **2.** US spelling of metre.

-meter Noun suffix denoting a measuring instrument (e.g. barometer, thermometer, voltmeter). Adjectival form: **-metric**.

metestrus (or **metestrum**) US forms of *metoestrus.

methacrylic acid $CH_2C(CH_3)COOH$ The traditional name for 2-methylpropenoic acid.

methanal HCHO The recommended name for the compound traditionally known as formaldehyde. Formaldehyde is acceptable in biological contexts for an aqueous solution of methanal used as a preservative.

methanamide $HCONH_2$ The recommended name for the compound traditionally known as formamide.

methanesulphonyl- (preferred to mesyl-) Prefix denoting the group CH_3SO_2-.

methanoate The recommended name for a salt of methanoic acid. The traditional name is formate.

Methanococcus (cap. M, ital.) A genus of methanogenic bacteria. Individual name: **methanococcus** (no cap., not ital.; pl. methanococci).

methanoic acid HCOOH The recommended name for the compound traditionally known as formic acid.

methanol CH_3OH The recommended name for the compound traditionally known as methyl alcohol.

Methanosarcina (cap. M, ital.) A genus of methanogenic bacteria. Individual name: **methanosarcina** (no cap., not ital.; pl. methanosarcinae).

methanoyl- *Prefix denoting the group HCO– derived from methanoic acid (e.g. methanoyl chloride).

methanoylation (preferred to formylation) *Chem.*

methionine Abbrev.: Met or met (no stop). Symbol: M *See* amino acid.

methoxy- Prefix denoting the methoxyl group CH_3O- (e.g. methoxybenzene, methoxyethanol).

methoxymethane $CH_3CH_2OCH_3$ The recommended name for the compound traditionally known as ethyl methyl ether.

methyl- *Prefix denoting the group CH_3- (e.g. methylamine, methyl chloride).

methyl alcohol CH_3OH The traditional name for methanol.

methylbenzene $C_6H_5CH_3$ The recommended name for the compound traditionally known as toluene.

methylbenzenesulphonyl- Prefix denoting the group $CH_3C_6H_4SO_2-$.

methylbenzoic acid $CH_3C_6H_4COOH$ The recommended name for the compound traditionally known as toluic acid.

methyl bromide CH_3Br The traditional name for bromomethane.

methylbuta-1,3-diene $H_2C=C(CH_3)CH=CH_2$ The recommended name for the compound traditionally known as isoprene.

3-methylbutanoic acid $(CH_3)_2CHCH_2COOH$ The recommended name for the compound traditionally known as isovaleric acid.

methyl chloride CH_3Cl The traditional name for chloromethane.

2-methyl-4-chlorophenoxyacetic acid Abbrev. and preferred form: MCPA (no stops).

methyl cyanide CH_3CN The traditional name for ethanenitrile.

methylene chloride CH_2Cl_2 The traditional name for dichloromethane.

(1-methylethyl)benzene $C_6H_5CH(CH_3)_2$ The recommended name for the compound traditionally known as cumene or isopropylbenzene.

methyl ethyl ketone $CH_3COCH_2CH_3$ Abbrev.: MEK (no stops). The traditional name for butanone.

methyl iodide CH_3I The traditional name for iodomethane.

methyl isobutyl ketone CH_3COCH_2-$CH(CH_3)_2$ The traditional name for 4-methyl-2-pentanone.

methylnitrobenzene $CH_3C_6H_2(NO_2)_3$ The recommended name for the compound traditionally known as nitrotoluene. The recommended name for *o*-nitrotoluene is methyl-2-nitrobenzene, etc.

Methylococcus (cap. M, ital.) A genus of methane-utilizing bacteria. Individual name: **methylococcus** (no cap., not ital.; pl. methylococci).

Methylomonas (cap. M, ital.) A genus of methane-utilizing bacteria. Individual name: **methylomonad** (no cap., not ital.).

4-methyl-2-pentanone CH_3COCH_2-$CH(CH_3)_2$ The recommended name for the compound traditionally known as methyl isobutyl ketone.

methylphenol $CH_3C_6H_4OH$ The recommended name for the compound traditionally known as cresol. The recommended name for *o*-cresol is 2-methylphenol, etc.

methylphenylamine $CH_3C_6H_4NH_2$ The recommended name for the compound traditionally known as toluidine. The recommended name for *o*-toluidine is 2-methylphenylamine, etc.

2-methylpropane $CH_3CH(CH_3)CH_3$ The recommended name for the compound traditionally known as isobutane.

2-methylpropanoic acid $(CH_3)_2$-$CHCOOH$ The recommended name for the compound traditionally known as isobutyric acid.

2-methylpropan-1-ol $(CH_3)_2CH$-CH_2OH The recommended name for the compound traditionally known as isobutyl alcohol.

2-methylpropan-2-ol $(CH_3)_3COH$ The recommended name for the compound traditionally known as *t*-butyl alcohol.

2-methylpropene $(CH_3)_2C=CH_2$ The recommended name for the compound traditionally known as isobutene or isobutylene.

2-methylpropenoic acid $CH_2=C(CH_3)COOH$ The recommended name for the compound traditionally known as methacrylic acid.

methyl-2,4,6-trinitrobenzene CH_3-$C_6H_2(NO_2)_3$ The recommended name for the compound traditionally known as 2,4,6-trinitrotoluene (TNT).

metoestrus US: **metestrus** or **metestrum**. *See* oestrus. Adjectival form: **metoestrous** (US: **metestrous**).

Meton (*fl.* 5th century BC) Greek astronomer.

Metonic cycle Also called lunar cycle.

metre US: **meter**. Symbol: m (no stop). The *SI unit of length. It is one of the seven SI base units, defined since 1983 as the length of the path travelled by light in a vacuum during a time interval of 1/299 792 458 of a second. In practice the kilometre is the largest multiple in use: in astronomy the astronomical unit, parsec, and light-year are used for distance measurements. At sea, the international nautical mile (1852 m) and sea mile are used. One metre = 39.3701 inches, 3.281 feet, 1.094 yards.

metric (not metrical) Involving measurement in general (associated noun: **metrology**) or a system of measurement based on the metre. The word is used more specifically as a noun (and adjective) in maths (e.g. metric space) and computing (e.g. font-metric file).

-metric *See* -meter; -metry.

metric carat A unit of mass, equal to 200 milligrams, used in trade in diamonds, fine pearls, and precious stones. It replaced the **carat**, equal to 3.17 grains or 205.3 milligrams, but its use is now deprecated.

metric horsepower *See* horsepower.

metric ton *See* tonne.

metro- (**metr-** before vowels) Prefix denoting **1.** measurement (e.g. metrology, metronome). **2.** *Med.* the uterus (e.g. metritis).

-metry Noun suffix denoting measurement or analysis (e.g. acidimetry, anthropometry, photogrammetry). Adjectival form: **-metric**.

Meyer, Julius Lothar (1830–95) German chemist.

Meyer, Victor (1848–97) German chemist.
Victor Meyer apparatus

Meyerhof, Otto Fritz (1884–1951) German-born US biochemist.
Embden–Meyerhof pathway (en dash) Another name for glycolysis.

MF Abbrev. for medium frequency.

mg (no stop) Symbol for milligram (or milligrams). *See* kilogram.

Mg Symbol for magnesium.

MHC Abbrev. and preferred form for major histocompatibility complex, a gene complex that encodes the most important antigens involved in rejection of foreign tissue. *See* H–2; HLA system.

MHD Abbrev. for magnetohydrodynamics.

mho Obsolete name for siemens.

MIBiol (or **M.I.Biol.**) Abbrev. for Member of the Institute of Biology. Members have the additional designation *CBiol (Chartered Biologist).

MIBK Abbrev. for methyl isobutyl ketone (4-methyl-2-pentanone).

Mic (no stop) *Astron.* Abbrev. for Microscopium.

mica (pl. micas) Adjectival form: **micaceous**.

MICE Abbrev. for Member of the Institution of Civil Engineers.

micelle (not micell, micella) *Biol., chem.* Adjectival form: **micellar**.

Michaelis, Leonor (1875–1949) German-born US biochemist.
Michaelis constant Symbol: K_m
Michaelis–Menten equation (en dash) Also named after Maud Lenore Menten (1879–1960).

Michell, John (1724–93) British geologist and astronomer.

Michelson, Albert Abraham (1852–1931) US physicist.
Michelson interferometer
Michelson–Morley experiment (en dash)

MIChemE (or **M.I.Chem.E.**) Abbrev. for Member of the Institution of Chemical Engineers.

MICR Abbrev. for magnetic-ink character recognition.

micRNA Abbrev. for mRNA-interfering complementary RNA.

micro- 1. Symbol: μ A prefix to a unit of measurement that indicates 10^{-6} times that unit, as in microsecond (μs), microamp (μA), or microfarad (μF). In medical work the prefix is always written out in full. *See also* SI units. **2.** (sometimes **micr-** before vowels) Prefix denoting **a.** small in size, range, or scale (e.g. microclimate, microglaciology, microinstruction, microorganism (one word), microspore). **b.** involving small quantities, objects, etc. (e.g. microaerophilic, microanalysis, microfiche, micrometer, microprocessor, microscope).

Microbacterium (cap. M, ital.) A genus of mainly rod-shaped aerobic bacteria. Individual name: **microbacterium** (no cap., not ital.; pl. microbacteria). *See also Aureobacterium.*

Microbispora (cap. M, ital.) A genus of *maduromycete bacteria. Individual name: **microbispora** (no cap., not ital.; pl. microbisporae).

Micrococcus (cap. M, ital.) A genus of Gram-positive aerobic coccoid bacteria. Individual name: **micrococcus** (no cap., not ital.; pl. micrococci). Use of the trivial name 'micrococci' should be restricted to members of the genus, and not extended to include other members of the family Micrococcaceae. Adjectival form: **micrococcal**.

Microcystis (cap. M, ital.) A former genus of *cyanobacteria, now of uncertain taxonomic status. Some bacterial taxonomists now consider it merely a cluster of morphologically similar strains. Italicization and capitalization are retained.

microfiche (pl. microfiche)

computer output on microfiche Abbrev.: COM (no stops).

micrometer 1. An instrument for measuring distance or angles. *Compare* micrometre. **2.** US spelling of micrometre.

micrometre US: **micrometer**. Symbol: μm One millionth of a metre, i.e. 10^{-6} metre. Former name: micron. *Compare* micrometer.

Micromonospora (cap. M, ital.) A genus of *actinoplanete bacteria. Individual name: **micromonospora** (no cap., not ital.; pl. micromonosporae).

micron Symbol: μ Use *micrometre.

microorganism (one word)

Micropolyspora (cap. M, ital.) A now defunct genus of bacteria that contained the causative agent of farmer's lung, *M. faeni*, now renamed **Faenia rectivirgula*.

microscope Adjectival form: **microscopic** (not microscopical). Care must be taken in using 'microscopic' to avoid confusion with its other, more general, meaning: 'very small'. Thus for parts of a microscope and other objects used in microscopy, use 'microscope' in an adjectival sense (e.g. microscope slide, microscope specimen); for more abstract senses, use 'microscopic' (e.g. microscopic examination, microscopic techniques). Derived noun: **microscopy**.

microscopical Use microscopic (*see* microscope) except in Royal Microscopical Society.

Microscopium A constellation. Genitive form: Microscopii. Abbrev.: Mic (no stop). *See also* stellar nomenclature.

microvillus (pl. microvilli; usually referred to in the pl.) *Anat.*

Microvirus (cap. M, ital.) Approved name for a genus of bacteriophages. Vernacular name: ϕX phage group (lower-case Greek phi).

mid- Prefix denoting the middle part or time. Terms incorporating 'mid-' are never written as two words; they are either hyphenated or written as one word.

Mid-Atlantic ridge (cap. M, A; hyphenated)

midbrain (one word)

Midgley, Thomas, Jr (1889–1944) US chemist.

midgut (one word)

mid-ocean ridge (hyphenated)

midpoint (one word)

MIEE Abbrev. for Member of the Institution of Electrical Engineers.

Mie scattering *Optics* Named after G. Mie.

Miescher, Johann Friedrich (1844–95) Swiss biochemist.

mil 1. One thousandth of an inch, i.e. 0.0254 millimetre. Also called thou. The use of both units is discouraged. **2.** A name formerly used in UK pharmacy to denote a millilitre.

Milankovich, Milutin (1879–1958) Yugoslav meteorologist and mathematician.

Milankovich radiation curves

mile 1. (or **statute mile**) A unit of length equal to 1.609 34 km, 1760 yards, 5280 feet. The mile is rarely used in scientific writing. **2.** *See* nautical mile, international. **3.** *See* sea mile.

millepede Use millipede.

Miller, Hugh (1802–56) British geologist.

Miller, Jacques Francis Albert Pierre (1931–) French-born Australian immunologist.

Miller, John Milton (1882–1962) US physicist.

Miller effect

Miller, Oskar von (1855–1935) German electrical engineer.

Miller, Stanley Lloyd (1930–) US chemist.

Miller–Urey experiment (en dash)

Miller, William Hallowes (1801–80) British mineralogist and crystallographer.

Miller indices Symbols: h, k, l or h_1, h_2, h_3

milli- 1. Symbol: m A prefix to a unit of measurement that indicates 10^{-3} times that unit, as in millimetre (mm), milliamp (mA), or millisecond (ms). In medical work the prefix is always written out in full. *See also* SI units. **2.** Prefix denoting a very large number (e.g. millipede).

Millikan, Robert Andrews (1868–1953) US physicist.
Millikan oil-drop experiment

millimicron Symbol: mμ Use nanometre.

millipede (preferred to millepede) *See* Diplopoda.

Millon, Auguste (1812–67) French chemist.
Millon's test (or **reaction**)

Mills, Bernard Yarnton (1920–) Australian physicist and radio astronomer.
Mills cross antenna

Mills, William Hobson (1873–1959) British chemist.

Milne, Edward Arthur (1896–1950) British mathematician and astrophysicist.

Milne, John (1850–1913) British seismologist.

Milstein, César (1927–) Argentinian-born British molecular biologist.

MIMD *Computing* Abbrev. for multiple instruction multiple data.

MIMechE (or **M.I.Mech.E.**) Abbrev. for Member of the Institution of Mechanical Engineers.

MIMinE (or **M.I.Min.E.**) Abbrev. for Member of the Institution of Mining Engineers.

mimosa (pl. mimosas) Any ornamental species of *Acacia*, especially *A. dealbata*, used by florists. Not to be confused with the genus *Mimosa* (cap. M, ital.), which includes the sensitive plant (*M. pudica*).

min 1. Symbol for minute. **2.** Abbrev. for minimum.

mineral nomenclature In geology, the common name of the mineral is used: bauxite, haematite, pyrites, etc. In chemical contexts it is recommended by IUPAC that the systematic chemical name should be used, followed by the traditional mineral name in brackets, e.g. hydrated aluminium oxide (bauxite), iron(III) oxide (haematite), and iron(II) disulphide (pyrites). *See also* inorganic chemical nomenclature.

mineralocorticoid (one word) *Endocrinol.*

mineralogy (not minerology)

mini- Prefix denoting very small or smallest (e.g. minichromosome, minicomputer, minimax).

minimum (pl. minima) Abbrev.: min (no stop). Adjectival forms: **minimum, minimal.** Verb form: **minimize.**

Ministry of Agriculture, Fisheries and Food Abbrev.: MAFF (no stops).

Minkowski, Hermann (1864–1909) Russian-born German mathematician.
Minkowski's inequality
Minkowski space–time (en dash)

minor planet (preferred to asteroid)

MInstP (or **M.Inst.P.**) Abbrev. for Member of the Institute of Physics.

minus sign Symbol: – An *en dash is usually used; avoid a hyphen if possible.

minute 1. Symbol: min A unit of time equal to 60 seconds. Although not an *SI unit, the minute may be used with the SI units. **2.** Symbol: ′ A unit of angle equal to 1/60 of a *degree, i.e. 0.291 milliradian. Also called minute of arc, arc minute.

Mio Abbrev. for Miocene.

mio- Prefix denoting reduction or less (e.g. Miocene, miosis). *See also* meio-.

Miocene Abbrev.: Mio (no stop). **1.** (adjective) Denoting the penultimate epoch of the Tertiary period. **2.** (noun; preceded by 'the') The Miocene epoch.

miosis *Ophthalmol.* Not to be confused with *meiosis.

mips *Computing* Acronym for million instructions per second.

miracidium (pl. miracidia) *Zool.* Adjectival form: **miracidial.**

Mis *Geol.* Abbrev. for Mississippian.

MIS Abbrev. for metal–insulator semiconductor.

misch metal (from German; no cap.)

miscible Noun form: **miscibility.**

mis-sense (hyphenated) *Genetics* Denoting a type of mutation.

Mississippian Abbrev.: Mis (no stop). **1.** (adjective) Denoting a US subperiod in the geological time scale corresponding approximately to the Lower *Carboniferous subperiod. **2.** (noun; preceded by 'the') The Mississippian subperiod.

mistral (no cap.) *Meteorol.*

MIT Abbrev. for Massachusetts Institute of Technology.

Mitchell, Peter Dennis (1920–) British biochemist.

Mitchell, R(eginald) J(oseph) (1895–1937) British aeronautical engineer.

mito- Prefix denoting threadlike (e.g. mitochondrion, mitosis).

mitochondrion (pl. mitochondria) *Cell biol.* Adjectival form: **mitochondrial**.

mitochondrial DNA Abbrev.: mtDNA (lower-case m, t).

mitosis (pl. mitoses) Cell division resulting in daughter nuclei identical to the parent nucleus. Adjectival form: **mitotic**. *Compare* meiosis.

mitral valve Use bicuspid valve.

Mitscherlich, Eilhardt (1794–1863) German chemist.

MIX Abbrev. for 1-methyl-3-isobutylxanthine.

mixed melting point Abbrev.: m.m.p. (stops, no caps.).

mks (or **m.k.s.**) **units** A metric system of units in which the metre, kilogram, and second are the fundamental units of length, mass, and time. When electric and magnetic properties are included a fourth fundamental property is required. Giovanni Giorgi proposed that one of the practical electrical units (*see* cgs units) be selected. In 1948 the unit of current, the ampere, was definitively chosen, leading to the **mksa system** (or Giorgi system); this contains all the practical electrical units of the cgs system. There are two types, rationalized and nonrationalized, which differ in the form of equations involving electrical and magnetic units. Rationalization changes certain equations so that those involving spheres contain a factor 4π, those with cylinders 2π, and linear systems have no factor involving π. *SI units are based on rationalized mksa units.

ml (no stop) Symbol for millilitre (or millilitres). *See* litre.

MLO Abbrev. for mycoplasma-like organism(s).

M–L relation (en dash) *Astron.* Short for mass–luminosity relation.

mm (no stop) Symbol for millimetre (or millimetres). *See* metre.

MMC Abbrev. for magnesium methyl carbonate.

mmf (or **m.m.f.**) Abbrev. for magnetomotive force.

mmH$_2$O *See* mmHg.

mmHg Symbol for millimetre of mercury, a unit of pressure expressed in terms of the height of a column of mercury. If the column contains water, pressure can be expressed in mmH$_2$O, i.e. millimetres of water. In terms of the SI unit of pressure, the pascal:

$$1\ \text{mmHg} = 13.5951\ \text{mmH}_2\text{O} = 133.322\ \text{Pa}.$$

If the column height is measured in inches, the corresponding pressure units are inHg and inH$_2$O. Use of these four units is now discouraged.

MMI *Computing* Abbrev. for man–machine interface.

m.m.p. Abbrev. for mixed melting point.

MMS Abbrev. for **1.** methyl methanesulphonate. **2.** multimission modular spacecraft.

MMT Abbrev. for Multiple Mirror Telescope (Arizona).

MMTS Abbrev. for methyl methylthiomethyl sulphoxide.

Mn Symbol for manganese.

MNNG Abbrev. for 1-methyl-3-nitro-1-nitrosoguanidine.

Mo Symbol for molybdenum.

MO Abbrev. for molecular orbital.

mobility Symbol: μ (Greek mu). The mean speed of an ion, electron, or hole in unit electric field.

Möbius, August Ferdinand (1790–1868) German mathematician.
Möbius net
Möbius strip

mod (no stop) *Maths, computing* Abbrev. for *modulo.

mode Adjectival form: **modal**. Derived noun: **modality**.

model Verb form: **models, modelling** (US: **modeling**), **modelled** (US: **modeled**).

modem Acronym for modulator–demodulator.

modular *See* module; modulus.

modulation transfer function *Image technol.* Abbrev.: MTF (no stops) or m.t.f. (stops).

module Adjectival form: **modular**.

modulo *Maths., computing* Abbrev.: mod, as in

$$x \bmod y$$

the result being the remainder of the division of integer x by integer y.

modulus (pl. moduli) **1.** *See* complex numbers. **2.** Another name for coefficient. Adjectival form: **modular**.

modulus of elasticity Another name for Young modulus.

modulus of impedance *See* impedance.

modulus of rigidity Another name for shear modulus.

Mohl, Hugo von (1805–72) German botanist.

Mohorovičić, Andrija (1857–1936) Yugoslav geologist.
Mohorovičić discontinuity Shortened form: **Moho** (no stop).

Mohs, Friedrich (1773–1839) German mineralogist.
Mohs scale

moiré (adjective; acute accent)

Moissan, Ferdinand Frédéric Henri (1852–1907) French chemist.

Moivre, Abraham De Usually alphabetized as *De Moivre.

mol 1. Symbol for mole. *See also* SI units. **2.** Abbrev. for molecule, molecular.

molality Symbol: m_B (of a solute B) or b_B (recently introduced) or, where a complicated formula is to be written, m(B) or b(B), as in $m(KNO_3)$. A measure of the molal concentration of a solute B, equal to the *amount of substance of solute B in a solution divided by the mass of the solvent. The *SI unit is the mole per kilogram. *Compare* molar concentration (*see* molar).

molar 1. Denoting a physical quantity measured for one mole of a substance. It is usually denoted by the subscript m in the symbol, e.g. *molar volume, symbol V_m, is the volume of one mole of a substance. Other examples include molar *heat capacity, C_m, and molar *enthalpy, H_m. **2.** Denoting the concentration of a solute, expressed as the *amount of substance of the solute divided by the mass of the solution (not the solvent, as in *molality). **3.** In the case of *molar conductivity and *molar absorption coefficient, molar means divided by concentration.

molar absorption coefficient Symbol: ϵ (Greek epsilon). A physical quantity relating to the absorption of electromagnetic radiation by a substance. It indicates the probability of an electronic transition in a chromophore. The SI unit is the square metre per mole; the unit normally used in ultraviolet and visible spectroscopy is 1000 square centimetres per mole (1000 $cm^2\ mol^{-1}$), but is by convention not expressed. *See also* absorbance.

molar conductivity Symbol: Λ (Greek cap. lambda). A physical quantity, *conductivity divided by *concentration (not amount of substance as in most *molar quantities). The *SI unit is the siemens square metre per mole ($S\ m^2\ mol^{-1}$).

molar enthalpy *See* enthalpy.

molar enthalpy change *See* enthalpy.

molar entropy *See* entropy.

molar extinction coefficient Former name for *molar absorption coefficient.

molar gas constant Symbol: R A fundamental constant equal to

$$8.314\,510\ J\ K^{-1}\ mol^{-1}.$$

It occurs in the gas law, which should be written in one of the following forms:

$$pV = nRT = (m/M)RT$$
$$pV_m = RT,$$

where p is the pressure, V is volume, V_m is molar volume, T is thermodynamic temperature, n is amount of substance (gas), m is mass, and M is molar mass. The symbol R was formerly used for the gas constant occurring in the equation $pV = RT$; it had the units $J\,K^{-1}$. This notation should no longer be used. The symbol R and the name gas constant now refer to the molar gas constant by international agreement.

molar Gibbs function *See* Gibbs function.

molar heat capacity *See* heat capacity.

molar Helmholtz function *See* Helmholtz function.

molar internal energy *See* internal energy.

molarity Another name for *concentration. The word should be avoided because of confusion with *molality.

molar mass Symbol: M (no subscript m). A physical quantity, the mass of one mole of a substance. The *SI unit is the kilogram per mole.
$M = 10^{-3} M_r$ kg mol^{-1} = M_r g mol^{-1},
where M_r is the *relative molecular mass.

molar volume Symbol: V_m A physical quantity, the volume of one mole of a substance. The *SI unit is normally the cubic metre per mole but the litre per mole may also be used. The molar volume of an ideal gas is 22.414 dm^3 mol^{-1}.

mole Symbol: mol The *SI unit of *amount of substance. It is one of the seven SI base units, defined as the amount of substance of a system that contains as many elementary entities as there are atoms in 0.012 kilogram of carbon-12. When the mole is used, the elementary entities must be specified: they may be atoms, molecules, ions, radicals, electrons, other particles, or specified groups of such particles; the entity may also be a reaction. To avoid ambiguity, a symbol or formula should be stated rather than a name, as in a mole of NaCl(s), a mole of C–C bonds, a mole of the reaction $N_2O_4 \rightleftharpoons 2NO_2$. One mole of a compound has a mass equal to its *relative molecular mass in grams. *See also* Avogadro constant.

molecular-beam epitaxy (hyphenated) Abbrev.: MBE (no stops).

molecular concentration Symbol: C_B (for a substance B) or, when a complicated formula is to be written, C(B), as in $C(H_2SO_4)$. A physical quantity, the number of molecules of a substance B divided by the volume of the mixture. The *SI unit is the reciprocal cubic metre (m^{-3}).

molecular orbital Abbrev.: MO (no stops). *See* orbital.

molecular weight Abbrev.: mol. wt. (stops). Former name for relative molecular mass.

molecule Adjectival form: **molecular**. Abbrev. (for both): mol (no stop).

mole fraction Symbol: x_B (for a substance B) or, when a complicated formula is to be written, x(B), as in $x(NaClO_3)$. A dimensionless physical quantity, the ratio of the *amount of substance of the substance B to the amount of substance of the mixture.

Molisch, Hans (1856–1937) Austrian chemist.

Molisch test *Biochem.* Also called α-naphthol test.

mollicute Any prokaryotic organism belonging to the class Mollicutes (cap. M). It has been proposed that 'mollicute' should replace *mycoplasma as the trivial name for members of the class as a whole, with the latter restricted to members of the genus *Mycoplasma*. Members of the remaining four genera – *Ureaplasma*, *Acholeplasma*, *Spiroplasma*, and *Anaeroplasma* – are thus referred to as, respectively, ureaplasmas, acholeplasmas, spiroplasmas, and anaeroplasmas. The genus *Thermoplasma*, formerly included with the mollicutes, is now classified as an *archaebacterium.

Mollusca (cap. M) A phylum of invertebrates including the bivalves, snails, squids, etc. Individual name: **mollusc** (US: **mollusk**; no caps.). Adjectival form: **molluscan** (not molluscous; US: **molluskan**).

Mollweide, Karl B. (died 1825) German mathematician and astronomer.
Mollweide (homalographic) projection

mol. wt. (preferred to MW) Abbrev. for molecular weight.

molybdenum Symbol: Mo *See also* periodic table; nuclide.

moment of force Symbol: ***M*** A physical quantity measuring the turning effect produced by a *force. It is a *pseudovector whose magnitude is *rF*; *F* is the magnitude of the force and *r* the perpendicular distance from the force's line of action to the axis of rotation. Anticlockwise moments are opposite in sign to clockwise moments. The *SI unit is the newton metre.

moment of inertia Symbol: *I* A scalar physical quantity, the sum Σmr^2 for all the particles of a body, where *m* is the mass of a particle and *r* its perpendicular distance from the axis under consideration. For a homogeneous body the sum is replaced by an integral. The *SI unit is the kilogram metre squared.

moment of momentum Another name for angular momentum.

momentum Symbol: ***p*** A physical quantity, the product of *mass and *velocity. The *SI unit is the kilogram metre per second. Momentum is a *vector quantity having the direction of the velocity. Also called linear momentum. *See also* angular momentum.

Mon (no stop) *Astron.* Abbrev. for Monoceros.

mon- *See* mono-.

Mond, Ludwig (1839–1909) German-born British industrial chemist.
Mond process

Monge, Gaspard (1746–1818) French mathematician.

mongolism Use Down's syndrome.

mono- (sometimes **mon-** before vowels) Prefix denoting one, single, or alone (e.g. monoamine, monoclonal, monohybrid, monosaccharide, monoecious, monoxide).

monoamine oxidase Abbrev.: MAO (no stops).
monoamine oxidase inhibitor Abbrev. and preferred form: MAO inhibitor.

monocarpellary Denoting a fruit formed from a single ovary. *Compare* monocarpic.

monocarpic Denoting a plant (known as a **monocarp**) that flowers and fruits only once in its lifetime. *Compare* monocarpellary.

Monoceros A constellation. Genitive form: Monocerotis. Abbrev.: Mon (no stop). *See also* stellar nomenclature.

monochloroethanoic acid CH_2-ClCOOH *See* chloroethanoic acid.

Monocotyledonae (cap. M) A subclass of flowering plants having embryos with one cotyledon. In some classifications these plants are regarded as a class, Liliopsida; in a recent classification they are regarded as a subclass, Liliidae, of the Magnoliopsida (*see* Angiospermae). Individual name: **monocotyledon** (no cap.; often shortened to **monocot**). Adjectival form: **monocotyledonous**. Note that this name and adjectival form apply irrespective of the classification used.

Monod, Jacques Lucien (1910–76) French biochemist.
Jacob–Monod hypothesis (or **model**; en dash) Also called operon model.

monoecious US: **monecious**. *Bot.* Noun form: **monoecy** (US: **monecy**).

monosodium glutamate HOOC-$(CH_2)_2CH(NH_2)COONa$ Abbrev.: MSG (no stops). Also called sodium hydrogen glutamate, sodium glutamate.

Monotremata (cap. M) The sole order of the *Prototheria, comprising the egg-laying mammals (spiny anteaters and duck-billed platypus). Individual name and adjectival form: **monotreme** (no cap.; not monotrematous).

Monro, Alexander (not Monroe) (1733–1817) British anatomist.
foramen of Monro

mons (not ital.; pl. montes) *Anat., astron.* In astronomy, the word, with an initial capital, is used in the approved Latin name of a mountain on a planet or satellite, as in

Olympus Mons on Mars and Maxwell Montes on Venus.

montbretia (not montbrietia; pl. montbretias) Any of several ornamental plants of the genera *Crocosmia* and *Tritonia* (formerly *Montbretia*), especially the hybrid *Crocosmia* × *crocosmiflora*.

month 1. The period of the moon's revolution around the earth, measured relative to a particular frame of reference in the sky. The length of the period depends on the frame of reference (*see* year). For example, the **sidereal month** has 27.321 66 days. the **synodic** (or **lunar**) **month** has 29.530 59 days. **2.** A time interval of 28, 29, 30, or 31 days, according to the calendar. In a leap year February has 29 rather than 28 days.

moon 1. (or **Moon**) The only satellite of the earth. Related adjective: **lunar**.

The mass of the moon, symbol M_M, and its radius, symbol R_M, have been adopted as astronomical constants:

$M_M = 7.35 \times 10^{22}$ kg

$R_M = 1736$ km.

See also month.

2. A satellite of a planet other than the earth.

Moore, Stanford (1913–82) US biochemist.

MOPS Abbrev. for 4-morpholinepropanesulphonic acid.

moraine *Geol.* Adjectival forms: **morainal, morainic.**

Moraxella (cap. M, ital.) A genus of bacteria of the family Neisseriaceae. Some authorities divide the genus into two subgenera, *Moraxella* (rod-shaped organisms) and **Branhamella* (coccal organisms). Species are designed in the form *Moraxella* (*Moraxella*) *bovis* and *Moraxella* (*Branhamella*) *ovis*, for example. Individual name: **moraxella** (no cap., not ital.; pl. moraxellae).

MORB *Geol.* Acronym for mid-ocean-ridge basalt.

morbilli (takes a sing. form of verb) The medical name for measles.

mordant (not mordent) *Chem.*

Mordell, Louis Joel (1888–1972) US-born British mathematician.

Morgan, Augustus De Usually alphabetized as *De Morgan.

Morgan, Thomas Hunt (1866–1945) US geneticist.

Morgan, William Wilson (1906–) US astronomer.

Morgan–Keenan classification (en dash) Also named after P. C. Keenan.

Morley, Edward William (1838–1923) US chemist and physicist.

Michelson–Morley experiment (en dash)

-morph 1. Noun suffix denoting a body or cell (e.g. ectomorph, polymorph, rhizomorph). Adjectival form: ***-morphic**. **2.** *See* -morpha.

-morpha Noun suffix denoting a *suborder in animal taxonomy (e.g. Hystricomorpha). Adjectival form (no initial cap.): **-morph.**

-morphic Adjectival suffix denoting form or shape (e.g. actinomorphic, ectomorphic, metamorphic, polymorphic). Noun form: **-morphism** (e.g. metamorphism, polymorphism) or **-morphy** (e.g. actinomorphy, ectomorphy). *See also* -morph.

-morphism *See* -morphic.

morpho- (**morph-** before vowels) Prefix denoting form, shape, or structure (e.g. morphogenesis, morphology, morphallaxis).

morphotype Use *morphovar.

morphovar A morpho(logical) var(iety): an unofficial category of classification used in microbiology and ranking below subspecies. Morphovars are strains that are distinguished by special morphological features.

-morphy *See* -morphic.

Morse, Samuel (1791–1872) US inventor.

Morse code

Morse telegraphy

morula (pl. morulas or morulae) The solid ball of cells resulting from cleavage of a fertilized animal egg cell: it develops into a *blastula.

MOS Acronym and preferred form for metal-oxide semiconductor. Used in combinations (e.g. CMOS, MOSFET, MOSRAM) or adjectivally (e.g. MOS capacitor, MOS circuit).

Mosander, Carl Gustav (1797–1858) Swedish chemist.

Moseley, Henry Gwyn Jeffreys (1887–1915) British physicist.
Moseley's law

MOSFET *See* MOS; FET.

MOSRAM *See* MOS; RAM.

Mössbauer, Rudolph Ludwig (1929–) German physicist.
Mössbauer effect
Mössbauer spectroscopy

mosses *See* Bryopsida.

MOST (pronounced mosst) Acronym for *MOS transistor. Also called MOSFET.

most significant bit *Computing* Abbrev.: MSB or msb (no stops).

most significant digit *Computing* Abbrev.: MSD or msd (no stops).

motor neurone (preferred to motor neuron and motoneuron)

Mott, Sir Nevill Francis (1905–) British physicist.
Gurney–Mott theory (en dash)

Mottelson, Benjamin Roy (1926–) US-born Danish physicist.

mould US: **mold.** Common name for any of various fungi, including *Mucor* (bread mould) and **Penicillium* (common mould).

Mountain Standard Time Abbrev.: MST (no stops).

moving-coil instrument (hyphenated)

m.p. Abbrev. for melting point.

M phase (cap. M) A phase of the *cell cycle.

MPhil (or **M.Phil.**) Abbrev. for Master of Philosophy.

MPPH Abbrev. for 5-(4-methylphenyl)-5-phenylhydantoin.

MRAeS (or **M.R.Ae.S.**) Abbrev. for Member of the Royal Aeronautical Society.

MRBS Abbrev. for Member of the Royal Botanic Society.

MRC Abbrev. for Medical Research Council.

MRCP Abbrev. for Member of the Royal College of Physicians.

MRCS Abbrev. for Member of the Royal College of Surgeons.

MRCVS Abbrev. for Member of the Royal College of Veterinary Surgeons.

MRGS Abbrev. for Member of the Royal Geographical Society.

MRI Abbrev. for **1.** magnetic resonance imaging. **2.** Member of the Royal Institution.

MRIC Abbrev. for Member of the Royal Institute of Chemistry, now *MRSC.

MRMetS (or **M.R.Met.S.**) Abbrev. for Member of the Royal Meteorological Society.

mRNA (lower-case m) Abbrev. for messenger RNA.

MRSC Abbrev. for Member of the Royal Society of Chemistry. Members have the additional designation *CChem (Chartered Chemist).

Ms Symbol sometimes used to denote the mesyl (methanesulphonyl) group in chemical formulae, e.g. CH_3SO_2Cl can be written as MsCl.

MS Abbrev. for **1.** mass spectrometry. **2.** Master of Science (US universities). **3.** Master of Surgery.

MSB (or **msb**) *Computing* Abbrev. for most significant bit.

MSc (or **M.Sc.**) Abbrev. for Master of Science.

MSD (or **msd**) *Computing* Abbrev. for most significant digit.

MS-DOS (hyphenated, caps.) A trade name for an operating system for microcomputers.

MSG Abbrev. for monosodium glutamate.

MSH Abbrev. for melanocyte-stimulating hormone.

MSH Abbrev. for mesitylenesulphonyl hydrazide.

M shell (not hyphenated) *Physics*

MSI *Electronics* Abbrev. and preferred form for medium-scale integration.

msl (no stops) Abbrev. for mean sea level.

MSNT Abbrev. for 1-(mesitylene-2-sulphonyl)-3-nitro-1,2,4-triazole.

MSS Abbrev. for **1.** Member of the Royal Statistical Society. **2.** multispectral scanner.

MST Abbrev. for Mountain Standard Time (in the USA).

MSTFA Abbrev. for *N*-methyl-*N*-(silyltrimethyl)trifluoroacetamide.

α-MT Abbrev. for DL-α-methyltyrosine.

MTBE Abbrev. for methyl *t*-butyl ether.

MTBF Abbrev. for mean time between failures.

MTBSTFA Abbrev. for *N*-methyl-*N*-(*t*-butyldimethylsilyl)trifluoroacetamide.

mtDNA Abbrev. for mitochondrial DNA.

MTech (or **M.Tech.**) Abbrev. for Master of Technology.

MTF (or **m.t.f.**) *Image technol.* Abbrev. for modulation transfer function.

MTPA Abbrev. for α-methoxy-α-trifluoromethylphenylacetic acid.

MTU *Computing* Abbrev. for magnetic tape unit.

mu Greek letter, symbol: μ (lower case), M (cap.).

μ Symbol for **1.** a bridging species in chemical formulae, e.g. $[Cl_2Al(\mu-Cl)AlCl_2]$. **2.** chemical potential (μ_B = that of substance B). **3.** friction coefficient. **4.** heavy chain of IgM (*see* immunoglobulin). **5.** Joule–Thomson coefficient. **6.** linear attenuation coefficient. **7.** magnetic dipole moment. **8.** magnetic moment, of a particle (μ_e, μ_p = electron, proton magnetic moments, respectively). **9.** micro-. **10.** mobility (μ_H = Hall mobility). **11.** muon. **12.** (bold type) permanent dipole of a molecule. **13.** permeability (μ_0 = permeability of vacuum, also called the magnetic constant, μ_r = relative permeability). **14.** Poisson ratio. **15.** *Astron.* proper motion. **16.** reduced mass. **17.** shear modulus.

μ_B Symbol for Bohr magneton.

μ_N Symbol for nuclear magneton.

muco- (**muc-** before vowels) Prefix denoting mucus or mucous membrane (e.g. mucopolysaccharide, mucoprotein, mucin).

mucosa (pl. mucosae) A mucous membrane. Adjectival form: **mucosal.**

mucro (pl. mucrones) *Biol.* A fine point projecting from certain organs or structures. Adjectival form: **mucronate.**

mucus A viscous secretion. Not to be confused with its adjectival form, **mucous,** which is most commonly applied to the secreting membrane (**mucous membrane,** not mucus membrane).

mudstone (one word) *Geol.*

Müller, Erwin Wilhelm (1911–77) German-born US physicist.

Muller, Hermann Joseph (1890–1967) US geneticist.

Müller, J.F.T. (1821–97) German zoologist.

Müllerian mimicry

Müller, Johannes Peter (1801–58) German physiologist.

***Müllerian duct**

Müller, K. Alex (1927–) Swiss physicist.

Müller, Otto Friedrich (1730–84) Danish microscopist.

Müller, Paul Hermann (1899–1965) Swiss chemist.

Müllerian duct US: **müllerian duct.** Also called paramesonephric duct. Named after J. P. *Müller.

Müllerian mimicry *Zool.* Named after J. F. T. *Müller.

Mulliken, Robert Sanderson (1896–1986) US physicist and chemist.

Mulliken symbols

multi- Prefix denoting many or several (e.g. multiaccess, multicellular, multinucleate, multiplex).

multimission modular spacecraft Abbrev.: MMS (no stops).

multiple instruction multiple data *Computing* Abbrev.: MIMD (no stops).

Multiple Mirror Telescope (in Arizona) Abbrev.: MMT (no stops).

multiplication *See* arithmetic operations.

multispectral scanner Abbrev.: MSS (no stops).

mu meson *See* muon.

Mumetal (cap. M) A trade name for a type of nickel-based ferromagnetic alloy.

Munk, Walter Heinrich (1917–) US geophysicist.

Munsell colour system *Physics, etc.* Named after Albert Munsell (died 1918).

Muntz, George Frederick (1794–1857) British metallurgist and political reformer.
Muntz metal

muon An unstable negatively charged elementary particle, denoted by μ or μ^- (Greek mu) in nuclear reactions, etc. Its antiparticle, the **positive muon** or **antimuon**, is denoted by $\overline{\mu}$ or μ^+. The muon mass is 206.77 times the electron mass. Former name: mu meson (now known to be a lepton, it was originally erroneously classified as a meson).

Murchison, Sir Roderick Impey (1792–1871) British geologist.

Murray, Sir John (1841–1914) British marine zoologist and oceanographer.

Mus (no stop) *Astron.* Abbrev. for Musca.

musa Acronym for multiple unit steerable aerial (or antenna).

Musca A constellation. Genitive form: Muscae. Abbrev.: Mus (no stop). *See also* stellar nomenclature.

Musci *See* Bryopsida.

mushroom Loosely, any fungus of the order Agaricales (*see* agaric), sometimes restricted to the edible species. The term has no taxonomic significance.

Muspratt, James (1793–1886) Irish industrial chemist.

Musschenbroek, Pieter van (1692–1761) Dutch physicist.

mutual inductance Symbol: M or L_m A property, Φ/I, of two interacting conducting loops, where I is the current in one loop that produces a *magnetic flux Φ in the second loop. The *SI unit is the henry. *See also* self-inductance.

mux Short for multiplexer.

MV Symbol for megavolt (or megavolts). *See* volt.

MW **1.** Symbol for megawatt (or megawatts). *See* watt. **2.** Abbrev. for medium-wave (or medium waves). **3.** Abbrev. for molecular weight. Use mol. wt.

Mx Symbol for maxwell.

my- *See* myo-.

myc (ital.) A viral *oncogene causing *my*elo*c*ytomatosis. Two homologous oncogenes found in certain human tumours are designated L-*myc* (cap. L, not ital.; from lung cancer) and N-*myc* (cap. N, not ital.; from neuroblastoma).

Mycalex (cap. M) A trade name for a form of mica bonded with glass.

mycelia sterilia *See* Agonomycetales.

mycelium (pl. mycelia) *Bot.* Adjectival form: **mycelial**.

-mycetes Noun suffix denoting a fungal class in plant taxonomy (e.g. Discomycetes, Hymenomycetes). Adjectival form (no initial cap.): **-mycete**.

Mycetozoa (cap. M) *See* Myxomycetes. Individual name and adjectival form: **mycetozoan** (no cap.).

myco- (**myc-** before vowels) Prefix denoting a fungus (e.g. mycobiont, mycology, mycotrophic, mycosis).

Mycobacterium (cap. M, ital.) A genus of actinomycete bacteria. Individual name: **mycobacterium** (no cap., not ital.; pl. mycobacteria).

mycobiont A fungal symbiont, especially the fungal constituent of a lichen. *Compare* Mycobionta.

Mycobionta (cap. M) Another name for *Eumycota, a division comprising the true fungi. *Compare* mycobiont; Mycobiota.

mycobiota (no cap.) The fungal flora of a habitat or area. *Compare* Mycobiota.

Mycobiota (cap. M) In a recent classification, a kingdom comprising the *fungi. *Compare* Mycobionta; mycobiota; Mycota.

mycoplasma (no cap., not ital.; pl. mycoplasmas) Any prokaryotic microorganism of the genus *Mycoplasma* (cap. M, ital.), class Mollicutes. The term has traditionally been applied to all members of the class, but many authorities now favour the term *mollicute as a general trivial name.

mycoplasma-like organism Abbrev.: MLO (no stops)

mycorrhiza (not mycorhiza; pl. mycorrhizae) *Bot.* Adjectival form: **mycorrhizal**.

Mycota In traditional classifications, a kingdom comprising the *fungi. *Compare* Mycobiota.

myelo- (myel- before vowels) Prefix denoting **1.** bone marrow (e.g. myelocyte, myeloid). **2.** the spinal cord (e.g. myelencephalon, myelitis).

Mylar (cap. M) *See* Melinex.

myo- (my- before vowels) Prefix denoting muscle (e.g. myofibril, myoglobin, myoneme, myenteron).

myriapod Any arthropod formerly regarded as a member of the class Myriapoda, now reclassified into separate classes (*Chilopoda and *Diplopoda). The term myriapod is still used for descriptive (rather than taxonomic) purposes.

myrmeco- Prefix denoting ants (e.g. myrmecochory, myrmecophily).

myxo- (myx- before vowels) Prefix denoting slime or mucus (e.g. myxomycete, myxamoeba, myxoedema).

myxobacter *See* myxobacteria.

myxobacteria (pl. noun, no cap.; sing. myxobacterium) Unicellular rod-shaped gliding bacteria belonging to the order Myxococcales. The trivial name **myxobacter** is used synonymously and interchangeably with myxobacterium.

Myxobionta *See* Myxomycota.

Myxococcus (cap. M, ital.) A genus of *myxobacteria. Individual name: **myxococcus** (no cap., not ital.; pl. myxococci).

myxoma virus (two words) Type species of the genus **Leporipoxvirus* (vernacular name: myxoma subgroup).

Myxomycetes (cap. M) A class comprising the slime moulds, traditionally classified as fungi (*see* Myxomycota) but placed by some authorities in the animal kingdom (as the subphylum Mycetozoa). Individual name and adjectival form: **myxomycete** (no cap.).

Myxomycota A division of fungi that do not possess a mycelium. Also called Myxobionta. It includes the classes *Acrasiomycetes (cellular slime moulds), *Myxomycetes (slime moulds), and *Plasmodiophoromycetes. In a recent classification these organisms are not regarded as fungi and the division is placed in the subkingdom Protistobionta (*see* Protista).

Myxophyta Obsolete name for the *cyanobacteria (blue-green algae).

Mz Abbrev. for Mesozoic.

N

n Symbol for **1.** nano-. **2.** neutron ($\bar{n}$ = antineutron). Abbrev. for n-type (semiconductor).

n Symbol (light ital.) for **1.** amount of substance. **2.** the *haploid chromosome number. **3.** number density. **4.** principal quantum number. **5.** refractive index. **6.** rotational frequency.

n_0 (light ital. n) Symbol for Loschmidt constant.

N 1. Symbol for **i.** asparagine. **ii.** guanosine, adenosine, thymidine (or uridine), or cytidine (unspecified). **iii.** newton. **iv.** nitrogen. **v.** nucleon. **2.** Abbrev. for **i.** *Elec. eng.* neutral. **ii.** *north or northern.

N Symbol (light ital.) for **1.** neutron number. **2.** number of molecules.

N_A (light ital. N) Symbol for Avogadro constant.

***n*-** (ital., always hyphenated) Prefix denoting normal, indicating an unbranched alkane chain (e.g. *n*-butane, *n*-propyl alcohol).

***N*-** (ital., always hyphenated) Prefix denoting substitution on a nitrogen atom in an organic compound (e.g. *N*-phenylhydroxylamine).

Na Symbol for sodium. [from Latin *natrium*]

NA Abbrev. for **1.** *Optics, etc.* numerical aperture. **2.** noradrenaline.

NAA Abbrev. for **1.** neutron activation analysis. **2.** naphthaleneacetic acid.

nabla Use *del.

NAD Abbrev. and preferred form for nicotinamide–adenine dinucleotide. The reduced form is designated **NADH** and the oxidized form $\mathbf{NAD^+}$. Former names: coenzyme I (CoI), cozymase, diphosphopyridine nucleotide (DPN).

NADP Abbrev. and preferred form for nicotinamide–adenine dinucleotide phosphate. The reduced form is designated **NADPH** and the oxidized form $\mathbf{NADP^+}$. Former names: coenzyme II, triphosphopyridine nucleotide (TPN).

Naegeli, Karl Wilhelm von (1817–91) Swiss botanist.

NAK (or **nak**) Short for negative acknowledgment.

named laws, effects, etc. *See* eponyms.

NAND *See* logic symbols; electronics, graphical symbols.

nano- 1. Symbol: n A prefix to a unit of a measurement that indicates 10^{-9} times that unit, as in nanometre (nm), nanosecond (ns), or nanofarad (nF). *See also* SI units. **2.** (**nan-** before vowels) Prefix denoting minute size (e.g. nanoplankton, nanandrous).

nanoplankton (preferred to nannoplankton)

Nansen, Fridtjof (1861–1930) Norwegian explorer and biologist.

naphthalen-1-amine $C_{10}H_7NH_2$ The recommended name for the compound traditionally known as α-naphthylamine.

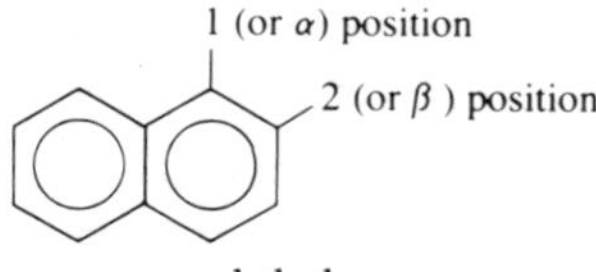

naphthalene

naphthalene (not napthalene)

naphthalene acetic acid Abbrev.: NAA (no stops).

naphthalen-1-ol $C_{10}H_7OH$ The recommended name for the compound traditionally known as α-naphthol.

naphthalen-2-ol $C_{10}H_7OH$ The recommended name for the compound traditionally known as β-naphthol.

α-naphthol $C_{10}H_7OH$ The traditional name for naphthalen-1-ol.

β-naphthol $C_{10}H_7OH$ The traditional name for naphthalen-2-ol.

α-naphthylamine $C_{10}H_7NH_2$ The traditional name for naphthalen-1-amine.

Napier (or **Neper**), **John** (1550–1617) Scottish mathematician.
Napierian logarithm Use natural logarithm.
Napier's analogies
Napier's rules
***neper**

nappe *Geol., maths.*

NAR *Bot.* Abbrev. for net assimilation rate.

narcissus (pl. narcissi) Generic name *Narcissus* (cap. N, ital.).

nares (pl. noun; sing. naris) The paired openings of the nasal cavity. There are two pairs: the external nares are the nostrils; the internal nares (or choanae) open into the pharynx.

NASA Acronym for National Aeronautics and Space Administration, a US government organization.

Nasmyth, Alexander (died 1848) British dentist.
Nasmyth's membrane

Nasmyth, James (1808–90) British engineer.
Nasmyth focus

naso- Prefix denoting the nose (e.g. nasolacrimal, nasopharynx).

nasturtium (pl. nasturtiums) Common name for the garden ornamental *Tropaeolum majus*. Not to be confused with the genus **Nasturtium*.

Nasturtium (cap. N, ital.) A genus of flowering plants that includes watercress. *Compare* nasturtium.

nasty A nondirectional movement of a plant organ in response to a stimulus. The word is often used in combination (*see* -nasty). Also called nastic movement. Adjectival form: **nastic**.

-nasty Noun suffix denoting a nastic movement in plants (e.g. nyctinasty, thermonasty). Adjectival form: **-nastic**.

Nathans, Daniel (1928–) US molecular biologist.

National Bureau of Standards (in the USA) Abbrev.: NBS (no stops).

National Engineering Laboratory (in East Kilbride) Abbrev.: NEL (no stops).

National Oceanic and Atmospheric Administration (in the USA) Abbrev.: NOAA (no stops).

National Optical Astronomy Observatories (headquarters Tucson, Arizona) Abbrev.: NOAO (no stops).

National Physical Laboratory (in Teddington, Middlesex) Abbrev.: NPL (no stops).

National Radio Astronomy Observatory (in Green Bank, West Virginia) Abbrev.: NRAO (no stops).

National Research Council Abbrev.: NRC (no stops).

National Rivers Authority Abbrev.: NRA (no stops).

National Science Foundation Abbrev.: *NSF (no stops).

Natta, Giulio (1903–79) Italian chemist. **Ziegler–Natta catalysts** (en dash)

Natural Environment Research Council Abbrev. or acronym: NERC (no stops).

natural killer cell Abbrev.: NK cell (no stops). A type of lymphocyte that kills tumour cells.

natural number Any member of the set of whole positive numbers and zero. As some authors exclude zero, the membership should be stated explicitly. *See also* sets (for denotation). *Compare* integer.

Nature Conservancy Council Abbrev.: NCC (no stops).

Naudin, Charles (1815–99) French experimental botanist and horticulturist.

nauplius (pl. nauplii) A crustacean larva. *Compare* nautilus.

nautical mile, international Abbrev.: n mile. A unit of distance used for precise scientific measurements at sea (but not in navigation) and defined as exactly 1852 metres. One n mile is equal to 1.1508 statute miles. The UK nautical mile, 6080 feet, 1853.18 metres, is now obsolete. *See also* sea mile.

nautilus (pl. nautiluses or nautili) A cephalopod mollusc of the genus *Nautilus* (cap. N, ital.; **pearly nautilus**) or the genus *Argonauta* (**paper nautilus**). *Compare* nauplius.

Nb Symbol for niobium.

NBS Abbrev. for **1.** National Bureau of Standards (US). **2.** *N*-bromosuccinimide.

nBu, n-Bu Use Bu^n (*see* Bu).

NCC Abbrev. for Nature Conservancy Council.

Nd Symbol for neodymium.

ND (or **n.d.**) Abbrev. for neutral density.

NDP Abbrev. for nucleoside diphosphate.

NDT Abbrev. for nondestructive testing.

Ne Symbol for neon.

NE Abbrev. and preferred form for northeast, north-eastern. For usage, *see* north.

ne- *See* neo-.

Neanderthal man (not Neandertal) Common name for a subspecies of fossil hominids (*see Homo*). Individuals are called **Neanderthalers** or **Neanderthals**, not Neanderthal men. Adjectival form: **Neanderthal** or **Neanderthaloid** (cap. N). Named after the site in the Neander Valley, near Düsseldorf, Germany, where the first fossils were identified in 1856.

near letter quality *Computing* Abbrev.: NLQ (no stops).

nebula (pl. nebulae; preferred to nebulas) *Astron.* Adjectival form: **nebular**.

necro- (**necr-** before vowels) Prefix denoting death or dissolution (e.g. necrobiosis, necrosis).

necrosis (pl. necroses) Adjectival form: **necrotic**.

Needham, Joseph (1900–) British biochemist, historian, and sinologist.

Néel, Louis Eugène Félix (1904–) French physicist.
Néel temperature (preferred to Néel point) Symbol: T_N

Ne'eman, Yuval (1925–) Israeli physicist.

Ng (no stop) Abbrev. for *Neogene.

Neisser, Albert Ludwig Siegmund (1855–1916) German bacteriologist. ****Neisseria***

Neisseria (cap. N, ital.) A genus of bacteria of the family Neisseriaceae. Individual name: **neisseria** (no cap., not ital.; pl. neisseriae). *See also* gonococcus; meningococcus.

NEL Abbrev. for National Engineering Laboratory (East Kilbride, Scotland).

NEM Abbrev. for *N*-ethylmaleimide.

nemato- (**nemat-** before vowels) Prefix denoting a threadlike form (e.g. nematocyst, nematic).

Nematoda (cap. N; not Nematoida) A phylum of invertebrates comprising the roundworms. Individual name and adjectival form: **nematode** (no cap.).

Nemertea (cap. N) A phylum of invertebrates comprising the ribbon worms. Also called Nemertina, Nemertini, Nemertinea, Rhynchocoela. Individual name and adjectival form: **nemertean** (no cap.; preferred to nemertine).

nemertine Use nemertean (*see* Nemertea).

neo- (sometimes **ne-** before vowels) Prefix denoting **1.** new, most recent, modern (e.g. neopallium, neoplasm, neoteny, Neotropical, Nearctic). **2.** *Chem.* an isomer having three substituent groups at the end of an otherwise unbranched alkane chain, with the end carbon atom linked to three other carbon atoms (e.g. neobutane, neopentane).

neo-Darwinism Named after C. *Darwin.

neodymium Symbol: Nd *See also* periodic table; nuclide.

Neogene Abbrev.: Ng (no stop). **1.** (adjective) Denoting the more recent subperiod of the Tertiary period, comprising the Miocene and Pliocene epochs. **2.** (noun; preceded by 'the') The Neogene subperiod. In some recent schemes the Neogene has the status of a period (*see* Tertiary).

neognathous Describing the palate structure of all birds except the *ratites (flightless birds). Such birds were formerly classified as the superorder Neognathae; the term neognathous is now used as an anatomical (rather than taxonomic) term. *See also* carinate. *Compare* palaeognathous.

neo-Lamarckism Named after J. B. *Lamarck.

neon Symbol: Ne *See also* periodic table; nuclide.

neopentyl alcohol $(CH_3)_3CCH_2OH$ The traditional name for 2,2-dimethylpropan-1-ol.

neoprene (no cap., not a trade name).

neper Symbol: Np A dimensionless unit used in telecommunications and acoustics to measure *power level differences and *sound pressure levels. It is defined in terms of the natural logarithm of the relevant ratio. One neper is approximately equal to 8.686 *decibels. Named after John *Napier.

nephelinite An igneous rock consisting substantially of the mineral **nepheline.** Normally the suffix *-ite indicates a mineral.

nephelo- (**nephel-** before vowels) Prefix denoting cloudiness (e.g. nepheline, nephelometric).

nepho- (**neph-** before vowels) Prefix denoting clouds (e.g. nephograph, nephanalysis).

nephridium (pl. nephridia) An excretory organ of invertebrates: may be a primitive **protonephridium** or a more advanced **metanephridium**. Adjectival form: **nephridial**.

nephro- (**nephr-** before vowels) Prefix denoting a kidney or kidneys (e.g. nephrogenic, nephridium).

Nepovirus (cap. N, ital.) Approved name for the *tobacco ringspot virus group. Individual name: **nepovirus** (no cap., not ital.). [from *ne*matode *po*lyhedral *virus*]

Neptune A planet. Adjectival form: **Neptunian.**

neptunium Symbol: Np *See also* periodic table; nuclide.

NEQ gate *Electronics* Abbrev. for non-equivalence gate.

NERC Abbrev. or acronym for Natural Environment Research Council.

Nernst, Walther Hermann (1864–1941) German physical chemist.
Nernst effect or **Nernst–Ettinghausen effect** (en dash) Also named after A. von Ettinghausen (died 1932).
Nernst equation
Nernst heat theorem

nerve Two-word terms in which the first word is 'nerve' (e.g. nerve cell, nerve cord, nerve fibre, nerve impulse) are not hyphenated unless used adjectivally (e.g. nerve-cell membrane).

Nessler, Julius (1827–1905) German agricultural chemist.
Nessler's reagent
Nessler tube

net assimilation rate Abbrev.: NAR (no stops). The rate of increase in dry weight of a plant divided by leaf area: used in assessing plant productivity.

Neumann, Carl Gottfried (1832–1925) German mathematician.
Neumann function
Neumann method
Neumann problem

Neumann, Franz Ernst (1798–1895) German physicist.
Faraday–Neumann law (en dash)
Neumann lines
Neumann's principle

Neumann, John von Usually alphabetized as *von Neumann.

neurilemma (preferred to neurolemma)

neuro- (sometimes **neur-** before vowels) Prefix denoting nerves or the nervous system (e.g. neuroanatomy, neuroendocrine, neurohormone, neurohypophysis, neuromuscular, neurotransmitter, neuralgia, neuritis).

neuroglia Use glia.

neuron *See* neurone.

neurone A nerve cell. The variant spelling **neuron**, originally the US spelling, is now widely used in British English. Adjectival forms: **neuronal, neuronic**.

neutral density Abbrev.: ND (no stops) or n.d. (stops).

neutrino An uncharged elementary particle, denoted by ν (Greek nu) in nuclear reactions, etc. Its antiparticle, the **antineutrino**, is denoted by $\bar{\nu}$. The three known types of neutrino, associated with the electron, muon, and tauon, are denoted by ν_e, ν_μ, and ν_τ, respectively.

neutron An uncharged particle, denoted n in nuclear reactions, etc. Its antiparticle, the **antineutron**, is denoted $\bar{n}$. Its rest mass, symbol m_n, is equal to

$$1.674\,929 \times 10^{-27}\,\text{kg}$$

It is 1.0014 times greater than the *proton mass. Two- or three-word terms in which the first word is 'neutron' are not hyphenated (e.g. neutron number, neutron star, neutron activation analysis).

neutron activation analysis (not hyphenated) Abbrev.: NAA (no stops).

neutron number Symbol: N A dimensionless physical quantity, the number of neutrons in an atomic nucleus. The sum $(N + Z)$ of neutron number and *proton number is called the *nucleon number, A.

névé (acute accents) *Glaciol.* Another name for firn.

new candle Obsolete name for candela.

Newcomb, Simon (1835–1909) US astronomer.

Newcomen, Thomas (1663–1729) British engineer.

New General Catalogue *Astron.* Abbrev.: *NGC (no stops).

Newlands, John Alexander Reina (1837–98) British chemist.
Newlands' law (of octaves)

New Style Abbrev.: *NS.

New Technology Telescope (at the *ESO) Abbrev.: NTT (no stops).

newton (no cap.) Symbol: N The SI unit of force.

$$1\ \mathrm{N} = 1\ \mathrm{kg\ m\ s^{-2}} = 1\ \mathrm{J\ m^{-1}}.$$

Named after Sir Isaac *Newton.

Newton, Alfred (1829–1907) British ornithologist.

Newton, Sir Isaac (1642–1727) English physicist and mathematician. Adjectival form: **Newtonian.**
***newton**
Newtonian fluid
Newtonian mechanics
Newtonian potential
Newtonian system
Newtonian telescope
Newton's formula
Newton's interpolation polynomial
Newton's law of gravitation
Newton's laws of motion
Newton's rings

Ng Abbrev. for Neogene.

NGC *Astron.* Abbrev. for New General Catalogue, generally prefixed to a number (as in NGC 224) to designate an object (nebula, cluster, or galaxy) in the Catalogue.

Ni Symbol for nickel.

niacin Use nicotinic acid.

Nicholas of Cusa (1401–64) German cardinal, mathematician, and philosopher.

Nicholson, William (1753–1815) British chemist.
Nicholson hydrometer

Nichrome (cap. N) A trade name for a group of high-resistivity heat-resistant nickel–chromium alloys.

nickel Symbol: Ni *See also* periodic table; nuclide.

nickel dimethylglyoxime $C_8H_{14}N_4NiO_4$ The traditional name for bis(butanedione dioximato)nickel(II).

nickelic Denoting compounds in which nickel has an oxidation state of +3. The recommended system is to use oxidation numbers, e.g. nickelic oxide, Ni_2O_3, has the systematic name nickel(III) oxide.

nickelous Denoting compounds in which nickel has an oxidation state of +2. The recommended system is to use oxidation numbers, e.g. nickelous oxide, NiO, has the systematic name nickel(II) oxide.

Nicol, William (1768–1852) British geologist and physicist.
Nicol prism

nicotinamide–adenine dinucleotide (en dash) Abbrev. and preferred form: *NAD (no stops).

nicotinamide–adenine dinucleotide phosphate (en dash) Abbrev. and preferred form: *NADP (no stops).

nicotinic acid (preferred to niacin) *See* vitamin B complex.

nidicolous Describing birds that hatch in a helpless state, requiring lengthy parental care. *Compare* nidifugous.

nidifugous Describing birds that hatch in a well-developed state and are soon independent of their parents. *Compare* nidicolous.

nidus (pl. nidi or niduses) **1.** The nest of an insect or spider. **2.** A focus of infection. Adjectival form: **nidal.**

nielsbohrium (one word) *See* element 105.

Niemann, Albert (1880–1921) German physician.
Niemann–Pick disease (en dash)

Niepce, Joseph-Nicéphore (hyphen) (1765–1833) French inventor.

Nieuwland, Julius Arthur (1878–1936) US chemist.

Ni-Fe (hyphenated, cap. N and F) A trade name for a nickel–iron-alkaline accumulator.

Nilson, Lars Fredrick (1840–99) Swedish chemist.

nimbostratus (one word; pl. nimbostrati) Abbrev.: *Ns* (cap. N, ital.).

niobium Symbol: Nb Former name: columbium. *See also* periodic table; nuclide.

Nirenberg, Marshall Warren (1927–) US biochemist.

Nissl, Franz (1860–1919) German neurologist.

Nissl body
Nissl degeneration
Nissl granules
Nissl stain
Nissl substance

nitric oxide NO The traditional name for nitrogen oxide.

nitrile Any of a class of organic compounds containing the group –CN joined directly to a carbon atom.
Nitriles are named in systematic nomenclature either by adding the suffix -nitrile to the name of the parent hydrocarbon, e.g. ethanenitrile, CH_3CN, and propenenitrile, CH_2CHCN, or as cyano derivatives of the parent hydrocarbons, e.g. 1,2-cyanoethane, CH_2CNCH_2CN.
In nonsystematic nomenclature nitriles are named either by replacing the ending -ic of the trivial name of the corresponding carboxylic acid by the suffix -nitrile (often with an 'o' added for euphony), e.g. acetonitrile, CH_3CN, and acrylonitrile, CH_2CHCN, or by naming the hydrocarbon group attached to the cyano group followed by the word cyanide, e.g. *t*-butyl cyanide, $(CH_3)_3CCN$.

nitro- Prefix denoting the group $-NO_2$ (e.g. nitroamine, nitrobenzene).

nitroaniline $H_2NC_6H_4NO_2$ The traditional name for nitrophenylamine.

Nitrococcus (cap. N, ital.) A genus of nitrifying bacteria that use nitrite as an energy source. Not to be confused with *Nitrosococcus*, a related genus of ammonia-oxidizing bacteria.

nitrogen Symbol: N *See also* periodic table; nuclide.

nitrogen oxide NO The recommended name for the compound traditionally known as nitric oxide.

nitroglycerine $CH_2(ONO_2)CH(ONO_2)CH_2(ONO_2)$ The traditional name for propane-1,2,3-triyl trinitrate.

nitronium Denoting a compound containing the ion $NO_2{}^+$, e.g. nitronium nitrate, $(NO_2)^+(NO_3)^-$. The recommended name is nitryl.

nitrophenol $OHC_6H_4NO_2$ The recommended name for the compound traditionally known as *o*-nitrophenol is 2-nitrophenol, etc.

nitrophenylamine $H_2NC_6H_4NO_2$ The recommended name for the compound traditionally known as nitroaniline. The recommended name for *o*-nitroaniline is 2-nitrophenylamine, etc.

nitroprusside A compound containing the ion $[Fe(CN)_5NO]^{2-}$, e.g. potassium nitroprusside, $K_2(Fe(CN)_5NO)$. The recommended name is pentacyanonitrosylferrate(II).

nitroso- Prefix denoting the group –NO (e.g. nitrosoamine, nitrosodimethylaniline).

Nitrosococcus (cap. N, ital.) *See Nitrococcus.*

nitrosyl- *Prefix denoting the group –NO bound to a metal atom through the nitrogen atom in inorganic compounds (e.g. nitrosyl chloride).

nitrotoluene $CH_3C_6H_4NO_2$ The traditional name for *methylnitrobenzene.

nitrous oxide N_2O The traditional name for dinitrogen oxide.

nitryl Denoting a compound containing the ion $NO_2{}^+$, e.g. nitryl nitrate, $(NO_2)^+(NO_3)^-$. The traditional name is nitronium.

NK cell Abbrev. for natural killer cell.

NLQ *Computing* Abbrev. for near letter quality.

n mile Abbrev. for international *nautical mile.

NMO Abbrev. for nonmotile *Oerskovia*-like strain. *See Oerskovia.*

NMR Abbrev. for nuclear magnetic resonance.

NNE Abbrev. and preferred form for north-north-east or north-north-eastern.

NNW Abbrev. and preferred form for north-north-west or north-north-western.

No Symbol for nobelium.

No. (or **no.**; pl. Nos. or nos.) Abbrev. for number.

NOAA Abbrev. for National Oceanic and Atmospheric Administration, a US organization.

NOAO Abbrev. for National Optical Astronomy Observatories (headquarters Tucson, Arizona).

Nobel, Alfred Bernhard (1833–96) Swedish chemist, engineer, and inventor.
***nobelium**
Nobel prize

nobelium Symbol: No The IUPAC name for element 102. The Soviet name, **joliotium**, is sometimes encountered. *See also* periodic table; nuclide.

Nobili, Leopoldo (1784–1835) Italian physicist.

noble gases (preferred to inert gases and rare gases) The elements of group 0 of the *periodic table.

Nocardia (cap. N, ital.) A genus of nocardioform actinomycete bacteria. Individual name: **nocardia** (no cap., not ital.; pl. nocardiae). Note that 'nocardiae' should not be used as a synonym for *nocardioform bacteria. Adjectival form: **nocardial**.

nocardioform Describing an actinomycete bacterium typically having a short-lived mycelium that breaks up into rods or cocci. The term is also used as a noun to mean a nocardioform bacterium. *See also Nocardia.*

nocti- (**noct-** before vowels) Prefix denoting night (e.g. noctilucent, noctambulation).

nodamura virus (two words) Type species of the single genus, *Nodavirus*, of the family Nodaviridae. Named after Nodamura, the Japanese village where the species was first isolated.

Nodaviridae Approved family name of the *nodamura virus group.

node of *Ranvier

Noguchi, (Seisako) Hideyo (1876–1928) Japanese bacteriologist.

no-load (hyphenated) *Elec. eng.*

nomo- Prefix denoting a law or laws (e.g. nomograph, nomology).

-nomy Noun suffix denoting a science or the laws of a scientific field (e.g. agronomy, astronomy). Adjectival form: **-nomical** or **-nomic**.

non- 1. (hyphenated before n and proper nouns) Prefix denoting lack or absence of or exclusion from a specified class (e.g. nondisjunction, nonessential, non-Euclidean, nonmetal, non-negative, non-Newtonian, nonporous, nonreducing). **2.** *See* nona-.

nona- (**non-** before vowels) Prefix denoting nine (e.g. nonagon, nonane).

nonanoic acid $CH_3(CH_2)_7COOH$ The recommended name for the compound traditionally known as pelargonic acid.

nonbonding orbital *See* orbital.

nondestructive testing Abbrev.: NDT (no stops).

nonequivalence gate Abbrev.: NEQ gate.

nonpaired spatial orbitals Abbrev.: NPSO (no stops)

Nor (no stop) *Astron.* Abbrev. for Norma.

NOR *See* logic symbols; electronics, graphical symbols.

nor- Prefix denoting a relationship, usually through alkylation or isomerization, between compounds (e.g. noradrenaline has one less methyl group than adrenaline; norleucine and leucine are isomers).

noradrenaline (not noradrenalin) US: **norepinephrine**. Abbrev.: NA (no stops). *See also* adrenergic.

norepinephrine US name for noradrenaline.

Norma A constellation. Genitive form: Normae. Abbrev.: Nor (no stop). *See also* stellar nomenclature.

normality *Chem.* Symbol: N Obsolete measurement of *concentration. Associated term: **normal solution**.

normal stress *See* stress.

normal temperature and pressure Abbrev.: N.T.P (stops) or NTP (no stops). Use *stp.

Norrish, Ronald George Wreyford (1897–1978) British physical chemist.

north Adjectival forms: **north, northern**. Abbrev.: N (no stop). Use the abbrev. only descriptively, not in place names or concepts (e.g. N London, N Canada, but Northern Ireland, North Atlantic Drift,

North Sea, *North Pole, magnetic north, true north, north-seeking pole). Use the same principle for **north-east(ern)** and **north-west(ern)** (e.g. NE London but North-West Passage).

North Atlantic Drift (initial caps.) An extension of the *Gulf Stream that flows north-eastwards from the Newfoundland Banks across the Atlantic Ocean.

north-east (hyphenated) Adjectival forms: **north-east, north-eastern**. Abbrev.: NE (no stops). For use of abbrev., *see* north.

Northern blotting (cap. N) A chromatographic technique used for the analysis of RNA. It was named by analogy to the similar technique of *Southern blotting.

north-north-east (hyphenated) Adjectival form: **north-north-eastern**. Abbrev. (for both): NNE (no stops).

north-north-west (hyphenated) Adjectival form: **north-north-western**. Abbrev. (for both): NNW (no stops).

north polar distance Abbrev.: NPD (no stops).

North Pole (preceded by 'the'; not N Pole) Use capitals in the case of the earth, otherwise use lower case (e.g. north celestial pole, north galactic pole, Jupiter's north pole).

Northrop, John Howard (1891–1987) US chemist.

north-west (hyphenated) Adjectival forms: **north-west, north-western**. Abbrev.: NW (no stops). For use of abbrev., *see* north.

Norton's theorem *Elec. eng.*

NOT *See* logic symbols.

noto- Prefix denoting **1.** the back or dorsal (e.g. notochord). **2.** south (e.g. Notogaea, notoungulate).

notochord (not notocord) The skeletal rod lying beneath the nerve cord of chordate animals. Adjectival form: **notochordal**.

nova (pl. novae; preferred to novas) *See also* stellar nomenclature.

Noyes, William Albert (1857–1941) US chemist.

Np Symbol for **1.** neptunium. **2.** neper.

NPD *Astron.* Abbrev. for north polar distance.

NPK Abbrev. for nitrogen, phosphorus, and potassium, used in fertilizers.

NPL Abbrev. for National Physical Laboratory (Teddington, Middlesex).

n-p-n transistor (or **npn transistor**) *See* n-type; p-type.

nPr, n-Pr Use Pr^n (*see* Pr).

NPSO Abbrev. for nonpaired spatial orbitals.

NQR Abbrev. for nuclear quadrupole resonance.

NRA Abbrev. for National Rivers Authority, a UK environmental agency.

NRAO Abbrev. for National Radio Astronomy Observatory (Green Bank, West Virginia).

NRC Abbrev. for National Research Council (in several countries).

Ns (ital.) Abbrev. for nimbostratus.

NS Abbrev. for New Style, used to indicate that a date has been calculated using the (current) Gregorian calendar. Italy, France, Spain, and Portugal switched from the Julian to the Gregorian calendars in 1582; other countries followed at different times. England and Wales and the American colonies switched in 1752, losing 11 days (3–13 Sept). *Compare* OS.

NSF Abbrev. for National Science Foundation, the US organization that promotes and supports research and education in science but not medicine.

N shell (not hyphenated) *Physics*

N-t-B Abbrev. for nitroso-*t*-butane.

NTP Abbrev. for nucleoside triphosphate.

N.T.P. (or **NTP**) Abbrev. for normal temperature and pressure. Use *stp.

NTT Abbrev. for New Technology Telescope (at the *ESO).

n-type (n not ital.) *Electronics* Denoting a doped semiconductor in which mobile negative-charge carriers (electrons) are the majority carriers. Abbrev. in combinations: n, as in n-p-n transistor, pnpn device.

nu Greek letter, symbol: ν (lower case), N (cap.).

ν Symbol for **1**. amount of substance. **2**. frequency. **3**. kinematic viscosity. **4**. neutrino ($\bar{\nu}$ = antineutrino). **5**. Poisson ratio. **6**. stoichiometric coefficient (ν_B = that of substance B).

Nu (ital.) Symbol for Nusselt number.

nuclear magnetic resonance (not hyphenated) Abbrev.: NMR (no stops). It is usual to prefix this term with the nucleus being investigated, e.g. ^{13}C nuclear magnetic resonance or ^{13}C NMR.

nuclear magneton *See* Bohr magneton.

nuclear quadrupole resonance (not quadropole) Abbrev.: NQR (no stops). *Physics*

nuclear reaction A reaction between *nuclides, expressed in the form:

$X_i(P_i, P_o)X_f$,

where X_i and X_f are the initial and final nuclides and P_i and P_o are the incoming and outgoing particle(s). Examples include:

$^{14}N(\alpha, p)^{17}O$
$^{59}Co(n, \gamma)^{60}Co$
$^{23}Na(\gamma, 3n)^{20}Na$
$^{31}P(\gamma, pn)^{29}Si$,

where p, n, α, and γ are symbols for the proton, neutron, alpha particle, and photon, respectively.

nucleo- (**nucle-** or **nucl-** before vowels) Prefix denoting **1**. a cell nucleus (e.g. nucleohistone, nucleoplasm). **2**. nucleic acid (e.g. nucleoprotein, nucleoside, nuclease). **3**. an atomic nucleus (e.g. nucleonics, nucleophilic, nucleon, nuclide).

nucleolus (pl. nucleoli) *Cell biol.* Adjectival form: **nucleolar**.

nucleon number Symbol: *A* A dimensionless physical quantity, the total number of nucleons (protons and neutrons) in an atomic nucleus. It is the sum of the *proton number, *Z*, and *neutron number, *N*, of the nucleus. In the specification of a particular *nuclide, the value of *A* is placed as a superscript to the left of the chemical symbol, as in ^{238}U, ^{14}N; alternatively, the name of the element or its symbol is hyphenated to A, as in uranium-238 or U-238. Also called mass number.

nucleophilic *See* reaction mechanisms.

nucleoside A molecule consisting of a purine or pyrimidine base linked to a pentose sugar. Symbols for nucleosides are typically formed from the initial letter (capitalized) of the name; for example, adenosine is A. To avoid confusion, the symbol may be prefixed by r to denote ribonucleosides or by d to denote 2′-deoxyribonucleosides; for example, 2′-deoxyribosyladenine (adenosine in DNA) is denoted by dA. Individual nucleosides and their symbols are listed alphabetically in the dictionary. *Compare* nucleotide.

nucleoside diphosphate Abbrev.: NDP (no stops).

nucleoside triphosphate Abbrev.: NTP (no stops).

nucleotide A compound consisting of a purine or pyrimidine base linked to a pentose sugar (i.e. a *nucleoside) and phosphoric acid. Free nucleotides (**mononucleotides**) include AMP and coenzyme A. **Polynucleotides** include DNA and RNA. Sequences in polynucleotides are denoted using nucleoside symbols: the linking *phosphate is denoted by a hyphen in known sequences and by a comma in unknown sequences; a terminal phosphate is indicated by a lower-case p. For example, the heptanucleotide

A-G-A(C,U,G)Gp

contains three nucleotides of known sequence followed by three of unknown sequence before the terminal nucleotide. Note that in designating the sequence or content of a polynucleotide, there must be an intervening hyphen or plus sign between the nucleoside symbols; e.g. G+C content, G-C sequence (not GC content, GC sequence). Repetitive sequences may be indicated by the prefix poly (not hyphenated, no intervening space), e.g. poly(U). *See also* poly(A).

nucleus (pl. nuclei) Adjectival forms: **nuclear, nucleate, nucleic**.

nuclide An atom as defined by its *proton number (i.e. atomic number) and *nucleon number (i.e. mass number). Different nuclides with the same proton number are called isotopes or isotopic nuclides. Different nuclides with the same nucleon number are called isobars or isobaric nuclides.

A particular nuclide is specified using information set in superscript and subscript positions to the left and right of the symbol of the element:

left superscript: nucleon number (e.g. ^{14}N, ^{16}O);

left subscript: proton number (e.g. $_{7}N$);

right superscript: a state of ionization (e.g. Pb^{2+}, Cl^{-}), an excited atomic state (e.g. He*, ^{4}He*), or, using a Roman numeral, an *oxidation number (e.g. $^{207}Pb^{II}$);

right subscript: number of atoms in the entity (e.g. $^{14}N_2$, $PO_4{}^{3-}$) or, in nuclear physics, number of neutrons in the nucleus (e.g. $^{235}U_{143}$).

Only the details required need be written: the left subscript is redundant and usually omitted. When it is not possible to set a right superscript immediately above the subscript, it can be printed to the immediate right of the subscript, as in $PO_4{}^{3-}$. In some cases a nuclide is specified by the name of the element or its symbol hyphenated to its nucleon number, e.g. uranium-238 or U-238.

nudi- Prefix denoting naked or bare (e.g. nudibranch, nudicaulous).

number Abbrev.: No. or no. (stops; pl. Nos. or nos.). In scientific usage, numbers are normally printed in roman (upright) type. An integer should never be terminated by a decimal sign. There should always be at least one digit before and after the decimal sign; when the number is less than unity, the sign should be preceded by a zero, e.g. 0.007 (not .007).

To facilitate the reading of long numbers, digits may be grouped in threes (counting to the right or left of the decimal sign), separated by a fixed space or thin space, e.g.

317 118.127 57.

In scientific writing a comma should not be used to separate groups of three digits as it could be confused with a decimal sign (as used in ISO publications in English). However, commas may be used to separate groups of three digits in sums of money (e.g. £5,234,567 but 5 234 567 metres).

Instead of a single final digit, e.g. 2.141 6, the last four digits may be grouped, e.g. 2.1416; with numbers between 1000 and 9999 it is usual to use a four-digit group with no spaces. *See also* Roman numerals.

number density Symbol: n A physical quantity, N/V, where N is the number of particles, molecules, etc., and V is the volume. The *SI unit is the reciprocal cubic metre (m^{-3}).

The **electron number density** and **hole number density**, are used in solid-state physics. The former is the number density of electrons in the conduction band, symbol n_n, n, or n_-. The latter is the number density of holes in the valence band, symbol n_p, p, or n_+. The **intrinsic number density**, symbol n_i, is given by $\sqrt{(n_n n_p)}$. The **donor number density**, symbol n_d, is the number density of donor impurities. The **acceptor number density**, symbol n_a, is the number density of acceptor impurities. The symbols n_n and n_p are also used for electron densities in n-type and p-type regions of a p-n junction, respectively.

numerical aperture *Optics, etc.* Abbrev.: NA (no stops).

nunatak (not nunatack) *Geol.* [from Eskimo via Danish]

Nusselt, E. K. Wilhelm (1882–1957) German engineer.

***Nusselt number**

Nusselt number Symbol: Nu A dimensionless quantity equal to hl/κ, where l is a characteristic length, κ the *thermal conductivity, and h the heat transfer coefficient (i.e. heat divided by the product of time, cross-sectional area, and temperature difference). *See also* parameter. Named after E. K. W. Nusselt.

NW Abbrev. and preferred form for northwest or north-western. For usage, *see* north.

nycto- (**nyct-** before vowels, **nycti-**) Prefix denoting night or darkness (e.g. nyctoperiod, nyctophobia, nyctanthous, nyctinasty).

Nyholm, Sir Ronald Sydney (1917–71) Australian-born British chemist.

nylon A generic name for a series of synthetic polyamide fibres produced under various trade names including Enkalon, Bri-Nylon, and Celon.

Nymphaeaceae (not Nympheaceae) A family of dicotyledonous plants including many water lilies.

Nyquist, Harry (1889–1976) Swedish-born US physicist.
Nyquist criterion
Nyquist diagram
Nyquist limit
Nyquist noise theorem

O

o *Computing, maths See* O.

O 1. A blood group (*see* ABO). **2.** Symbol for **i.** (or o) *Computing, maths* order (followed in brackets by limiting value of a function). **ii.** oxygen. **3.** Abbrev. for Ordovician.

O (light ital.) *Genetics* Symbol for *operator (O^c (superscript lower-case c) = operator-constitutive mutant).

***o*-** (ital., always hyphenated; preferred to *ortho-*) *Chem.* Prefix denoting *ortho*, indicating 1,2-substitution on a benzene ring (e.g. *o*-cresol, *o*-dihydroxybenzene). The use of numbered substitution positions is recommended (e.g. 1,2-cresol). *Compare m-*; *p-*.

***O*-** (cap. O, ital., always hyphenated) *Chem.* Prefix denoting substitution on an oxygen atom in an organic compound (e.g. *O*-ethyl *S*-methyl-3-p-tolyl-2-butenoate).

oasis (pl. oases)

ob- Prefix denoting **1.** inverse, reversed, turned about (e.g. obcordate, obduction, obovate). **2.** towards, over (e.g. obtect).

***ob*-** (ital., always hyphenated) *Chem.* Prefix denoting geometric isomerization in octahedral tris(chelate) structures in which the carbon-to-carbon bonds of, for example, ethylenediamine are oblique to the three-fold axis of the complex (e.g. *ob*-tris (ethylenediamine)cobalt(III) chloride). *Compare lel-*.

Oberth, Hermann Julius (1894–) German rocket scientist.

OBM Abbrev. for Ordnance benchmark.

obs. Abbrev. for observed.

Occam (cap. O) A programming language devised specifically for use with transputer-based systems.

Occhialini, Giuseppe Paolo Stanislao (1907–) Italian physicist.

occiput The back of the head. Adjectival form: **occipital**.

occlusion *Meteorol., chem., zool.* Adjectival form: **occluded**.

Oceanospirillum (cap. O, ital.) A genus of spirally shaped bacteria (*see* spirillum). Individual name: **oceanospirillum** (no cap., not ital.; pl. oceanospirilla).

Oceanus *See* mare.

ocellus (pl. ocelli) *Biol.* **1.** The simple eye of many invertebrates. **2.** Any eyelike marking. Adjectival forms: **ocellar, ocellate** (or **ocellated**).

Ochoa, Severo (1905–) Spanish-born US biochemist.

OCR Abbrev. for optical character recognition (or reader).

Oct (no stop) *Astron.* Abbrev. for Octans.

oct- *See* octo-.

octa- Prefix denoting eight (e.g. octadecanoate, octagon, octahedron, octahydrate, octavalent). *See also* octo-.

octadecanoic acid $CH_3(CH_2)_{16}$-COOH The recommended name for the compound traditionally known as stearic acid.

***cis*-octadec-9-enoic acid** $CH_3(CH_2)_7$-$CH{=}CH(CH_2)_7COOH$ The recommended name for the compound traditionally known as oleic acid.

octanoic acid $CH_3(CH_2)_6COOH$ The recommended name for the compound traditionally known as caprylic acid.

Octans A constellation. Genitive form: Octantis. Abbrev.: Oct (no stop). *See also* stellar nomenclature.

octo- (**oct-** before vowels) Prefix denoting eight (e.g. octopod, octane). *See also* octa-.

octyl- *Prefix denoting the group $C_8H_{17}-$ (e.g. octylbenzoic acid, octyl bromide).

oculo- Prefix denoting an eye (e.g. oculomotor).

OD Abbrev. for Ordnance datum, the standard sea level of the Ordnance Survey.

ODA Abbrev. for 4,4′-oxydianiline.

(*Z*)-13-ODAL Abbrev. for *cis*-13-octadecenal.

-ode Noun suffix denoting **1.** resemblance (e.g. geode, staminode). **2.** path, direction (e.g. cathode, dynode).

Odling, William (1829–1921) British chemist.

-odont Adjectival suffix denoting teeth (e.g. diphyodont, heterodont, polyphyodont).

odonto- (**odont-** before vowels) Prefix denoting a tooth (e.g. odontoblast, odontoid).

odour US: **odor**. Adjectival form: **odorous** (not odourous). Derived noun: **odorimetry** (not odourimetry).

Oe 1. Symbol for oersted. **2.** *Astron. See* spectral types.

oedema US: **edema**. Adjectival form: **oedematous** (US: **edematous**).

OEM Abbrev. for original equipment manufacturer.

Oerskovia (cap. O, ital.; not *Ørskovia*) A genus of nocardioform actinomycete bacteria. Individual name: **oerskovia** (no cap., not ital.; pl. oerskoviae). A distinction is made between true oerskoviae and similar organisms referred to as 'nonmotile *Oerskovia*-like strains' (NMOs).

oersted Symbol: Oe The *cgs (electromagnetic) unit of magnetic field strength, now discouraged. In *SI units, magnetic field strength is measured in amperes per metre: $1\ \mathrm{Oe} = 10^3/4\pi\ \mathrm{A\ m^{-1}}$. Named after H. C. *Oersted.

Oersted, Hans Christian (not Ørsted) (1777–1851) Danish physicist.
***oersted**

oesophagus (pl. oesophagi or oesophaguses) US: **esophagus**. Adjectival form: **oesophageal** (US: **esophageal**).

oestrogen US: **estrogen**. Adjectival form: **oestrogenic** (US: **estrogenic**).

oestrus US: **estrus** or **estrum**. The period of sexual receptivity (heat) of female mammals. Not to be confused with its adjectival form, **oestrous** (US: **estrous**), which is most commonly applied to the cycle of reproductive activity (**oestrous cycle**) of which oestrus is a phase (other phases are pro-oestrus, metoestrus, and dioestrus). The adjective also occurs in combination (e.g. monoestrous, polyoestrous).

off-centre (hyphenated)

offlap (one word) *Geol.*

offline (one word) *Computing*

offset (noun and verb; one word)

offshore (adjective and adverb; one word)

off the film *Photog.* Abbrev.: OTF (no stops).

ogive *Stats.* Adjectival form: **ogival.**

ohm (no cap.) Symbol: Ω (Greek cap. omega). The *SI unit of electric *resistance.

$$1\ \Omega = 1\ \mathrm{V\ A^{-1}}.$$

Named after Georg *Ohm.

Ohm, Georg Simon (1787–1854) German physicist.
***ohm**
ohmic contact
ohmic loss
ohmmeter
Ohm's law

O horizon A generally wet o(rganic) *soil horizon.

-oic Noun suffix denoting the carboxyl group –COOH (e.g. benzoic acid, propanoic acid).

-oid 1. Noun suffix denoting like or similarity (e.g. asteroid, ellipsoid). Adjectival form: **-oidal. 2.** Adjectival suffix denoting similarity (e.g. amoeboid, coccoid).

-oidea Noun suffix denoting a *superfamily in animal taxonomy (e.g. Bovoidea).

-oideae Noun suffix denoting a *subfamily in plant taxonomy (e.g. Rosoideae).

oil Two-word terms in which the first word is oil (e.g. oil gland, oil hardening, oil rig, oil slick, oil switch) should not be hyphenated unless they are used adjectivally (e.g. oil-immersion objective, oil-slick pollution). Adjectives such as oil-cooled, oil-fired, and oil-insulated are always hyphenated. *See also* oleo-.

oilcake (one word) *Agric.*

oilfield (one word)

oilgas (one word) *Chem.*

Okazaki fragment *Molecular biol.* Named after Reiji Okazaki (1930–75).

-ol Noun suffix denoting alcohol (e.g. ethanol, phenol).

Olbers, Heinrich W. M. (1758–1840) German astronomer.
Olbers's paradox

Oldham, Richard Dixon (1858–1936) British seismologist and geologist.

Old Style Abbrev.: *OS.

olefin (or **olefine**) The traditional name for an alkene.

oleic acid $CH_3(CH_2)_7CH=CH(CH_2)_7$-COOH The traditional name for *cis*-octadec-9-enoic acid.

oleo- (**ole-** before vowels) Prefix denoting oil (e.g. oleoplast, oleoresin).

Oli Abbrev. for Oligocene.

oligo- (**olig-** before vowels) Prefix denoting few or little (e.g. oligopeptide, oligosaccharide, oligotrophic).

Oligocene Abbrev.: Oli (no stop). **1.** (adjective) Denoting the third epoch of the Tertiary period. **2.** (noun; preceded by 'the') The Oligocene epoch.

Oligochaeta (cap. O) A class of annelid worms including the earthworms. Individual name and adjectival form: **oligochaete** (no cap.).

Oliphant, Sir Mark (or **Marcus**) **Laurence Elwin** (1901–) Australian physicist.

omasum (pl. omasa) The third compartment of a ruminant's stomach. Adjectival form: **omasal**.

OMC-1 Abbrev. for *Orion molecular cloud.

-ome Noun suffix denoting a mass, group, or part of a specified kind (e.g. biome, genome, rhizome).

omega Greek letter, symbol: ω (lower case), Ω (cap.).

ω Symbol for **1.** angular frequency (ω_L, ω_D = Larmor, Debye angular frequency, respectively). **2.** angular velocity. **3.** solid angle.

Ω Symbol for **1.** ohm. **2.** omega particle (Ω^-). **3.** solid angle.

omicron Greek letter, symbol: o (lower case), O (cap.).

ommatidium (pl. ommatidia) A unit of the compound eye of an arthropod. Adjectival form: **ommatidial**.

OMR Abbrev. for optical mark reading.

-on Noun suffix denoting **1.** an elementary particle or particles (e.g. electron, lepton, proton). **2.** a noble gas (e.g. argon, neon).

onco- Prefix denoting a tumour (e.g. oncology, oncotic).

oncogene A gene capable of converting a normal cell into a cancerous cell. Individual oncogenes are designated by an italicized lower-case three-letter label (*see* gene); viral oncogenes are specified by the hyphenated prefix v- (not ital.), e.g. v-*myb*; cellular oncogenes are prefixed by c-, e.g. c-*myc*. In certain families of oncogenes, hyphenated prefixes with initial capitals (not ital.) are part of the accepted designation. For instance, the *ras* oncogenes (from *rat s*arcoma), occurring in humans and rats, are designated K-*ras* (or Ki-*ras*; after the Kirsten sarcoma virus), Ha-*ras* (or H-*ras*; after the Harvey sarcoma virus), and N-*ras* (after neuroblastoma). In these cases any c- or v- prefix comes first, e.g. v-K-*ras*, v-Ha-*ras*.

oncogenic Causing tumour formation; the term is applied to some viruses, chemicals, and environmental factors. Genes responsible for the transformation of a normal cell into a cancer cell are called *oncogenes. Noun form: **oncogenesis**.

oncosphere (preferred to onchosphere) A tapeworm larva. Also called hexacanth.

oncovirus (one word) Any *retrovirus of the subfamily Oncovirinae. They are classified into four types, designated A, B, C, and D.

ondo- Prefix denoting a wave (e.g. ondometer, ondoscope).

-one Noun suffix denoting ketone (e.g. acetophenone, propanone).

one gene–one enzyme hypothesis (en dash)

online (one word) *Computing*

Onnes, Heike Kammerlingh Usually alphabetized as *Kammerlingh Onnes.

Onsager, Lars (1903–76) Norwegian-born US chemist.
Onsager equation

onycho- (**onych-** before vowels) Prefix denoting a nail or claw (e.g. onychomycosis, Onychophora).

oo- Prefix denoting an ovum or egg (e.g. oogamy, oogenesis, oogonium, oosphere, oospore).

Oomycetes (cap. O) A class of fungi in the subdivision *Mastigomycotina. In a recent classification these organisms are regarded as algae and classified in the division *Chromophyta. Individual name and adjectival form: **oomycete** (no cap.).

Oort, Jan Hendrik (1900–) Dutch astronomer.
Oort cloud
Oort constants

OP Abbrev. for osmotic pressure.

op-amp (or **opamp**) Short for operational amplifier.

Oparin, Aleksandr Ivanovich (1894–1980) Soviet biochemist.

open circuit (two words) Hyphenated when used adjectivally (e.g. open-circuit voltage).

open reading frame *Genetics* Abbrev.: ORF (no stops). A base sequence in a polynucleotide that has no stop codons and so is ‘open’ for translation.

open systems interconnection (not hyphenated) *Computing* Abbrev.: OSI (no stops).

operational taxonomic unit Abbrev.: OTU (no stops). An entity studied in numerical taxonomy.

operations research (preferred to operational research) Abbrev.: OR (no stops).

operator *Genetics* Symbol: *O* (ital.). Operators for leftwards or rightwards transcription units are designated by a subscript capital L or R (not ital.) plus, where appropriate, an identifying number, e.g. $O_{\mathrm{R}}1$, $O_{\mathrm{L}}3$.

operator-constitutive mutant *Genetics* Symbol: O^{c} (cap. ital. O, superscript lower-case c, not ital.)

operculum (pl. opercula) *Biol.* Adjectival forms: **opercular**, **operculate**.

Oph (no stop) *Astron.* Abbrev. for Ophiuchus.

ophio- (**ophi-** before vowels) Prefix denoting a snake or snakelike (e.g. Ophioglossales, ophiolite, Ophiuroidea).

Ophioglossales (cap. O) An order of ferns including the genus *Ophioglossum* (cap. O, ital.; the adder’s tongues). Adjectival form: **ophioglossoid**.

Ophiuchus A constellation. Genitive form: Ophiuchi. Abbrev.: Oph (no stop). *See also* stellar nomenclature.
Ophiuchids (meteor shower)

Ophiuroidea (cap. O) A class of echinoderms comprising the brittlestars. Individual name and adjectival form: **ophiuroid** (no cap.).

ophthalmic (not opthalmic)

ophthalmo- (**ophthalm-** before vowels) Prefix denoting the eye (e.g. ophthalmology, ophthalmoscope).

opistho- *Anat., zool.* Prefix denoting behind or posterior (e.g. opisthosoma, opisthotonos).

Oppel zones *Geol.* Named after Albert Oppel.

Oppenheimer, Julius Robert (1904–67) US physicist.
Born–Oppenheimer approximation (en dash)

optical character reader (or **recognition**) (not hyphenated) Abbrev.: OCR (no stops).

optical density Former name for *internal transmission density.

optical mark reading (not hyphenated) Abbrev.: OMR (no stops).

optical rotary dispersion (not hyphenated) Abbrev.: ORD (no stops).

optical rotation, angle of Symbol: α (Greek alpha). The angle through which the plane of plane-polarized light is rotated by a substance. The *SI unit is the *radian. The *molar optical rotatory power*, symbol α_n, has the unit radian square metre per mole.

optic axis (not optical axis) *Optics, crystallog.*

opto- Prefix denoting **1.** vision or the eye (e.g. optometer). **2.** optical phenomena or signals (e.g. optoelectronics, optoisolator).

OR 1. Abbrev. for operations research. **2.** *See* logic symbols; electronics, graphical symbols.

Oracle (cap. O) A trade name for the IBA's teletext system.

oral Relating to or administered by mouth (*see also* os). *Compare* aural.

orbital *Chem.* A region in space in which there is a high probability of finding an electron in an atom or molecule. The shapes of atomic and molecular orbitals are obtained from solution of the Schrödinger wave equation.

Atomic orbitals are designated s, p, d, f, g, . . . and correspond to values of the orbital angular momentum quantum number l of 0, 1, 2, 3, 4, ... The magnetic quantum number m_l is related to angular momentum along a chosen axis, usually the z axis by convention, and can take integer values $-l$ to $+l$. Thus, for $l = 1$, m can equal $-1, 0$, or $+1$, and there are three p orbitals possible, p_x, p_y, and p_z. Similarly, for $l = 2$, $m_l = -2, -1, 0, +1, +2$, giving the five d orbitals d_{z^2}, $d_{x^2-y^2}$, d_{xy}, d_{xz}, and d_{yz}. The principal quantum number n allows only values of l from zero to $n-1$; therefore, in the nth energy level there are n types of orbital. The spin quantum number s is allowed to take only the values $-\frac{1}{2}$ and $+\frac{1}{2}$ and each orbital can thus only contain two electrons with opposing spins.

The electronic configuration of an atom can be described symbolically by:

$$(nl)^K \ldots,$$

where n is the principal quantum number, l is the term symbol for the orbital angular momentum quantum number, and K the number of electrons in the orbital(s), e.g.

$$(1s)^2(2s)^2(2p)^6(3s)^2(3p)^6(4s)^2(3d)^{10}.$$

The electronic state of an atom is derived from the orbital angular momentum quantum number L, which is obtained by combining the orbital angular momenta of the individual electrons. Thus, for $L = 0, 1, 2, 3, 4, \ldots$ the corresponding term symbols are denoted by the upper case letters S, P, D, F, G, . . . , which parallels the term symbols for values of l.

A right subscript attached to the term symbol indicates the total angular momentum quantum number j or J. A left superscript indicates the spin multiplicity $2s+1$ or $2S+1$. Examples are $p_{1/2}$ and $^3D_2$.

Molecular orbitals are formed by the overlap of atomic orbitals. For linear molecules the lower-case Greek letters σ, π, δ, ϕ, . . . are used as term symbols designating the molecular orbitals corresponding to the atomic orbitals s, p, d, f, . . . ; similarly, the Greek capital letters Σ, Π, Δ, Φ, . . . correspond to the atomic term symbols S, P, D, F, ... For nonlinear molecules the term symbols a, b, e, t, . . . and A, B, E, T, . . . (from group theory) are used.

The combination of atomic orbitals with amplitudes of the same sign generates a **bonding orbital** (e.g. σ and π bonds); combination of atomic orbitals with amplitudes of opposite signs generates an **antibonding orbital**, denoted by an asterisk (e.g. σ^* and π^*).

The molecular term symbols can have superscripts and subscripts. A plus (+) or minus (−) sign as a right superscript indicates the symmetry as regards reflection in any plane through the axis of symmetry of the molecule; g or u as a right subscript indicates *gerade or *ungerade, respectively; and a left superscript indicates the spin multiplicity. A right subscript numeral

has different meanings for the term symbols of linear and nonlinear molecules: for linear molecules it indicates the total angular momentum quantum number; for nonlinear molecules it indicates a symmetry property. Examples are σ_g, $^3\Sigma^+_u$, $^3\Pi_2$, a_{2u}, and $^2T_{2g}$.
Atomic orbitals of different types but similar energies may combine to form **hybrid orbitals**. Common examples of hybrid orbitals are sp, p^2, sp^2, p^3, sp^3, dsp^3, and d^2sp^3. The superscript numerals indicate the number of each type of orbital involved, e.g. d^2sp^3 hybridization involves two d orbitals, one s orbital, and three p orbitals.
Heteroatomic molecules have electrons in orbitals that are not involved in the bonding system: such orbitals are designated **nonbonding** and given the symbol n, e.g. the 2p orbital of the oxygen atom in the carbonyl group.

orbital angular momentum quantum number Symbol: *l* (individual entity) or *L* (whole system). An integer that characterizes the orbital angular momentum of a particle, atom, nucleus, etc.
In the case of electrons within an atom, for each value of *n*, the *principal quantum number, *l* can only have the values:

$$0, 1, 2, 3, \ldots, (n-1).$$

Each value of *l* defines a sub-shell. In atomic spectroscopy, the electron states – atomic orbitals – corresponding to these values of *l* are respectively designated by the term symbols:

s, p, d, f, g, h, i, k, l, m, n, ...

Capital letters, S, P, D, F, . . . , are used for equivalent values of *L*. *See* orbital.

orbivirus (one word) Any virus belonging to the genus *Orbivirus* (cap. O, ital.). *See also* reovirus.

orchid Any plant of the family Orchidaceae. The word is used both specifically, with the appropriate qualifier (e.g. lady's slipper orchid), and more generally, for any member of the family. Adjectival form: **orchidaceous** (usually referring to the family).

ORD Abbrev. for optical rotary dispersion.

order A category used in biological classification (*see* taxonomy) consisting of a number of similar families (sometimes only one family). Names of orders are printed in roman (not italic) type with an initial capital letter and typically end in -ales for plants (e.g. Coniferales, Rosales). There is no uniform ending for animal orders, although within some classes endings tend to be consistent (e.g. -ptera for insect orders, -iformes for bird orders). *See also* suborder.

Ordnance Survey (not Ordinance) Abbrev.: OS (no stops).
Ordnance benchmark Abbrev.: OBM (no stops).
Ordnance datum Abbrev.: OD (no stops).

Ordovician Abbrev.: O (no stop). **1.** (adjective) Denoting the second period of the Palaeozoic era. **2.** (noun; preceded by 'the') The Ordovician period.

ORF *Genetics* Abbrev. for *open reading frame.

organic chemical nomenclature Both systematic and nonsystematic names are widely used in the nomenclature of organic compounds. The choice of which system to use depends on the context; strict systematic names should always be used in chemical contexts.
A systematic name comprises a root, a suffix, and one or more prefixes. The root is generally derived from the name of an aliphatic or cyclic hydrocarbon. If the root is an aliphatic hydrocarbon the longest carbon chain possible is chosen, e.g. 2-methylpentane, $CH_3CH(CH_3)CH_2CH_2CH_3$. The suffix is derived from the functional group. When a compound contains more than one functional group it is necessary to designate the principal one. The normal order for choosing the principal function is (in decreasing priority): carboxylic acid, sulphonic acid, acyl halide, amide, nitrile, aldehyde, ketone, alcohol, pheonol, thiol, amine. Thus CH_3COCH_2COOH (which contains both ketonic and carboxyl groups) is systematically named 3-oxobutanoic acid.
The names of substituents that replace hydrogen atoms in a compound are prefixed directly to the root, e.g. 2-chloromethylbenzene. However, if a compound is named by a general term then each part of the name

remains separate, e.g. ethyl ethanoate, $CH_3CH_2OOCCH_3$.
The names of substituents used as prefixes are generally arranged alphabetically, ignoring any associated numerical prefixes (e.g. mono-, di-, tri-, tetra-, etc.), e.g. trichloronitromethane, CCl_3NO_2. A composite prefix is treated as a unit (and numerical prefixes are therefore considered), e.g. 1-dimethylamino-2-ethylnaphthalene. If two or more prefixes start with the same letter, the shorter one precedes the longer, e.g. 2-methyl-1-methylaminonaphthalene.
The position of a substituent is indicated by a numeral (locant), obtained by numbering the carbon atoms in the parent compound (i.e. the root): the appropriate locant, followed by a hyphen, precedes each substituent name. The numbering is chosen to give the lowest locants possible to the substituents, and when a series of locants is required, the numbering is chosen to give the lowest first locant, e.g. $CH_3CH_2CH(CH_3)CH(CH_3)CH_2CH_2CH_2CH_2CH(CH_3)_2$ is named 2,7,8-trimethyldecane and not 3,4,9-trimethyldecane. Locants are often omitted when the structure of a compound can be deduced unambiguously without them, e.g. ethanol, CH_3CH_2OH, and butanone, $CH_3CH_2COCH_3$.
There are numerous methods of less systematic nomenclature for organic compounds, many specific only to certain classes of compound. The most common nomenclature is similar to that described for systematic nomenclature above, but with the following differences:
(1) Trivial names are used with the structural prefixes, *n*-, *s*-, *t*-, iso-, *o*-, *m*-, *p*-, etc., e.g. isobutyl chloride, $(CH_3)_2CHCH_2Cl$.
(2) The order of hydrocarbon substituents is sometimes cited in order of increasing complexity, e.g. 3-methyl-4-ethylheptane, $CH_3CH_2CH_2CH(CH_2CH_3)CH(CH_3)CH_2CH_3$.
(3) Greek letters are sometimes used to denote successive carbon atoms in the hydrocarbon skeleton, e.g. β-naphthalene. If a functional group is present, the carbon atom directly attached to the functional group is designated as α, e.g. α,β,β-tribromopropionic acid, $Br_2CHCHBrCOOH$.
More specific information on organic nomenclature is given under the names of types of compound (e.g. alcohol, aldehyde, etc.). *See also* alicyclic hydrocarbon; heterocyclic.

organic sulphide US: **organic sulfide.** Any of a class of organic compounds containing a sulphur atom linking two carbon atoms.
Organic sulphides are systematically named by adding the prefix thio- to the name of the root hydrocarbon, e.g. ethylthioethane, $CH_3CH_2SCH_2CH_3$.
In nonsystematic nomenclature organic sulphides are named by adding either of the words sulphide or thioether after the names of the hydrocarbon groups attached to the sulphur atom, e.g. diethyl thioether, $CH_3CH_2SCH_2CH_3$.

organo- Prefix denoting **1.** *Biol.* an organ or organs (e.g. organogenesis, organotrophic). **2.** *Chem.* a compound containing an organic group (e.g. organochlorine, organometallic, organophosphorus, organosilicone).

organoboron compound Any organic compound containing one or more carbon-to-boron bonds.
Several systems of nomenclature are in use and simple organoboron compounds are usually named by one of the two following methods. In the first method organoboron compounds are named as derivatives of borane, BH_3, e.g. trimethylborane, $(CH_3)_3B$, and dichloroethoxyborane, $Cl_2BOCH_2CH_3$. In the second method organoboron compounds are named as derivatives of boron or, if the compound contains oxygen, of the boron oxo acids, i.e. boric acid, $B(OH)_3$, boronic acid, $HB(OH_2)$, borinic acid, $H_2B(OH)$, e.g. methylboron trichloride, CH_3BCl_2, and ethyl dichloroborinate, $Cl_2BOCH_2CH_3$. The first system of nomenclature is recommended.

organ of *Corti Also called spiral organ.

organophosphorus compound Any organic compound containing one or more carbon-to-phosphorus bonds; however, ester and amide derivatives are often included.

Organophosphorus compounds are commonly named as derivatives of phosphorus hydrides (phosphine, PH_3, phosphorane, PH_5) or of phosphorus oxo acids (phosphorous acid, H_3PO_4, phosphonous acid, $HP(OH)_2$, phosphinous acid, $H_2P(OH)$, phosphoric acid, $(HO)_3PO$, phosphonic acid, $HPO(OH)_2$, phosphinic acid, $H_2PO(OH)$, and phosphine oxide, H_3PO). Compounds with carbon-to-phosphorus bonds are named as substitution products of the parent compound, e.g. diphenylphosphorane, $(C_6H_5)_2PH_3$, and methylphosphonic acid, $CH_3PO(OH)_2$. Esters of the oxo acids are named by the usual rules for ester nomenclature, e.g. triethyl phosphate, $(CH_3CH_2O)_3PO$. Compounds with phosphorus-to-halogen bonds or phosphorus-to-nitrogen bonds are named as acyl halides or amides, respectively, e.g. phenylphosphonous dichloride, $(CH_3)_2PCl$.

Ori (no stop) *Astron.* Abbrev. for Orion.

orient (verb; preferred to orientate) To align. Noun form: **orientation**. Adjectival form: **oriented** (preferred to orientated, especially in computing, e.g. object-oriented, machine-oriented, problem-oriented).

original equipment manufacturer Abbrev.: OEM (no stops).

O-ring (hyphenated) A rubber or metal ring used as a vacuum or pressure seal.

Orion A constellation. Genitive form: Orionis. Abbrev.: Ori (no stop). *See also* stellar nomenclature.

Orionids (meteor shower)

Orion molecular cloud Abbrev.: OMC-1 (hyphenated)

Orion nebula

Ornithischia (cap. O) An order of herbivorous dinosaurs comprising the bird-hipped dinosaurs. Individual name and adjectival form: **ornithischian** (no cap.).

ornitho- (**ornith-** before vowels) Prefix denoting a bird or birds (e.g. ornithology, ornithophily).

oro- Prefix denoting **1.** the mouth (e.g. oroanal, oronasal, oropharynx). **2.** a mountain or mountains (e.g. orocline, orographic).

orogenesis The process by which mountains are formed. The term **orogeny** is sometimes used synonymously, but this word is better reserved for a mountain-building period (e.g. the Caledonian orogeny). Adjectival form: **orogenic**.

Ørskovia Use **Oerskovia*.

Ørsted, Hans Christian Use *Oersted.

Ortelius, Abraham (1527–98) Flemish cartographer.

ortho- (**orth-** before vowels) Prefix denoting **1.** straight, correct, or upright (e.g. orthostatic, orthostichy, orthotropism). **2.** perpendicular or perpendiculars (e.g. orthocentre, orthogonal, orthorhombic). **3.** the most highly hydrated form of an oxo acid (e.g. orthophosphoric acid). The use of oxidation numbers is now recommended (e.g. orthophosphoric acid is phosphoric(V) acid). **4.** a diatomic molecule in which both nuclei have parallel spins (e.g. orthodeuterium, orthohydrogen).

ortho- (ital., always hyphenated) *Chem. See o-.*

orthophosphate Denoting a compound containing the ion PO_4^{3-}, e.g. sodium orthophosphate, Na_3PO_4. The recommended name is phosphate(V).

orthophosphoric acid H_3PO_4 The traditional name for phosphoric(V) acid.

orthophosphorous acid H_3PO_3 The traditional name for phosphonic acid.

Orthopoxvirus (cap. O, ital.) Approved name for a genus of *poxviruses. Vernacular name: vaccinia subgroup. Individual name: **orthopoxvirus** (no cap., not ital.).

Orthoptera (cap. O) An order of insects including grasshoppers, locusts, and crickets; formerly included cockroaches and mantids (now classified as *Dictyoptera). Individual name: **orthopteran** (no cap.). Adjectival form: **orthopterous** or **orthopteran**.

orthotropic Describing or relating to plant organs that exhibit **orthotropism**, a growth response directly towards or away from the stimulus. *Compare* orthotropous.

orthotropous Describing a plant ovule that develops in an upright position within the ovary. *Compare* orthotropic.

os 1. (pl. ossa) A bone. Adjectival form: **osseous. 2.** (pl. ora) A mouth or mouthlike part. Adjectival form: **oral. 3.** (pl. osar) *Geol.* An esker.

Os Symbol for osmium.

OS Abbrev. for **1.** Ordnance Survey. **2.** Old Style, used to indicate that a date has been calculated using the Julian calendar rather than the current Gregorian calendar. *Compare* NS.

-osan Noun suffix characteristic of polysaccharides (e.g. hexosan).

Osborn, Henry Fairfield (1857–1935) US palaeontologist.

oscillate To fluctuate; not to be confused with *osculate or ocellate (*see* ocellus). Noun form: **oscillation**. Adjectival form: **oscillatory**. Derived nouns: **oscillator, oscilloscope**.

Oscillatoriales (cap. O) An order of *cyanobacteria. Individual name and adjectival form: **oscillatorian** (no cap.).

osculate (of geometric figures) To touch at a point; not to be confused with *oscillate. Noun form: **osculation**. Adjectival form: **osculatory**.

osculum (pl. oscula) *Zool.* A mouthlike aperture. Adjectival forms: **oscular, osculate**.

-ose Noun suffix denoting a simple sugar (e.g. aldose, fructose, ribose).

O shell (not hyphenated) *Physics*

OSI *Computing* Abbrev. for open systems interconnection.

-oside Noun suffix denoting glycoside (e.g. methylglucoside, rhamnoside).

osmium Symbol: Os *See also* periodic table; nuclide.

osmo- Prefix denoting **1.** water or liquid movement or uptake (e.g. osmosis, osmoregulation). **2.** osmosis (e.g. osmometer).

osmosis (pl. osmoses) Adjectival form: **osmotic**.

osmotic potential Symbol: Ψ_o Also called solute potential (symbol: Ψ_s). *See* water potential.

osmotic pressure Symbol: Π (Greek cap. pi). Abbrev.: OP (no stops). The pressure required to prevent the flow of a pure solvent into a solution across a semipermeable membrane. The *SI unit is the *pascal. In biology, it was formerly used in calculations of water relations, but these are now expressed in terms of osmotic potential (*see* water potential).

Osteichthyes (cap. O) A class of vertebrates comprising the bony fishes. Individual name and adjectival form: **osteichthyan** (no cap.), although the common name is more widely used for individuals.

osteo- (sometimes **oste-** or **ost-** before vowels) Prefix denoting bone or bones (e.g. osteoarthritis, osteoblast, osteocyte, osteitis, osteoma, ostectomy).

ostiole A small pore in the fruiting bodies of some algae and fungi. *Compare* ostium.

ostium (pl. ostia) *Anat., biol.* A small hole or opening, as in a sponge or an arthropod heart. *Compare* ostiole.

Ostracoda (cap. O) A subclass of small bivalved crustaceans. Individual name and adjectival form: **ostracod** (no cap.).

Ostwald, Friedrich Wilhelm (1853–1932) German chemist.
Ostwald's dilution law
Ostwald process
Ostwald viscometer

ot- *See* oto-.

OTF *Photog.* Abbrev. for off the film.

oto- (**ot-** before vowels) Prefix denoting the ear (e.g. otocyst, otolith, otitis).

Otto, Nikolaus August (1832–91) German engineer.
Otto cycle
Otto engine

OTU Abbrev. for operational taxonomic unit.

Ouchterlony, O. T. G. (1914–) Swedish microbiologist and immunologist.

Oughtred, William (1575–1660) English mathematician.

ounce Symbol: oz A unit of mass used in the UK and USA, equal to $^1/_{16}$ pound or 28.3495 grams. This is the avoirdupois ounce and must be differentiated from the troy ounce (equal to 480 grains, 31.1035 grams) and the apothecaries' ounce (equal to the troy ounce). The apothecaries' ounce and its submultiples – the scruple (20 grains) and the drachm (60 grains) – are no longer used in the UK. *See also* fluid ounce.

ounce troy *See* ounce.

-ous Adjectival suffix denoting **1.** relating to (e.g. igneous, mucous, oestrous). **2.** having or containing (e.g. hydrous). **3.** *Chem.* traditionally, combined in the form having the lower of two possible oxidation numbers (e.g. chromous, cuprous); use of the oxidation number in the name is now recommended.

outcrop (one word)

outgassing (one word)

output (noun and adjective) Verb form: **outputs, outputting, output**.

overpotential (one word) Symbol: η (Greek eta). The difference between an electrode potential under practical conditions (e.g. in an electrolytic cell) and the value under reversible conditions. The *SI unit is the volt.

overwinter (verb; one word)

ovi- Prefix denoting **1.** an egg or ovum (e.g. oviduct, oviparity, ovipositor). *See also* ovo-. **2.** sheep (e.g. ovibovine).

ovo- Prefix denoting an egg or ovum (e.g. ovotestis, ovoviviparity). *See also* ovi-.

ovum (pl. ova) A mature female gamete (egg cell) of animals and humans. In botany, the word is sometimes used as a synonym for 'oosphere' (the female gamete of lower plants); for seed plants the term 'ovule' is used to include the egg cell together with its protective and nutritive tissue.

Owen, Sir Richard (1804–92) British anatomist and palaeontologist.

ox (no stop) Abbrev. for the dibasic oxalate ion often used in the formulae of coordination compounds, e.g. $[Cr(ox)_3]^{3-}$.

oxa- (**ox-** before vowels) Prefix denoting a heterocyclic compound in which the hetero atom is oxygen (e.g. oxabicycloheptane, oxetane).

oxalate *See* ox.

oxalic acid HOOCCOOH The traditional name for ethanedioic acid.

oxamide $H_2NCOCONH_2$ The traditional name for ethanediamide.

oxbow (one word)

oxidation (preferred to oxidization) Derived nouns: **oxidant, oxidate**.

oxidation number The number of electrons that an atom in a chemical compound is considered to have more or less than the number in the free atom. In ionic compounds, the oxidation number is the charge. Thus, in calcium chloride ($CaCl_2$) the calcium has an oxidation number of +2 and the chlorine an oxidation number of −1, $Ca^{2+}2Cl^-$. In covalent compounds the oxidation number is obtained by assigning the electrons to the more electronegative element of the pair in a bond. For example, in ammonia (NH_3) it is assumed that the compound is $N^{3-}3H^+$; nitrogen has an oxidation number of −3 and hydrogen an oxidation number of +1. In coordination complexes, the coordinate bonds with neutral ligands are considered to have no effect on the oxidation state. Thus, $Ni(CO)_4$ contains nickel with an oxidation number of 0. The ion $[Fe(CN)_6]^{4-}$ is regarded as an iron ion Fe^{2+} coordinated to six CN^- ions. The iron thus has an oxidation number of +2. Oxidation numbers are used in *inorganic chemical nomenclature.

oxidation–reduction (en dash) Abbrev.: redox (no stops). *Chem.*

oxidize (not oxydize) Noun form: ***oxidation** (preferred to oxidization). Derived noun: **oxidizer**.

oxidoreductase (one word) *See* enzyme nomenclature.

oxime Any of a class of organic compounds containing the group =CNOH, where the carbon atom is joined to two hydrocarbon groups or to a hydrocarbon group and a hydrogen atom.

Oximes are systematically named by adding the word oxime after the name of the corresponding aldehyde or ketone, e.g. ethanal oxime, CH_3CHNOH, and propanone oxime, $(CH_3)_2CNOH$.

In nonsystematic nomenclature oximes are named as in systematic nomenclature but using the trivial names of the corresponding aldehydes or ketones, e.g. acetaldehyde oxime, CH_3CHNOH. Alternatively, the respective endings -ehyde and -one of the trivial names of the aldehydes and ketones are replaced by -oxime, e.g. acetaldoxime and acetoxime.

oxirane Use epoxide.

oxo- (**ox-** before vowels) Prefix denoting hydroxy- (e.g. oxoheptanoic acid, oxime). Hydroxy- is recommended in most contexts.

oxo acid An acid in which the acidic hydrogen atom(s) are bound to oxygen atoms. The alternative generic name 'oxy acid' is not recommended.

Oxo acids are systematically named by naming the acids as polyoxo- derivatives, e.g. tetraoxosulphuric(VI) acid, H_2SO_4, and trioxosulphuric(VI) acid, H_2SO_3.

In nonsystematic nomenclature oxo acids have the suffix -ic or -ous, denoting a higher or lower oxidation state, respectively, e.g. sulphuric acid, H_2SO_4, and sulphurous acid, H_2SO_3.

The trivial names sulphuric acid, sulphurous acid, nitric acid, HNO_3, nitrous acid, HNO_2, and thiosulphuric acid, $H_2S_2O_3$, are recommended for general use.

For elements that form numerous oxo acids, e.g. phosphorus, certain oxo acids are given specific systematic names, e.g. $HOPH_2O$ has the trivial name phosphorous acid and the recommended name phosphonic acid.

3-oxobutanoic acid CH_3COCH_2-COOH The recommended name for the compound traditionally known as acetoacetic acid.

oxoethanoic acid CHOCOOH The recommended name for the compound traditionally known as glyoxalic acid.

Oxon (no stop) Abbrev. for Oxoniensis (Latin: of Oxford), used with academic awards.

2-oxopropanoic acid $CH_3COCOOH$ The recommended name for the compound traditionally known as pyruvic acid.

oxy- Prefix denoting **1.** oxygen (e.g. oxyacetylene, oxycyanogen, oxymercuration). **2.** the hydroxyl group (e.g. oxyacetic acid); the alternative hydroxy- is recommended in all contexts.

-oxy Noun suffix characteristic of alkoxy groups (e.g. butoxy, ethoxy).

oxy acid Use *oxo acid.

oxygen Symbol O The names dioxygen and trioxygen can be used for the forms O_2 and O_3. The traditional name for O_3, ozone, is still in use.

oxygen difluoride OF_2 The recommended name for the compound traditionally known as fluorine monoxide.

oxyhaemoglobin US: **oxyhemoglobin**. Symbol: HbO_2 (no stops).

-oyl Noun suffix characteristic of acyl groups (e.g. benzoyl, ethanoyl).

oz Symbol for ounce.

ozone O_3 The traditional name for trioxygen.

P

p **1.** Symbol for **i.** electron state $l = 1$ (*see* orbital angular momentum quantum number). **ii.** *Biochem.* a terminal *phosphate in a polynucleotide. **iii.** pico-. **iv.** proton ($\bar{p}$ = antiproton). **2.** Abbrev. for p-type (semiconductor).

p Symbol for **1.** (bold ital.) electric dipole moment. **2.** (light ital.) momentum (in nonvector equations; p_i = generalized momentum); in vector equations it is printed in bold italic type (***p***; includes generalized momentum). **3.** (bold ital.) permanent dipole moment of a molecule ($\boldsymbol{p}_i$ = induced dipole moment of a molecule). **4.** (light ital.) pressure, including sound pressure (p_B = partial pressure of substance B). **5.** (light ital.) pyranose (*see* sugars).

p. (stop) Abbrev. for page. *See also* Appendix 9 (for style of page numbers in references).

P 1. Symbol for **i.** *Genetics* parental generation. **ii.** perianth (in a *floral formula). **iii.** peta-. **iv.** *Biochem.* *phosphate (P_i = orthophosphate). **v.** phosphorus. **vi.** *phytochrome (P_{fr} = phytochrome absorbing far red light; P_r = phytochrome absorbing red light). **vii.** poise. **viii.** proline. **2.** Abbrev. for Permian.

P Symbol (light ital.) for **1.** parity. **2.** power, including radiant power, sound power (also P_a) (P_s = active power, P_q = reactive power). **3.** *Genetics* *promoter (P_{RE} = repressor establishment promoter, P_{RM} = repressor maintenance promoter).

***p*-** (ital., always hyphenated; preferred to *para-*) *Chem.* Prefix denoting *para*, indicating 1,4-substitution on a benzene ring (e.g. *p*-cresol, *p*-dihydroxybenzene). The use of numbered substitution positions is recommended (e.g. 1,4-cresol). *Compare m-*; *o-*.

P680 Chlorophyll *a* with a light absorption peak of 684 nm. *See* photosystem.

P700 Chlorophyll *a* with a light absorption peak of 700 nm. *See* photosystem.

Pa Symbol for **1.** pascal. **2.** protactinium.

PA Abbrev. for public-address (system).

PAABA Abbrev. for *p*-acetamidobenzoic acid.

PABA Abbrev. for *p*-aminobenzoic acid.

PABP Abbrev. for poly(A)-binding protein.

PABX *Telephony* Abbrev. for private automatic branch exchange.

pachy- Prefix denoting thick (e.g. pachycaul, pachytene).

Pacific Standard Time Abbrev.: PST (no stops).

Pacini, Filippo (1812–83) Italian anatomist.

Pacinian (US **pacinian**) **corpuscles**

packet assembler/disassembler (solidus) *Computing* Abbrev.: PAD (no stops).

PAD *Computing* Acronym for packet assembler/disassembler.

paedo- (paed- before vowels) US: **pedo- (ped-)**. Prefix denoting **1.** children (e.g. paediatrics). **2.** larval or immature forms (e.g. paedogenesis).

paedogenesis US: **pedogenesis**. Sexual reproduction in a larval or immature form of an animal. *Compare* pedogenesis.

paeony Use *peony.

page Abbrev.: p. (stop; pl. pp.). *See also* Appendix 9 (for style of page numbers in references).

Page, Sir Frederick Handley (1885–1962) British aircraft designer.

PAGE Abbrev. for polyacrylamide gel electrophoresis.

Pal Abbrev. for Palaeocene.

PAL Acronym for **1.** *Television* phase alternation line, the colour TV system generally adopted in Europe. **2.** present atmospheric level.

Palade, George Emil (1912–) Romanian-born US physiologist and cell biologist.

palaeo- (sometimes **palae-** before vowels) US: **paleo- (pale-)**. Prefix denoting old or ancient (e.g. palaeobotany, palaeoclimatology, palaeoecology, palaeontology). The US spelling is often used in Britain in the names of geological time units and rock formations. However, this dictionary follows the recommendation of the Geological Society in retaining the British spelling for these names (e.g. Palaeocene, not Paleocene).

Palaeocene US: **Paleocene** (*see* palaeo-). Abbrev.: Pal (no stop). **1.** (adjective) Denoting the earliest epoch of the Tertiary period. **2.** (noun; preceded by 'the') The Palaeocene epoch. In some older schemes it was included in the *Eocene epoch.

Palaeogene US: **Paleogene** (*see* palaeo-). Abbrev.: Pg (no stop). **1.** (adjective) Denoting the older subperiod of the Tertiary period, comprising the Palaeocene, Eocene, and Oligocene epochs. **2.** (noun; preceded by 'the') The Palaeogene subperiod. In some recent schemes the Palaeogene has the status of a period (*see* Tertiary).

palaeognathous US: **paleognathous.** Describing the palate structure of the *ratites (flightless birds). These birds were formerly classified as the superorder Palaeognathae; the term palaeognathous is now used as an anatomical (rather than taxonomic) term. *Compare* neognathous.

Palaeozoic US: **Paleozoic** (*see* palaeo-). Abbrev.: Pz (no stop). **1.** (adjective) Denoting an era in the geological time scale. **2.** (noun; preceded by 'the') The Palaeozoic era. In some schemes this time is separated into two eras: the Lower Palaeozoic (initial caps.), comprising the Cambrian, Ordovician and Silurian periods; and the Upper Palaeozoic (initial caps.), comprising the Devonian, Carboniferous, and Permian periods.

paleo- US spelling of *palaeo-.

palladium Symbol: Pd *See also* periodic table; nuclide.

pallium (pl. pallia) The cerebral cortex, which is divided into several regions, the most evolutionary advanced of which is the **neopallium.** Adjectival form: **pallial.**

Palmae (not Palmaceae) A family of monocotyledonous plants comprising the palms. Alternative name: Arecaceae. Individual name and adjectival form: **palm.**

palmitic acid $CH_3(CH_2)_{14}COOH$ The traditional name for hexadecanoic acid.

Palus *See* mare.

PAM (or **p.a.m.**) Abbrev. for pulse amplitude modulation.

2-PAM Abbrev. for 2-pyridinealdoxime methiodide.

pampas (takes a pl. or sing. form of verb)

PAN Abbrev. for 1-(2-pyridylazo)-2-naphthol.

pan- Prefix denoting all (e.g. panchromatic, pandemic, pangenesis, panmixis). *See also* panto-.

pancreozymin Use *cholecystokinin.

pandemic 1. (adjective) Describing a widespread *epidemic disease affecting large numbers of people in different countries simultaneously. The term should be reserved for human diseases; the equivalent term for animal diseases is panzootic. **2.** (noun) A pandemic disease. *Compare* endemic.

Paneth, Friedrich Adolf (1887–1958) Austrian chemist.
Paneth reaction

Paneth, Josef (1857–90) Austrian physiologist.
Paneth cells

Pangaea (preferred to Pangea) *Geol.*

Panofsky, Wolfgang Kurt Hermann (1919–) German-born US physicist.

panto- Prefix denoting all (e.g. pantocolpate, pantograph, pantonematic). *See also* pan-.

panzootic 1. (adjective) Describing a widespread *epizootic disease, affecting large numbers of animals in several countries simultaneously. **2.** (noun) A panzootic disease. *See also* pandemic. *Compare* enzootic.

Papanicolaou, George Nicholas (not Papanicolau) (1883–1962) Greek-born US anatomist.
Pap test

papilla (pl. papillae) Adjectival forms: **papillary, papillate, papillose.**

Papillomavirus (cap. P, ital.) A genus of *papovaviruses. Individual name: **papillomavirus** (no cap., not ital.).

Papin, Denis (1647–*c.* 1712) French physicist and inventor.

papovavirus (one word) Any member of the family Papovaviridae (vernacular name: papovavirus group). [from *pa*pilloma, *po*lyoma, and *va*cuolating agent – an early name for *SV40]

Pappenheimer, Alwin M. (1908–) US biochemist.
Pappenheimer bodies

pappus (pl. pappi) A ring of modified sepals, usually in the form of fine hairs. Adjectival form: **pappose.**

Pappus of Alexandria (*fl.* early 4th century AD) Greek mathematician.

Pappus' rules
Pappus' theorems

Pap test (cap. P) *Med.* Named after G. N. *Papanicolaou.

PAR Abbrev. for 4-(2-pyridylazo)-resorcinol.

para- (**par-** before vowels and h) Prefix denoting **1.** beside (e.g. parameter, parapodium, parathyroid, paraxial, parhelion). **2.** resembling (e.g. parautochthonous, paraphysis). **3.** opposing, contrary, or abnormal (e.g. paraphase, paraprotein, parasexual, parasympathetic, *parenteral). **4.** a polymer or isomer of a compound (e.g. paraldehyde). **5.** a diatomic molecule in which both nuclei have opposite spins (e.g. paradeuterium, parahydrogen).

para- (ital., always hyphenated) *Chem. See p-.*

-para *Med.* Noun suffix denoting a woman in relation to the number of children she has borne (e.g. multipara, nullipara). Adjectival form: ***-parous**.

parabola (pl. parabolas or parabolae) Adjectival form: **parabolic**. Derived noun: **paraboloid**. Adjectival form: **paraboloid** or **paraboloidal**.

A parabola is a geometric curve, a paraboloid is a curved surface. The adjective parabolic is often used in optics, astronomy, etc., to describe a paraboloid surface (as in parabolic mirror, parabolic dish), since it is the parabolic sections of the surface that are relevant.

Paracelsus, Philippus Aureolus (1493–1541) German physician, chemist, and alchemist. Also called Theophrastus Bombastus von Hohenheim.

paraffin (not parrafin) **1.** The traditional name for an alkane. **2.** (or **paraffin oil** or **kerosene**) A mixture of hydrocarbons obtained from petroleum, with 11–12 carbon atoms and a boiling-point range of 160–250°C.

paraformaldehyde $(CH_2O)_n{\cdot}xH_2O$, where $n > 6$. The traditional name for poly(methanal).

parainfluenza virus (two words) Abbrev.: PI (no stops). Any of the certain *paramyxoviruses. Most serovars are designated by an Arabic numeral, e.g. parainfluenza virus 2.

paraldehyde The traditional name for *ethanal trimer.

parallax Symbol: π (Greek pi). The angular displacement in the apparent position of a body when viewed from two widely separated points. In astronomy, it is measured in arc seconds. Adjectival form: **parallactic**.

parallel (not paralel)

parallelepiped (not parallelopiped)

parallel input/output *Computing* Abbrev.: PIO (no stops).

parallelogram (not paralellogram)

parameter A combination of physical quantities that is useful in characterizing the behaviour or properties of a system. An example is Grüneisen parameter. A dimensionless parameter is usually described as a number, as in Mach number or Reynolds number.

The symbol for a dimensionless parameter may be a two-letter combination; this is printed in italic type as with single-letter symbols of physical quantities. The use of roman type to distinguish a two-letter symbol from the product of two single-letter symbols is now discouraged. When occurring in a product, a two-letter symbol should be separated from other symbols by a thin space, a multiplication sign, or brackets.

paramyxovirus (one word) Any virus belonging to the genus *Paramyxovirus* (cap. P, ital.). Authors should avoid extending usage to include any other member of the family Paramyxoviridae, which also contains the genera *Morbillivirus* and *Pneumovirus*. *See also* parainfluenza virus.

Paranthropus (cap. P, ital.) The generic name originally given to fossil remains of hominids now usually classified as *Australopithecus robustus* (*see Australopithecus*). *Paranthropus* is retained as a separate genus by some authorities.

parapodium (pl. parapodia) A paired appendage in polychaete worms, bearing chaetae. Adjectival form: **parapodial**.

Parapoxvirus (cap. P, ital.) Approved name for a genus of *poxviruses. Vernacular name: orf subgroup. Individual name: **parapoxvirus** (no cap., not ital.).

paraquat (not cap. P) A substance containing 1,1′-dimethyl-4,4′-dipyridylium salts, used as a herbicide.

parathyroid hormone (preferred to parathormone) Abbrev.: PTH (no stops).

Parazoa (cap. P) A subkingdom of invertebrates comprising the sponges (phylum *Porifera). Individual name and adjectival form: **parazoan** (no cap.).

parental generation (in breeding experiments) Abbrev.: P (no stop). *Compare* filial generation.

parenteral *Med.* Denoting any route of administration of drugs, etc., other than by mouth. Not to be confused with **parental** (relating to parentage).

parenthesis (pl. parentheses) *See* brackets.

parhelion (pl. parhelia) *Meteorol.*

pari- Prefix denoting even or equal (e.g. paripinnate, parity).

parity Symbol: P The property of a wave function determining its symmetry under spatial reflection. If the wave function changes sign then $P = -1$, otherwise $P = +1$. *See also* gerade; ungerade.

Parkes, Alexander (1813–90) British chemist and inventor.
Parkes process

Parkinson, James (1755–1824) British surgeon and palaeontologist.
***parkinsonism**
Parkinson's disease
Parkinson's syndrome Use parkinsonism.

parkinsonism (no cap.; preferred to Parkinson's syndrome) A neurological condition, characterized by tremor, rigidity, etc., with many causes, including psychoactive drug therapy, encephalitis, and **Parkinson's disease** (a degenerative disorder of the basal ganglia of the brain). Adjectival form: **parkinsonian.** Named after J. *Parkinson.

-parous Adjectival suffix denoting giving birth or bearing (e.g. multiparous, viviparous). Noun form: **-parity**. *Compare* -para.

parse (**parsing, parsed**) *Computing, etc.* Derived noun: **parser**.

parsec Symbol: pc A unit of length used in astronomy. Although not an *SI unit, it may be used with the SI units and SI prefixes can be attached to it, e.g. megaparsec (Mpc). One parsec is approximately equal to 3.0857×10^{16} m, 2.0626×10^{5} AU, 3.2616 l.y. *See also* astronomical unit.

Parsons, Sir Charles Algernon (1854–1931) British engineer.

Parsons, William *See* Rosse, Lord.

partheno- Prefix denoting the absence of fertilization (e.g. parthenocarpy, parthenogenesis).

partial pressure of a gas Symbol: p_B (for a gas B) or, when a complicated formula is to be written, p(B), as in $p(N_2O_4)$. A physical quantity, the product of the *mole fraction, x_B, of gas B and the total *pressure, p, of the mixture. The *SI unit is the pascal.

particle 1. *Physics* A system of definite mass and spin: a hydrogen molecule or a helium nucleus, each in a definite energy level, is just as much a particle as an electron or proton. Particles that are not compound, such as the electron (but not the proton), are called *fundamental or elementary particles; protons and other hadrons, which are composed of quarks, should be referred to as subatomic particles. **2.** Any minute quantity of matter.

particle fluence Symbol: Φ (Greek cap. phi). A physical quantity associated with nuclear reactions and ionizing radiation. It is the number of particles that, within a time interval, fall on a small sphere at a given point, divided by the cross-sectional area of the sphere. The *SI unit is the reciprocal of the square metre (m^{-2}). For a specified particle, such as a proton or electron, the term used is **proton fluence**, **electron fluence**, etc. The particle fluence rate (or particle flux density), symbol ϕ (lower-case phi), is equal to $d\Phi/dt$; it is measured in $m^{-2}\ s^{-1}$.

By analogy, **energy fluence** at a given point and within a time interval is the sum of the energies (excluding rest energies) of all incident particles divided by the cross-sectional area of the sphere. The SI unit is the joule per square metre and the symbol is Ψ (Greek cap. psi). The energy fluence rate or (or energy flux density), symbol ψ (lowercase psi), is equal to $d\Psi/dt$; it is measured in $J\ m^{-2}\ s^{-1}$).

particle flux density Another name for particle fluence rate. *See* particle fluence.

Partington, James Riddick (1886–1965) British chemist.

parvovirus (one word) Any virus belonging to the genus *Parvovirus* (cap. P, ital.). Authors should avoid extending usage to include any other member of the family Parvoviridae, which contains two other genera.

pascal (no cap.) Symbol: Pa The *SI unit of pressure and stress.

$$1\ Pa = 1\ N\ m^{-2} = 1\ J\ m^{-3}.$$

Fluid pressure is often measured in bars: one pascal = 10^{-5} bar. Named after Blaise *Pascal.

Pascal (or **PASCAL**) A computer language, named after Blaise *Pascal.

Pascal, Blaise (1623–62) French mathematician, physicist, and religious philosopher.

***pascal**
***Pascal**
Pascal's law *Physics*
Pascal's theorem *Maths.*
Pascal's triangle

Paschen, L. C. H. Friedrich (1865–1947) German physicist.

Paschen–Back effect (en dash) Also named after Ernst Back (1881–1959).
Paschen curve
Paschen series
Paschen's law

Passeriformes (cap. P) An order comprising the perching birds. Individual name and adjectival form: **passerine** (no cap.).

Pasteur, Louis (1822–95) French chemist and microbiologist.

Pasteur effect
******Pasteurella***
******Pasteuria***
pasteurization

Pasteurella (cap. P, ital.) A genus of bacteria of the family Pasteurellaceae.

P. pestis: now reclassified as *Yersinia pestis*.
P. tularensis: now reclassified as *Francisella tularensis*.
Individual name: **pasteurella** (no cap., not ital.; pl. pasteurellae).

Pasteuria (cap. P, ital.) A genus of Gram-positive mycelium-forming bacteria. The name *P. ramosa* has been applied to two quite different organisms: (1) a bacterial parasite of cladocerans, named by Metchnikoff in 1888; (2) a budding bacterium subsequently found on *Daphnia* spp. and mistaken for Metchnikoff's bacterium. The latter bacterium has now been assigned to the **Blastocaulis-Planctomyces* group and renamed *Planctomyces staleyi*. The other *Pasteuria* species were for a long time erroneously classified as protozoa (i.e. *Duboscqia penetrans*); they are now assigned to two species – *P. penetrans* and *P. thornei*.

patella (pl. patellae) The kneecap. Adjectival forms: **patellar, patellate**.

patera (not ital.; pl. paterae) *Astron.* A volcanic structure with a relatively shallow profile and a central depression. The word, with an initial capital, is used in the approved Latin name of such a feature on a planet or satellite, as in Alba Patera on Mars and Loki Patera on Io.

patho- Prefix denoting disease (e.g. pathogen).

pathotype Use *pathovar.

pathovar (preferred to pathotype) Abbrev.: pv. (stop). A patho(logical) var(iety): an unofficial category of classification used in microbiology and ranking below subspecies. Pathovars are strains having distinct pathogenic attributes for specific hosts.

Pattinson, Hugh Lee (1796–1858) British metallurgical chemist.

Pattinson process

Pauli, Wolfgang (1900–58) Austrian-born Swiss physicist.

Pauli exclusion principle

Pauli matrices
Pauli paramagnetism

Pauling, Linus Carl (1901–) US chemist.

-pause Noun suffix denoting a boundary (e.g. gravipause, magnetopause, stratopause).

Pav (no stop) *Astron.* Abbrev. for Pavo.

Pavlov, Ivan Petrovich (1849–1936) Soviet physiologist. Adjectival form: **Pavlovian.**

Pavo A constellation. Genitive form: Pavonis. Abbrev.: Pav (no stop). *See also* stellar nomenclature.

Payen, Anselme (1795–1871) French chemist.

payload (one word)

Pb *Chem.* Symbol for lead. [from Latin *plumbum*]

PBX *Telephony* Abbrev. for private branch exchange.

pc Symbol for parsec. *See also* SI units.

PC Abbrev. for **1.** personal computer. **2.** printed circuit.

PCB Abbrev. for **1.** polychlorinated biphenyl. **2.** printed circuit board.

PCC Abbrev. for pyridinium chlorochromate.

PCM (or **pcm**) Abbrev. for pulse code modulation.

PCR Abbrev. for polymerase chain reaction.

Pd Symbol for palladium.

p.d. Abbrev. for potential difference.

PDC Abbrev. for pyridinium dichromate.

PDGF Abbrev. for platelet-derived growth factor.

pdl Symbol for poundal.

PDT Abbrev. for 3-(2-pyridyl)-5,6-diphenyl-1,2,4-triazine.

Pe (italic type) Symbol for Peclet number.

PE Abbrev. for potential energy.

Peachey, (Eleanor) Margaret *See* Burbidge, (Eleanor) Margaret.

Peano, Giuseppe (1858–1932) Italian mathematician and logician.
Peano curve
Peano's axioms

Pearson, Karl (1857–1936) British statistician.
Pearson correlation Also called product-moment correlation.

peck *See* bushel.

Peclet, J. C. E. (not Péclet) (1793–1857) French physicist.
Peclet number

Peclet number Symbol: *Pe* A dimensionless quantity equal to the product of the *Reynolds number and the *Prandtl number. *See also* parameter. Named after J. C. E. Peclet.

pecten *Zool.* **1.** (pl. pectines) A comblike structure, especially a vascular structure in the eyes of reptiles and birds. Adjectival form: **pectinate**. **2.** (pl. pectens) Any scallop of the genus *Pecten* (cap. P, ital.). *Compare* pectin.

pectin *Biochem.* A polysaccharide occurring in plant cell walls, used as a gelling agent. This and related plant polysaccharides are called **pectic substances**. *Compare* pecten.

ped- *See* pedi-; pedo-.

-ped (or **-pede**) Noun suffix denoting a foot or feet (e.g. biped, centipede). Adjectival form: **-pedal**.

pedi- (**ped-** before vowels) Prefix denoting a foot or footlike (e.g. pedicel, pedipalp).

Pediococcus (cap. P, ital.) A genus of lactic acid-producing coccoid bacteria. Individual name: **pediococcus** (no cap., not ital.; pl. pediococci).

pedo- (**ped-** before vowels) **1.** Prefix denoting soil (e.g. pedocal, pedogenesis, pedalfer). **2.** US spelling of *paedo-.

pedogenesis 1. Soil formation. Adjectival form: **pedogenic**. **2.** US spelling of paedogenesis.

Pedomicrobium (cap. P, ital.) A genus of prosthecate bacteria. Individual name: **pedomicrobium** (no cap., not ital.; pl. pedomicrobia).

Peg (no stop) *Astron.* Abbrev. for Pegasus.

PEG Abbrev. for polyethylene glycol.

Pegasus A constellation. Genitive form: Pegasi. Abbrev.: Peg (no stop). *See also* stellar nomenclature.

Peierls, Sir Rudolph Ernst (1907–) German-born British theoretical physicist.

Peirce, Charles Sanders (1839–1914) US philosopher, logician, and mathematician.
Peirce's law

Peking man *See Homo.*

pelargonic acid $CH_3(CH_2)_7COOH$ The traditional name for nonanoic acid.

pelargonium (pl. pelargoniums) Any plant of the genus *Pelargonium* (cap. P, ital.), especially any of the horticultural varieties and hybrids popularly known as *geraniums.

P element (cap. P) *Genetics* A transposable element (*see* transposon) occurring in *Drosophila melanogaster*. They are so named because the deleterious effects of such elements are only apparent when a so-called P (paternally contributing) male is crossed with an M (maternally contributing) female, not vice versa.

Peligot, Eugene Melchior (1811–90) French chemist.

Pell, John (1610–85) British mathematician.
Pellian equation

Pelletier, Pierre Joseph (1788–1842) French chemist.

Peltier, Jean Charles Athanase (1785–1845) French physicist.
Peltier coefficient Symbol: Π_{ab} (Greek cap. pi, substances a and b)
Peltier effect
Peltier element

pelvis (pl. pelves) **1.** The bony structure formed by the two hip bones (the **pelvic girdle**), the sacrum, and the coccyx. The term is also used loosely to denote this region of the body. **2.** Any basin-shaped anatomical structure, as in the kidney (renal pelvis). Adjectival form: **pelvic**.

PEMA Abbrev. for 2-phenyl-2-ethylmalonamide.

Pen (no stop) *Geol.* Abbrev. for Pennsylvanian.

Penck, Albrecht (1858–1945) German geographer and geologist.

pene- (pen- before vowels) Prefix denoting almost (e.g. peneplain, peninsula, penumbra).

penicillin An antibiotic; not to be confused with its source, **Penicillium*.

Penicillium (cap. P, ital.) A genus of moulds (*see* Hyphomycetes) important as the source of penicillin.

peninsula (pl. peninsulas) Adjectival form: **peninsular**.

Penney, William George, Lord (1909–91) British mathematician.
Kronig–Penney model (en dash) Also named after R. de L. Kronig.

Pennsylvanian Abbrev.: Pen (no stop). **1.** (adjective) Denoting a US subperiod in the geological time scale corresponding approximately to the Upper Carboniferous subperiod. **2.** (noun; preceded by 'the') The Pennsylvanian subperiod.

Penrose, Roger (1931–) British mathematician and theoretical physicist.
Penrose tiling

Pensky–Martens test (en dash) *Chem.*

penta- Prefix denoting five (e.g. pentadactyl, pentagon, pentamerous, pentavalent). *See also* quinque-.

pentacyanonitrosylferrate(II) A compound containing the ion $Fe(CN)_5NO^{2-}$, e.g. potassium pentacyanonitrosylferrate(II), $K_2Fe(CN)_5NO$. The traditional name is nitroprusside.

pentaerythritol triacrylate Abbrev.: PETA (no stops).

pentane-2,4-dione $CH_3COCH_2COCH_3$ The recommended name for the compound traditionally known as acetylacetone.

pentanoic acid $CH_3(CH_2)_3COOH$ The recommended name for the compound traditionally known as *n*-valeric acid.

Pentaphane *See* Penton.

Penton (cap. P) A trade name for a chlorinated polyether with a high chemical resistance. The film form is called **Pentaphane**.

pentose phosphate pathway Abbrev.: PPP (no stops). Also called hexose monophosphate shunt, phosphogluconate pathway.

pentyl- *Prefix denoting the group $C_5H_{11}-$. Isomers include *n*-pentyl, CH_3-$(CH_2)_3CH-$, isopentyl, $(CH_3)_2CHCH_2$-CH_2-, neopentyl, $(CH_3)_3CCH_2-$, and *t*-pentyl, $(CH_3)_2(CH_3CH_2)C-$. Examples are *n*-pentyl alcohol, neopentyl alcohol, isopentylbenzene.

penumbra (pl. penumbras; preferred to penumbrae) Adjectival form: **penumbral.**

Penzias, Arno Allan (1933–) US astrophysicist.

peony (not paeony) Generic name: *Paeonia* (cap. P, ital.; not *Peonia*).

PEP Abbrev. for phosphoenolpyruvate or phosphoenolpyruvic acid.
PEP carboxylase Abbrev. and preferred form for phosphoenolpyruvate carboxylase.

pepo (pl. pepos) A type of fleshy fruit characteristic of the Cucurbitaceae.

peptide A compound consisting of two or more amino acids linked together by covalent bonds. The number of constituent amino acids in individual peptides is indicated by an appropriate prefix, which may be exact, e.g. dipeptide (two), pentapeptide (five), decapeptide (ten); or approximate, e.g. oligopeptide (few), polypeptide (many; proteins are polypeptides). The three-letter abbreviations for amino acids are used in depicting peptide sequences. In a known sequence, the abbreviations are joined by hyphens; if the sequence is not known, they are separated by commas. For example, the heptapeptide

Gly-Phe(Gly,Tyr,Ser)Val-Ala

contains two amino acids of known sequence, then three of unknown sequence, followed by two of known sequence. The glycine on the left carries the free amino group; the alanine on the right carries the free carboxyl group. The position of an amino acid on a peptide chain is indicated by a numerical suffix (e.g. Ser135, Tyr35).

Peptococcus (cap. P, ital.) A genus of anaerobic coccoid bacteria. Individual name: **peptococcus** (no cap., not ital.; pl. peptococci). The species *P. asaccharolyticus*, *P. prevotii*, *P. magnus*, and *P. indolicus* have recently been transferred to the genus **Peptostreptococcus*, leaving *P. niger* as the sole member.

Peptostreptococcus (cap. P, ital.) A genus of anaerobic coccoid bacteria. Individual name: **peptostreptococcus** (no cap., not ital.; pl. peptostreptococci). *See also Peptococcus.*

Per (no stop) *Astron.* Abbrev. for Perseus.

per- Prefix denoting **1.** through, throughout, or thoroughly (e.g. perennial, perfoliate). **2.** *Chem.* **a.** a compound containing an excess of a specified element (e.g. peroxide). **b.** traditionally, an element in a higher oxidation state (e.g. perchlorate, permanganate); systematic names are now preferred. *Compare* peroxy-.

peracetic acid ($CH_3C(O)OOH$) The traditional name for peroxyethanoic acid.

peracids Oxo acids containing either of the groups $(O-O)^{2-}$ or $(O-OH)^-$, e.g. perdisulphuric acid, $H_2S_2O_8$. The recommended name is peroxoacids.

perbenzoic acid $C_6H_5C(O)OOH$ The traditional name for peroxybenzoic acid.

perbromic acid $HBrO_4$ The traditional name for peroxobromic(VII) acid.

perceived noise decibal Symbol: PNdB (no stops).

per cent (two words) US: **percent.** Symbol: %
percentage (one word)
percentile (one word)

perchloric acid $HClO_4$ The traditional name for peroxochloric(VII) acid.

perdisulphuric acid $H_2S_2O_8$ US: **perdisulfuric acid.** The traditional name for peroxodisulphuric(VI) acid.

Perey, Marguerite Catherine (1909–75) French nuclear chemist.

perfect fungi Fungi that undergo sexual reproduction at some stage of their life cycle. *Compare* imperfect fungi.

peri- Prefix denoting **1.** about or surrounding (e.g. perianth, pericarp, perimeter, peristome). **2.** near or nearest (e.g. perigee, perihelion).

peri- (ital., always hyphenated) Prefix sometimes used to denote 1,8-substitution on the naphthalene ring (e.g. *peri*-dinitronaphthalene). *Compare amphi-; ana-; epi-; kata-; pros-.*

pericardium (pl. pericardia) *Anat.* Adjectival form: **pericardial** (not pericardiac).

perihelion (pl. perihelia) *Astron.*

period Symbol: T A physical quantity, the duration of one cycle or oscillation of a periodic phenomenon, i.e. the reciprocal of *frequency. The *SI unit is the *second. Also called periodic time.

periodic table The table of elements arranged in order of increasing proton number. In the *short form* (given in Appendix 6), the *lanthanoids and *actinoids are shown separately. In the vertical **groups**, numbered IA to VIIB, and 0 for the noble gases, the elements all have the same number of electrons in the outer shell. The horizontal **periods** contain elements with the same number of electron shells with an increasing number of electrons in the outer shell, moving from left to right. The first three periods are called **short periods**, the next four are **long periods**.

period–luminosity relation (en dash) *Astron.* Abbrev.: P–L relation.

peripheral nervous system Note that this is not abbreviated to PNS; it should always be spelt out in full.

Perissodactyla (cap. P) An order of hoofed mammals comprising the odd-toed ungulates (e.g. rhinos, horses). Individual name and adjectival form: **perissodactyl** (no cap.; not perissodactylous).

Perkin, Sir William Henry (1838–1907) British chemist.

Perkin reaction

Perkin's mauve Also called mauveine.

Perkin, William Henry, Jr (1860–1929) British chemist, son of Sir William Perkin.

Perkin reaction

Perkin synthesis

permanganate A compound containing the ion MnO_4^-, e.g. potassium permanganate, $KMnO_4$. The recommended name is manganate(VII).

permeability Symbol: μ (Greek mu). A property of a medium equal to the *magnetic flux density in it divided by the *magnetic field strength, $\boldsymbol{B}/\boldsymbol{H}$. The *SI unit is the henry per metre or newton per square ampere.

The permeability of vacuum, symbol μ_0, is a fundamental constant:

$$\mu_0 = 4\pi \times 10^{-7}\,\mathrm{H\,m^{-1}}$$
$$= 12.566\,370\,614\,4 \times 10^{-7}\,\mathrm{H\,m^{-1}}.$$

μ_0 is also called the magnetic constant. It is related to the *permittivity of vacuum, ϵ_0:

$$c^2 = 1/\mu_0\epsilon_0,$$

where c is the speed of light in vacuum.

The relative permeability, symbol μ_r, is a dimensionless quantity equal to μ/μ_0. The magnetic susceptibility, symbol χ (Greek chi) or χ_m, is the dimensionless quantity $(\mu_r - 1)$.

In anisotropic media, permeability and magnetic susceptibility are second-rank *tensors and component notation for the symbols should then be used, e.g. μ_{ij}.

permeability coefficient A physical quantity, the volume of fluid of unit viscosity flowing per second through unit area in a porous substance with a unit pressure gradient. The *SI unit is the square metre.

permeability of vacuum Also called permeability of free space. *See* permeability.

permeance Symbol: Λ (Greek cap. lambda). A physical quantity, the reciprocal of *reluctance. The *SI unit is the henry.

Permian Abbrev.: P (no stop). **1.** (adjective) Denoting the final period of the Palaeozoic era. **2.** (noun; preceded by 'the') The Permian period. The period is divided into the Lower (or Early) Permian and Upper (or Late) Permian epochs (initial caps.); abbrevs. (respectively): P_1 and P_2 (subscript numerals). *See also* Permo-Triassic.

permittivity Symbol: ϵ (Greek epsilon). A property of a medium equal to the *electric flux density in it divided by *electric field strength, D/E. The *SI unit is the farad per metre.

The **permittivity of vacuum**, symbol ϵ_0, is a fundamental constant:

$$\epsilon_0 = 1/(c^2\mu_0) = 8.854\,187\,817 \times 10^{-12}\,\mathrm{F\,m^{-1}},$$

where c is the speed of light in vacuum and μ_0 the *permeability of vacuum. ϵ_0 is also called the electric constant.

The **relative permittivity**, symbol ϵ_r, is the dimensionless quantity ϵ/ϵ_0 (also called the dielectric constant). The electric susceptibility, symbol χ (Greek chi) or χ_e, is the dimensionless quantity $(\epsilon_r - 1)$.

In anisotropic media, permittivity and electric susceptibility are second-rank *tensors and component notation for the symbols should be used, e.g. ϵ_{ij}.

permittivity of vacuum Also called permittivity of free space. *See* permittivity.

permonosulphuric acid H_2SO_5 US: **permonosulfuric acid.** The traditional name for peroxosulphuric(VI) acid.

Permo-Triassic (hyphenated) **1.** (adjective) Denoting the geological time during which there was deposition of New Red Sandstone. This occurred during the *Permian and *Triassic periods, but because of their nature such strata cannot be dated more precisely. **2.** (noun; preceded by 'the') The Permo-Triassic time. Also called Permo-Trias.

Permutit (cap. P) A trade name for a natural or synthetic zeolite.

Pérot, Alfred (1863–1925) French physician.

Fabry–Pérot interferometer *Physics* (en dash)

peroxoacids Oxo acids containing either of the groups $(O-O)^{2-}$ or $(O-OH)^{-}$, e.g. peroxodisulphuric(VI) acid, $H_2S_2O_8$. The traditional name is peracids.

peroxobromic(VII) acid $HBrO_4$ The recommended name for the compound traditionally known as perbromic acid.

peroxochloric(VII) acid $HClO_4$ The recommended name for the compound traditionally known as perchloric acid.

peroxodisulphuric(VI) acid $H_2S_2O_8$ US: **peroxodisulfuric(VI) acid.** The recommended name for the compound traditionally known as perdisulphuric acid.

peroxosulphuric(VI) acid H_2SO_5 US: **peroxosulfuric(VI) acid.** The recommended name for the compound traditionally known as Caro's acid or permonosulphuric acid.

peroxy- (not per-) Prefix used to denote an organic peroxide (e.g. peroxyethanoic acid).

peroxybenzoic acid $C_6H_5C(O)OOH$ The recommended name for the compound traditionally known as perbenzoic acid.

peroxyethanoic acid $CH_3C(O)OOH$ The recommended name for the compound traditionally known as peracetic acid.

Perrault, Claude (1613–88) French anatomist, engineer, and architect.

Perrin, Jean Baptiste (1870–1942) French physicist.

Perseus A constellation. Genitive form: Persei. Abbrev.: Per (no stop). *See also* stellar nomenclature.

Perseids (meteor shower)

personal computer Abbrev.: PC (no stops).

Perspex (cap. P) US: **Plexiglass**. A trade name for a transparent poly(methylmethacrylate).

PERT Acronym for project (or programme) evaluation and review technique.

Perthes, Georg Clemens (1869–1927) German surgeon.

Perthes' disease (or **Legg–Calvé–Perthes disease**)

Perutz, Max Ferdinand (1914–) Austrian-born British biochemist.

PES Abbrev. for photoelectron spectroscopy.

PETA Abbrev. for pentaerythritol triacrylate.

peta- Symbol: P A prefix to a unit of measurement that indicates 10^{15} times that unit, as in petajoule (PJ). *See also* SI units.

-petalous Adjectival suffix denoting petals (e.g. gamopetalous, polypetalous). Noun form: **-petaly**.

Peters, Sir Rudolph Albert (1889–1982) British biochemist.

Petersen, Carl Georg Johan (1860–1928) Danish marine biologist.
Petersen grab

Petersen, Julius (1839–1910) Danish mathematician.
Petersen graph

Peters projection *Cartog.* Named after Arno Peters.

Petit, Alexis Thérèse (1791–1820) French physicist.
Dulong and Petit's law

Petri, Richard Julius (1852–1921) German bacteriologist.
Petri dish (preferred to petri dish)

petro- Prefix denoting **1.** rock or stone, or rocklike (e.g. petrogenesis, petrography, petrology). **2.** petroleum (e.g. petrochemical).

Petzval curvature *Optics* Named after J. M. Petzval (1807–91).

Peyer, Johann Konrad (1653–1712) Swiss anatomist. Adjectival form: **Peyerian**.
Peyer's patches

Pezizales An order of ascomycete fungi (*see* Ascomycotina) containing some of the cup fungi, a group formerly included in the class Discomycetes.

Pfeffer, Wilhelm Friedrich Philipp (1845–1920) German botanist.

Pfeiffer, Richard Friedrich Johannes (1858–1945) German bacteriologist.
Pfeiffer's bacillus Old name for *Haemophilus influenzae*.

Pfund, A. Herman (1879–1949) US physicist.
Pfund series

Pg Abbrev. for Palaeogene.

PGA Abbrev. for **1.** programmable gate array. **2.** pin grid array. **3.** pteroylglutamic acid. **4.** phosphoglyceric acid.

PGAL Abbrev. for phosphoglyceraldehyde.

pH A dimensionless physical quantity expressing the acidity or alkalinity of a solution. To a first approximation,

$$\mathrm{pH} = -\log_{10}[\mathrm{H}^+],$$

where $[\mathrm{H}^+]$ is the *concentration of hydrogen ions measured in moles per cubic decimetre (mol dm^{-3}). A pH below 7 indicates an acid solution, one above 7 an alkaline solution. The formal definition is based on measurements of *electromotive force. pH is short for potential of hydrogen.
The dimensionless quantity $\mathrm{p}K_\mathrm{a}$ can be defined analogously:

$$\mathrm{p}K_\mathrm{a} = -\log_{10}K_\mathrm{a},$$

where K_a is the acid dissociation constant measured in mol dm^{-3}.

Ph (no stop) **1.** Abbrev. for Phanerozoic. **2.** Symbol often used to denote the phenyl group in chemical formulae, e.g. C_6H_5OH can be written as PhOH.

PHA Abbrev. for phytohaemagglutinin.

Phaeophyta A division comprising the brown algae. In a recent classification it is reduced to the status of a class, Phaeophyceae, in the division *Chromophyta, subkingdom Chromobionta.

phage *See* bacteriophage.

-phage Noun suffix denoting an eater or engulfer (e.g. bacteriophage, macrophage). Adjectival form: **-phagous**.

phage λ (Greek lambda) Abbrev. and usual form for coliphage λ, the type species of the λ phage group, sole genus of the family Styloviridae.

phage T2 Abbrev. and usual form for coliphage T2, the type species of the *T-even phage group.

phage T7 Abbrev. and usual form of coliphage T7, type species of the T7 phage group, sole genus of the family Podoviridae.

phago- (**phag-** before vowels) Prefix denoting eating or engulfing (e.g. phagocyte).

phagocyte Adjectival form: **phagocytic**. Derived noun: **phagocytosis** (pl. phagocytoses; adjectival form: **phagocytotic**). Derived verb: **phagocytose**.

phagotype Use *phagovar.

phagovar (preferred to phagotype) A phago(cytotic) var(iety): an unofficial category of classification used in microbiology and ranking below subspecies. Phagovars are strains capable of being lysed by certain bacteriophages.

phalanges (sing. phalanx; not phalange) The bones of the digits. *See also* hallux; pollex. Adjectival form: **phalangeal**.

phanero- (phaner- before vowels) Prefix denoting visible or obvious (e.g. phanerocrystalline, phanerophyte, Phanerozoic).

phanerogam In former plant classification schemes, any plant with easily recognizable reproductive structures (e.g. a gymnosperm or angiosperm), placed in the taxon Phanerogamia. The term now has no taxonomic significance. *Compare* cryptogam.

Phanerozoic Abbrev.: Ph (no stop). **1.** (adjective) Denoting the eon comprising the Palaeozoic, Mesozoic, and Cenozoic eras. **2.** (noun; preceded by 'the') The Phanerozoic eon.

pharynx (pl. pharynges) In mammals it is divided into a **nasopharynx, oropharynx**, and **laryngopharynx**. Adjectival form: **pharyngeal** (not pharyngal).

phase Two-word terms in which the first word is 'phase' (e.g. phase modulation, phase rule, phase shift, phase splitter, phase velocity) are not hyphenated unless used adjectivally (e.g. phase-contrast microscope, phase-shift keying).

phase angle *See* phase difference.

phase difference A physical quantity associated with two sinusoidally varying quantities, *u* and *i*, of the same angular frequency, ω:

$$\text{if } u = u_m \cos \omega t$$
$$i = i_m \cos(\omega t - \phi),$$

then ϕ (Greek phi) is the phase difference. The letters *u* and *i* indicate instantaneous values, u_m and i_m being maximum values. Phase difference may be expressed as an angle – the phase angle – or as a time. The *SI units are therefore the radian, degree, or second.

phase modulation Abbrev.: PM (no stops) or p.m. (stops).

PhD (or **Ph.D.**) Abbrev. for Doctor of Philosophy.

Phe (no stop) **1.** (or **phe**) Abbrev. for phenylalanine. *See* amino acid. **2.** *Astron.* Abbrev. for Phoenix.

phellem The outermost layer of the outer protective tissue of woody plants, consisting of cork cells. It should be distinguished from the **phelloderm**, the innermost layer; and the **phellogen**, the cork cambium that lies between them.

phen (no stop) Abbrev. for 1,10-phenanthroline often used in the formulae of coordination compounds, e.g. $[Cr(phen)_3]^{3+}$.

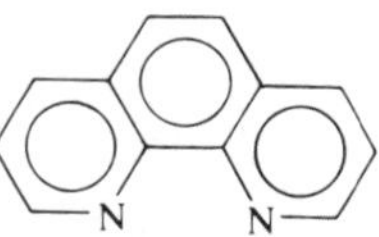

4,5-phenanthroline
(*o*-phenanthroline)

4,5-phenanthroline The recommended name for the compound traditionally known as *o*-phenanthroline.

phenetole $C_6H_5OC_2H_5$ The traditional name for ethoxybenzene.

pheno- (phen- before vowels) Prefix denoting **1.** showing or manifesting (e.g. phenocryst, phenotype). **2.** a molecule containing one or more benzene rings (e.g. phenobarbitone).

phenotype The sum of the observable characteristics of an organism. In prokaryotes, such phenotypic characters as metabolic deficiencies are designated by the abbreviation for the particular metabolite involved; it has an initial capital letter and is not italicized. Hence, a strain of bacterium unable to utilize lactose is designated Lac^- (superscript minus); one unable to synthesize tryptophan is designated Trp^-. The corresponding wild-type phenotypes are designated Lac^+ and Trp^+, respectively. When several characteristics are involved, the phenotype may be indicated in the form $Met^+Leu^-Lac^-$ (no intervening spaces).

In eukaryotes, phenotypic characters are usually given standard labels, written out in full. Thus in *Drosophila melanogaster* the family of mutant phenotypes relating to eye colour include 'white', 'vermilion', 'brown', 'apricot', 'eosin', etc. *See also* gene; genotype.

phenoxide *See* alkoxide.

phenoxy- Prefix denoting the group C_6H_5O- (e.g. phenoxyethanol).

phenyl- *Prefix denoting the group C_6H_5- (e.g. phenyl benzoate, phenylethanoic acid).

phenylalanine Abbrev.: Phe or phe (no stop). Symbol: F *See* amino acid.

phenylamine $C_6H_5NH_2$ The recommended name for the compound traditionally known as aniline.

phenylazobenzene $C_6H_5N{=}NC_6H_5$ The recommended name for the compound traditionally known as azobenzene.

(phenylazo)phenylamine $C_6H_5N{=}NC_6H_4NH_2$ The recommended name for the compound traditionally known as aminoazobenzene. The recommended name for *o*-aminoazobenzene is 4-(phenylazo)phenylamine.

***N*-(phenylazo)phenylamine** $C_6H_5N{=}NNHC_6H_5$ The recommended name for the compound traditionally known as diaminobenzene.

***o*-phenylenebis(dimethylarsine)** *See* diars.

phenylenediamine $C_6H_4(NH_2)_2$ The traditional name for *benzenediamine.

***N*-phenylethanamide** $C_6H_5NHCOCH_3$ The recommended name for the compound traditionally known as acetanilide.

phenylethanone $C_6H_5COCH_3$ The recommended name for the compound traditionally known as acetophenone.

phenylethene $C_6H_5CH{=}CH_2$ The recommended name for the compound traditionally known as styrene.

phenylketonuria Abbrev.: PKU (no stops).

phenylmethanol $C_6H_5CH_2OH$ The recommended name for the compound traditionally known as benzyl alcohol.

(phenylmethyl)amine $C_6H_5CH_2NH_2$ The recommended name for the compound traditionally known as benzylamine.

3-phenylpropenoic acid $C_6H_5CH{=}CHCOOH$ The recommended name for the compound traditionally known as cinnamic acid.

phi Greek letter, symbol: ϕ (lower case), Φ (cap.).

ϕ Symbol for **1.** electric potential. **2.** fluidity. **3.** geographical latitude. **4.** particle fluence rate. **5.** phase difference. **6.** the phenyl group used in formulae of coordination compounds, e.g. $[Re(S_2C_2\phi_2)_3]$. **7.** a plane angle. **8.** a polar coordinate. **9.** radiant flux density.

ϕX *See* bacteriophage; *Microvirus*.

Φ Symbol for **1.** heat flow rate. **2.** (or Φ_v) luminous flux. **3.** magnetic flux. **4.** particle fluence. **5.** potential energy. **6.** quantum yield. **7.** (or Φ_e) radiant flux. **8.** work function.

Phillips, Sir David Chilton (1924–) British biophysicist.

Phillips, John (1800–74) British geologist.

phlebo- (**phleb-** before vowels) Prefix denoting a vein (e.g. phlebothrombosis, phlebitis).

phlogiston (no cap.)

phloroglucinol $C_6H_3(OH)_3$ The traditional name for benzene-1,3,5-triol.

Phoenix A constellation. Genitive form: Phoenicis. Abbrev.: Phe (no stop). *See also* stellar nomenclature.

Phoenicids (meteor shower)

Pholidota (cap. P) An order of toothless mammals comprising the pangolins. Individual name and adjectival form: **pholidote** (no cap.; not pholidotan).

phon Symbol: phon A unit of loudness level, judged subjectively. The loudness level of a sound in phons is:

$$20\log_{10}(p/p_0),$$

where p is the effective (r.m.s.) *sound pressure (in *pascals) of a standard pure tone of frequency 1 kHz that is judged by a normal observer under standardized conditions as being equally loud as the sound; p_0 is a pressure of 20 micropascals.

-phone Noun suffix denoting sound (e.g. microphone, telephone).

-phore Noun suffix denoting carrying, supporting, or producing (e.g. chromatophore, chromophore, gametangiophore, spermatophore).

-phoresis Noun suffix denoting transmission or motion of (e.g. electrophoresis).

phosgene Cl_2CO The traditional name for carbonyl chloride.

phosphate In biochemistry, the symbol P is used to denote a phosphate group: P_i (subscript i) denotes orthophosphate, and PP_i pyrophosphate. In depicting polynucleotide sequences (*see* nucleotide), linking phosphate groups are denoted by hyphens in known sequences and by commas in unknown sequences. A terminal phosphate is denoted by p (lower case): to the left of a nucleoside this indicates a 5′-phosphate; to the right it denotes a 3′-phosphate.

phosphate(V) Denoting a compound containing the ion PO_4^{3-}, e.g. sodium phosphate(V), Na_3PO_4. The traditional name is orthophosphate.

phosphatide Use phospholipid.

phosphatidylcholine (one word) The systematic name for the phospholipid traditionally known as lecithin.

phosphatidylethanolamine (one word) The systematic name for the phospholipid traditionally known as kephalin (or cephalin).

phosphinate Denoting a compound containing the ion $H_2PO_2^-$, e.g. potassium phosphinate, KH_2PO_2. The traditional name is hypophosphite.

phosphite Denoting a compound containing the ion HPO_3^{2-}, e.g. sodium phosphite (Na_2HPO_3). The recommended name is phosphonate.

phospho- (**phosph-** before vowels) Prefix denoting phosphorus (e.g. phospholipid, phosphazene).

phosphoenolpyruvate Abbrev.: PEP (no stops).

phosphoenolpyruvate carboxylase Abbrev. and preferred form: PEP carboxylase.

phosphoglyceraldehyde Abbrev.: PGAL (no stops). Former names: glyceraldehyde 3-phosphate, triose phosphate.

phosphoglyceric acid Abbrev.: PGA (no stops).

phospholipid (preferred to phosphatide)

phosphonate Denoting a compound containing the ion HPO_3^{2-}, e.g. sodium phosphonate, Na_2HPO_3. The traditional name is phosphite.

phosphonic acid H_3PO_3 The recommended name for the compound traditionally known as orthophosphorous acid or phosphorous acid.

phosphor Derived noun: **phosphorescence**. Adjectival form: **phosphorescent**. Verb form: **phosphoresce**.

phosphoric acid H_3PO_4 The traditional name for phosphoric(V) acid.

phosphorous acid H_3PO_3 The traditional name for phosphonic acid.

phosphorus (not phosphorous) Symbol: P Allotropes may be distinguished as phosphorus (red) or polyphosphorus, phosphorus (white) or tetraphosphorus, and phosphorus (black). *See also* periodic table; nuclide.

phosphorus(III) chloride PCl_3 The recommended name for the compound traditionally known as phosphorus trichloride.

phosphorus(V) chloride PCl_5 The recommended name for the compound traditionally known as phosphorus pentachloride.

phosphorus(III) chloride oxide $POCl_3$ The recommended name for the compound traditionally known as phosphorus oxychloride or phosphoryl chloride.

phosphorus oxychloride $POCl_3$ The traditional name for phosphorus(III) chloride oxide.

phosphorus pentachloride PCl_5 The traditional name for phosphorus(V) chloride.

phosphorus trichloride PCl_3 The traditional name for phosphorus(III) chloride.

phosphoryl chloride $POCl_3$ The traditional name for phosphorus(III) chloride oxide.

phot A *cgs unit of illumination (now called *illuminance) equal to one lumen per square centimetre. The *SI unit of illuminance is the lux: 1 phot = 10^4 lux.

photo- Prefix denoting **1.** light or radiation lying mainly in the visible region (e.g. photoassimilate, photoautotroph, photocell, photochemistry, photoelectric, photoionization, photoreceptor, photosynthesis). **2.** photography (e.g. photogrammetry, photolithography).

photoelectron spectroscopy *Chem.* Abbrev.: PES (no stops).

photon The quantum of electromagnetic radiation, denoted γ (Greek gamma) in nuclear reactions, etc. It transfers energy $h\nu$, momentum $h\nu/c$, and mass $h\nu/c^2$, where h is the Planck constant, ν the frequency of the radiation, and c the speed of light in vacuum.

Photostat (cap. P) A trade name for a make of photocopier. The name has been used generically for any type of photocopy. Use photocopier (for the machine), photocopy (for the document produced). *See also* Xerox.

photosystem (or **pigment system**) Either of two photochemical systems, designated photosystems (or pigment systems) I and II (PSI and PSII), that carry out the light reactions of photosynthesis. PSI contains pigments including P700, a form of chlorophyll *a* with a light absorption peak of 700 nm. PSII includes P680, chlorophyll *a* with a light absorption peak of 684 nm.

phreatic *Geol.* Denoting ground water occurring below the water table. [from Greek *phrear*: well] *Compare* vadose.

Phthiraptera (cap. P) In some classifications, an order of insects comprising the lice. In other schemes these are split into two orders: *Anoplura and *Mallophaga. Individual name and adjectival form: **phthirapteran** (no cap.).

-phyceae Noun suffix denoting an algal *class in plant taxonomy (e.g. Chlorophyceae).

phyco- Prefix denoting algae (e.g. phycobiont, phycocyanin, phycoerythrin, phycology).

Phycomycetes (cap. P) In older classifications, a class of fungi corresponding to the subdivisions *Mastigomycotina plus *Zygomycotina. The trivial name, **phycomycetes** (no cap.), is still used as a synonym for the lower fungi.

-phyll Noun suffix denoting a leaf (e.g. chlorophyll, sporophyll).

phyllo- (**phyll-** before vowels) Prefix denoting a leaf or leaflike (e.g. phylloclade, phyllopodium, phyllosilicates, phyllode).

Phyllobacterium (cap. P, ital.) A genus of bacteria of the family Rhizobiaceae. Individual name: **phyllobacterium** (no cap., not ital.; pl. phyllobacteria).

phyllotaxis (preferred to phyllotaxy) The arrangement of leaves on a plant stem.

phylogeny (preferred to phylogenesis) The evolutionary history of an organism or group of organisms. Adjectival form: **phylogenetic**.

phylum (pl. phyla) A category used in animal classification (*see* taxonomy) consisting of a number of similar classes (sometimes only one class). Names of phyla are printed in roman (not italic) type with an initial capital letter; they often end in -a (e.g. Annelida, Arthropoda, Chordata) but there is no uniform ending. Large phyla may be divided into subphyla. *Compare* division.

physical quantities, symbols Single capital or lower-case letters of the Roman or Greek alphabet are almost always used, sometimes with a subscript, prime ('), or other modifying sign. (The two-letter symbols used for dimensionless *parameters are an exception.) In general, the symbols should be printed in italic type. *Vectors however should be set in bold italic type and *tensors in bold slanting sanserif type, when such type is available; this distinguishes the vector (or tensor) as an entity from its components and from other *scalar quantities. Different roman (upright), italic, and bold italic types for Greek letters are not often available, and the roman type is then used for Greek symbols.

A symbol can be made more specific by, for example, adding a subscript. Descriptive or numerical subscripts should be printed in roman type, as in E_{k} (kinetic energy), μ_{r} (relative permeability), S_{m} (molar entropy), or x_1, x_2, x_3. Subscripts that are themselves symbols for physical quantities or for numerical variables should, if possible, be set in italic (or if necessary bold italic) type, as in C_p (heat capacity at constant pressure), a_{ij} (matrix element), Σa_n. A symbol can also be made more specific by using both lower- and upper-case forms of a letter, as in *r* and *R* for radius, if there is no ambiguity with other symbols.

Symbols for particular physical quantities are given at the entries for those quantities, and are also listed at the beginning of each letter of the alphabet or in appropriate alphabetical order.

physisorption *See* adsorption.

-phyta Noun suffix denoting a *division in plant taxonomy (e.g. Bryophyta, Tracheophyta). Adjectival form (no initial cap.): **-phyte** (e.g. bryophyte).

-phyte Noun suffix denoting a plant or plants (e.g. halophyte, hydrophyte, xerophyte). Adjectival form: **-phytic**.

phyto- Prefix denoting a plant, plants, or of plant origin (e.g. phytoalexin, phytochrome, phytoplankton, phytotron).

Phytobiota In a recent classification, a kingdom containing all the plants; it corresponds to the traditional plant kingdom, *Plantae, excluding the fungi. *See also* Chlorobionta; Chromobionta.

phytochrome A proteinaceous plant pigment that regulates such light-dependent processes as flowering and greening of young leaves. It exists in two interchangeable forms, designated P_{r} (subscript r), absorbing red light (660 nm wavelength), and P_{fr}, absorbing far-red light (730 nm wavelength).

phytohaemagglutinin US: **phytohemagglutinin**. Abbrev.: PHA (no stops).

pi Greek letter, symbol: π (lower case), Π (cap.).

π Symbol for **1.** the irrational number 3.141 592 653 . . . . **2.** pion. **3.** solar parallax.

Π Symbol for **1.** mathematical product (*see also* limits). **2.** osmotic pressure. **3.** Peltier coefficient (Π_{ab} = that for substances a and b).

PI Abbrev. for *parainfluenza virus.

Piazzi, Giuseppe (1746–1826) Italian astronomer.

pi bond Usually written **π bond**. *See* orbital.

Pic (no stop) *Astron.* Abbrev. for Pictor.

pica *See* point.

Picard, Jean (1620–82) French astronomer.

Piccard, Auguste (1884–1962) Swiss physicist.

Pick, Arnold (1851–1924) Austrian psychiatrist and neurologist.
Pick's disease (a form of dementia)

Pick, Friedel (1867–1926) Austrian physician.
Pick's disease (a form of multiple sclerosis)

Pick, Ludwig (1868–*c.* 1944) German physician.
Niemann–Pick disease (en dash)

Pickering, Edward Charles (1846–1919) US astronomer.

pickup (noun and adjective; one word) Verb form: **pick up** (two words).

pico- Symbol: p A prefix to a unit of measurement that indicates 10^{-12} times that unit, as in picosecond (ps) or picofarad (pF). *See also* SI units.

Pictet, Raoul Pierre (1846–1929) Swiss chemist and physicist.

Pictor A constellation. Genitive form: Pictoris. Abbrev.: Pic (no stop). *See also* stellar nomenclature.

piedmont (no cap.) *Geol.* Situated at the foot of a mountain.

Pierce, John Robinson (1910–) US communications engineer.

piezo- Prefix denoting pressure or mechanical strain (e.g. piezochemistry, piezoelectric).

pileus (pl. pilei) The cap of a mushroom. Adjectival form: **pileate**. *Compare* pilus.

pillotina (pl. pillotinas) Any of various spirochaete bacteria inhabiting the hindguts of termites. The term is derived from *Pillotina* (cap. P, ital.) – a name proposed (but not approved) for a genus of such organisms.

pilo- (or **pili-**) Prefix denoting hair (e.g. pilomotor, piliferous).

Piltdown man A fraudulent fossil hominid shown to consist of a human cranium and an ape's jawbone. Named after the site in Piltdown, Sussex, where it was found in 1911.

pilus (pl. pili; usually referred to in the pl.) A fine hair, especially any of the hairlike projections (also called fimbriae) of the cell wall of certain bacteria. Adjectival forms: **piliferous, piliform, pilose**. *Compare* pileus.

pimelic acid $HOOC(CH_2)_5COOH$ The traditional name for heptanedioic acid.

pi meson *See* pion.

pinacol $(CH_3)_2C(OH)C(OH)(CH_3)_2$ The traditional name for 2,3-dimethylbutane-2,3-diol.

pinacolone $CH_3C(O)C(CH_3)_3$ The traditional name for 3,3-dimethyl-2-butanone.

Pinales *See* Coniferales.

p-i-n diode (hyphens) *See* p-type; i-type; n-type.

pin grid array (not hyphenated) Abbrev.: PGA (no stops).

Pinicae A subdivision of the *Pinophyta containing the orders Ginkgoales (comprising *Ginkgo*), Cordaitales (extinct), *Coniferales, and *Taxales. Individual name: **coniferophyte**.

pinna (pl. pinnae) **1.** A part of the outer ear of mammals. **2.** The leaflet of a compound leaf: may be subdivided into second-order leaflets (**pinnules**). Adjectival form: **pinnate**.

pinnati- *Bot.* Prefix denoting feather-like (e.g. pinnatifid).

pinniped 1. (noun) A member of the order *Pinnipedia. **2.** (adjective) Relating to or describing the Pinnipedia. **3.** (adjective) Having feet in which the digits are joined by a membrane; fin-footed.

Pinnipedia (cap. P) An order of mammals containing the seals, formerly classified as a suborder of the Carnivora. Individual name and adjectival form: **pinniped** (no cap.; not pinnipedian).

Pinophyta (cap. P) A division comprising the gymnosperms, used in certain classifications. It contains the subdivisions *Cycadicae (cycadophytes), *Pinicae (coniferophytes), and *Gneticae (gnetophytes). Individual name: **pinophyte** (no cap.). *See also* Gymnospermae.

Pinopsida *See* Coniferopsida; Gymnospermae.

pint Symbol: pt   A UK and US unit of volume, defined differently in the two countries and not used for scientific purposes. *See* gallon; bushel.

PIO *Computing* Abbrev. for parallel input/output.

pion An unstable elementary particle. The uncharged form is denoted by π^0 in nuclear reactions, etc.; the charged forms are denoted by π^+ and π^-. Also called pi meson.

pipelining (one word) *Computing*

piperazine The traditional name for *hexahydropyrazine.

piperidine The traditional name for *hexahydropyridine.

PIPES Abbrev. for 1,4-piperazinebis-(ethanesulphonic acid).

Pippard, Sir Alfred Brian (1920–) British physicist.

Pirani gauge *Physics* Named after M. S. von Pirani (1880–).

Pirie, Norman Wingate (1907–) British biochemist.

Pirquet, Clemens von (1874–1929) Austrian paediatrician.
Pirquet test

Pisces 1. *Astron.* A constellation. Genitive form: Piscium. Abbrev.: Psc (no stop). *See also* stellar nomenclature. **2.** *Zool.* A taxon that was formerly regarded as a class comprising: (1) originally, all fishes and fishlike vertebrates (now split into the classes *Agnatha, Placodermi (extinct jawed fishes), *Chondrichthyes, and *Osteichthyes); (2) later, all jawed fishes (i.e. as above but excluding the Agnatha); (3) later, all bony fishes (i.e. Osteichthyes). In some older schemes it is still used as a superclass comprising the four classes listed in (1) above.

pisci- Prefix denoting fish (e.g. pisciculture, piscivore).

Piscis Austrinus A constellation. Genitive form: Piscis Austrini. Abbrev.: PsA (no stops). *See also* stellar nomenclature.

pisi- (**piso-**) Prefix denoting a pea or pealike (e.g. pisiform, pisolitic).

Pithecanthropus (cap. P, ital.) The generic name originally given to fossil remains of hominids found at Java (Java man) and later at Peking (Peking man), both now classified as *Homo erectus*. The name **pithecanthropine** (not ital., no cap.) is sometimes still used as a general term for any *Homo erectus* fossil. *See Homo.*

Pitot, Henri (1695–1771) French scientist.
pitot head (no cap.)
pitot meter
pitot-static (hyphen)
pitot tube

Pitzer, Kenneth Sanborn (1914–) US theoretical chemist.

pivalic acid $(CH_3)_3CCOOH$ The traditional name for 2,2-dimethylpropanoic acid.

pixel *Computing* [from picture element]

pK_a *See* pH.

PKU Abbrev. for phenylketonuria.

PL/I (or **PL/1**) Programming language 1; a multipurpose programming language.

PLA Abbrev. for programmable logic array.

placenta (pl. placentae or placentas) Adjectival form: **placental**. Derived noun: **placentation**.

Placentalia Use *Eutheria.

plagio- Prefix denoting inclining or oblique (e.g. plagioclase, plagioclimax, plagiotropism).

plain 1. (noun) A level tract of land. **2.** (adjective) Level. **3.** (adjective) Not complicated. *Compare* plane.

Planck, Max Karl Ernst Ludwig (1858–1947) German physicist. Adjectival form: **Planckian**.
***Planck constant**
Planck function Symbol: Y
Planck's law
Planck's radiation formula

Planck constant Symbol: h A fundamental constant equal to

$$6.626\,076 \times 10^{-34}\,\mathrm{J\,s}.$$

Former name: Planck's constant. The ratio $h/2\pi$, symbol $\hbar$ (pronounced 'h bar'), is sometimes called the Dirac constant. It is equal to

$$1.054\,572 \times 10^{-34}\,\mathrm{J\,s}.$$

The **Planck mass**, symbol m_P, is given by $\sqrt{(hc/2\pi G)}$ and is equal to

$$2.176\,71 \times 10^{-8}\,\mathrm{kg}.$$

The **Planck length**, symbol l_P, is given by $\sqrt{(hG/2\pi c^3)}$ and is equal to

$$1.616\,05 \times 10^{-35}\,\mathrm{m}.$$

The **Planck time**, symbol t_P, is given by $\sqrt{(hG/2\pi c^5)}$ and is equal to

$$5.390\,56 \times 10^{-44}\,\mathrm{s}.$$

Planctomyces (cap. P, ital.) *See Blastocaulis-Planctomyces* group.

plane 1. *Maths.* A flat surface. Adjectival form: **planar**. **2.** Short for aeroplane. *Compare* plain.

plane angle *See* angle.

plani- Prefix denoting a plane, plane figure, or flatness (e.g. planimetric, planisphere). *See also* plano-.

planitia (not ital.; pl. planitiae) *Astron.* A plain. The word, with an initial capital, is used in the approved Latin name of such a feature on a planet or satellite, as in Chryse Planitia and Hellas Planitia, both on Mars.

plano- Prefix denoting **1.** a plane, plane figure, or flatness (e.g. planoconcave, planoconvex, planosol). **2.** wandering or motile (e.g. planogamete). *See also* plani-.

Planococcus (cap. P, ital.) A genus of motile Gram-positive coccoid bacteria. Individual name: **planococcus** (no cap.; pl. planococci). Adjectival form: **planococcal**.

plan-position indicator (hyphenated) *Radar* Abbrev.: PPI (no stops).

Plantae The kingdom comprising all the plants. Originally it included all organisms except animals, but *bacteria are now excluded and most taxonomists now place *fungi in a separate kingdom. In some classifications it is divided into two subkingdoms: *Thallophyta and *Embryophyta. In a recent five-kingdom classification, the plant kingdom is designated Phytobiota.

Planté, R. L. Gaston (1834–89) French physicist.
Planté battery

plant taxonomy *See* taxonomy.

planum (not ital.; pl. plana) *Anat., astron.* In astronomy, the word, with an initial capital, is used in the approved Latin name of a plateau on the surface of a planet or satellite, as in Solis Planum and Lunae Planum on Mars.

Plaskett, John Stanley (1865–1941) Canadian astronomer.
Plaskett's star

-plasm Noun suffix denoting the material forming cells (e.g. protoplasm, cytoplasm). Adjectival form: **-plasmic**.

plasma (or **blood plasma**) The fluid constituent of blood, in which the blood cells and platelets are suspended. *Compare* serum.

plasma- Prefix denoting cytoplasm or protoplasm (e.g. plasmagel, plasmagene, plasmalemma, plasmasol). *See also* plasmo-.

plasmid An independently replicating extrachromosomal DNA molecule present in many bacterial cells. Designations of individual plasmids usually consist of a lowercase p (for plasmid), followed by two capital letters and a number, e.g. pBR322 (no spaces), pER2, pRN3. However, some nonstandard designations, predating this convention, are still in use, e.g. RP1 (*see* R plasmid), ColE1 (*see* Col plasmid), and F (*see* F plasmid). Plasmids are also assigned to various incompatibility groups; members of any particular group are unable to coexist in the same cell. Designations vary, but are always prefixed by Inc (cap. I), with no intervening space; e.g. IncFI, IncI1, IncP, etc.

plasmo- Prefix denoting cytoplasm or protoplasm (e.g. plasmogamy, plasmolysis). *See also* plasma-.

plasmodesma (pl. plasmodesmata; usually referred to in the pl.) Any of the cytoplasmic strands that connect the protoplasm of adjacent plant cells.

Plasmodiophoromycetes A class of fungi of the division *Myxomycota, sometimes classified as an order, Plasmodiophorales, of the *Myxomycetes (slime moulds), that are parasites of plants. In a recent classification these parasites are regarded as a division, Plasmodiophoromycota, of the subkingdom Protistobionta.

plasmodium (pl. plasmodia) **1.** A motile multinucleate mass of protoplasm, especially forming the vegetative phase of the slime moulds. *Compare* coenocyte; syncytium. *See also* acellular. **2.** Any sporozoan protozoan of the genus *Plasmodium* (cap. P, ital.). Adjectival form: **plasmodial**.

-plast Noun suffix denoting **1.** a plastid (e.g. amyloplast, chloroplast, chromoplast, elaioplast). **2.** a unit of protoplasm (e.g. protoplast). Adjectival form: **-plastic**.

plasto- Prefix denoting chloroplasts (e.g. plastogene, plastoglobuli, plastoquinone).

plateau (pl. plateaux; preferred to plateaus)

Plateau, Joseph Antoine Ferdinand (1801–83) Belgian physicist.
Talbot–Plateau law (en dash) Use Talbot's law.

platelet A particle found in mammalian blood that is essential for blood coagulation. It is not a blood cell, therefore the synonym 'thrombocyte' (implying cellular form) should be avoided; however, such terms as 'thrombocytosis' and 'thrombocytopenia' are used for certain conditions affecting platelets.

platelet-derived growth factor Abbrev.: PDGF (no stops).

platinum Symbol: Pt *See also* periodic table; nuclide.

Plato (*c.* 428 BC–347 BC) Greek philosopher.
Platonic year

platy- Prefix denoting broad or flat (e.g. platyhelminth).

Platyhelminthes (cap. P) A phylum of invertebrates comprising the flatworms. Individual name: **platyhelminth** (no cap.). Adjectival form: **platyhelminthic**.

platyrrhine (not platyrhine) A member of the Platyrrhini, a division of the Primates comprising the New World monkeys. Adjectival form: **platyrrhine**. *Compare* catarrhine.

Playfair, John (1748–1819) Scottish mathematician and geologist.
Playfair's law

Playfair, Lyon, Lord (1818–98) British chemist and politician.

Ple Abbrev. for Pleistocene.

plecto- Prefix denoting twisted or interwoven (e.g. plectostele).

Plectomycetes A former class of ascomycete fungi (*see* Ascomycotina).

pleio- *See* pleo-.

pleiotropy (preferred to pleiotropism) The control of two or more apparently unrelated characteristics by a single gene. Adjectival form: **pleiotropic**.

pleisto- Prefix denoting most (e.g. Pleistocene).

Pleistocene Abbrev.: Ple (no stop). **1.** (adjective) Denoting the first epoch of the Quaternary period. **2.** (noun; preceded by 'the') The Pleistocene epoch.

Pleistogene Abbrev.: Ptg (no stop). Another name for *Quaternary (noun).

pleo- (or **pleio-, plio-**) Prefix denoting greater in number, size, degree, etc.; more (e.g. pleochroic, pleomorphic, pleiotropy, Pliocene).

plesiosaur Any extinct marine reptile of the order Plesiosauria, which includes the genus *Plesiosaurus*. Adjectival form: **plesiosaurian**.

plesiosaurus (pl. plesiosauruses or plesiosauri) Any *plesiosaur of the genus *Plesiosaurus*. The name should not be used for plesiosaurs of other genera.

pleura (pl. pleurae) The membrane that covers each lung (visceral pleura) and lines the chest wall (parietal pleura) in mammals, forming a sac. Adjectival form: **pleural**.

pleuro- (**pleur-** before vowels) Prefix denoting **1.** to one side (e.g. pleurocarpous). **2.** the pleura (e.g. pleuropneumonia).

Pleurocapsa (cap. P, ital.) A group of strains of *cyanobacteria belonging to the order Pleurocapsales. The group, currently of undefined taxonomic status, contains members assigned in the botanical literature to various genera, including *Pleurocapsa*. Hence, the botanical *Pleurocapsa* and the bacteriological *Pleurocapsa* are different entities. The latter may be distinguished by the epithet '*Pleurocapsa* group'.

Pleurocapsales An order of *cyanobacteria. Adjectival form: **pleurocapsalean**.

pleuropneumonia-like organism Abbrev.: PPLO (no stops). Obsolete name for a mycoplasma, in its broadest sense (*see* mollicute).

-plex Noun suffix denoting a link, connection, network (e.g. complex, duplex, multiplex).

Pli Abbrev. for Pliocene.

plio- *See* pleo-.

Pliocene (not Pleiocene) Abbrev.: Pli (no stop). **1.** (adjective) Denoting the most recent epoch of the Tertiary period. **2.** (noun; preceded by 'the') The Pliocene epoch.

-ploid Adjectival suffix denoting a complete set of chromosomes (e.g. diploid, haploid, polyploid, tetraploid). Noun form: **-ploidy**.

P–L relation (en dash) *Astron.* Short for period–luminosity relation.

Plücker, Julius (1801–68) German mathematician and physicist.

plumb- Prefix denoting lead (e.g. plumbic, plumbiferous, plumbous); in chemistry systematic names are now preferred.

plumbane PbH_4 The traditional name for lead(IV) hydride.

plumbate Denoting a compound containing the ion PbO_3^{2-}, e.g. potassium plumbate, K_2PbO_3. The recommended name is plumbate(IV).

plumbic Denoting compounds in which lead has an oxidation state of +4. The recommended system is to use oxidation numbers, e.g. plumbic oxide, PbO_2, has the systematic name lead(IV) oxide.

plumbous Denoting compounds in which lead has an oxidation state of +2. The recommended system is to use oxidation numbers, e.g. plumbous oxide, PbO, has the systematic name lead(II) oxide.

Pluto A planet. Adjectival form: **Plutonian**.

plutonium Symbol: Pu *See also* periodic table; nuclide.

Pm Symbol for promethium.

p.m. Abbrev. for post-mortem.

PM (or **p.m.**) Abbrev. for phase modulation.

PMBX *Telephony* Abbrev. for private manual branch exchange.

PMDA Abbrev. for pyromellitic dianhydride.

PMHS Abbrev. for polymethylhydrosilane.

PMS Abbrev. for phenazine methosulphate.

PMSF Abbrev. for phenylmethylsulphonyl fluoride.

PMSG Abbrev. for pregnant mare's serum gonadotrophin.

pn (no stop) Abbrev. for propylenediamine often used in the formulae of coordination compounds, e.g. $[Co(pn)_3]^{3+}$.

p.n. Abbrev. for proton number.

PNdB Symbol for perceived noise decibel.

pneumato- Prefix denoting air, gas, or respiration (e.g. pneumatolysis, pneumatophore). *See also* pneumo-; pneumono-.

pneumo- Prefix denoting the lungs (e.g. pneumococcus, pneumocyte, pneumovirus). *See also* pneumato-; pneumono-.

pneumococcus (no cap., not ital.; pl. pneumococci) Any strain of bacterium belonging to the species *Streptococcus pneumoniae*. Pneumococci cause pneumonias. Adjectival form: **pneumococcal**.

pneumono- (**pneumon-** before vowels) *Med.* Prefix denoting the lungs (e.g. pneumonectomy, pneumonitis). *See also* pneumato-; pneumo-.

p-n junction (or **pn junction**) *See* p-type; n-type.

PNL Abbrev. for polymorphonuclear leucocyte.

pnpn device (or **p-n-p-n device**) *See* p-type; n-type.

p-n-p transistor (or **pnp transistor**) *See* p-type; n-type.

PNSB *See* purple nonsulphur bacteria.

Po Symbol for polonium.

Poaceae (cap. P) *See* Gramineae. The adjectival form, **poaceous**, is rarely used.

4-POBN Abbrev. for α-(4-pyridyl 1-oxide)-*N*-*t*-butylnitrone.

Pockels, Friedrich Karl Alwin (1865–1913) German physicist.
Pockels cell
Pockels effect

-pod (not -pode) Noun suffix denoting a foot (e.g. cephalopod, tripod).

podo- (**pod-** before vowels) Prefix denoting a foot (e.g. *Podocarpus, Podophyllum*, podagra).

podzol (preferred to podsol) *Geol.* The word is printed with an initial capital when used as the name of a specific category in a *soil classification scheme; e.g. the world class Podzols in the RAO/UNESCO scheme. Adjectival form: **podzolic**. Verb form: **podzolize**. [from Russian: ash ground]

Poggendorff, Johann Christian (1796–1877) German physicist, biographer, and bibliographer.
Poggendorff compensation method
Poggendorff illusion

Pogson, Norman Robert (1829–91) British astronomer.
Pogson ratio

Pohl interferometer *Physics* Named after Robert W. Pohl (1884–1976).

-poiesis Noun suffix denoting the making or production of (e.g. haemopoiesis). Adjectival form: **-poietic**.

poikilo- (**poikil-** before vowels) Prefix denoting variation or irregularity (e.g. poikiloblastic, poikilocyte, poikilothermy, poikilosmotic).

poikilothermic (preferred to poikilothermal) Describing animals (**poikilotherms**) whose body temperature fluctuates with that of the environment. The term is often incorrectly used as a synonym for *ectothermic. Noun form: **poikilothermy**. *Compare* homoiothermic.

Poincaré, (Jules) Henri (1854–1912) French mathematician and philosopher of science.
Poincaré conjecture
Poincaré recurrence theorem

poinsettia (no cap., not ital., not poinsetia; pl. poinsettias) A popular ornamental shrub, *Euphorbia* (formerly *Poinsettia*) *pulcherrima*.

point Symbol: pt A unit of measurement used in printing, equal to approximately $^{1}/_{72}$ inch, 0.351 mm. In the Anglo-American point system there are 12 points in one pica, and 1 pica is equal to $^{35}/_{83}$ cm, i.e. about 0.166 inch (almost $^{1}/_{6}$ inch) or 4.2169 mm. The em, traditionally equal to the pica, has now been defined as $^{1}/_{6}$ inch exactly, i.e. 4.23 333 mm.

point spread function (not hyphenated) *Image technol.* Abbrev.: PSF (no stops) or p.s.f. (stops).

poise Symbol: P A *cgs unit of viscosity (dynamic viscosity), now discouraged. In *SI units dynamic viscosity is measured in newton seconds per square metre ($N\ s\ m^{-2}$): $1\ P = 0.1\ N\ s\ m^{-2}$. Named after J. L. M. *Poiseuille. *See also* stokes.

poiseuille A name used in France for the SI unit of dynamic viscosity, the newton second per square metre or the pascal second. Named after J. L. M. *Poiseuille. *Compare* poise.

Poiseuille, Jean Louis Marie (1799–1869) French physicist.
***poise**
***poiseuille**
Poiseuille flow
Poiseuille's equation

Poisson, Siméon Denis (1781–1840) French mathematician and mathematical physicist.
Poisson distribution
Poisson parentheses
Poisson process
Poisson ratio Symbol: μ or ν (Greek mu or nu)
Poisson's differential equation

Polanyi, John Charles (1929–) Canadian chemist and physicist.

Polanyi, Michael (1891–1976) Hungarian-born British physical chemist and philosopher.

polar coordinates *See* coordinates.

polarize Noun form: **polarization**.

Polaroid (cap. P) A trade name for a range of photographic and optical products.

Polenske number *Food technol.*

poliovirus (one word) Any of a subgroup of human *enteroviruses. There are three serovars, designated 1, 2, and 3.

polje (pl. poljes) *Geol.*

pollex (pl. pollices) The first (innermost) digit of a forelimb; the thumb in humans. *Compare* hallux.

polonium Symbol: Po *See also* periodic table; nuclide.

poly- Prefix denoting **1.** several to many (e.g. polyandrous, polycyclic, polygene, polymorphism, polyploidy). **2.** a polymer (e.g. polyester, polyethene, polypeptide).

poly(A) (no space) Used to indicate a sequence of repeated adenosine units (*see* nucleoside). In molecular genetics, the epithet 'poly(A)' is used to refer to the 3′ 'tail' of eukaryotic mRNA following the addition of repeated adenylated nucleotides to the primary transcript.
poly(A)-binding protein Abbrev.: PABP (no stops)

Pólya counting formula *Maths.* Named after G. Pólya.

polyacrylamide gel electrophoresis Abbrev.: PAGE (no stops).

Polychaeta (cap. P) A class of annelid worms comprising the bristleworms. Individual name and adjectival form: **polychaete** (no cap.; not polychaetous).

polychlorinated biphenyl Abbrev.: PCB (no stops).

polydioxoboric(III) acid *See* boric acid.

polyethylene glycol Abbrev.: PEG (no stops).

polymer Any chemical compound consisting of repeated molecular units.

Polymers are systematically named by enclosing the name of the repeated unit (the monomer) within parentheses and adding the prefix poly-, e.g. poly(ethene), $(-CH_2CH_2-)_n$, poly(ethenol), $(-CHOHCH_2-)_n$, and poly(tetrafluoroethene), $(-CF_2CF_2-)_n$.

In nonsystematic nomenclature polymers are named as in systematic nomenclature but the trivial names of the monomers are used and the parentheses are usually omitted, e.g. polyethylene, $(-CH_2CH_2-)_n$, polyvinyl alcohol, $(-CHOHCH_2-)_n$, and polytetrafluoroethylene, $(-CF_2CF_2-)_n$.

polymerase chain reaction Abbrev.: PCR (no stops).

poly(methanal) $(CH_2O)_n{\cdot}xH_2O$, where $n > 6$. The recommended name for the compound traditionally known as paraformaldehyde.

polymorphonuclear leucocyte Abbrev.: PNL (no stops). Also called polymorph.

Polyomavirus (cap. P, ital.) A genus of *papovaviruses. Individual name: **polyomavirus** (no cap., not ital.).

Polypodiales *See* Filicales.

Polypodiopsida *See* Filicopsida.

Polyporales *See* Aphyllophorales.

poly(tetrafluoroethene) Abbrev.: PTFE (no stops).

poly(vinyl acetate) Abbrev.: PVA (no stops).

poly(vinyl chloride) Abbrev.: PVC (no stops).

Polyzoa *See* Bryozoa.

POMC Abbrev. for proopiomelanocortin.

Pomeranchuk, Isaak Yakovlevich (1913–66) Russian physicist.
Pomeranchuk cooling
Pomeranchuk pole
pomeron

Poncelet, Jean-Victor (hyphen) (1788–1867) French mathematician.

Pond, John (1767–1836) British astronomer.

Ponnamperuma, Cyril Andrew (1923–) Ceylonese-born US chemist.

pons Varolii US: **pons varolii.** *Anat.* Usually shortened to **pons**. Named after C. *Varolio.

Pontecorvo, Guido (1907–) Italian-born British geneticist.

Pope, Sir William Jackson (1870–1939) British chemist.

Popov, Aleksandr Stepanovich (1859–1906) Russian physicist and electrical engineer.

Popper, Sir Karl Raimund (1902–) Austrian-born British philosopher. Adjectival form: **Popperian**.

p orbital *See* orbital.

Porifera (cap. P) A phylum of invertebrates comprising the sponges. Individual name and adjectival form: **poriferan** (no cap.); the adjective 'poriferous' refers to the possession of many pores.

porous Noun form: **porosity**.

porphyry *Geol.* Adjectival form: **porphyritic**.

Porter, Sir George (1920–) British chemist.

Porter, Keith Roberts (1912–) Canadian biologist.

Porter, Rodney Robert (1917–85) British biochemist.

position vector Symbol: $\boldsymbol{r}$ A *vector quantity expressing the position of a moving point P relative to a fixed point O. The *coordinates (x, y, z) or (r, θ, ϕ) of P referred to orthogonal axes through O are the components of $\boldsymbol{r}$ referred to these axes. The *SI unit is the metre.

positive muon *See* muon.

positive tauon *See* tauon.

positron *See* electron.

post- Prefix denoting **1.** after (e.g. post-Darwinian (hyphenated), postglacial, postpartum, post-transcriptional (hyphenated)). **2.** posterior to or behind (e.g. postganglionic, postsynaptic).

post-mortem (noun and adjective; hyphenated) Abbrev.: p.m.

Post's correspondence problem *Computing* Named after E. L. Post (1897–1954).

pot (no stop) Short for potentiometer.

potassium Symbol: K *See also* periodic table; nuclide.

potassium bicarbonate $KHCO_3$ The traditional name for potassium hydrogencarbonate.

potassium bromate $KBrO_3$ The traditional name for potassium bromate(V).

potassium chromate K_2CrO_4 The traditional name for potassium chromate(VI).

potassium dichromate $K_2Cr_2O_7$ The traditional name for potassium dichromate(VI).

potassium ferricyanide $K_3Fe(CN)_6$ The traditional name for potassium hexacyanoferrate(III).

potassium hexacyanoferrate(III) $K_3Fe(CN)_6$ The recommended name for the compound traditionally known as potassium ferricyanide.

potassium hydrogencarbonate $KHCO_3$ The recommended name for the compound traditionally known as potassium bicarbonate.

potassium iodate KIO_3 The traditional name for potassium iodate(V).

potassium manganate K_2MnO_4 The traditional name for potassium manganate(VI).

potassium manganate(VII) $KMnO_4$ The recommended name for the compound traditionally known as potassium permanganate.

potassium nitrate KNO_3 The recommended name for the compound traditionally known as nitre or saltpetre.

potassium permanganate $KMnO_4$ The traditional name for potassium manganate(VII).

potassium sodium 2,3-dihydroxybutanedioate NaOOCCH(OH)CH(OH)COOK The recommended name for the compound traditionally known as Rochelle salt.

potassium tetracyancuprate(I) $K_2Cu(CN)_4$ The recommended name for the compound traditionally known as cuprous potassium cyanide.

potato virus X Abbrev.: PVX (no stops). Type member of the **Potexvirus* group (vernacular name: potato virus X group).

potato virus Y Abbrev.: PVY (no stops). Type member of the **Potyvirus* group (vernacular name: potato virus Y group).

potential *See* electric potential; potential difference.

potential difference Symbol: U, V, or sometimes ΔV; u or v is often used for the instantaneous value of potential difference in alternating current technology. Abbrev.: p.d. A physical quantity, the difference in *electric potential between two points. It is a *scalar equal to the line integral of the *electric field strength, $\int \boldsymbol{E}_s d\boldsymbol{s}$, between the points. The *SI unit is the volt.

potential energy *See* energy.

Potexvirus (cap. P, ital.) Approved name for the *potato virus X group. Individual name: **potexvirus** (no cap., not ital.). [from *pot*ato *X virus*]

Potyvirus (cap. P, ital.) Approved name for the *potato virus Y group. Individual name: **potyvirus** (no cap., not ital.). [from *pot*ato *Y virus*]

Poulsen, Valdemar (1869–1942) Danish electrical engineer.
Poulsen arc

pound Symbol: lb The traditional UK and US unit of mass, now defined as exactly 0.453 592 37 kilogram. It was the fundamental unit of mass in the obsolete *fps system of units but has been superseded in scientific measurements by the kilogram (*see* SI units).

1 pound = 16 ounces = 16 × 16 drams = 7000 grains

Multiples used in the UK include the stone (14 lb), the hundredweight (112 lb), and the ton (2240 lb). In the US the short hundredweight (100 lb) and the short ton (2000 lb) are used; the equivalent UK units are sometimes called the long hundredweight and the long ton in the US.

The pound and its multiples and submultiples as defined above are avoirdupois units; they can be written pound (avoir), ounce (avoir), etc. When the pound and its multiples and submultiples are not qualified, it is assumed that they are avoirdupois units.

Compare pound troy.

poundal Symbol: pdl A unit of *force in the obsolete *fps system of units. The *SI unit, the newton, should now be used:

1 pdl = 1 lb ft/s^2 = 0.138 255 N.

Compare pound-force.

pound-force Symbol: lbf A unit of *force in the obsolete *fps system of units; *see also* psi (*or* p.s.i.). The *SI unit, the newton, should now be used:

1 lbf = 32.1740 lb ft/s^2 = 4.448 22 N.

32.1740 ft/s^2 is the value of the standard *acceleration of free fall (*compare* poundal). Associated units include the ounce-force and ton-force (*see also* kilogram-force).

The pound-weight, symbol lb wt, has also been used as an fps unit of force: 1 lb wt = g poundals; g is the acceleration of free fall and, like the pound-weight, varies with locality. The 'pound' in everyday usage is strictly the pound-weight.

pound troy A unit of mass used in the USA (but not in the UK), equal to 5760 grains or 0.373 242 kilogram. It contains 12 ounces troy. *Compare* pound.

pound-weight *See* pound-force.

Poupart, François (1661–1708) French surgeon.

Poupart's ligament

Powell, Cecil Frank (1903–69) British physicist.

power Symbol: P A physical quantity, the rate of doing *work or of *heat transfer. The *SI unit is the watt.

In the fields of electricity and magnetism, power is the product of current and potential difference. In an a.c. circuit, the **active power** is $IV \cos \phi$; I and V are the r.m.s. values of current and potential difference and ϕ is the phase angle between them; the SI unit of active power is the watt. The product IV is the **apparent power** and is measured in volt amperes. The **reactive power** is $IV \sin \phi$ and is measured in vars. The recommended symbols are P (active power), P_s or S (apparent power), and P_q or Q (reactive power).

power level difference Symbol: L_P or L_W A dimensionless quantity equal to

$$\tfrac{1}{2}\log_e(P_1/P_2)$$

in nepers or

$$\tfrac{1}{2}\log_{10}(P_1/P_2)$$

in decibels, where P_1 and P_2 are two powers. When P_1 and P_2 are *sound powers (P_2 being a reference power), the term 'sound power level' is employed.

poxvirus (one word) Any member of the family Poxviridae (vernacular name: poxvirus group). Vernacular names of species are named after the corresponding disease (e.g. buffalopox, canary pox, goat pox, rabbitpox, sheep pox, etc.), and therefore have the form buffalopox virus, canary pox virus, goat pox virus, rabbitpox virus, sheep pox virus, etc.

Poynting, John Henry (1852–1914) British physicist.

Poynting–Robertson effect (en dash) Also named after Howard Robertson.

Poynting's theorem

***Poynting vector**

Poynting vector Symbol: $\boldsymbol{S}$ The *vector product $\boldsymbol{E} \times \boldsymbol{H}$ of the electric and magnetic fields of an electromagnetic wave. The *SI unit is the joule per square metre per second (J m^{-2} s^{-1}).

pp. (stop) Abbrev. for pages. *See also* Appendix 9 (for style of page numbers in references).

PP$_i$ Symbol for pyrophosphate (in biochemistry).

PPI *Radar* Abbrev. for plan-position indicator.

PPLO Abbrev. for *pleuropneumonia-like organism(s).

ppm (no stops) Abbrev. for parts per million.

PPP Abbrev. for pentose phosphate pathway.

ppt. Abbrev. for precipitate.

Pr Symbol for **1.** praseodymium. **2.** the propyl group in chemical formulae, e.g. $CH_3CH_2CH_2OH$ can be written as PrOH. The following symbols are similarly used:
Pr^i (not *i*-Pr, iPr, or isoPr) for isopropyl group, e.g. Pr^iOH is $(CH_3)_2CHOH$;
Pr^n (not *n*-Pr or nPr) for *n*-propyl group, e.g. Pr^nOH is $CH_3CH_2CH_2OH$.

Pr (ital.) Symbol for Prandtl number.

practical electrical system *See* cgs units.

practical units *See* cgs units.

practice (noun) Verb form: **practise** (US: **practice**).

prae- Use *pre-.

Prandtl, Ludwig (1875–1953) German physicist.
***Prandtl number**
Prandtl's momentum transfer theory

Prandtl number Symbol: *Pr* A dimensionless quantity equal to ν/a, where ν is the kinematic *viscosity and *a* the *thermal diffusivity. *See also* parameter.

praseodymium Symbol: Pr *See also* periodic table; nuclide.

Pratt, John Henry (1809–71) British geophysicist.
Pratt hypothesis of isostasy

pre- (not prae-) Prefix denoting **1.** before (e.g. preamplifier, Precambrian, pre-eclampsia (hyphenated), pre-emphasis (hyphenated), prenatal). **2.** anterior to or in front of (e.g. prefrontal, preganglionic, premolar, presynaptic).

Precambrian (capital P, one word) **1.** (adjective) Denoting geological time before the *Phanerozoic eon, i.e. preceding the Cambrian period. **2.** (noun; preceded by 'the') Precambrian time. Due to the lack of consensus regarding stratigraphic classification of the Precambrian, its status is variable: in some recent systems the term 'Precambrian' is discarded in favour of the terms *Proterozoic and *Archaean.

precipitate *Chem.* Abbrev.: ppt. (stop).

precursor messenger RNA Abbrev.: pre-mRNA.

prefixes Many scientific prefixes, and general prefixes used in scientific terminology, are listed in this dictionary, together with examples of prefixed words. In general, such words are not hyphenated between the prefix and the ending; exceptions include (1) some (but not all) words in which the last letter of the prefix and the first letter of the ending are the same (e.g. anti-inflammatory, intra-atomic); (2) words in which the prefix is followed by a word derived from a proper noun (e.g. neo-Darwinism, non-Euclidean).
In chemistry, many prefixes are used to form the names of compounds. The convention is that if the prefix indicates replacement of hydrogen in a specific compound then it is run on to the compound name, e.g. ethylbenzene (one word). If it is used with a general term for a type of compound, then two words are used, e.g. ethyl alcohol.

Pregl, Fritz (1869–1930) Austrian chemist.

pregnant mare's serum gonadotrophin Abbrev.: PMSG (no stops).

Prelog, Vladimir (1906–) Swiss chemist.

pre-mRNA (hyphenated) Abbrev. and preferred form for precursor messenger RNA.

preproinsulin (one word)

pressure Symbol: *p* A physical quantity, the *force acting perpendicular to an element of area, divided by the element of area. The term is best limited to liquids and gases and to solids in uniform isotropic compression. (In general for solids one is concerned with *stress, which is not usually isotropic or uniform.) The *SI unit is the pascal but the bar may also be used for fluid pressure. Use of the atmosphere, millimetre of mercury, and the torr is discouraged; these units should only be used for rough comparisons.

The standard pressure for gases was formerly 101 325 pascals. It is now recommended by IUPAC that the standard pressure for reporting thermodynamic data should be 10^5 Pa (1 bar). 'Normal boiling points' are still reported with reference to a pressure of 101 325 Pa. *See also* stp.

pressure coefficient Symbol: β (Greek beta). A physical quantity equal to the rate of change of pressure p with temperature T at constant volume V:

$$\beta = (\partial p/\partial T)_V.$$

The *SI unit is the pascal per kelvin (Pa K^{-1}). The relative pressure coefficient, symbol α (Greek alpha), is equal to β/p. The SI unit is K^{-1}.

pressure potential Symbol: Ψ_p Also called turgor potential. *See* water potential.

pressurized-water reactor (hyphenated) Abbrev.: PWR (no stops).

Prestel (cap. P) A trade name for a UK *videotex system.

Prestwich, Sir Joseph (1812–96) British geologist.

preventive (not preventative)

Prévost, Pierre (1751–1839) Swiss physicist.
Prévost's theory of exchanges

Pribnow box (cap. P) *Genetics* A *consensus sequence of bases – TATAAT – in prokaryote DNA. Also called '–10 sequence' (minus sign) because it is ten base pairs upstream of the initiation point.

Priestley, Joseph (1733–1804) British chemist and Presbyterian minister.

Prigogine, Ilya (1917–) Belgian chemist.

Primates (cap. P) The order of mammals containing lemurs, monkeys, apes, and man. *See also* Scandentia. Individual name and adjectival form: **primate** (no cap.).

primula (pl. primulas) Generic name: *Primula* (cap. P, ital.).

principal 1. (adjective) Most important. **2.** (noun) A person in charge. **3.** (noun) Capital as opposed to income. *Compare* principle.

principal quantum number Symbol: n An integer used in the specification of electronic structure of an atom. It can take the values 1, 2, 3, . . . Each value corresponds to a particular electron shell (called K, L, M, . . . shells), $n = 1$ being the lowest and innermost shell. *See also* orbital angular momentum quantum number.

principle (noun) A general rule or law. *Compare* principal.

Pringsheim, Ernst (1859–1917) German physicist.

Pringsheim, Nathanael (1823–94) German botanist.

printed circuit Abbrev.: PC (no stops).
printed circuit board (not hyphenated) Abbrev.: PCB (no stops).

printout (noun; one word) Verb form: **print out** (two words).

prion protein Abbrev. and preferred form: *PrP (no stop).

private automatic branch exchange Abbrev.: PABX (no stops).

private branch exchange Abbrev.: PBX (no stops).

private manual branch exchange Abbrev.: PMBX (no stops).

Pro (or **pro**; no stop) Abbrev. for proline. *See* amino acid.

pro- Prefix denoting **1.** before (e.g. procambium, pro-oestrus (hyphenated), prophase, protandry). **2.** precursor of (e.g. prohormone, prothrombin, provitamin). **3.** anterior to or in front of (e.g. pronephros, proventriculus).

Proactinomyces (cap. P, ital.) A name used by certain authors for the genus **Nocardia*. It is not on the Approved Lists of Bacterial Names.

proactinomycete (one word) Any actinomycete bacterium in which the mycelial growth phase is only transitory or sporadic. *Compare* sporoactinomycete.

Proboscidea (cap. P) An order of mammals that contains the elephants. Individual name and adjectival form: **proboscidean** (no cap.).

proboscis (pl. probosces)

Procaryotae (not Prokaryotae) *See* prokaryote.

procaryote US spelling of *prokaryote.

proctodaeum (pl. proctodaea) US: **proctodeum** (pl. proctodea; the US spelling is now often used in British English). The embryonic anus. Adjectival form: **proctodaeal** (US: **proctodeal**).

product Symbol Π (Greek cap. pi). *See also* limits.

proglottis (pl. proglottides; preferred to proglottid, pl. proglottids) Any of the segments of a tapeworm.

program US spelling of programme. This spelling is used in the UK and elsewhere for all computing senses. Verb form: **programs, programming, programmed** (preferred to programing, programed). Adjectival form: **programmable** (preferred to programable). Derived noun: **programmer**.

programmable gate array (not hyphenated) Abbrev.: PGA (no stops).

programmable logic array (not hyphenated) Abbrev.: PLA (no stops).

programmable read-only memory Acronym: PROM (no stops).

programme *See* program.

programming support environment Abbrev.: PSE (no stops).

prokaryote US: **procaryote**. Any organism belonging to the kingdom Procaryotae (not Prokaryotae), which comprises the bacteria and blue-green algae. Adjectival form: **prokaryotic**.

Prokhorov, Aleksandr Mikhaylovich (1916–) Soviet physicist.

proline Abbrev.: Pro or pro (no stop). Symbol: P *See* amino acid.

Prolog (or **PROLOG**) *Computing* [from pro(gramming in) log(ic)]

PROM Acronym for programmable read-only memory.

promethium Symbol: Pm *See also* periodic table; nuclide.

promoter *Genetics* Symbol: *P* (ital.). Promoters are designated by *P* followed by subscript capital letters (not ital.); for example, repressor establishment promoter is P_{RE}, repressor maintenance promoter is P_{RM}. Promoters for leftwards or rightwards transcription units are denoted by a subscript capital L or R (not ital.) plus (as appropriate) an identifying number, e.g. P_{L}, $P_{\mathrm{R}}2$, etc.

proopiomelanocortin (one word) Abbrev.: POMC (no stops).

propadiene $CH_2{=}C{=}CH_2$ The recommended name for the compound traditionally known as allene.

propagate (not propogate) Noun form: **propagation**.

propanal CH_3CH_2CHO The recommended name for the compound traditionally known as propionaldehyde.

propanedioic acid $CH_2(COOH)_2$ The recommended name for the compound traditionally known as malonic acid.

propane-1,2,3-diol $CH_2(OH)CH(OH)CH_2OH$ The recommended name for the compound traditionally known as glycerol.

propanenitrile CH_3CH_2CN The recommended name for the compound traditionally known as ethyl cyanide.

propane-1,2,3-triyl trinitrate $CH_2(ONO_2)CH(ONO_2)CH_2(ONO_2)$ The recommended name for the compound traditionally known as nitroglycerine.

propanoic acid CH_3CH_2COOH The recommended name for the compound traditionally known as propionic acid.

propanone CH_3COCH_3 The recommended name for the compound traditionally known as acetone.

propanoyl- *Prefix denoting the group $CH_3CH_2CO{-}$ derived from propanoic acid (e.g. propanoyl chloride).

propel (**propels, propelling, propelled**) Adjectival form: **propellent.** Noun forms: **propellant** (not propellent), **propeller** (not propellor), **propulsion** (adjectival form: **propulsive**).

propenal $CH_2{=}CHCHO$ The recommended name for the compound traditionally known as acrolein.

propenenitrile CH_2CHCN The recommended name for the compound traditionally known as acrylonitrile.

propenoic acid $CH_2CHCOOH$ The recommended name for the compound traditionally known as acrylic acid.

prop-2-en-1-ol $CH_2=CHCH_2OH$ The recommended name for the compound traditionally known as allyl alcohol.

proper motion Symbol: μ (Greek mu). The apparent angular motion of a star, perpendicular to the line of sight, in a year. It is measured in arc seconds per year.

propiolic alcohol $HC{\equiv}CCH_2OH$ The traditional name for 2-propyn-1-ol.

propionaldehyde CH_3CH_2CHO The traditional name for propanal.

Propionibacterium (cap. P, ital.) A genus of typically rod-shaped Gram-positive bacteria that produce propionic acid during fermentation. Individual name: **propionibacterium** (no cap., not ital.; pl. propionibacteria).
Members are traditionally divided into two groups, according to their habitat: the 'classical propionibacteria', from cheese and dairy products; and the 'acnes group' (or 'cutaneous propionibacteria'), typically found on skin and named after *P. acnes* (formerly *Corynebacterium acnes*). Members of the acnes group are also commonly, but erroneously, referred to as 'anaerobic coryneforms' or 'anaerobic diphtheroids'.

propionic acid CH_3CH_2COOH The traditional name for propanoic acid.

propionyl- *Prefix denoting the group CH_3CH_2CO- derived from propanoic (propionic) acid (e.g. propionylacetone, propionyl chloride). Propanoyl- is recommended in all contexts.

Propliopithecus (cap. P, ital.) A genus of fossil apes. **Aegyptopithecus* fossils are sometimes included in this genus.

proportional Symbol: $\propto$ (not Greek alpha). Noun form: **proportionality**.

propulsion *See* propel.

propyl Prefix denoting the group C_3H_7-. Isomers are *n*-propyl, $CH_3CH_2CH_2-$, and isopropyl, $(CH_3)_2CH_2-$. Examples are *n*-propylbenzene, isopropylbenzene, isopropyl chloride.

propylenediamine *See* pn.

2-propyn-1-ol $HC{\equiv}CCH_2OH$ The recommended name for the compound traditionally known as propiolic alcohol.

pros- (ital., always hyphenated) *Chem.* Prefix sometimes used to denote 2,7-substitution on the naphthalene ring (e.g. *pros*-dimethylnaphthalene). *Compare amphi-*; *ana-*; *epi-*; *kata-*; *peri-*.

prostaglandin Any one of a large group of hormone-like substances with a wide range of pharmacological activity. The nine classes of prostaglandins are designated by PG followed by a capital letter (A–I); individual prostaglandins within these classes are denoted by a subscript Arabic numeral (1 or 2) indicating the number of C:C bonds in their fatty-acid chains (e.g. PGE_1, PGE_2, PGF_1, PGG_2, PGH_2, etc.).

prostheca (pl. prosthecae) Any of the conical or threadlike extensions of the cell wall found in certain bacteria. Adjectival form: **prosthecate**.

prosthecobacterium (no cap.; pl. prosthecobacteria) Trivial name for any prosthecate bacterium, including any of the genus *Prosthecobacter* (cap. P, ital.).

Prosthecomicrobium (cap. P, ital.) A genus of prosthecate bacteria. Individual name: **prosthecomicrobium** (no cap., not ital.; pl. prosthecomicrobia).

prot- *See* proto-.

protactinium Symbol: Pa *See also* periodic table; nuclide.

protein For depicting amino-acid sequences of proteins, *see* peptide. For denoting proteins as products of genes, *see* gene products.

Proterozoic (cap. P) **1.** (adjective) Denoting the eon preceding the Phanerozoic eon (*see* Precambrian). **2.** (noun; preceded by 'the') The Proterozoic eon. In some schemes the Proterozoic is divided into three eras: the Palaeoproterozoic, the Mesoproterozoic, and the Neoproterozoic (the most recent).

prothallus (pl. prothalli) *Bot.* Adjectival form: **prothallial**.

prothorax (pl. prothoraces) *Zool.* Adjectival form: **prothoracic**.

Protista (cap. P) A kingdom proposed by Haeckel in 1866 to include organisms that could not be classified as plants or animals (e.g. bacteria, protozoans, algae, and fungi). Although this grouping is now recognized as artificial, the taxon is still used for simple organisms of uncertain affinities. For example, in some classifications it is a kingdom containing all unicellular organisms (bacteria, unicellular algae and fungi, and protozoans), and in a recent classification it has been renamed **Protistobionta** and includes the protozoans, certain algae (e.g. dinoflagellates), and certain fungi (e.g. slime moulds and oomycetes). Individual name: **protist** (no cap.). Adjectival form: **protistan**.

protium *See* hydrogen.

proto- (usually **prot-** before vowels) Prefix denoting first in time, order, etc.; primitive, ancestral, or precursor of (e.g. protogyny, protonephridium, proto-oncogene (hyphenated), protostar, protostele, protoxylem, protactinium, protandry).

protochordate Any invertebrate chordate animal, sometimes classified as a division (Protochordata) of the phylum Chordata containing the *Cephalochordata and *Urochordata. Adjectival form: **protochordate**.

proton A positively charged particle, denoted p in nuclear reactions, etc. Its antiparticle, the **antiproton**, is denoted $\bar{p}$. The magnitude of the charge of the proton is equal to that of the *electron. Its rest mass, symbol m_p, is

$$1.672\,6231 \times 10^{-27}\,\text{kg}.$$

The ratio of proton mass to electron mass, m_p/m_e, is equal to

$$1.836\,152\,701 \times 10^3.$$

Two- or three-word terms in which the first word is 'proton' are not hyphenated (e.g. proton number, proton synchrotron).

protonema (pl. protonemata) The thread-like structure produced by a germinating bryophyte spore. Adjectival form: **protonemal**.

proton number Symbol: Z Abbrev.: p.n. (stops). A dimensionless physical quantity, the number of protons in an atomic nucleus. In specifying a particular *nuclide, the value of Z is placed as a subscript to the left of the chemical symbol, e.g. ${}_{12}\text{Mg}$. Also called atomic number.

Prototheria (cap. P) A subclass of primitive mammals containing a single order, *Monotremata. Individual name and adjectival form: **prototherian** (no cap.).

Protozoa (cap. P) A phylum or subkingdom comprising single-celled microscopic organisms usually classified as animals. Individual name: **protozoan** (pl. protozoans or protozoa; no caps.). Adjectival form: **protozoan**.

Proudman, Joseph (1888–1975) British mathematician and oceanographer.

Proust, Joseph Louis (1754–1826) French chemist.

Prout, William (1785–1850) British chemist and physiologist.
Prout's hypothesis

PrP (lower-case r) Abbrev. and preferred form for prion protein, the glycoprotein component of the scrapie agent (prion).

prussic acid HCN A traditional name for hydrogen cyanide.

Przewalski, Nikolai Mikhailovich (1839–88) Russian explorer.
Przewalski's horse (*Equus przewalskii*)

PSI, PSII (Roman numerals I and II) *Bot.* *See* photosystem.

PsA *Astron.* Abbrev. for Piscis Austrinus.

Psc *Astron.* Abbrev. for Pisces.

PSCM index *Meteorol.* Short for progression, southerly, cyclonicity, and meridionality index.

PSE *Computing* Abbrev. for **1.** programming support environment. **2.** project support environment.

pseudo- Prefix denoting false or closely resembling (e.g. pseudoacid, pseudoallele, pseudoendosperm, pseudopodium, pseudorandom).

pseudocowpox (one word) A disease of bovines.

pseudomonad Any bacterium of the family Pseudomonadaceae, which includes the genus *Pseudomonas* (cap. P, ital.) and related genera. Usage of the term 'pseudomonad' should avoid possible confusion between members of this genus and those of the family as a whole.

pseudouridine Symbol: Ψ (Greek cap. psi). A ribonucleoside. Also called 5-ribosyluracil. *See* nucleoside.

pseudovector A *vector whose sign is unchanged when the signs of all its components are changed. In general, vectors associated with rotations can be classed as pseudovectors, examples being angular velocity, angular momentum, and torque. Symbols of pseudovectors are printed, like other vectors, in bold italic type. *See also* vector product.

PSF (or **p.s.f.**) *Image technol.* Abbrev. for point spread function.

psi Greek letter, symbol: ψ (lower case), Ψ (cap.).

ψ Symbol for **1.** energy fluence rate. **2.** geographical latitude.

Ψ Symbol for **1.** electric flux. **2.** energy fluence. **3.** pseudouridine. **4.** water potential. **5.** wave function.

psi (or **p.s.i.**) Abbrev. for pound-force per square inch.

-psida Noun suffix denoting a *class of higher plants in taxonomy (e.g. Lycopsida). Adjectival form (no initial cap.): **-psid**.

Psilophytopsida A class of extinct pteridophytes. In some classifications these plants are combined with those of the class *Psilotopsida as the class Psilopsida.

Psilotopsida A class of pteridophytes containing the genera *Psilotum* and *Tmesipteris*. In some classifications these plants are combined with those of the class Psilophytopsida as the class **Psilopsida** (whisk ferns; individual name: **psilopsid**). In a recent classification these plants are classified as a superclass, Psilotatinae, in the infradivision Cormatae (vascular plants).

PST Abbrev. for Pacific Standard Time (in the USA).

psycho- Prefix denoting mental processes or the mind (e.g. psychoanalysis, psychogalvanic, psychometrics, psychomotor).

psychro- Prefix denoting cold (e.g. psychrometer).

psychrophilic Denoting microorganisms (known as **psychrophiles**) that thrive in temperatures below 20 °C.

pt 1. Symbol for pint. For UK usage it should be qualified as UKpt and for US usage as liq pt or dry pt. *See* gallon; bushel. **2.** Symbol for point.

Pt Symbol for platinum.

PTA Abbrev. for *p*-terephthalic acid.

-ptera Noun suffix denoting a wing; used especially in animal taxonomy to denote certain insect *orders (e.g. *Coleoptera, *Diptera, *Lepidoptera). Adjectival form (no initial cap.): **-pteran** (e.g. coleopteran).

Pteridophyta (cap. P) A division of plants comprising the classes *Filicopsida (ferns), *Sphenopsida (horsetails), *Lycopsida (clubmosses), *Psilophytopsida, and *Psilotopsida (the latter two are often classed together as Psilopsida. Individual name and adjectival form: **pteridophyte** (no cap.). Some authors place the pteridophytes in the division *Tracheophyta (vascular plants) together with seed plants.

Pteridospermales (cap. P) An extinct order of cycads (*see* Cycadopsida) comprising the seed ferns. Also called Cycadofilicales. Individual name and adjectival form: **pteridosperm** (no cap.).

ptero- (**pter-** before vowels) Prefix denoting a wing or winglike (e.g. pterodactyl, pterosaur).

Pteropsida A class of plants that in some classifications includes only the ferns (*see* Filicopsida) but in others includes, in addition, the gymnosperms and angiosperms.

Pterosauria (cap. P) An order of extinct flying reptiles containing the pterodactyls. Individual name: **pterosaur** (no cap.). Adjectival form: **pterosaurian**.

pteroylglutamic acid Abbrev.: PGA (no stops).

PTFE Abbrev. for poly(tetrafluoroethene).

Ptg Abbrev. for Pleistogene.

PTH Abbrev. for parathyroid hormone.

Ptolemy (*fl.* 2nd century AD) Egyptian astronomer. Also called Claudius Ptolemaeus.
Ptolemaic system
Ptolemy's theorem

ptomaine (preferred to ptomain) *Biochem.*

PTT Short for Postal, Telegraph, and Telephone Administration, a national governmental organization in many countries.

p-type (p not ital.) *Electronics* Denoting a doped semiconductor in which mobile positive-charge carriers (holes) are the majority carriers. Abbrev. in combinations: p, as in p-n junction, p-n-p transistor, pnpn device.

Pu Symbol for plutonium.

public-address system Abbrev.: PA system.

publications, citing *See* references; Appendix 9.

puffball (one word) *See* Lycoperdales.

pulsatance Another name for *angular frequency.

pulse Two-word terms in which the first word is 'pulse' (e.g. pulse frequency, pulse generator, pulse jet) should not be hyphenated unless used adjectivally (e.g. pulse-frequency modulation, pulse-height analyser, pulse-repetition frequency).

pulse-amplitude modulation (hyphenated) Abbrev.: PAM (no stops) or p.a.m. (stops).

pulse code modulation (not hyphenated) Abbrev.: PCM or pcm (no stops).

Punnett, Reginald Crundall (1875–1967) British geneticist.
Punnett square

Pup (no stop) *Astron.* Abbrev. for Puppis.

pupa (pl. pupae) Adjectival form: **pupal**.

Puppis A constellation. Genitive form: Puppis. Abbrev.: Pup (no stop). *See also* stellar nomenclature.

Purbach, Georg von (1423–61) Austrian astronomer and mathematician.

Purcell, Edward Mills (1912–) US physicist.

Purkinje, Johannes Evangelista (1787–1869) Czech physiologist.
Purkinje cell
Purkinje fibre
Purkinje phenomenon

purple nonsulphur bacteria US: **nonsulfur**. Trivial name for a group of photosynthetic bacteria. The term 'Purple Nonsulfur Bacteria' (US spelling, initial caps.; abbrev.: PNSB) has been proposed as a taxonomic name but as yet has no official status.

putrescine $H_2N(CH_2)_4NH_2$ The traditional name for 1,4-diaminobutane.

pv. (stop) Abbrev. for pathovar.

PVA Abbrev. for poly(vinyl acetate) or poly(vinyl alcohol).

PVC Abbrev. for poly(vinyl chloride).

PVP Abbrev. for poly(vinyl pyrrolidone).

PVPDC Abbrev. for poly(4-vinyl pyridinium dichromate).

PVP/I (solidus) Abbrev. for poly(vinyl pyrrolidone)–iodine complex (en dash).

PVX Abbrev. for potato virus X.

PVY Abbrev. for potato virus Y.

PWR Abbrev. for pressurized-water reactor.

py (no stop) Abbrev. for pyridine often used in the formulae of coordination compounds, e.g. $[Co(acac)_2(py)_2]^{2+}$.

pycno- (**pycn-** before vowels, **pykno-**, **pykn-**) Prefix denoting thickness or density (e.g. pycnometer, pycnoxylic, pycnidium, pyknosis).

pykno- *See* pycno-.

pylorus (pl. pylori) *Anat.* Adjectival form: **pyloric**.

pyr- *See* pyro-.

pyrazine The traditional name for *1,4-diazine.

Pyrenomycetes (cap. P) A former class of ascomycete fungi (*see* Ascomycotina) containing the flask fungi. The individual name and adjectival form, **pyrenomycete** (no cap.), may still be used for descriptive purposes.

Pyrex (cap. P) A trade name for a type of heat-resistant borosilicate glass.

pyridine *See* py.

pyridoxine Permitted synonym: vitamin B_6.

pyro- (sometimes **pyr-** before vowels or h) Prefix denoting **1.** high temperature or fire (e.g. pyroclastic, pyroelectric, pyrometry, pyrites, pyroxene, pyrheliometer). **2.** *Chem.* traditionally, an acid or salt with a water content between that of the ortho- and meta- compound (e.g. pyrophosphoric acid); systematic names are now recommended.

pyrogallol $C_6H_3(OH)_3$ The traditional name for benzene-1,2,3-triol.

pyrophosphoric acid $H_4P_2O_7$ The traditional name for heptaoxodiphosphoric(V) acid.

pyrrolidine The traditional name for *tetrahydropyrrole.

Pyrrophyta (not Pyrophyta) *See* Dinophyta.

pyruvic acid $CH_3COCOOH$ The traditional name for 2-oxopropanoic acid.

Pythagoras (*c.* 580 BC–*c.* 500 BC) Greek mathematician and philosopher.
Pythagoras's theorem

Pyx (no stop) *Astron.* Abbrev. for Pyxis.

Pyxis A constellation. Genitive form: Pyxidis. Abbrev.: Pyx (no stops). *See also* stellar nomenclature.

Pz Abbrev. for Palaeozoic.

Q

q Symbol for quartet (in nuclear magnetic resonance spectroscopy).

q Symbol for **1.** (light ital.) density of heat flow rate. **2.** (light ital.) electric charge. **3.** (bold ital.) generalized coordinate. **4.** (light ital.) *Chem.* partition function (individual entity). **5.** (light ital.) specific humidity.
q_i (light ital. q) generalized coordinate.

Q **1.** Symbol for glutamine. **2.** Abbrev. for Quaternary.

Q Symbol (light ital.) for **1.** *Chem.* partition function (whole system). **2.** *Elec. eng.* quality factor. **3.** quantity of electricity (i.e. electric charge). **4.** quantity of heat. **5.** (or Q_v) quantity of light. **6.** reactive *power. **7.** *Chem. eng.* throughput.

Q_{10} Symbol for temperature coefficient (in biology).

QA Abbrev. for quality assurance.

QAM (or **qam**) Abbrev. for quadrature-amplitude modulation.

Q band (not hyphenated) *Telecom.*, *radar* Superseded (approximately) by the designation Ka band. *See* K band.

QC Abbrev. for quality control.

QCD *Physics* Abbrev. for quantum chromodynamics.

QED Abbrev. for **1.** quod erat demonstrandum (Latin: which was to be proved). **2.** *Physics* quantum electrodynamics.

QI Abbrev. for quartz iodine.

QSO Abbrev. for quasistellar object; the contraction quasar is now preferred.

qt Symbol for quart. For UK usage it can be qualified as UKqt and for US usage as liq qt or dry qt. *See* gallon; bushel.

Quadrans Murali A former constellation, close to *Bootes.
Quadrantids (meteor shower)

quadrature-amplitude modulation (hyphenated) Abbrev.: QAM or qam (no stops).

quadri- (**quadr-** before vowels) Prefix denoting four (e.g. quadrifoliate, quadrilateral). *See also* tetra-.

quadruped Any vertebrate animal, especially a mammal, with four legs. *Compare* tetrapod.

quadrupole (not quadropole)

quality assurance Abbrev.: QA (no stops).

quality control Abbrev.: QC (no stops).

quality factor Symbol: Q A dimensionless quantity associated with electric circuits and equal to the magnitude of the reactance divided by the resistance, $|X|/R$ (*see* impedance).

quantity of electricity Another name for electric charge.

quantity of heat *See* heat.

quantity of light Symbol: Q or Q_v (v stands for visible). A physical quantity, the time integral of the *luminous flux, $\int \Phi dt$. The *SI unit is the lumen second.

quantum (pl. quanta) Verb form: **quantize.**

quantum chromodynamics Abbrev.: QCD (no stops).

quantum electrodynamics Abbrev.: QED (no stops).

quark Any of six elementary particles that have a fractional charge. The names of the quarks and their symbols are identical: u, d, c, s, t, and b. The terms 'up', 'down', 'charm', 'strange', 'top' (or 'truth'), and 'bottom' (or 'beauty') are mnemonics for these symbols. The quarks u, c, and t have a positive charge ⅔ that of the *electron; d, s, and b have a negative charge $-\frac{1}{3}e$. The quark antiparticles are denoted by ū, d̄, c̄, s̄, t̄, and b̄.

Different combinations of quarks, held together by the strong interaction, form the baryons (three quarks), such as the *proton and *neutron, and the mesons (quark plus antiquark), such as the *pions.

quart Symbol: qt A UK and US unit of volume, defined differently in the two countries and not used for scientific purposes. *See* gallon; bushel.

quarter *Maths.* US: **fourth**. *See also* fractions.

quasar Contraction of quasistellar object.

quasi- Prefix denoting closely resembling, mimicking (e.g. quasigeostrophic, quasirandom).

Quaternary (not Quarternary) Abbrev.: Q (no stop). **1.** (adjective) Denoting the second and most recent period of the Cenozoic era. **2.** (noun; preceded by 'the') The Quaternary period. Also called (the) Pleistogene.

Queckenstedt, Hans Heinrich (1876–1918) German physician.

Queckenstedt test

Quetelet, (Lambert) Adolphe (Jacques) (1796–1874) Flemish astronomer, mathematician, and sociologist.

quicklime CaO A traditional name for calcium oxide.

quillworts *See* Isoetales.

quino- (quin- before vowels and h) Prefix denoting derivation of a chemical compound from quinone (cyclohexadiene-1,4-dione) (e.g. quinolone, quinhydrone).

quinol $C_6H_4(OH)_2$ The traditional name for benzene-1,4-diol.

quinone The traditional name for *cyclohexadiene-1,4-dione.

quinque- Prefix denoting five (e.g. quinquefoliate, quinquepartite). *See also* penta-.

quintal Symbol: q A unit of mass equal to 100 kilograms.

qwerty keyboard (preferred to QWERTY)

R

r Symbol for ribonucleoside (preceding the *nucleoside symbol).

r Symbol for **1.** (light ital.) a polar coordinate. **2.** (bold ital.) position vector. **3.** (light ital.) radius.

R **1.** Symbol for **i.** arginine. **ii.** an organic group, often an alkyl group, in chemical formulae, e.g. ROH. **iii.** resonance effect (*see* electron displacement). **iv.** roentgen. **2.** Abbrev. for *restriction (point).

R Symbol (light ital.) for **1.** molar gas constant. **2.** radius ($R_\odot$, R_M, $R_\oplus$ = solar, lunar, terrestrial radius, respectively). **3.** resistance.

R_F (light ital. R) *Chromatog.* The distance travelled by the solvent front divided by the distance travelled by a given component of the solution.

R_H (light ital. R) Symbol for Hall coefficient.

R_m (light ital. R) Symbol for reluctance.

R$_\infty$ (light ital. R) Symbol for Rydberg constant.

(*R*)- (ital., parentheses, always hyphenated) *Chem.* Prefix denoting *rectus*, indicating an optically active isomer in which the priority (obtained using the Cahn–Ingold–Prelog sequence rules) of the substituent groups on the chiral centre decreases in a clockwise direction (e.g. (*R*)-butan-2-ol). If more than one asymmetric carbon atom is present, the configuration at each is specified together with the carbon atom number (if necessary) (e.g. (*R*,*R*)-dichlorobutane, (2*R*,3*S*)-2,3-dichloropentane, (2*R*,3*R*)-tartaric acid). *Compare* (*S*)-.

Ra Symbol for radium.

Ra (ital.) Symbol for Rayleigh number.

RA *Astron.* Abbrev. for right ascension.

rabbitpox (one word) *See also* poxvirus.

Rabi, Isidor Isaac (1898–1988) Austrian-born US physicist.

raceme *Bot.* A type of inflorescence. Adjectival form: **racemose** (*compare* racemic).

racemic *Chem.* Racemic compounds are denoted by the prefix (±)- (parentheses, always hyphenated), e.g. (±)-lactic acid. The alternative DL- (but not *dl*-) may be used, depending on the context.

racemic acid HOOCCH(OH)CH(OH)COOH The traditional name for (±)-2,3-dihydroxybutanedioic acid.

racemize *Chem.* Noun form: **racemization.**

rachi- (or **rachio-**; not rhachi-, rhachio-) Prefix denoting a spine or a shaft or axis (e.g. rachianaesthesia, rachiodont).

rachis (preferred to rhachis; pl. rachides) A central axis, for example of a feather or of an inflorescence. Adjectival form: **rachial** or **rachidial**.

rad 1. Symbol for radian. *See also* SI units. **2.** A former unit of *absorbed dose of ionizing radiation, equal to 0.01 joule per kilogram (originally 100 erg/g). The *SI unit now in use is the gray.

radar Acronym for radio detecting and ranging.

radian Symbol: rad The *SI unit of plane angle. It is a dimensionless derived unit, known as a supplementary unit:

$$2\pi \text{ rad} = 360°$$
$$1 \text{ rad} = 57.2958°.$$

Because plane angle, defined as a ratio of two lengths, is dimensionless, the unit of plane angle is the number 1; it is often convenient, however, to use the special name radian instead of the number 1. *See also* steradian.

radiance Symbol: L or L_e (e stands for energetic). A physical quantity that, for a point on a surface emitting or receiving radiation, corresponds to *radiant flux divided by area and solid angle. For emission, it is *radiant exitance per unit solid angle (equivalent to *radiant intensity per unit area); for incident radiation, it is *irradiance per unit solid angle. In both cases the *SI unit is the watt per steradian per square metre.

radiant emittance Former name for radiant exitance.

radiant energy Symbol: Q, Q_e (e stands for energetic), or W A physical quantity, the *energy of any form of electromagnetic radiation. The *SI unit is the joule.

radiant energy density Symbol: w A physical quantity, the *radiant energy per unit volume. The *SI unit is the joule per cubic metre.

radiant energy flux Another name for radiant flux.

radiant exitance Symbol: M or M_e (e stands for energetic). A physical quantity that, at a point on a surface, is the *radiant flux leaving an element of the surface divided by the area of that element. The *SI unit is the watt per square metre. *Compare* luminous exitance.

radiant flux Symbol: P, Φ, or Φ_e (Greek cap. phi; e stands for energetic). A physical quantity, the rate of flow of energy emitted or propagated as electromagnetic radiation. The *SI unit is the watt. Also called radiant energy flux, radiant power. *Compare* luminous flux.

radiant flux density Symbol: ϕ (Greek phi). A physical quantity, the *radiant flux per unit area. The *SI unit is the watt per square metre.

radiant intensity Symbol: I or I_e (e stands for energetic). A physical quantity, the energy emitted per second in a given direction in unit solid angle by a source of electromagnetic radiation. The *SI unit is the watt per steradian. *Compare* luminous intensity.

radiant power Another name for radiant flux.

radical **1.** (noun) *Chem.* (not free radical) Do not use in place of the word 'group'. A single centred dot is used to represent the unpaired electron of a radical, e.g. $CH_3\cdot$. **2.** (noun) *Maths.* A root of a number or quantity, such as $\sqrt{5}$, $\sqrt[3]{x}$. **3.** (adjective) *Bot.* Relating to a *radicle.

radicle The embryonic root of seed plants. Adjectival form: ***radical** (note that this also has noun senses in chemistry and maths).

radio Two-word terms in which the first word is 'radio' (e.g. radio receiver, radio source, radio telescope, radio wave) should not be hyphenated unless used adjectivally (e.g. radio-source catalogue). Some 'radio' terms formerly written as two words are now more usually written as one word; these, together with other words prefixed by 'radio-', are listed under the entry *radio-.

radio- (**radi-** before o) Prefix denoting **1.** a ray or radius (e.g. Radiolaria, radiospermic, radiosymmetrical). **2.** electromagnetic radiation, often specifically radio waves (e.g. radioastronomy, radiofrequency, radiometer, radiosonde, radiotelegraphy, radiotelephone). **3.** ionizing radiation (e.g. radioactivity, radiobiology, radiocarbon, radioimmunoassay, radioisotope, radiometric, radiopaque). *See also* radio.

radioactivity (one word) Adjectival form: **radioactive**. *See also* activity.

radiofrequency (one word) Abbrev.: r.f. (stops) or RF (no stops). See Appendix 1.

radiograph An image produced by X-rays. Such images are commonly referred to as 'X-rays' but this should be avoided in technical usage. Derived noun: **radiography**.

radioisotope (one word) Adjectival form: **radioisotope** (preferred to radioisotopic).

radiolabelled compounds It is usual to prefix the name of a radiolabelled compound with the symbol for the labelling isotope with the position of substitution, if it is known. This information is enclosed in square brackets, e.g. [^{3}H] glucose, [6-^{3}H] glucose.

Radiolaria (cap. R) An order of marine protozoans. Individual name and adjectival form: **radiolarian** (no cap.).

radiopaque (preferred to radio-opaque)

radium Symbol: Ra *See also* periodic table; nuclide.

radius (pl. radii) **1.** *Maths.* Symbol: r or R *See also* length. **2.** *Anat., zool.* A bone of the forelimb. Adjectival form: **radial.**

radix (pl. radices) *Maths., biol.*

radon Symbol: Rn *See also* periodic table; nuclide.

radula (pl. radulae) A feeding organ of molluscs. Adjectival form: **radulate**.

rain Terms incorporating the word 'rain' are usually written as one word (e.g. rainbow, raindrop, rainfall, rainforest, rainsplash, rainstorm, rainwater); two-word terms (e.g. rain day, rain gauge, rain shadow) should not be hyphenated unless used adjectivally (e.g. rain-gauge measurements).

Rainwater, (Leo) James (1917–86) US physicist.

RAL Abbrev. for Rutherford Appleton Laboratory, at Harwell, Oxfordshire.

RAM **1.** Acronym and preferred form for random-access memory. **2.** Acronym for radar-absorbing material.

r.a.m. Abbrev. for *relative atomic mass.

Raman, Sir Chandrasekhara Venkata (1888–1970) Indian physicist.
Raman effect
Raman spectroscopy

Ramanujan, Srinivasa Aaiyangar (1887–1920) Indian mathematician.

Ramapithecus (cap. R, ital.) The generic name originally given to fossil apes from India and Pakistan, now usually included in the genus **Sivapithecus*.

ramjet (one word)

Ramón y Cajal, Santiago (1852–1934) Spanish histologist.

Ramsauer, Carl Wilhelm (1879–1955) German physicist.
Ramsauer effect (preferred to Ramsauer–Townsend effect)

Ramsay, Sir William (1852–1916) British chemist.

Ramsden, Jesse (1735–1800) British instrument maker.
Ramsden circle
Ramsden eyepiece

ramus (pl. rami) *Anat., zool.* A branch or branchlike structure. Adjectival forms: **ramal, ramose.**

Ranales Use *Magnoliidae.

R&D (no spaces) Abbrev. for research and development.

random-access memory (hyphenated) Acronym and preferred form: RAM (no stops).

Raney, Murray (1885–1966) US engineer.
Raney nickel

rangefinder (one word) *Photog.*

Rankine, William John Macquorn (1820–72) British engineer and physicist.
***degree Rankine**
Rankine cycle
Rankine temperature scale

Ranunculaceae A large family of dicotyledonous plants, commonly known as the buttercup family. Adjectival form: **ranunculaceous.**

Ranvier, Louis-Antoine (hyphen) (1835–1922) French histologist.
node of Ranvier

Raoult, François-Marie (1830–1901) French physical chemist.
Raoult's law (or **rule**)

rapid eye movement Abbrev.: *REM (no stops).

rare earth elements Use *lanthanoids.

rare gases Use *noble gases.

RAS Abbrev. for **1.** Royal Astronomical Society. **2.** Royal Agricultural Society.

Raschig, Friedrich (1863–1928) German industrial chemist.
Raschig process
Raschig rings

Ratcliffe, John Ashworth (1902–87) British physicist.

rate coefficient Symbol: k A physical quantity associated with chemical reactions. The *SI unit depends on the order of reaction, which is defined in terms of the rate of increase of the *concentration of a component, i.e. dc/dt. For a particular substance B, if dc_B/dt depends on the concentrations of some other species C, D, etc., according to the expression

$$dc_B/dt = \nu_B k(c_C)^x (c_D)^y \ldots,$$

then the reaction is of order x with respect to C, of order y with respect to D and so on. (The presence of the *stoichiometric coefficient ν_B means that k is the same whichever substance is chosen as B.) The SI units of k are then as follows:
zero order: mol/(m^3 s) or mol/(L s)
first order: s^{-1}
second order: m^3/mol s) or L/(mol s)
third order: m^6/(mol^2 s) or L^2/(mol^2 s)
The rate coefficient is also called 'rate constant' but this is correctly used only for reactions involving elements.

rate constant *See* rate coefficient.

rate of conversion *See* stoichiometric coefficients.

rate of formation *See* stoichiometric coefficients.

rate of reaction *See* stoichiometric coefficients.

rationalized units *See* mks units.

ratite Any flightless running bird, such as an emu, kiwi, or ostrich. Ratites lack a keel on the sternum and were formerly classified as the subclass or superorder Ratitae; the term ratite is now used for descriptive (rather than taxonomic) purposes. *See also* palaeognathous. *Compare* carinate.

Ratzel, Friedrich (1844–1904) German geographer and ethnographer.

Raunkiaer, Christen (1860–1938) Danish botanist.
Raunkiaer's classification (or **system**)

Raunkiaer's life forms

rauwolfia 1. Any plant of the genus *Rauvolfia* (cap. R, ital.; not *Rauwolfia*). 2. A drug obtained from *R. serpentina*. Named after L. Rauwolf (16th century).

Ray, John (1627–1705) English naturalist and taxonomist.
Ray's bream

Rayet, Georges Antoine Pons (1839–1906) French astronomer.
Wolf–Rayet stars (en dash)

Rayleigh, John William Strutt, Lord (1842–1919) British physicist.
Rayleigh criterion
Rayleigh disc
Rayleigh–Jeans formula (en dash)
Rayleigh limit
***Rayleigh number**
Rayleigh scattering
Rayleigh waves

Rayleigh number Symbol: *Ra* A dimensionless quantity equal to the product of the *Grashof number and the *Prandtl number. *See also* parameter.

Raynaud, Maurice (1834–81) French physician.
Raynaud's disease
Raynaud's phenomenon

Rb Symbol for rubidium.

RBC Abbrev. for red blood cell.

RC (not ital.) Abbrev. for resistance-capacitance, used in adjectival form to describe a circuit, device, etc.

r.d. Abbrev. for relative density.

r determinant (lower-case r) *See* R plasmid.

rDNA Abbrev. for ribosomal DNA, a region of DNA found in eukaryotes that contains gene clusters coding for ribosomal RNA.

RDP (or **RuDP**) Abbrev. for ribulose diphosphate, former name for *ribulose 1,5-bisphosphate.

Re Symbol for rhenium.

Re (ital.) Symbol for Reynolds number.

re- (hyphenated before e or to clarify sense) Prefix denoting **1.** again; return or restoration to; response to (e.g. reaction, recombination, re-entry, re-format, renaturation, repolarization). **2.** back (e.g. recurvate, reflection, revolute).

reactance *See* impedance.

reaction mechanisms In chemistry, various shorthand forms are used to designate different types of reaction mechanism. An initial capital letter (roman type) shows the type of reaction as follows:
A – acid-catalysed;
B – base-catalysed;
E – elimination;
S – substitution.
In addition, a subscript capital letter (or letters) can be used:
AC – acyl-oxygen cleavage;
AL – alkyl-oxygen cleavage;
N – nucleophilic;
E – electrophilic.
A number (not subscript) is used to show the molecularity of the reaction (1 = unimolecular, 2 = bimolecular). A reaction that occurs with rearrangement is indicated by a prime mark after the number. Examples of reaction notations are:
$\mathrm{A_{AC}1}$ – acid-catalysed, acyl-oxygen cleavage, unimolecular;
$\mathrm{B_{AL}2}$ – base-catalysed, alkyl-oxygen cleavage, bimolecular;
E1 – elimination, unimolecular;
$\mathrm{S_E1}$ – substitution, electrophilic, unimolecular;
$\mathrm{S_N2}$ – substitution, nucleophilic, bimolecular;
$\mathrm{S_N2'}$ – substitution, nucleophilic, bimolecular, with rearrangement.
Other notations are:
$\mathrm{E_\alpha}$ – 1,1′-elimination;
E1cB – elimination, unimolecular, from conjugate base;
Ei – elimination, intramolecular.

reactive power *See* power.

Read, Herbert Harold (1889–1970) British geologist.

Read diode *Elec. eng.* Named after W. T. Read.

read-only memory (hyphenated) Acronym and preferred form: *ROM (no stops).

readout (one word)

read–write head (en dash; preferred to read/write head) *Computing*

Réaumur, René Antoine Ferchault de (1683–1757) French entomologist, physicist, and metallurgist.
Réaumur temperature scale

Reber, Grote (1911–) US radio astronomer.

Recent (cap. R) *Geol.* Use *Holocene.

reciprocal Derived nouns: **reciprocation**, **reciprocity**.

reciprocal ohm Obsolete name for siemens.

Recklinghausen, Friedrich von Usually alphabetized as *von Recklinghausen.

recryst. Abbrev. for recrystallized.

recti- (**rect-** before vowels) Prefix denoting straight or right (e.g. rectilinear, rectirostral, rectangle).

rectrix (pl. rectrices) A tail feather of a bird. *Compare* remex.

rectum (pl. recta or rectums) The terminal portion of the alimentary canal. Adjectival form: **rectal**. *Compare* rectus.

rectus (pl. recti) Any of several muscles. *Compare* rectum.

red algae *See* Rhodophyta.

red blood cell (preferred to red blood corpuscle) Abbrev.: RBC (no stops). Also called erythrocyte.

Redfield, William C. (1789–1857) US meteorologist.

Redi, Francesco (1626–97) Italian biologist, physician, and poet.
redia (pl. rediae) *Zool.*

red lead Pb_3O_4 The traditional name for dilead(II) lead(IV) oxide.

redox (no stops) Abbrev. for reduction–oxidation.

redshift (one word) *Astron.*

reduced-instruction-set computer (hyphenated) Acronym: RISC (no stops).

reductio ad absurdum (not ital.; from Latin)

reduction–oxidation (en dash) Abbrev.: redox.

re-entry (hyphenated) Adjectival form: **re-entrant**.

refer (**refers, referring, referred**) Noun form: **reference**.

references A references section consists of a list of works mentioned in the text. (It should be distinguished from a bibliography, which includes the works consulted or otherwise used by the author in the preparation of the text but not necessarily referred to directly there.) Reference sections should contain a source for every published work and all other publicly available material (e.g. unpublished theses) mentioned in the text. Most publishers have their own preferred system for citing references, so authors and volume editors should ascertain in detail what system their publisher is proposing to use before starting to prepare the reference section.

A reference should provide sufficient detail for the reader to be able to find the original work easily. Each entry should contain the following elements:

Reference to a book:
Name(s) of author(s) or editor(s);
Year of publication;
Title of chapter or paper (if the reference is to a contribution in a multiauthor book);
Full title of the book, including subtitle (if any);
Series, if any (optional);
Edition (if not the first);
Volume number (if any);
Translator (if any);
Editor (if the reference is to a contribution in a multiauthor book);
Page numbers (if the reference is not to the whole book);
Name of the publisher;
Town or city of publication (or address for obscure publishers, e.g. local societies).

Reference to an article in a periodical:
Name(s) of author(s);
Year of publication;
Title of article or paper (optional, but adopt a consistent policy about these);
Title of journal or periodical;
Volume number;
Issue number (only if the pagination of the journal is by issue rather than by volume);
Page numbers.

Reference lists should be typed with double spacing, since these sections usually need very detailed marking up for the typesetter. The two systems preferred by Oxford University Press for their science books (a version of the Harvard (author-date) system and a numbered system) are described in detail in Appendix 9.

reflect Noun form: **reflection** (not reflexion). Derived nouns: **reflector, reflectivity**.

reflectance Symbol: ρ (Greek rho). A dimensionless physical quantity equal to the ratio Φ_r/Φ_0, where Φ_0 is the radiant (or luminous) flux incident on a body or substance and Φ_r is the flux reflected from it.

reflection coefficient *See* acoustic absorption coefficient.

reflexion Use reflection.

refract *Physics* Noun form: **refraction**. Adjectival form: **refractive**. Derived noun: **refractor**.

refractive index Symbol: n A dimensionless physical quantity, the ratio of the sine of the angle of incidence of electromagnetic radiation on a medium to the sine of its angle of refraction in the medium.

For radiation passing from a medium of index n_1 to one of index n_2, the ratio n_2/n_1 is written n_{21}; this is the **relative refractive index** of the two media. If the first medium is a vacuum or air so that $n_1 = 1$ (or very nearly 1), it is common to write n_2 and n_{21} as simply n.

refractory (not refactory) *Eng., physiol.*

refrigerant (not refridgerant) Derived nouns: **refrigeration, refrigerator** (shortened form: **fridge**).

Regge, T. E. (1931–) Italian physicist.
Regge pole
Regge trajectory

regio (not ital.; pl. regiones) *Astron.* A large area distinguished by shading or colour. The word, with an initial capital, is used in the approved name of such a feature on the surface of a planet or satellite, as in Beta Regio on Venus or Chalybes Regio on Io.

Regiomontanus (1436–76) German astronomer and mathematician. German name: Johann Müller.

Registry of Toxic Effects of Chemical Substances Abbrev.: RTECS. Compounds on this register are assigned a number, prefixed by a two-letter code, and are listed with the relevant information. For example, lead(IV) oxide is registered as RTECS OG 0700000.

Regnault, Henri Victor (1800–78) French physicist and chemist.
Regnault's hygrometer

Reichstein, Tadeus (1897–) Polish-born Swiss biochemist.

Reid, Harry Fielding (1859–1944) US seismologist and glaciologist.
Reid's theory

Reines, Frederick (1918–) US physicist.

Reissner, Ernst (1824–78) German anatomist.
Reissner's membrane

Reiter, Hans (1881–1969) German bacteriologist.
Reiter's disease (or **syndrome**)

relative activity Another name for *activity.

relative atomic mass Symbol: A_r, followed when necessary by the chemical symbol in brackets, as in A_r(Cl). Abbrev.: r.a.m. (stops). A dimensionless physical quantity, the ratio of the average mass of an atom of an element to $^1/_{12}$ of the mass of an atom of carbon-12. A_r depends on the isotopes present in the element. The natural isotopic composition is assumed unless otherwise stated, in which case **relative isotopic** (or **nuclidic**) **mass** is the preferred term. *See also* mole.

relative density Symbol: d Abbrev.: r.d. (stops). A dimensionless quantity, the ratio of the *density of a substance to the density of a reference substance under specified conditions. For a gas, the reference substance is usually air, both being at stp. For a solid or liquid at a specified temperature (often 20 °C), the reference substance is water at 4 °C (when it has its maximum density, 1000 kg m^{-3}). Formerly called specific gravity. *See also* specific.

relative isotopic mass *See* relative atomic mass.

relative molecular mass Symbol: M_r, followed when necessary by the chemical formula, as in M_r(NaCl). Abbrev.: r.m.m. (stops) or RMM (no stops). A dimensionless physical quantity, the ratio of the average mass of a molecule or other molecular entity to $^1/_{12}$ of the mass of an atom of carbon-12. Natural isotopic composition of the entity is assumed unless otherwise stated. The concept is not restricted to entities that are strictly molecular; it can refer, for example, to a particular isotope or to a particular formula.

relative nuclidic mass *See* relative atomic mass.

relative permeability *See* permeability.

relative permittivity *See* permittivity.

relative pressure coefficient *See* pressure coefficient.

relative refractive index *See* refractive index.

relativistic coordinates *See* coordinates.

relativity *Physics.* Adjectival form: **relativistic**.

relaxation time Symbol: τ (Greek tau). A physical quantity that for some function of time $f(t)$ is expressed as:

$$f(t) = \exp(-t/\tau).$$

The *SI unit is the second. The ratio T/τ is the logarithmic decrement, symbol Λ (Greek cap. lambda), where T is the *period. *Compare* damping coefficient.

release factor *Genetics* Abbrev.: *RF (no stops).

reluctance Symbol: R_m A physical quantity corresponding to magnetic resistance, equal to the *magnetic potential difference divided by magnetic flux, U_m/Φ. The *SI unit is the reciprocal of the henry (H^{-1}).

rem Symbol: rem A unit of *dose equivalent that is being superseded by the sievert, an *SI unit equal to 100 rem. The word 'rem' is an acronym for roentgen equivalent man (or mammal).

REM Abbrev. for rapid eye movement, used especially to describe a stage of sleep (REM sleep).

Remak, Robert (1815–65) Polish-born German embryologist and anatomist.

remex (pl. remiges) A flight feather of a bird's wing. *Compare* rectrix.

renin An enzyme, released by the kidney, that catalyses the formation of angiotensin. *Compare* rennin.

rennin An enzyme, secreted by the stomach, that coagulates milk. *Compare* renin.

reovirus (one word) Any virus belonging to the genus *Reovirus* (cap. R, ital.). Usage should not be extended to include any other member of the family Reoviridae, which contains several other genera (including *Orbivirus* and *Rotavirus*).

repeat segment *Genetics* Abbrev. and preferred form: *R segment.

repetency Another name for *wavenumber.

repetition frequency Symbol: f *Genetics* The ratio of chemical complexity to kinetic complexity of a given sample of DNA. *See* complexity.

Reptilia (cap. R) The class of vertebrates comprising the reptiles. Adjectival form: **reptilian**.

research and development Abbrev.: R&D (no stops or spaces).

reseau (preferred to réseau; pl. reseaux) *Astron., photog.* A grid on a star map, etc. [from French: net]

resistance Symbol: R A physical quantity indicating the extent to which an object resists the flow of an *electric current. The resistance to direct current is equal to the potential difference applied divided by the resulting current, V/I, when there is no emf in the object. In an alternating-current circuit, the resistance is the real part of the *impedance. The SI unit of resistance is the ohm. *See also* RC.

resistance plasmid *See* R plasmid.

resistance transfer factor Abbrev.: RTF (no stops). *See* R plasmid.

resistivity Symbol: ρ (Greek rho). A property of a substance equal to the *resistance of a uniform conductor of the substance with unit length and unit cross-sectional area. It is also the *electric field strength, *E*, divided by the current density *J* when there is no emf in the conductor. The *SI unit is the ohm metre (Ω m); in some contexts (Ω mm^2)/m is used.

resistor An electrical component introducing a known resistance; formerly called a 'resistance'.

resonance effect *See* electron displacement.

resorcinol $C_6H_4(OH)_2$ The traditional name for benzene-1,3-diol.

respiratory quotient Abbrev.: RQ (no stops).

rest mass Symbol: m_0 (subscript zero). In particle physics it is usually expressed in *electronvolts, i.e. in terms of the equivalent energy.

restriction enzyme (or **restriction endonuclease**) An enzyme that cleaves a molecule of foreign DNA internally. These enzymes are produced by many bacteria and there are three types: type I, type II, and type III (Roman numerals). The name of each enzyme consists of an italic three-letter prefix plus an identifying letter and/or numbers (no intervening spaces). The prefix identifies the species of bacterium from which the enzyme derives, comprising the initial letter of the generic name and the first two letters of the specific name. Subsequent characters indicate the host strain (if applicable) and the particular enzyme (in cases where a single host produces several restriction enzymes). Hence, *Eco*B is derived from *E. coli* strain B, *Mbo*I is enzyme I from *Moraxella bovis*, and *Bam*HI is enzyme I from *Bacillus amyloliquefaciens* strain H (note that strain designation always precedes enzyme number). Correspondence between the strain label and the form incorporated in the name of the restriction enzyme apparently follows no rules and can be highly variable between different enzymes.

restriction fragment length polymorphism (no hyphens) Abbrev.: RFLP (no stops). Variation in the lengths of restriction fragments of DNA obtained from different individuals due to genetic polymorphism of the restriction sites in the chromosomes. Such sites can be used as markers, e.g. for loci determining genetic diseases.

restriction point Abbrev.: R or R point. The point during the G_1 phase of the *cell cycle at which the cycle may be halted. Cells in this resting state are sometimes said to have entered to G_0 (subscript zero) phase.

Ret (no stop) *Astron.* Abbrev. for Reticulum.

reticulo- Prefix denoting a net or network (e.g. reticulocyte, reticulodromous, reticuloendothelial).

reticulum (pl. reticula) **1.** A network of fibres, blood vessels, tubules, etc.: often specified (e.g. endoplasmic reticulum, sarcoplasmic reticulum). **2.** The second compartment of a ruminant's stomach. Adjectival forms: **reticular, reticulate**.

Reticulum A constellation. Genitive form: Reticuli. Abbrev.: Ret (no stop). *See also* stellar nomenclature.

retina (pl. retinas or retinae) Adjectival form: ***retinal** (note that this also has a noun sense).

retinal 1. (adjective) Relating to the retina of the eye. **2.** (noun) The aldehyde of vitamin A (retinol), which combines with the protein opsin to form the light-sensitive pigment rhodopsin. Also called retinene.

retinol The chemical name for vitamin A. *Compare* retinal.

retro- Prefix denoting backwards, back, or behind (e.g. retrobulbar, retroflexion, retrograde, retrotransposon, retroversion).

retrovirus (one word) Any virus belonging to the family Retroviridae.

Reye, Ralph Douglas Kenneth (1912–78) Australian paediatrician.
Reye's syndrome

Reynolds, Osborne (1842–1912) Irish engineer and physicist.
***Reynolds number**
Reynolds's law

Reynolds number Symbol: *Re* A dimensionless quantity equal to vl/ν, where v is a characteristic speed, l a characteristic length, and ν the kinematic *viscosity. The **magnetic Reynolds number**, symbol *Rm*, is equal to $vl\mu\sigma$, where μ is magnetic *permeability and σ is electric *conductivity. *See also* parameter.

Rf Symbol for rutherfordium. *See* element 104.

RF *Genetics* Abbrev. for release factor. In *E. coli* there are two, designated RF1 and RF2. In eukaryotes there is only one; the abbreviation is prefixed by e (for eukaryote), i.e. eRF.

r.f. (or **RF**) *Abbrev. for* radiofrequency.

R factor Use *R plasmid.

RFLP Abbrev. for *restriction fragment length polymorphism.

R-form (hyphenated) Abbreviation and preferred form for rough form: used to describe bacterial colonies with a jagged perimeter. *Compare* S-form.

RFT Abbrev. for richest-field telescope.

RGB *Elec. eng., etc.* Abbrev. for red green blue.

RGO Abbrev. for Royal Greenwich Observatory.

Rh 1. Symbol for rhodium. **2.** Abbrev. for *rhesus factor.

rhachi- Use *rachi-.

rhenium Symbol: Re *See also* periodic table; nuclide.

rheo- Prefix denoting **1.** flow (e.g. rheology, rheopexy). **2.** electric current (e.g. rheostat).

rhesus factor (lower-case r) Abbrev.: Rh (cap. R, no stop). A complex group of antigens that may or may not be present on the surface of human erythrocytes, so named because they also occur in rhesus monkeys. Individuals or blood possessing the factor are described as rhesus-positive (hyphenated, often abbreviated to Rh-positive or Rh +); those without the factor are rhesus-negative (Rh-negative or Rh –). The most important rhesus antigen is the D antigen; cells with the antigen are described as RhD + ; those without as RhD –. The corresponding antibody is anti-D (hyphenated).

Rheticus (or Rhäticus) (1514–76) Austrian astronomer and mathematician. Austrian name: Georg Joachim von Lauchen.

rhino- (rhin- before vowels) Prefix denoting the nose or sense of smell (e.g. rhinovirus, rhinencephalon).

rhizo- (rhiz- before vowels) Prefix denoting a root or rootlike (e.g. rhizomorph, rhizophore, rhizopodium, rhizome).

Rhizobium (cap. R, ital.) A genus of nitrogen-fixing bacteria of the family Rhizobiaceae. Individual name: **rhizobium** (no cap., not ital.; pl. rhizobia). Some authors use 'rhizobia' to denote members of the closely related genus **Bradyrhizobium*, but this should be avoided.

Rhizopoda (cap. R) A subclass of protozoans (class Sarcodina) containing the amoebae and foraminiferans. Individual name and adjectival form: **rhizopod** (no cap.); 'rhizopodia' (sing. rhizopodium) are thin branching pseudopodia possessed by some rhizopods.

rho Greek letter, symbol: ρ (lower case), Ρ (cap.).

ρ Symbol for **1.** density (ρ_l = linear density, ρ_A = surface density). **2.** mass concentration (ρ_B = that of substance B). **3.** reflectance. **4.** reflection coefficient (acoustic). **5.** resistivity. **6.** volume density of charge.

Rhodesian man *See Homo.*

rhodium Symbol: Rh *See also* periodic table; nuclide.

rhodo- (rhod- before vowels) Prefix denoting a red colour (e.g. rhodocrosite, Rhodophyta, rhodopsin).

Rhodococcus (cap. R, ital.) A genus of nocardioform actinomycete bacteria. It includes the species formerly classified as *Corynebacterium fascians*. Individual name: **rhodococcus** (no cap., not ital.; pl. rhodococci). Adjectival form: **rhodococcal**.

rhododendron (pl. rhododendrons) Any evergreen shrub of the genus *Rhododendron* (cap. R, ital.). Deciduous species and hybrids of this genus are called azaleas; they were formerly classified in the genus *Azalea*, now subsumed into *Rhododendron*.

Rhodophyta (cap. R) A division comprising the red algae. In a recent classification the red algae are classified as a subkingdom, Rhodobionta. Individual name and adjectival form: **rhodophyte** (no cap.).

rho factor Sometimes written **ρ factor** (Greek rho). *Genetics* A protein required for the termination of transcription at certain sites. This type of termination is described as rho-dependent; termination not requiring a rho factor (called self-termination) is described as rho-independent (note that 'rho' is written out in full in these cases).

rhombo- (**rhomb-** before vowels) Prefix denoting a rhombus (e.g. rhombochasm, rhombohedron, rhombencephalon).

rhombus (pl. rhombuses) Adjectival form: **rhombic**.

R horizon A *soil horizon consisting of (bed)r(ock).

rhyncho- Prefix denoting a snout or beak (e.g. Rhynchocephalia).

Rhynchocoela Use *Nemertea.

rhyolite *Geol.* Adjectival form: **rhyolitic**.

Ri (ital.) Symbol for Richardson number.

ria (pl. rias) *Geol.*

Rib (no stop) Abbrev. for ribose.

ribbon worm (two words) *See* Nemertea.

riboflavin (preferred to riboflavine) Permitted synonym: vitamin B_2.

ribonuclease Abbrev. and preferred form: RNAase (no stops).

ribonucleic acid Abbrev. and preferred form: *RNA (no stops).

ribonucleoprotein (one word) Abbrev.: *RNP (no stops).

ribose Abbrev.: Rib (no stop). *See* sugars.

ribosomal protein Abbrev.: r-protein (lower-case r, hyphenated).

ribosomal RNA Abbrev.: rRNA (no stops).

ribosome These ribonucleoproteins are described in terms of their *sedimentation coefficients (*see also* Svedberg unit). The prokaryote (70S) ribosome comprises a large (50S) subunit and a small (30S) subunit; the eukaryote (80S) ribosome has 60S and 40S subunits (note that there is no space between the number and S). The constituent proteins of the subunits are accordingly designated L (large) or S (small), respectively, and numbered. Hence, in the prokaryotic ribosome, the large subunit proteins are designated L1 to L34 (no space); the small subunit proteins are given as S1 to S21. The S prefix for small subunit proteins must not be confused with the symbol for Svedberg unit, e.g. distinguish between 'the S16 protein' and 'a 16S subunit'. Adjectival form: **ribosomal**.

ribosylthymine Symbol: T A ribonucleoside (*see* nucleoside). *Compare* thymidine.

ribulose 1,5-bisphosphate Abbrev.: RuBP or RUBP (no stops). Former name: ribulose diphosphate (abbrev.: RDP or RuDP), superseded to emphasize the fact that the two phosphate groups are attached to different carbon atoms.
ribulose bisphosphate carboxylase Abbrev.: RUBP carboxylase.

Ricci, Matteo (1552–1610) Italian astronomer and mathematician.

Ricciolo, Giovanni Battista (1598–1671) Italian astronomer.

Richard of Wallingford (*c.* 1291–1336) English astronomer and mathematician.

Richards, Theodore William (1868–1928) US chemist.

Richardson, Lewis Fry (1881–1953) British meteorologist and physicist.
Richardson number Symbol: *Ri* (*see* parameter).

Richardson, Sir Owen Willans (1879–1959) British physicist.
Richardson constant Symbol: *A*
Richardson's equation (preferred to Richardson–Dushman equation).

richest-field telescope (hyphenated) Abbrev.: RFT (no stops).

Richet, Charles Robert (1850–1935) French physiologist.

Richter, Burton (1931–) US physicist.

Richter, Charles Francis (1900–85) US seismologist.
Richter scale

Richthofen, Ferdinand von Usually alphabetized as *von Richthofen.

Ricketts, Howard Taylor (1871–1910) US pathologist.
***rickettsia**

rickettsia (not ricketsia; pl. rickettsiae) Any parasitic microorganism of the order Rickettsiales, often classified as bacteria. The order contains several genera, including *Rickettsia* (cap. R, ital.). Adjectival form: **rickettsial**. Named after H. T. *Ricketts.

Riddle, Oscar (1877–1968) US biologist.

Rideal, Sir Eric (Keightley) (1890–1974) British chemist.

Riemann, (Georg Friedrich) Bernhard (1826–66) German mathematician.
Cauchy–Riemann integral (en dash)
Riemann–Christoffel tensor (en dash)
Riemannian geometry
Riemannian space
Riemann integral
Riemann's hypothesis
Riemann surface
Riemann zeta function

Righi, Augusto (1850–1920) Italian physicist.
Righi–Leduc effect (en dash) Also named after S. Leduc.

right ascension Symbol: α (Greek alpha); abbrev.: RA (no stops). A coordinate used with *declination to give the position of an astronomical object with respect to the celestial equator. It is the angular distance measured eastwards along the equator from the vernal equinox (strictly a catalogue *equinox) to the object's hour circle; it is the equivalent of terrestrial longitude. It is generally expressed in hours (0–24 hours), where one hour equals 15° of arc.

rille (preferred to rill) *Astron.*

Ringer, Sydney (1834–1910) British physician.
Ringer's solution

RISC Acronym for reduced-instruction-set computer.

Ritchey, George Willis (1864–1945) US astronomer.
Ritchey–Chrétien telescope (en dash) Also named after Henri Chrétien.

Ritter, Johann Wilhelm (1776–1810) German scientist.

Ritz's combination principle *Spectroscopy* Named after Walther Ritz (1878–1909).

Rm (ital.) Symbol for magnetic Reynolds number.

r.m.m. (or **RMM**) Abbrev. for relative molecular mass.

rms Abbrev. for root mean square.

Rn Symbol for radon.

RNA Abbrev. and preferred form for ribonucleic acid. When prefixed by one or more lower-case letters, it indicates a particular type of RNA. Thus:
hnRNA (heterogeneous nuclear RNA)
mRNA (messenger RNA)
micRNA (mRNA-interfering complementary RNA)
pre-mRNA (hyphenated; precursor messenger RNA)
rRNA (ribosomal RNA)
scRNA (small cytoplasmic RNA)
snRNA (small nuclear RNA)
***tRNA** (transfer RNA).
The RNA of RNA viruses is prefixed by a bracketed plus or minus sign (or both), preceded by a space, according to type:
(+) RNA The RNA of plus-strand (or positive-strand) RNA viruses, which codes directly for viral proteins.
(–) RNA The RNA of minus-strand (or negative strand) RNA viruses, which serves as a template for mRNA synthesis.
(±) RNA The RNA of double-stranded RNA viruses.

RNAase Abbrev. and preferred form for ribonuclease. Note that this is the recommended form; RNase, though commonly used, is deprecated.

RNA polymerase An enzyme that catalyses transcription during protein synthesis. Different types are designated by Roman numerals, e.g. RNA polymerase I. *See also* sigma factor.

RNP Abbrev. for ribonucleoprotein. When prefixed by one or more lower-case letters, it indicates a particular type of RNP. Thus:
hnRNP (heterogeneous nuclear RNP)
scRNP (small cytoplasmic RNP)
snRNP (small nuclear RNP)
snoRNP (small nucleolar RNP)

Roberts, John D. (1918–) US chemist.

Robertson, John Monteath (1900–) British X-ray crystallographer.

Robertson, Sir Robert (1869–1949) British chemist.

Robinson, Sir Robert (1886–1975) British chemist.

Roche, Édouard Albert (1820–83) French mathematician.
Roche limit
Roche lobe

Rochelle salt NaOOCCH(OH)CH(OH)COOK The traditional name for potassium sodium 2,3-dihydroxybutanedioate.

roche moutonée (pl. roches moutonées) *Geol.* A rounded rock. [from French: fleecy rock]

Rochon prism *Optics*

rockrose (one word) A shrub of the genus *Helianthemum* or *Cistus* (family Cistaceae, not *Rosaceae).

Rodentia (cap. R) The order of mammals comprising the rodents; it does not include rabbits and hares (*see* Lagomorpha).

ROE Abbrev. for Royal Observatory, Edinburgh.

Roemer, Ole Christensen Usually alphabetized as *Romer.

roentgen (not röntgen) Symbol: R A unit of *exposure to X-rays or gamma rays, equal to the amount of radiation that produces ions with a total charge of 2.58×10^{-4} coulomb per kilogram of air. The roentgen is being superseded by the *SI unit of coulomb per kilogram. Named after W. C. *Roentgen.

Roentgen (or **Röntgen**), **Wilhelm Conrad** (1845–1923) German physicist.
***roentgen**
Roentgen rays Now called X-rays.

Rohrer, Heinrich (1933–) Swiss physicist.

Rolando, Luigi (1773–1831) Italian anatomist. Adjectival form: **Rolandic.**
fissure of Rolando

Rolle's theorem *Maths.* Named after Michel Rolle (1652–1719).

ROM Acronym and preferred form for read-only memory. Used in combinations (e.g. CD-ROM, PROM, EPROM, romware (lower case)) or adjectivally (e.g. ROM cartridge).

Roman numerals The system of numbers originally used by the Romans and based on letters of the alphabet:

$$I = 1, V = 5, X = 10, L = 50,$$
$$C = 100, D = 500, M = 1000.$$

Roman numerals are used in science for several purposes, including:

1. Numbering a series of objects, such as the satellites of a planet.
2. Specifying the *oxidation number of a compound (in small capitals following the name of an element, etc., within parentheses with no space); e.g. copper(II) sulphate, chlorate(V) ion.
3. Specifying the state of ionization of a z-fold ionized atom ($z = 0, 1, 2, \ldots$) (in small capitals, preceded by a thin space) corresponding to $z + 1$; e.g. H I, Al III.
4. Specifying a luminosity class of stars (*see* spectral types).

Romanowsky, Dmitriy Leonidovitch (1861–1921) Russian physician.
Romanowsky stain

Romberg, Moritz Heinrich (1795–1873) German physician.
Romberg's sign
Romberg's test

Romer, Alfred Sherwood (1894–1973) US palaeontologist.

Rømer (or **Römer, Roemer**), **Ole Christensen** (1644–1710) Danish astronomer.

romware *See* ROM.

röntgen Use *roentgen.

Röntgen, Wilhelm Conrad Usually alphabetized as *Roentgen.

rood An obsolete unit of area, equal to 1/4 acre, 1210 square yards.

root Two-word terms in which the first word is 'root' are not hyphenated; e.g. root cap, root hair, root nodule.

root mean square Abbrev.: rms (no stops). Hyphenated when used adjectivally (e.g. root-mean-square value) although the abbreviated form (e.g. rms power) is usually preferred. The rms value of a quantity x is usually denoted x_{rms} or $\tilde{x}$.

rootstock (one word)

Roque de los Muchachos Observatory *See* La Palma Observatory.

Rosaceae A large family of dicotyledonous plants, commonly known as the *rose family. Adjectival form: **rosaceous**.

rose An unspecified plant of the genus *Rosa* (family *Rosaceae), especially any of the cultivated hybrids; the word is qualified for individual species, e.g. dog rose (*R. canina*). Note that the word is also used in the common names of plants of other genera and families, e.g. Christmas rose (*Helleborus niger*), *rockrose.

Rose, William Cumming (1887–1985) US biochemist.

Ross, Sir Ronald (1857–1932) British physician and bacteriologist.

Rossby, Carl-Gustaf Arvid (1898–1957) Swedish-born US meteorologist.

Rossby waves Also called upper-air waves.

Rossby number Symbol: *Ro* (*see* parameter).

Rosse, William Parsons, Lord (1800–67) Irish astronomer and telescope builder.

Rossi, Bruno Benedetti (1905–) Italian-born US physicist.

Rossi–Forel scale (en dash) *Seismol.*

rot Another symbol for curl. *See* vector.

rotational frequency Symbol: n A physical quantity, the number of revolutions within an interval of time divided by that time. The *SI unit is the reciprocal of the second (s^{-1}). *See also* rpm.

rotavirus (one word) Any virus belonging to the genus *Rotavirus* (cap. R, ital.). *See also* reovirus.

Rotifera (cap. R) A phylum of minute aquatic invertebrates comprising the wheel animalcules. Individual name: **rotifer** (no cap.). Adjectival form: **rotiferal**.

Rot value (cap. R) A measure of RNA kinetic *complexity based on its hybridization ability. The notation is a modification of R_0t, where R_0 (subscript zero) is the initial RNA concentration and t is time. It is analogous to the *Cot value for DNA. $\text{Rot}_{1/2}$ (subscript 1/2) is the value corresponding to half complete association between complementary nucleotides.

Rouget, Charles Marie Benjamin (1824–1904) French physiologist.

Rouget cell Use pericyte.

roundworm (one word) *See* Nematoda.

Rous sarcoma virus Abbrev.: RSV (no stops). Also called avian sarcoma virus (abbrev.: ASV). Named after Francis Peyton Rous (1879–1970).

Roux, Paul Émile (1853–1933) French microbiologist.

Rowland, Henry Augustus (1848–1901) US physicist.

Rowland circle

Rowland ghost

Royal Agricultural Society Abbrev.: RAS (no stops).

Royal Astronomical Society Abbrev.: RAS (no stops).

Royal Greenwich Observatory Abbrev.: RGO (no stops). Now in Cambridge, previously at Herstmonceaux, Sussex, originally at Greenwich.

Royal Observatory, Edinburgh Abbrev.: ROE (no stops).

Royal Society of Chemistry Abbrev.: RSC (no stops).

R plasmid (preferred to R factor) Abbrev. and preferred form for resistance *plasmid (formerly factor). Most comprise two segments: a resistance transfer factor (RTF) and the r determinant (lower-case roman r). R plasmids are generally designated by a capital R in various combinations with other characters, e.g. RP1, R6K, RSF2124, pBR322, etc.

rpm (or **r.p.m.**) Abbrev. for revolutions per minute, used to indicate *rotational frequency. The unit of rotational frequency is the reciprocal of the second.

r-protein (lower-case r, hyphenated) Abbrev. for ribosomal protein.

RQ Abbrev. for respiratory quotient.

-rrhoea US: **-rrhea**. Noun suffix denoting flow or discharge (e.g. diarrhoea).

rRNA (lower-case r) Abbrev. for ribosomal RNA.

RSC Abbrev. for Royal Society of Chemistry.

R segment (cap. R) Abbrev. and preferred form for repeat segment, either of the segments that occur at the ends of retroviral RNA and its complementary DNA.

***r*-strategist** (hyphenated) An organism that colonizes unstable habitats, expending much of its resources on reproduction. The intrinsic rate of increase, *r*, of such organisms is high. *Compare* *K*-strategist.

RSV Abbrev. for Rous sarcoma virus.

RTECS Abbrev. for *Registry of Toxic Effects of Chemical Substances.

RTF Abbrev. for resistance transfer factor. *See* R plasmid.

Ru Symbol for ruthenium.

Rubbia, Carlo (1934–) Italian physicist.

rubella The medical name for German measles. *Compare* rubeola.

rubeola A former medical name for *measles. *Compare* rubella.

Rubiaceae A large family of dicotyledonous plants, commonly known as the madder family. Adjectival form: **rubiaceous**.

rubidium Symbol: Rb *See also* periodic table; nuclide.

RuBP (or **RUBP**) Abbrev. for ribulose 1,5-bisphosphate.

Rudbeck, Olof (1630–1702) Swedish naturalist.

rudbeckia or, as generic name (*see* genus), ***Rudbeckia***

RuDP (or **RDP**) Abbrev. for ribulose diphosphate, former name for ribulose 1,5-bisphosphate.

Ruhmkorff, Heinrich Daniel (1803–77) German-born French inventor.

Ruhmkorff coil

rumen (pl. rumens or rumina) The first compartment of a ruminant's stomach.

Rumford, Benjamin Thompson, Count (1753–1814) US-born British physicist.

Ruminantia (cap. R) A suborder of artiodactyl mammals that possess a complex stomach (*see* rumen) and chew the cud. Individual name and adjectival form: **ruminant** (no cap.).

Ruminococcus (cap. R, ital.) A genus of coccoid fermentative bacteria, typically inhabiting the rumen or colon. Individual name: **ruminococcus** (no cap., not ital.; pl. ruminococci).

Runcorn, Stanley Keith (1922–) British geophysicist.

Runge, Carl David Tolme (1856–1927) German mathematician.

Runge–Kutta method (en dash) Also named after W. M. Kutta (1867–1944).

Runge, Friedlieb Ferdinand (1795–1867) German chemist.

runoff (noun; one word) *Meteorol.*

Ruska, Ernst August Friedrich (1906–88) German physicist.

Russell, Bertrand Arthur William, Earl (1872–1970) British philosopher and mathematician.

Russell's paradox

Russell, Sir Edward John (1872–1965) British agricultural scientist.

Russell, Sir Frederick Stratten (1897–1984) British marine biologist.

Russell, Henry Norris (1877–1957) US astronomer.

Hertzsprung–Russell diagram (en dash)

Russell–Saunders coupling (en dash) Also named after F. A. Saunders (1875–1963). Also called LS coupling.

Russell–Vogt theorem (en dash; preferred to Vogt–Russell theorem) Also named after H. Vogt.

rust fungi *See* Urediniomycetes.

ruthenium Symbol: Ru *See also* periodic table; nuclide.

Rutherford, Ernest, Lord (1871–1937) New Zealand physicist.
Rutherford Appleton Laboratory Abbrev.: RAL (no stops).
Rutherford atom
rutherfordium *See* element 104.
Rutherford scattering

Ružička, Leopold (1887–1976) Croatian-born Swiss chemist.

Rydberg, Johannes Robert (1854–1919) Swedish physicist and spectroscopist.
***Rydberg constant**
Rydberg series
Rydberg's formula
Rydberg spectroscopy

Rydberg constant Symbol: R_∞ (subscript infinity). A fundamental constant equal to

$$10.973\,731\,534 \times 10^6\,\mathrm{m}^{-1}.$$

Former name: Rydberg's constant.

Ryle, Sir Martin (1918–84) British radio astronomer.

S

s Symbol for **1.** electron state $l = 0$ (*see* orbital angular momentum quantum number). **2.** second. **3.** singlet (in nuclear magnetic resonance spectroscopy). **4.** *solid state. *See also* state symbols. **5.** strong absorption (used in infrared spectroscopy).

s Symbol (light ital.) for **1.** path length. **2.** sedimentation coefficient. **3.** specific entropy. **4.** spin quantum number.

S **1.** Symbol for **i.** guanosine or cytidine (unspecified). **ii.** serine. **iii.** siemens. **iv.** solar mass. **v.** spiral *galaxy (followed by letter (s)). **vi.** substitution (*see* reaction mechanisms). **vii.** sulphur. **viii.** *Svedberg unit. **2.** Abbrev. for **i.** Silurian. **ii.** *south or southern.

S Symbol for **1.** (light ital.) apparent *power. **2.** (light ital.) entropy (S_m = molar entropy). **3.** (bold ital.) Poynting vector. **4.** (light ital.) spin quantum number (of a system). **5.** (light ital.) strangeness quantum number.
S_{ab} (light ital. S) Symbol for Seebeck coefficient (for substances a and b).

***s*-** (ital., always hyphenated) *Chem.* **1.** Prefix denoting secondary (e.g. *s*-butyl alcohol, *s*-butylamine). The alternative *sec-* is not recommended. **2.** Use **sym-*.

***S*-** (ital., always hyphenated) Prefix denoting substitution on a sulphur atom in an organic compound (e.g. *O*-ethyl *S*-methyl-3-ρ-tolyl-2-butenoate).

(*S*)- (ital., parentheses, always hyphenated) *Chem.* Prefix denoting *sinister*, indicating an optically active isomer in which the priority (obtained using the Cahn–Ingold–Prelog sequence rules) of the substituent groups on the chiral centre decreases in an anticlockwise direction (e.g. (*S*)-butan-2-ol). If more than one asymmetric carbon atom is present, the configuration at each is specified together with the carbon atom number (if necessary) (e.g. (*S*,*S*)-dichlorobutane, (2*S*,3*R*)-2,3-dichloropentane, (2*S*,3*S*)-tartaric acid. *Compare* (*R*)-.

SA12 (no space) Abbrev. for simian agent 12, a member of the genus **Polyomavirus*.

Sabatier, Armand (1834–1910) French scientist.
Sabatier effect

Sabatier, Paul (1854–1941) French chemist.

Sabin, Albert Bruce (1906–) Polish-born US microbiologist.
Sabin vaccine

Sabine, Sir Edward (1788–1883) British explorer, soldier, and geophysicist.
Sabine's gull (*Larus sabini*)

Sabine, Wallace Clement Ware (1868–1919) US physicist.
Sabine reverberation formula

saccharide Any carbohydrate, especially a sugar. The word is usually used in combination (e.g. monosaccharide, disaccharide, polysaccharide). *See also* sugars.

Saccharopolyspora (cap. S, ital.) A genus of nocardioform actinomycete bacteria. Individual name: **saccharopolyspora** (no cap., not ital.; pl. saccharopolysporas).

sac fungi *See* Ascomycotina.

Sachs, Julius von (1832–97) German botanist.

Sadron, Charles Louis (1902–) French physical chemist and biophysicist.

Sagan, Carl Edward (1934–) US astronomer.

Sagitta A constellation. Genitive form: Sagittae. Abbrev.: Sge (no stop). *See also* stellar nomenclature. *Compare* Sagittarius.

sagittal (not saggital) *Anat., physics*

Sagittarius A constellation. Genitive form: Sagittarii. Abbrev.: Sgr (no stop). *See also* stellar nomenclature. *Compare* Sagitta.

Saha, Meghnad N. (1894–1956) Indian astrophysicist.
Saha ionization equation

SAIDS Acronym for simian acquired immune deficiency syndrome.

Sainte-Claire Deville, Henri Étienne (hyphen) (1818–81) French chemist.

Saint-Hilaire, Étienne Geoffroy Usually alphabetized as *Geoffroy Saint-Hilaire.

Salam, Abdus (1926–) Pakistani physicist.

salic *Geol.* Denoting the silicon- and aluminium-rich minerals, calculated by the CIPW classification. *Compare* femic.

salicyl alcohol $C_6H_5CH_2OH$ The traditional name for 2-hydroxyphenylmethanol.

salicylaldehyde HOC_6H_4CHO The traditional name for 2-hydroxybenzaldehyde.

salicylamide $HOC_6H_4CONH_2$ The traditional name for 2-hydroxybenzamide.

salicylic acid HOC_6H_4COOH The traditional name for 2-hydroxybenzoic acid.

Salisbury, Sir Edward James (1886–1978) British botanist.

Salk, Jonas Edward (1914–) US microbiologist.
Salk vaccine

Salmon, Daniel Elmer (1850–1914) US pathologist.
***Salmonella**
salmonellosis

Salmonella (cap. S, ital.) A genus of bacteria of the family *Enterobacteriaceae that cause a variety of diseases in animals and man. Individual name: **salmonella** (no cap., not ital.; pl. salmonellas or salmonellae).
Taxonomically the genus is generally considered as comprising a single species containing numerous serovars. There is no officially endorsed system of subdividing the species or naming the different serovars. However, one widely accepted scheme divides the species into five subgroups ('subgenera'), designated I to V (Roman numerals).
Serovars of subgenus I are designated by a specific-like epithet, e.g. *S. choleraesuis*, *S. typhi*, and continuation of this practice for new serovars is recommended. Existing serovars of the other subgenera (except subgenus III) have been similarly designated, e.g. *S.* II *salamae*, *S.* IV *flint*, *S.* V *bongor*. Serovars of subgenus III, which is also known as the '*S. arizonae* subgroup' or simply the 'Arizona group', are designated only by their antigenic formulae (it is recommended that this practice be extended to subgenera II, IV, and V). The antigenic formula represents the somatic (O) antigens, the phase 1 flagellar (H) antigens, and the phase 2 flagellar (H) antigens. Broadly it takes the form 4,12:c:1,6 (groups of antigen labels separated by colons, with no intervening spaces).

SALR *Meteorol.* Abbrev. for saturated adiabatic lapse rate.

salt A chemical compound formed by reaction of an acid with a base, comprising a cation and an anion.
Simple salts are systematically named from their constituent ions as binary compounds (*see* inorganic chemical nomenclature), e.g. iron(III) chloride, $FeCl_3$, and copper(I) chloride, CuCl.
Salts containing acidic hydrogen are systematically named by incorporating the prefix hydrogen- in the name of the anion,

e.g. sodium hydrogensulphate, $NaHSO_4$, and lithium dihydrogenphosphate, LiH_2PO_4. (Note: in the USA it is common practice to separate the two components of the anion name, e.g. sodium hydrogen sulphide.)

The systematic nomenclature of salts of oxo acids is based on specific names of certain anions, which are often the same as the trivial names except for the addition of the oxidation number where necessary, e.g. carbonate, CO_3^{2-}, and phosphate(V), PO_4^{3-}. Related anions are named in a rational manner indicating structure using these simple names, e.g. diphosphate(V), $P_2O_7^{4-}$, and polytrioxophosphate, $(PO_3)^{n-}{}_n$. Examples are calcium carbonate, $CaCO_3$, sodium chlorate(I), NaOCl, and sodium chlorite(III), $NaClO_2$.

In nonsystematic nomenclature salts are named by naming the anions and cations, e.g. ferric chloride, $FeCl_3$, cuprous chloride, CuCl, sodium hypochlorite, NaOCl, and sodium chlorite, $NaClO_2$. Salts containing acidic hydrogen are named by incorporating the prefix bi- in the name of the anion, e.g. sodium bisulphate, $NaHSO_4$. Many common salts have trivial names, e.g. quicklime, $CaCO_3$.

The trivial anion names sulphate, SO_4^{2-}, sulphite, SO_3^{2-}, nitrate, NO_3^-, nitrite, NO_2^-, and thiosulphate, $S_2O_3^{2-}$, are recommended for general use.

saltpetre KNO_3 US: **saltpeter**. A traditional name for potassium nitrate.

samarium Symbol: Sm *See also* periodic table; nuclide.

Samuelsson, Bengt Ingemar (1934–) Swedish biochemist.

Sandage, Allan Rex (1926–) US astronomer.

sandbank (one word)

Sandmeyer reaction *Chem.* Named after Traugott Sandmeyer (1854–1922).

sandstone (one word)

Sanger, Frederick (1918–) British biochemist.
Sanger method (for gene sequencing)

Sänger, Eugen (1905–64) Austrian rocket scientist.

SA node Abbrev. for sinoatrial node.

sanserif (one word) A style of typeface (without serifs) used, for example, to denote a *tensor.

sapphire (not saphire)

sapro- (sapr- before vowels) Prefix denoting decaying matter or decomposition (e.g. saprolite, saprophyte, saprozoic).

Saprospira (cap. S, ital.) A genus of gliding bacteria. Individual name: **saprospira** (no cap., not ital.).

SAR Abbrev. for synthetic aperture radar.

Sarcina (cap. S, ital.) A genus of coccoid bacteria. The trivial name **sarcinae** (no cap., not ital.) is used to refer to similar cocci of the genera **Methanosarcina* and **Sporosarcina*, as well as members of *Sarcina*.

sarco- (sarc- before vowels) Prefix denoting muscle or fleshy tissue (e.g. sarcolemma, sarcomere, sarcoplasm, sarcoma).

Sarcopterygii *See* Choanichthyes.

SAS Abbrev. for Small Astronomical Satellite (for X-ray and γ-ray studies).

satellite laser ranging (not hyphenated) Abbrev.: SLR (no stops).

saturated adiabatic lapse rate (not hyphenated) Abbrev.: SALR (no stops).

saturated vapour pressure (not hyphenated) Abbrev.: SVP (no stops).

Saturn A planet. Adjectival form: **Saturnian.**

Saunders, Frederick Albert (1875–1963) US physicist.
Russell–Saunders coupling (en dash) Also called LS-coupling.

-saur Noun suffix denoting a lizard, usually used in names of extinct reptiles (e.g. dinosaur, plesiosaur, pterosaur). The suffix *-saurus* (ital.) is frequently used in generic names of extinct reptiles (e.g. *Ichthyosaurus*, *Tyrannosaurus*) and in the individual names derived from them (not ital.); it should not be used as a variant of -saur.

saurian Any reptile formerly classified in the suborder Sauria (now called Lacertilia), which comprises the lizards. The term saurian is still used as an adjective for descriptive (rather than taxonomic) purposes.

Saurischia (cap. S) An order of carnivorous and herbivorous dinosaurs. Individual name and adjectival form: **saurischian** (no cap.).

sauro- (**saur-** before vowels) Prefix denoting a lizard (e.g. Sauropterygia, Saurischia).

Saussure, Horace Bénédict de (1740–99) Swiss physicist and geologist.

SAV Abbrev. for simian *adenovirus.

Savart, Félix (1791–1841) French physicist.
Biot–Savart law (en dash)

Savery, Thomas (*c.* 1650–1715) British engineer and inventor.

SAW Abbrev. for surface acoustic wave.

sb Symbol for stilb.

Sb Symbol for antimony. [from Latin *stibium*]

SB (followed by a, b, c, ab, or bc) Symbol for a barred spiral *galaxy.

sBu, s-Bu Use Bu^s (*see* Bu).

Sc Symbol for scandium.

Sc (ital.) **1.** Symbol for Schmidt number. **2.** Abbrev. for stratocumulus.

scalar A *physical quantity, such as mass or temperature, that has magnitude but not direction. Symbols of scalars are printed in italic type.

scalar product A *scalar quantity involving two *vectors, $\boldsymbol{a}$ and $\boldsymbol{b}$. It is denoted $\boldsymbol{a} \cdot \boldsymbol{b}$ (centred bold dot) and has a magnitude $ab \sin \theta$, where θ is the angle between the two vectors. Also called dot product.

Scandentia An order of mammals comprising the tree shrews, formerly classified as a family (Tupaiidae) of the order Primates.

scandium Symbol: Sc *See also* periodic table; nuclide.

scanning electron microscope (not hyphenated) Abbrev.: SEM (no stops).
scanning transmission electron microscope Abbrev.: STEM (no stops).
scanning tunnelling microscope Abbrev.: STM (no stops).

scapho- Prefix denoting boat-shaped (e.g. scaphocephaly, Scaphopoda).

scato- Prefix denoting dung or excrement (e.g. scatology, scatophagous).

Schally, Andrew Victor (1926–) Polish-born US physiologist.

Schawlow, Arthur Leonard (1921–) US physicist.

Scheele, Karl Wilhelm (1742–86) Swedish chemist.
Scheele's green

Scheiner, Christoph (1575–1650) German astronomer.

Scheiner, Julius (1858–1913) German astrophysicist.
Scheiner number

Schering, Harald Ernst Malmsten (1880–1959) German engineer.
Schering bridge

Schiaparelli, Giovanni Virginio (1835–1910) Italian astronomer.

Schick, Bela (1877–1967) Hungarian-born US paediatrician.
Schick test

Schiff, Hugo (1834–1915) German-born Italian chemist.
Schiff base
Schiff reagent
Schiff test

schiller *Geol.* [from German: iridescence]

Schilling, Robert Frederick (1919–) US physician.
Schilling test

Schilling, Victor (1883–1960) German haematologist.
Schilling haemogram

Schimper, Andreas Franz Wilhelm (1856–1901) German plant ecologist.

schist *Geol.* Adjectival form: **schistose**. Derived noun: **schistosity**.

schistosome A fluke of the genus *Schistosoma* (formerly *Bilharzia*), which causes the disease **schistosomiasis** (former names: bilharzia, bilharziasis).

schizo- (**schiz-** before vowels) Prefix denoting a split, division, or cleavage (e.g. schizocarp, schizogony, schizont).

Schizomycetes An obsolete class of fungi comprising the bacteria, used when the latter were thought to have affinities to fungi.

Schleiden, Matthias Jakob (1804–81) German botanist.

Schlemm, Friedrich (1795–1858) German anatomist.
Schlemm's canal

schlieren (no cap.) *Physics, geol.* Often used adjectivally (e.g. **schlieren technique, schlieren photography**). [from German: streaks]

Schmidt, Bernhard Voldemar (1879–1935) Estonian-born German telescope maker.
Schmidt–Cassegrain telescope (en dash)
Schmidt corrector
Schmidt–Maksutov telescope (en dash)
Schmidt telescope (or **camera**) *See also* UKST.

Schmidt, Ernst Heinrich Wilhelm (1892–) German engineer.
***Schmidt number**

Schmidt, Karl Friedrich (1887–1988) German chemist.
Claisen–Schmidt condensation (en dash)
Schmidt reaction

Schmidt, Maarten (1929–) Dutch-born US astronomer.

Schmidt number Symbol: *Sc* A dimensionless quantity equal to ν/D, where ν is the kinematic *viscosity and D the *diffusion coefficient. *See also* parameter. Named after E. H. W. *Schmidt.

Schmitt, Otto Herbert (1913–) US biophysicist and electronic engineer.
Schmitt trigger

Schoenflies, Arthur Mortiz (1853–1928) German mathematician.
Schoenflies symbols

Schönbein, Christian Friedrich (1799–1868) German chemist.

Schönberg–Chandrasekhar limit (en dash) *Astrophysics.* Use Chandrasekhar limit. Named after Mario Schönberg (1916–) and S. *Chandrasekhar.

Schönlein, Johann Lucas (1793–1864) German physician.
Henoch–Schönlein purpura (en dash)

Schottky, Walter (1886–1976) German physicist.
Schottky barrier
Schottky defect
Schottky diode
Schottky effect
Schottky noise
Schottky TTL

Schrieffer, John Robert (1931–) US physicist. *See also* BCS theory.

Schrödinger, Erwin (1887–1961) Austrian physicist.
Schrödinger (or **Schrödinger's**) **equation**

Schrötter, Anton (1802–75) Austrian chemist.

Schuler pendulum *Physics* Named after Max Schuler.

Schüller, Artur (1874–1958) Austrian neurologist.
Hand–Schüller–Christian disease or **Schüller–Christian disease** (en dashes)

Schultz, Werner (1878–1948) German physician.
Schultz–Charlton test (en dash) Also named after W. Charlton (1889–).

Schultze, Max Johann Sigismund (1825–74) German zoologist.

Schultze's solution Chlor-zinc iodide (CZI), a stain used to detect the presence of cellulose. Named after Ernst Schultze (1860–1912).

Schuster, Sir Arthur (1851–1934) British physicist and spectroscopist.

Schwann, Theodor (1810–82) German physiologist.
Schwann cell
schwannoma
sheath of Schwann Use neurilemma.

Schwartz, Melvin (1932–) US physicist.

Schwarz, Hermann Amadeus (1843–1921) German mathematician.
Schwarz inequality

Schwarzschild, Karl (1873–1916) German astronomer.

Schwarzschild black hole
Schwarzschild radius

Schwinger, Julian Seymour (1918–) US physicist.

Science and Engineering Research Council Abbrev.: SERC (no stops).

Scl (no stop) *Astron.* Abbrev. for Sculptor.

sclera The white fibrous outer layer of the eyeball. Adjectival forms: **scleral, sclerotic**; the latter is often used (as a noun) as a synonym for sclera but this is not recommended.

sclero- (scler- before vowels) Prefix denoting hardness or thickness (e.g. sclerometer, scleroprotein, sclerenchyma, sclerosis).

sclerotic 1. Affected with sclerosis; hardened. **2.** Relating to the outer layer of the eyeball (sclera); the word is also used (as a noun) as a synonym for sclera but this is not recommended.

sclerotium (pl. sclerotia) A structure produced by certain fungi as a means of surviving adverse conditions. Not to be confused with *Sclerotium* (cap. S, ital.), a genus of fungi. *See also* ergot.

Sco (no stop) *Astron.* Abbrev. for Scorpius.

scolex (pl. scolices, not scoleces) The head of a tapeworm.

-scope Noun suffix denoting an instrument for observing or examining (e.g. microscope, oscilloscope, telescope). Adjectival form: **-scopic**. Derived noun form: **-scopy**.

Scorpius A constellation. Genitive form: Scorpii. Abbrev.: Sco (no stop). *See also* stellar nomenclature.

scoto- Prefix denoting darkness or low illumination (e.g. scotometer, scotophor, scotopic).

Scott, Charles F. (1864–1944) US electrical engineer.
Scott connection

Scott, Dukinfield Henry (1854–1934) British palaeobotanist.

SCP Abbrev. for single-cell protein.

SCR Abbrev. for silicon-controlled rectifier.

scRNA Abbrev. for small cytoplasmic RNA. It exists as *scRNP.

scRNP Abbrev. for small cytoplasmic ribonucleoprotein. Such molecules are known colloquially as 'scyrps'.

Scrophulariaceae A large family of dicotyledonous plants, commonly known as the figwort family. Adjectival form: **scrophulariaceous**.

scruple *See* ounce.

Sct (no stop) *Astron.* Abbrev. for Scutum.

scuba (pl. scubas) Acronym for self-contained underwater breathing apparatus.

Sculptor A constellation. Genitive form: Sculptoris. Abbrev.: Scl (no stop). *See also* stellar nomenclature.

scutellum (pl. scutella) *Bot., zool.*

scutum (pl. scuta) *Zool.*

Scutum A constellation. Genitive form: Scuti. Abbrev.: Sct (no stop). *See also* stellar nomenclature.

scyphistoma (pl. scyphistomae or scyphistomas; not scyphistomata) The sedentary stage of scyphozoans.

Scyphozoa (cap. S) A class of coelenterates (phylum Cnidaria) comprising the true jellyfish. Also called Scyphomedusae. Individual name and adjectival form: **scyphozoan** (no cap.).

scyrps *See* scRNP.

SD Abbrev. for standard deviation.

SDD *Telecom.* Abbrev. for subscriber direct dialling.

SDS Abbrev. for sodium dodecyl sulphate.

Se Symbol for selenium.

SE (no stops) Abbrev. and preferred form for south-east or south-eastern. For usage, *see* south.

sea Two-word terms in which the first word is 'sea' (e.g. sea breeze, sea fog, sea horse, sea level) should not be hyphenated unless used adjectivally (e.g. sea-floor spreading, sea-level pressure).

sea anemone (two words) *See* Cnidaria.

Seaborg, Glenn Theodore (1912–) US nuclear chemist.

sea cucumber (two words) *See* Holothuroidea.

seagull (one word) Avoid: use 'gull' for unidentified species; qualify this when the species is known (e.g. herring gull, black-backed gull).

sea lily (two words) *See* Crinoidea.

sealion (one word)

sea mile Symbol: M The length of one minute of latitude, measured along the meridian, in the latitude of the position. It varies slightly with latitude but is approximately 1853 metres. It is the principal means of expressing distance on Admiralty charts, including metric charts. A cable is 1/10 of a sea mile. *See also* nautical mile, international.

seamount (one word) *Geol.*

sea squirt (two words) *See* Urochordata.

sea urchin (two words) *See* Echinoidea.

seaweed (one word)

sebacic acid $HOOC(CH_2)_8COOH$ The traditional name for decanedioic acid.

sec Symbol for secant (reciprocal of cosine), written with a space or thin space before an angle or variable:

sec 30°, sec (−θ), sec *x*, sec (*a* + *b*).

The inverse function of $y = \sec x$ is denoted by:

$$x = \operatorname{arcsec} y$$

or

$$x = \sec^{-1} y \text{ (no space).}$$

sec- (ital., always hyphenated) *Chem. See s-*.

Secchi, (Pietro) Angelo (1818–78) Italian astronomer.
Secchi classification
Secchi disc

sech Symbol for hyperbolic secant, written with a space or thin space before a variable as in sech *x*, sech (*a* + *b*). The inverse function of $y = \operatorname{sech} x$ is denoted using the prefix ar-:

$$x = \operatorname{arsech} y$$

or

$$x = \operatorname{sech}^{-1} y \text{ (no space).}$$

second 1. Symbol: s (no stop). The *SI unit of time. It is one of the SI base units, defined since 1967 as the duration of 9 192 631 770 periods of the radiation corresponding to the transition between the two hyperfine levels of the ground state of the caesium-133 atom. **2.** Symbol: ″ A measure of angle equal to 1/60 of a minute, i.e. 4.848 14 microradian. Also called arc second, arcsec. *See also* degree.

section (in plant *taxonomy) Abbrev.: sect. (stop).

Sedgwick, Adam (1785–1873) British geologist and mathematician.

sediment Adjectival form: **sedimentary**. Derived noun: **sedimentation**.

sedimentation coefficient Symbol: s The rate of sedimentation of a particle in an ultracentrifuge. Sedimentation coefficients corrected to 20 °C in water are specified as $s_{20,w}$; sedimentation coefficients at zero concentration are specified as s^0 or $s^0_{20,w}$. Sedimentation coefficients of macromolecules and cellular particles are normally expressed in *Svedberg units.

Seebeck, Thomas Johann (1770–1831) Estonian-born German physicist.
Seebeck coefficient Symbol: S_{ab} (for substances a and b)
Seebeck effect

seed ferns *See* Pteridospermales.

Seeliger–Donker-Voet scheme (en dash, second name hyphenated) *Bacteriol. See Listeria.*

Segrè, Emilio Gino (1905–89) Italian-born US physicist.
Segrè chart

Seidel aberrations *Optics* Named after Ludwig von Seidel (1821–96).

seif dune *Geol.* [from Arabic: sword]

seismo- (**seism-** before vowels) Prefix denoting **1.** an earthquake or earthquakes (e.g. seismograph, seismology). **2.** shock, especially mechanical shock (e.g. seismonasty).

Selachii (cap. S; not Selachi) An order of elasmobranch fishes comprising the sharks. Individual name and adjectival form: **selachian** (no cap.).

Selaginella (cap. S, ital.) A genus of clubmosses, the sole extant genus of the order Selaginellales (*see* Lycopsida). Individual name: **selaginella** (no cap., not ital.; pl. selaginellas).

selenium Symbol: Se *See also* periodic table; nuclide.

seleno- (selen- before vowels) Prefix denoting the moon or moonlike; crescent-shaped (e.g. selenography, selenium).

self- (always hyphenated) Prefix denoting the same individual, component, material, etc.; lack of external involvement, control, etc. (e.g. self-fertilization, self-inductance, self-pollination, self-sterility).

self-inductance (hyphenated) Symbol: L A property of a single conducting loop, Φ/I, where I is the current in it and Φ is the *magnetic flux through it caused by this current. The *SI unit is the henry. *See also* mutual inductance.

Seliwanoff's test *Biochem.* Named after F. F. Seliwanoff.

SEM Abbrev. for scanning electron microscope.

Semenov, Nikolay Nikolaevich (1896–1986) Russian chemist.

semi- (hyphenated before i) Prefix denoting **1.** half (e.g. semicircular, semitone). **2.** partial or partially; intermediate (e.g. semiconductor, semigroup, semiparasite, semipermeable).

semicarbazone Any of a class of organic compounds containing the group $=CNNCONH_2$, in which the carbon atom is joined to two hydrocarbon groups or to a hydrocarbon group and a hydrogen atom. Semicarbazones are systematically named by adding the word semicarbazone after the name of the corresponding aldehyde or ketone, e.g. ethanal semicarbazone, $CH_3CHNNC(O)NH_2$, and propanone semicarbazone, $(CH_3)_2CNNC(O)NH_2$. In nonsystematic nomenclature semicarbazones are named as in systematic nomenclature but the trivial names of the corresponding aldehydes or ketones are used, e.g. acetaldehyde semicarbazone, $CH_3CHNNC(O)NH_2$, and acetone semicarbazone, $(CH_3)_2CNNC(O)NH_2$.

semiconductor *See* i-type; n-type; p-type.

semimetal Use metalloid.

-sepalous Adjectival suffix denoting sepals (e.g. gamosepalous, polysepalous). Noun form: **-sepaly**.

septi- (sept- before vowels) Prefix denoting **1.** seven (e.g. septifolious, septivalent). *See also* hepta-. **2.** a partition (e.g. septicidal).

septicaemia US: **septicemia**.

septum (pl. septa) *Anat., biol.* Adjectival forms: **septal, septate**.

Ser (no stop) **1.** (or **ser**) Abbrev. for serine. *See* amino acid. **2.** *Astron.* Abbrev. for Serpens.

seral Relating to a sere.

SERC Abbrev. or acronym for Science and Engineering Research Council. Former name: Science Research Council (SRC).

sere An ecological community at a particular stage in a succession. Different types are often denoted by prefixes (e.g. halosere, hydrosere, microsere). Adjectival form: **seral**.

serine Abbrev.: Ser or ser (no stop). Symbol: S *See* amino acid.

sero- Prefix denoting serum (e.g. serology, serotaxonomy).

serosa (pl. serosae) A serous membrane. Adjectival form: **serosal**.

serotonin Preferred to 5-hydroxytryptamine, but note that serotonin receptors are designated 5-HT_1 (subdivided into 5-HT_{1A} and 5-HT_{1B}) and 5-HT_2.

serotype Use *serovar.

serous Relating to, resembling, or producing *serum.

serovar Abbrev.: sv. (stop). A sero(logical) var(iety): an unofficial category of classification used in microbiology and ranking below subspecies. It is preferred to serotype, although the latter term is very widely used. Serovars are strains distinguished by their antigenic properties.

Serpens A constellation. Genitive form: Serpentis. Abbrev.: Ser (no stop). *See also* stellar nomenclature.

Serre, Jean-Pierre (hyphen) (1926–) French mathematician.

Sertoli, Enrico (1842–1910) Italian histologist.
Sertoli cell

serum (pl. sera) The fluid that separates from clotted blood, similar to *plasma but lacking coagulation factors. The adjectival form, **serous**, is most commonly applied to a type of membrane (**serous membrane**).

servomechanism (one word) Often shortened to **servo** (pl. servos).

servomotor (one word)

sesqui- Prefix denoting a ratio of 3 : 2 or a value of 1½ (e.g. sesquioxide).

seta (pl. setae) A bristle-like structure in plants and invertebrates, such as a *chaeta of an annelid worm or the stalk supporting a bryophyte capsule. Adjectival forms: **setaceous, setose**.

SETI *Astron.* Acronym for search for extraterrestrial intelligence.

sets The symbols used in set theory are shown in the table. There is usually a space or thin space on one or both sides of the symbol when used in an expression.

Sewall Wright effect (not hyphenated) Another name for genetic drift. Named after S. *Wright.

Seward, Albert Charles (1863–1941) British palaeobotanist.

Sex (no stop) *Astron.* Abbrev. for Sextans.

sex- Prefix denoting six (e.g. sexagesimal, sextile). *See also* hexa-.

sex chromosome A chromosome associated with the determination of sex. In most organisms there are two types, the normal-sized X chromosome and the small Y chromosome; in the diploid state the homogametic sex (the female sex in humans and many mammals) is denoted by XX, the heterogametic sex (male) by XY. In organisms that lack a Y chromosome the heterogametic sex is denoted by XO (capital O). In birds and lepidopterans the larger chromosome is called the Z chromosome and the smaller, the W chromosome; in these animals the homogametic sex (WW) is male, the heterogametic sex (WZ) female.

Symbols in set theory

is an element of: $x \in A$	$\in$
is not an element of: $x \notin A$	$\notin$
contains as element: $A \ni x$	$\ni$
set of elements	$\{a_1, a_2, \cdots\}$
empty set	$\varnothing$
the set of positive integers and zero	$\mathbb{N}$, **N**
the set of all integers, $\{\ldots, -2, -1, 0, 1, 2, \ldots\}$	$\mathbb{Z}$, **Z**
the set of rational numbers	$\mathbb{Q}$, **Q**
the set of real numbers	$\mathbb{R}$, **R**
the set of complex numbers	$\mathbb{C}$, **C**
set of elements of A for which $p(x)$ is true	$\{x \in A \mid p(x)\}$
is included in, subset of: $B \subseteq A$	$\subseteq$, $(\subset)$
contains: $A \supseteq B$	$\supseteq$, $(\supset)$
is properly contained in	$\subset$
contains properly	$\supset$
union: $A \cup B = \{x \mid (x \in A) \vee (x \in B)\}$	$\cup$
intersection: $A \cap B = \{x \mid (x \in A) \wedge (x \in B)\}$	$\cap$
difference: $A \setminus B = \{x \mid (x \in A) \wedge (x \notin B)\}$	$\setminus$
complement of: $\complement A = \{x \mid x \notin A\}$	$\complement$

Sextans A constellation. Genitive form: Sextantis. Abbrev.: Sex (no stop). *See also* stellar nomenclature.

Seyfert, Carl Keenan (1911–60) US astronomer.
Seyfert galaxy

S-form (hyphenated) Abbrev. and preferred form for smooth form: used to describe bacterial colonies with a smooth appearance. *Compare* R-form.

Sge (no stop) *Astron.* Abbrev. for Sagitta.

Sgr (no stop) *Astron.* Abbrev. for Sagittarius.

sh Short for sinh.

SHA Abbrev. for sidereal hour angle.

Shannon, Claude Elwood (1916–) US mathematician.
Shannon diagram
Shannon model
Shannon's theorems

Shapley, Harlow (1885–1972) US astronomer.

Sharpey, William (1802–80) British anatomist.
Sharpey's fibres

Sharpey-Schafer, Sir Edward Albert (hyphen) (1850–1935) British physiologist.

Shaw, Sir William Napier (1854–1945) British meteorologist.

shear modulus Symbol: *G* or μ (Greek mu). A physical quantity equal to the ratio of shear *stress to shear *strain, τ/γ. The *SI unit is the newton per square metre per radian (N m^{-2} rad^{-1}). Also called modulus of rigidity.

shear strain *See* strain.

shear stress *See* stress.

sheep pox (two words) *See also* poxvirus.

Shen Kua (*c.* 1031–*c.* 1095) Chinese scientist and scholar.

Shepard, Francis Parker (1896–1985) US marine geologist.

Sheppard, Phillip Macdonald (1921–76) British geneticist.

Sherrington, Sir Charles Scott (1857–1952) British physiologist.

SHF Abbrev. for superhigh frequency.

Shiga, Kiyoshi (1870–1957) Japanese bacteriologist.
****Shigella***
shigellosis

Shigella (cap. S, ital.) A genus of bacteria of the family *Enterobacteriaceae that are responsible for bacillary dysentery in primates (including humans). The four species – *S. dysenteriae*, *S. flexneri*, *S. boydii*, and *S. sonnei* – may alternatively be referred to as subgroups A, B, C, and D, respectively. Individual name: **shigella** (no cap., not ital.; pl. shigellae).

Shine–Dalgarno sequence (en dash) *Cell biol.*

SHM Abbrev. for simple harmonic motion.

Shockley, William Bradford (1910–89) British-born US physicist.
Shockley diode

Shoemaker, Eugene Merle (1928–) US geologist.

shoran Acronym for short-range navigation.

short Two-word terms in which the first word is 'short' are hyphenated when used adjectivally (e.g. short-circuit impedance, short-range force, short-wave radio).

Short, Charles W. (1794–1863) US botanist.
shortia (pl. shortias) or, as generic name (*see* genus), ***Shortia***

short circuit (noun; two words) Verb form: **short-circuit** (hyphenated).

short-day plant (hyphenated)

short hundredweight *See* pound.

short-sighted (hyphenated) US: **nearsighted** (one word). Noun form: **short-sightedness** (US: **nearsightedness**). Medical name: myopia.

Shortt clock *Astron., Physics* Named after W. H. Shortt.

short ton *See* pound.

short wave (hyphenated when used adjectivally) Abbrev. (for both) SW (no stops).

Si Symbol for silicon.

SI Abbrev. for Système International (d'Unités). *See* SI units.

sickle-cell anaemia, trait (hyphenated)

sideband (one word) *Telecom.*

sidereal (not siderial) Denoting, involving, or measured with reference to a star or stars.

sidereal day *See* day.

sidereal hour angle (not hyphenated) Abbrev.: SHA (no stops).

sidereal month *See* month.

sidereal year *See* year.

sidero- (**sider-** before vowels) Prefix denoting **1.** iron (e.g. siderocyte, siderolite, siderophile, siderite). **2.** the stars (e.g. siderostat).

Sidgwick, Nevil Vincent (1873–1952) British chemist.

Siebold, Karl Theodor Ernst von (1804–85) German zoologist and parasitologist.

Siegbahn, Kai (1918–) Swedish physicist, son of Karl Siegbahn.

Siegbahn, Karl Manne Georg (1886–1978) Swedish physicist, father of Kai Siegbahn.

Siegbahn unit

siemens (no cap.; pl. siemens) Symbol: S The *SI unit of electric *conductance.

$$1\,\mathrm{S} = 1\,\Omega^{-1}.$$

Named after Werner von *Siemens. Former names: mho, reciprocal ohm.

Siemens, (Ernst) Werner von (1816–92) German electrical engineer and inventor.

***siemens**

Siemens electrodynamometer Also named after his brother Karl von Siemens (1829–1906).

Siemens relay

Siemens, Sir (Charles) William (originally Karl Wilhelm von Siemens; 1823–83) German-born British engineer and inventor.

Siemens furnace Also called regenerative furnace.

Siemens–Martin process (en dash) Also named after P. E. *Martin. Now called open-hearth process.

Siemens process Also named after his brother Friedrich von Siemens (1826–1904).

sierra (pl. sierras) Adjectival form: **sierran.** [from Spanish: saw]

sieve (not seive)

sieve element *Bot.* A phloem cell that transports nutrients. The two types are **sieve cells**, having sieve areas; and **sieve-tube elements** or **members** (hyphenated), having sieve plates. A **sieve tube** is a series of sieve-tube elements.

sievert (no cap.) Symbol: Sv The *SI unit of *dose equivalent.

$$1\,\mathrm{Sv} = 1\,\mathrm{J\,kg^{-1}}.$$

The sievert has recently replaced the rem: one rem is equal to 10^{-2}Sv. Named after R. M. Sievert, 20th-century Swedish radiologist. *Compare* gray.

sigma Greek letter, symbol: σ (lower case), Σ (cap.).

σ Symbol for **1.** conductivity (electric). **2.** cross section. **3.** normal stress. **4.** *Genetics* sigma factor. **5.** Stefan–Boltzmann constant. **6.** surface density of charge. **7.** surface tension. **8.** symmetry number. **9.** wavenumber.

Σ Symbol for **1.** sigma particle (Σ^+, Σ^-, or Σ^0). **2.** sum of (*see also* limits).

sigma bond Usually written **σ bond** (Greek sigma). *See* orbital.

sigma factor (or **subunit** or **polypeptide**) Usually written **σ factor, subunit,** or **polypeptide** (Greek sigma). The subunit of RNA polymerase that determines promoter specificity during gene transcription. Different sigma factors are designated by a superscript number, which corresponds to the molecular mass in kilodaltons, e.g. σ^{55}, σ^{37}, etc.

sigma replication Sometimes written **σ-replication** (Greek sigma, hyphenated). A mode of DNA replication in which a structure resembling the letter σ is formed. Also called rolling circle replication.

signal-to-noise ratio (hyphenated; preferred to signal/noise ratio) Abbrev.: S/N ratio or SNR (no stops). It is often measured in *decibels.

Sikorsky, Igor Ivan (1889–1972) Russian-born US aeronautical engineer.

SIL *Electronics* Acronym for single in-line.

sila- (**sil-** before vowels) Prefix denoting a heterocyclic compound in which the hetero atom is silicon (e.g. silabicycloheptane, silolane).

silane Any of a class of compounds containing silicon and hydrogen atoms in an arrangement analogous to that of alkanes.

Silanes are named by using numerical prefixes to indicate the number of silicon atoms present (the first homologue, silane, SiH_4, has no prefix), e.g. disilane, SiH_3-SiH_3, and trisilane, $SiH_3SiH_2SiH_3$.

Organic silane derivatives (organosilanes) are named by prefixing the name of the parent silane by the names of the substituent groups, e.g. ethylsilane, CH_3CH_2-SiH_3, and 1,2-dimethyldisilane, CH_3-$SiH_2SiH_2CH_3$.

Important derivatives of silanes and organosilanes are halosilanes, silanols, siloxanes, silazanes, and silyl esters, analogous to haloalkanes, alcohols, ethers, secondary amines, and esters, respectively.

Halosilanes are named by prefixing the name of the parent silane or organosilane by the name of the substituent halogen, e.g. dichlorosilane, SiH_2Cl_2, and trichloro (methylsilane), CH_3SiCl_3. Halosilanes are traditionally known as silyl halides and named accordingly, e.g. methylsilyl trichloride, CH_3SiCl_3.

Silanols are named by adding the suffix -ol (along with any multiplying prefix) to the name of the parent silane or organosilane, e.g. silanol, SiH_3OH, and methyldisilane-1,2-diol, $CH_3SiH(OH)SiH_2OH$.

Siloxanes and silazanes are named by using numerical prefixes to indicate the number of silicon atoms attached to the linking oxygen and nitrogen atoms, respectively, along with the name of any substituent groups, e.g. disiloxane, SiH_3OSiH_3, methydisiloxane, $CH_3SiH_2OSiH_3$, and 3-methyltrisilazane, $SiH_3NHSiH(CH_3)NHSiH_3$. Poly (siloxanes) are known as silicones.

Silyl esters are named as salts of the parent carboxylic acid, e.g. trimethylsilyl ethanoate, $(CH_3)_3SiOOCCH_3$.

silanol *See* silane.

silazane *See* silane.

silica SiO_2 A traditional name for silicon(IV) oxide.

silicon Symbol: Si *See also* periodic table; nuclide.

silicon-controlled rectifier (hyphenated) Abbrev.: SCR (no stops).

silicon dioxide SiO_2 A traditional name for silicon(IV) oxide.

silicone Any of a class of organic polymers containing silicon.

silicon(IV) oxide SiO_2 The recommended name for the compound traditionally known as silica or silicon dioxide.

silicula (preferred to silicle; pl. siliculae) A type of dry fruit, produced by some members of the Cruciferae, that resembles a short broad pod. It is similar to, but should not be confused with, a *siliqua.

siliqua (preferred to silique; pl. siliquae) A type of dry fruit, produced by some members of the Cruciferae, that resembles a long narrow pod. It is similar to, but should not be confused with, a *silicula.

Silliman, Benjamin (1779–1864) US chemist.

silo- (**sil-** before vowels) Prefix denoting silicon (e.g. silane).

siloxane *See* silane.

Silurian Abbrev.: S (no stop). **1.** (adjective) Denoting the third period in the Palaeozoic era. **2.** (noun; preceded by 'the') The Silurian period.

silver Symbol: Ag *See also* periodic table; nuclide.

silver acetylide Ag_2C_2 The traditional name for silver(I) dicarbide.

silver(I) dicarbide Ag_2C_2 The recommended name for the compound traditionally known as silver acetylide.

SIMD *Computing* Abbrev. for single instruction multiple data.

simian adenovirus Abbrev.: SAV (no stops). *See* adenovirus.

simian immunodeficiency virus Abbrev. and preferred form: *SIV (no stops).

simian virus 40 Abbrev. and preferred form: SV40.

Simmonds, Morris (1855–1925) German pathologist.
Simmonds' disease (apostrophe)

simple harmonic motion (not hyphenated) Abbrev.: SHM (no stops).

Simplexvirus (cap. S, ital.) A genus of *alphaherpesviruses containing the *herpes simplex viruses.

Simpson, Sir George Clark (1878–1965) British meteorologist.

Simpson, George Gaylord (1902–84) US palaeontologist.

Simpson, Sir James Young (1811–70) British obstetrician.

Simpson, Thomas (1710–61) British mathematician.
Simpson's rule

simulation The imitation of some or all aspects of the behaviour of a system by another system, often in the form of a model and usually involving a computer. A computer device or program performing a simulation is called a **simulator**. *Compare* assimilation; emulation.

sin Symbol for sine, written with a space or thin space before an angle or variable:

$\sin 30°, \sin (-\theta), \sin x, \sin (a + b).$

The inverse function of $y = \sin x$ is denoted by:

$$x = \arcsin y$$

or

$$x = \sin^{-1}y \text{ (no space).}$$

Sinanthropus (cap. S, ital.) The generic name originally given to fossil remains found at Peking (Peking man), now classified as *Homo erectus*. *See Homo.*

single-cell protein (hyphenated) Abbrev.: SCP (no stops).

single in-line (hyphenated) *Electronics* Acronym: SIL (no stops).

single instruction multiple data *Computing* Abbrev.: SIMD (no stops).

single-lens reflex (hyphenated) *Photog.* Abbrev.: SLR (no stops).

single-sideband transmission Abbrev.: SST or SSB (no stops).

single-strand binding protein *Genetics* Abbrev.: SSB protein (no stops).

singlet state *See* electronic states.

sinh Symbol for hyperbolic sine, written with a space or thin space before a variable as in $\sinh x$, $\sinh (a + b)$. The shortened form sh is also permitted.
The inverse function of $y = \sinh x$ is denoted using the prefix ar-:

$$x = \text{arsinh}\, y \text{ (or arsh } y)$$

or

$$x = \sinh^{-1}y \text{ (no space).}$$

sinistro- (**sinistr-** before vowels) Prefix denoting on or towards the left (e.g. sinistrorse).

sinoatrial (one word)
sinoatrial node Abbrev.: SA node.

sinter (not sintre) Derived noun: **sintering**.

sinus (not ital.) **1.** *Anat., zool.* (pl. sinuses or, in human anatomical nomenclature, sinus) A cavity, channel, bulge, or indentation. **2.** *Astron.* (pl. sini) A small *mare or a semienclosed break in a scarp. The word, with an initial capital, is used in the approved name of such features, as in Sinus Medii on the moon and Sinus Meridiani on Mars.

sinusoid (noun) *Anat., maths., physics, etc.* Adjectival form: **sinusoidal**.

siphon (not syphon)

Siphonaptera (cap. S) An order of insects comprising the fleas. Also called Aphaniptera. Individual name and adjectival form: **siphonapteran** (no cap.).

Siphonophora (cap. S) An order of colonial hydrozoan coelenterates including the Portuguese man-of-war. Individual name: **siphonophore** (no cap.). Adjectival form: **siphonophorous**.

Siphunculata Use *Anoplura.

Sipunculida (cap. S; not Sipunculoidea) A phylum of marine burrowing wormlike invertebrates, formerly regarded as a class of annelids. Individual name: **sipunculid** (no cap.). Adjectival forms: **sipunculid**, **sipunculoid**.

Sirenia (cap. S) An order of aquatic mammals containing the dugongs and manatees. Individual name and adjectival form: **sirenian** (no cap.).

sirocco (not scirocco) *Meteorol.*

Site of Special Scientific Interest Abbrev.: SSSI (no stops).

Sitter, Willem de Usually alphabetized as *de Sitter.

SI units (Système International d'Unités; International System of Units) The system of units of measurement that was derived from the *mks system and is in use for all scientific and technical purposes. SI units have displaced *cgs units and *fps units. By international agreement seven physical quantities are regarded as being dimensionally independent; these are known as **base quantities**. Appendix 7, Table 7.1 lists these quantities together with the seven **base units** (and their symbols) on which the SI system is founded. The base units are all arbitrarily defined.

Derived units are expressed algebraically in terms of the base units, for example the SI unit of velocity is metres per second. Some of the derived SI units have special names and symbols. There are also two named **supplementary units**, the radian and steradian, which are now regarded as dimensionless derived units. Named derived units, together with their symbols, are given in Appendix 7, Table 7.3.

The full name of a unit is printed in lower case roman (upright) type, even when named after a person: the newton not the Newton; the degree Celsius is an exception. The plural forms of units usually end with an s, with the exception of hertz, lux, siemens, kelvin (kelvin or kelvins), tesla (tesla or teslas), and henry (henrys or henries). The singular form is used when the numerical value of a quantity is less than 1, as in 0.5 second.

The prefixes shown in Appendix 7, Table 7.2 are used to form names and symbols of decimal multiples and submultiples of SI units (*see also* kilogram). A prefix name is attached directly to the unit name, usually without a hyphen and without a loss of vowel, as in picoampere. However, megohm and mega-ampere are also used, although megaampere is preferred. The symbol of a prefix is combined with the single unit symbol to which it is directly attached, forming with it a new symbol; for example 1 cm^3 is $(10^{-2}\ m)^3$, i.e. $10^{-6}\ m^3$. Compound prefixes, such as kMHz, should not be used. The multiple or submultiple can usually be chosen so that the numerical value of a quantity lies between 0.1 and 1000, as in 4.677 kilovolts or 25 milliseconds. In some cases it may be better to use the same multiple or submultiple, for example in a table of values. It is recommended that only one prefix be used in forming a multiple or submultiple of a compound unit.

Symbols for units and prefixes should be printed in roman type, irrespective of the type used in the rest of the text. Unit symbols should remain unchanged in the plural (i.e. 20 kg not 20 kgs), should be written without a final full stop, except for normal punctuation, and should be placed after the complete numerical value for a quantity, leaving a space between value and unit symbol. No space should be left between the symbols for a unit and its prefix, as in mN, kHz. When the name of the unit is derived from a proper name, the unit symbol (or its first letter) is a capital letter, for example J for joule or Hz for hertz. Almost all other unit symbols use only lowercase letters, for example m for metre or lx for lux; the *litre, symbol L or l, has recently been made an exception.

A compound unit formed by multiplication of two or more units may be written as follows:

Pa s Pa·s

A fixed small (thin) space is usually used between the symbols when printing the former. In the case of the latter, a centred dot is preferred to a dot on the line.

A compound unit formed by dividing one unit by another may be indicated as follows:

$Pa\ s^{-1}$ Pa/s

or by any other way of writing the product of Pa and s^{-1}. Not more than one solidus should be used in an expression, for example J/(mol K) rather than J/mol/K. In complicated cases negative powers or parentheses should be used.

Certain units outside the SI system have been retained because of their practical importance (e.g. minute, hour, day, degree, minute, and tonne) or their use in specialized fields (e.g. electronvolt, atomic mass unit (unified), astronomical unit, parsec, and bar). SI prefixes may be attached to many of these units, as in millibar, megaparsec, kilotonne. In some cases compound units can be formed using these non-SI units and SI units, as in kilometre per hour.

SIV Abbrev. and preferred form for simian immunodeficiency virus. Isolates of the virus are designated by a subscript term (caps.) derived from the name of the host monkey: e.g. SIV_{MAC} is isolated from macaques; SIV_{AGM} is isolated from African green monkeys.

Sivapithecus (cap. S, ital.) A genus of fossil apes related to orang-utans; includes fossils originally classified as *Ramapithecus*.

Skinner, Burrhus Frederic (1904–90) US behavioural psychologist.

Skraup, Zdenko Hans (1850–1910) Austrian chemist.
Skraup synthesis

SLAC *See* SLC.

slaked lime $Ca(OH)_2$ The traditional name for calcium hydroxide.

SLC Abbrev. for Stanford Linear Collider, at the Stanford Linear Accelerator Center (SLAC), California.

slime moulds (preferred to slime fungi) *See* Myxomycota.

Slipher, Vesto Melvin (1875–1969) US astronomer.

SLR Abbrev. for **1.** single-lens reflex (camera). **2.** satellite laser ranging.

slug An *fps unit of mass equal to 32.1740 pounds or 14.5939 kilograms. One pound-force acting on this mass produces an acceleration of one foot per second per second.

Sm Symbol for samarium.

Small Astronomical Satellite Abbrev.: SAS (no stops).

SMC *Astron.* Abbrev. for Small Magellanic Cloud.

Smith, Hamilton Othanel (1931–) US molecular biologist.

Smith, Henry John (1826–83) British mathematician.

Smith, John Maynard Usually alphabetized as *Maynard Smith.

Smith, Theobald (1859–1934) US bacteriologist.

Smith, William (1769–1839) British surveyor and geologist.

Smithson, James (1765–1829) British geologist.

SMR Abbrev. for standard metabolic rate.

smut fungi *See* Ustilaginomycetes.

Sn Symbol for tin. [from Latin *stannum*]

SN Abbrev. for supernova.

Snell, Willebrord van Roijen (1591–1626) Dutch mathematician and physicist.
Snell's law

SNG Abbrev. for substitute natural gas.

snoRNP Abbrev. for small nucleolar ribonucleoprotein. Such molecules are known colloquially as 'snorps'.

snorps *See* snoRNP.

snow Terms incorporating the word 'snow' are usually written as one word (e.g. snowblind, snowcap, snowdrift, snowdrop, snowfall, snowflake); two-word terms (e.g. snow bunting, snow cover, snow goose, snow leopard, snow line) should not be hyphenated unless used adjectivally (e.g. snow-line variations).

SNR Abbrev. for **1.** supernova remnant. **2.** signal-to-noise ratio.

S/N ratio (or **SNR**) Abbrev. for signal-to-noise ratio.

snRNA Abbrev. for small nuclear RNA. There are several varieties, designated U1, U2, U4, U5, and U6 (cap. U (for uracil), no space). snRNAs exist as components of *snRNP.

snRNP Abbrev. for small nuclear ribonucleoprotein. These molecules comprise proteins and *snRNAs; they are known colloquially as 'snurps'.

snurps *See* snRNP.

Soddy, Frederick (1877–1966) British chemist.
Fajans–Soddy laws (en dash)

sodium Symbol: Na *See also* periodic table; nuclide.

sodium bicarbonate $NaHCO_3$ The traditional name for sodium hydrogencarbonate.

sodium bisulphate $NaHSO_4$ US: **sodium bisulfate.** The traditional name for sodium hydrogensulphate.

sodium bisulphite $NaHSO_3$ US: **sodium bisulfite.** The traditional name for sodium hydrogensulphite.

sodium borohydride $NaBH_4$ The traditional name for sodium tetrahydridoborate(III).

sodium chlorate(III) $NaClO_2$ The recommended name for the compound traditionally known as sodium chlorite.

sodium chlorite $NaClO_2$ The traditional name for sodium chlorate(III).

sodium disulphate(IV) $Na_2S_2O_5$ US: **sodium disulfate(IV).** The recommended name for the compound traditionally known as sodium metabisulphite.

sodium dodecyl sulphate US: **sodium dodecyl sulfate.** Abbrev.: SDS (no stops).

sodium heptaoxotetraborate(III)-10-water $Na_2B_4O_7{\cdot}10H_2O$ The recommended name for the compound traditionally known as borax.

sodium hydrogencarbonate $NaHCO_3$ The recommended name for the compound traditionally known as sodium bicarbonate.

sodium hydrogensulphate $NaHSO_4$ US: **sodium hydrogensulfate.** The recommended name for the compound traditionally known as sodium bisulphate.

sodium hydrogensulphite $NaHSO_3$ US: **sodium hydrogensulfite.** The recommended name for the compound traditionally known as sodium bisulphite.

sodium metabisulphite $Na_2S_2O_5$ US: **sodium metabisulfite.** The traditional name for sodium disulphate(IV).

sodium orthophosphate Na_3PO_4 The traditional name for sodium phosphate(V).

sodium phosphate(V) Na_3PO_4 The recommended name for the compound traditionally known as sodium orthophosphate.

sodium sulphate Na_2SO_4 US: **sodium sulfate.** The recommended name for the compound traditionally known as Glauber's salt.

sodium tetrahydridoborate(III) $NaBH_4$ The recommended name for the compound traditionally known as sodium borohydride.

software (one word) *Computing*

soil classification There is no single universally adopted system of classifying soils. Many countries employ national schemes that suit local requirements, although these often borrow from international systems. Traditional systems have been greatly influenced by the concept of zonal, intrazonal (transitional), and azonal soils, developed in Russia in the early 20th century. Zonal soils correspond to a particular climatic or vegetational zone; intrazonal soils are influenced by local factors that modify the prevailing climate (e.g. topology); and azonal soils are immature soils that are too young to reflect the effects of climate. These three principal categories accommodated the so-called 'great soil groups', with each group containing soils of a similar nature. These groups (e.g. sierozems, chestnut soils, chernozems, etc.) are still referred to in the literature, although for classificatory purposes they have been superseded by modern empirical definitional systems.

One of the most influential modern classifications is the *Soil Taxonomy* system, published by the US Department of Agriculture in 1975. It is a hierarchical system, with 10 orders subdivided into 47 suborders, 185 great groups, 970 subgroups, 4500 families, and over 10 000 series; the names of these groups have initial capital letters. Each order name has the ending '-sols', e.g. Entisols, Spodosols, etc. Names of suborders incorporate as a suffix an identifying element from the order name; hence the Aquents form a suborder of *Ent*isols, and Orthods are a suborder of the Sp*od*osols. This identifying element is retained in names of the lower ranks. For example, the suborder Orthods contains the great group Fragiorth*od*s, which itself contains the subgroup Typic Fragiorth*od*s. FAO and UNESCO have produced another important system, originally published as a legend to their *Soil Map of the World*. This draws from *Soil Taxonomy* as well as more traditional nomenclatures. It comprises 106 soil units grouped into 26 world classes. Most of the class names have the ending '-sols' (e.g. Fluvisols, Lithosols) but some traditional names have been retained as class names, e.g. Podzols (*see* podzol), Rendzinas, Phaeozems, Chernozems (initial caps.).

soil horizon Any of the various horizontal layers in a soil profile. Horizons are designated by capital letters: *see* A horizon; B horizon; C horizon; E horizon; F horizon; G horizon; H horizon; L horizon; O horizon; R horizon. Subdivisions of horizons are designated by an Arabic numeral; e.g. the subdivisions of the B horizon are designated B1, B2, and B3. Each of these subdivisions may be further subdivided, again denoted by a suffixed Arabic numeral; hence a B2 horizon can be differentiated into B21 and B22 (no spaces). Transitional or intermediate horizons are designated AB, BC, etc. (no space).
Specific characteristics of a particular horizon can be indicated by a lower-case suffix or suffixes. Often the suffix is the initial or other letter of a key word in the full descriptive gloss. Hence Ap is a ploughed A horizon, Bg is a gleyed B horizon, etc. For subdivisions, the descriptive suffix either follows the numerals (e.g. B21t, B22t) or precedes them (e.g. Ap1, Ap2: here the ploughed nature of the A horizon is considered the principal defining feature and subdivisions 1 and 2 are subsidiary features).
Layers of different lithology, i.e. from different parent materials, are denoted by prefixed Arabic or Roman numerals; horizons of the uppermost layer in such a profile are designated '1' or 'I', although this is usually omitted from the notation. For example, Cg, 2Cg, 3Cg indicates a gleyed C horizon from three sources. A buried soil horizon is indicated by an additional lower-case 'b' prefix.

sol. (stop) *Chem.* Abbrev. for **1.** solution. **2.** soluble.

Solanaceae A large family of dicotyledonous plants, commonly known as the nightshade family. Adjectival form: **solanaceous.**

solar luminosity *See* luminosity.

solar mass Symbol: S or $M_\odot$ The mass of the sun (1.9891×10^{30} kg), adopted as an astronomical constant and used as the unit of mass for astronomical bodies, e.g. 'a star of 7 solar masses'.

solar parallax Symbol: π (Greek pi). An astronomical constant adopted in 1976, equal to 8.794 148 arc seconds.

solar radius Symbol: $R_\odot$ The radius of the sun (696 000 km), adopted as an astronomical constant.

soleno- (**solen-** before vowels) Prefix denoting a tube (e.g. solenocyte, solenostele, solenoid).

solid It is recommended that the solid state of a substance X be denoted X(s), where X may be a chemical name or formula.

solid angle Symbol: Ω or ω (Greek cap. or lower-case omega). A dimensionless quantity indicating the region subtended by any area at a point. The *SI unit is the steradian.

solid-state (adjective; hyphenated)

soln. Abbrev. for solution.

Solo man Common name for a subspecies of fossil hominids (*see Homo*). Named after the River Solo, in Java, where the first fossils were found in 1931–32.

solonchak *Geol.* A soil type. [from Russian: salt marsh]

solonetz (preferred to solonets) *Geol.* A soil type. [from Russian: salt]

solstice Adjectival form: **solstitial.**

solubility product Symbol: K_s A physical quantity, the product of the *concentrations of ions in a saturated solution. The concentrations, c(sat), are expressed in mol m^{-3} or mol L^{-1} so that, for a binary electrolyte, the unit of K_s will be (mol $m^{-3})^2$ or (mol $L^{-1})^2$.

soluble *Chem.* Abbrev.: sol. (stop).

soluble RNA Abbrev.: sRNA Use transfer RNA.

solution *Chem.* Abbrev.: sol. or soln. (stops).

Solvay, Ernest (1838–1922) Belgian industrial chemist.
Solvay process

solvent (not solvant)

somato- (**somat-** before vowels) Prefix denoting the body (e.g. somatomedin, somatopleure, somatotrophin, somatotype).

somatotrophin US: **somatotropin.** Also called somatotrophic hormone (US: somatotropic hormone; abbrev.: STH). The term is synonymous with *growth hormone, but the two terms tend to be used in different contexts: somatotrophin in veterinary and agricultural contexts, growth hormone in human medicine.

-some Noun suffix denoting a body (e.g. acrosome, chromosome, kinetosome). Adjectival form: **-somal.**

Somerville, Mary (1780–1872) British astronomer and physical geographer.

Sommerfeld, Arnold Johannes Wilhelm (1868–1951) German physicist.
Bohr–Sommerfeld theory (en dash)
Fermi–Dirac–Sommerfeld law (en dashes)

Somogyi, Michael (1883–1971) Hungarian-born US biochemist.
Somogyi unit

sonar Acronym for sound navigation and ranging. Former name: asdic.

Sondheimer, Franz (1926–81) British chemist.

sono- (son- before vowels) Prefix denoting sound (e.g. sonogram, sonometer, sonoprobe).

Sopwith, Sir Thomas Octave Murdoch (1888–1989) British aircraft designer.

sorbic acid $CH_3CH=CHCH=CHCOOH$ The traditional name for 2,4-hexadienoic acid.

s orbital *See* orbital.

Sorby, Henry Clifton (1826–1908) British geologist.

Sørensen, Søren Peter Lauritz (1868–1939) Danish chemist.

Soret, Jacques-Louis (hyphen) (1827–90) Swiss physicist.
Soret band
Soret effect

sorghum (not sorgum) Generic name: *Sorghum* (cap. S, ital.).

-sorption Noun suffix denoting a taking up (e.g. adsorption, chemisorption). Adjectival form: **-sorptive.**

sorus (pl. sori) A collection of sporangia in ferns and certain algae and fungi. Adjectival forms: **soral, sorose.**

Sosigenes (*fl.* 1st century BC) Egyptian astronomer.

SOS response (or **SOS repair**) The enhanced DNA repair capability induced in *E. coli* by treatments that damage DNA. The response involves so-called SOS genes, in particular a base sequence known as the **SOS box.**

sound intensity Symbol: L or J A physical quantity, the *sound power (unidirectional) through an area perpendicular to the direction of propagation divided by the area. The *SI unit is the watt per square metre.

sound power Symbol: P or P_a A physical quantity, the sound energy transferred in a certain time interval divided by the length of the time interval. The *SI unit is the watt. Also called sound energy flux, acoustic power (applies to ultrasound and infrasound as well as audible sound). *See also* power level difference.

sound power level *See* power level difference.

sound pressure Symbol: p A physical quantity, the difference between the instantaneous total pressure and the static pressure, i.e. the pressure that would exist in the absence of sound waves. The *SI unit is normally the pascal but the bar may also be used. The term **acoustic pressure** is more general, applying to ultrasound and infrasound as well as audible sound, but it has the same symbol and units as sound pressure.
The **sound pressure level,** symbol L_p, is a dimensionless quantity equal to:

$$\log_e(p/p_0) = \log_e 10.\log_{10}(p/p_0),$$

in *nepers, or

$$20\log_{10}(p/p_0),$$

in *decibels, where p_0 is a reference pressure that must be explicitly stated. The sound pressure level is twice the sound power level.

south Adjectival forms: **south, southern.** Abbrev.: S (no stop). Use the abbrev. only descriptively, not in place names or concepts (e.g. S London, S Canada, but Southern Cross, South Atlantic, *South Pole, southern oscillation, magnetic south, true south, south-seeking pole). Use the same principle for **south-east(ern)** and **south-west(ern)** (e.g. SE London but South-East Asia).

south-east (hyphenated) Adjectival forms: **south-east, south-eastern.** Abbrev.: SE (no stops). For use of abbrev., *see* south.

Southern blotting (cap. S) A chromatographic technique for analysing DNA restriction fragments. Named after E. M. Southern. The similar techniques *Northern blotting and *Western blotting are named by analogy, not after their inventors.

South Pole (preceded by 'the'; not S Pole) Use initial capitals in the case of the earth, otherwise use lower case (e.g. south celestial pole, south galactic pole, Jupiter's south pole).

south polar distance Abbrev.: SPD (no stops).

south-south-east (hyphenated) Adjectival form: **south-south-eastern.** Abbrev. (for both): SSE (no stops).

south-south-west (hyphenated) Adjectival form: **south-south-western.** Abbrev. (for both): SSW (no stops).

south-west (hyphenated) Adjectival forms: **south-west, south-western.** Abbrev.: SW (no stops). For use of abbrev., *see* south.

sp. (stop; pl. spp.) Abbrev. for *species.

space Adjectival form: **spatial.**
Two-word terms in which the first word is 'space' (e.g. space charge, space group, space platform, space probe, space shuttle, space station) should not be hyphenated unless used adjectivally (e.g. space-charge limited).

spacecraft (one word)

spacelab (one word)

spacesuit (one word)

space–time (en dash)

spacewalk (one word)

spadix (pl. spadices) A type of inflorescence typical of the Araceae.

SPADNS Abbrev. for 2-(4-sulphophenylazo)-1,8-dihydroxy-3,6-naphthalenesulphonic acid.

Spallanzani, Lazzaro (1729–99) Italian biologist.

spatial (not spacial)

SPD *Astron.* Abbrev. for south polar distance.

speciality US: **specialty.** The US form is often used in British English, especially in medical contexts.

special relativity Abbrev.: SR (no stops).

species (pl. species) Abbrev.: sp. (stop; pl. spp.). The fundamental unit of biological classification (*see* taxonomy). All species have a two-part Latin name according to the system of *binomial nomenclature devised by Linnaeus; for example the brown rat is *Rattus norvegicus*. An unidentified species of rat would be designated *Rattus* sp. (or *Rattus sp.*). Many plant and animal species also have a common name; this should not be capitalized unless it is derived from a proper name. For example, lesser spotted woodpecker and wood anemone, but Père David's deer and Norway spruce. Adjectival form: **specific.**

specific 1. *Physics* A word that when placed before a physical quantity is now restricted in meaning to 'divided by mass', as in specific entropy, specific heat capacity. When the quantity is represented symbolically by an italic capital letter, the specific quantity is represented by the corresponding italic lower-case letter. Terms in which the word specific has some other meaning have generally been renamed; for example specific gravity is now called relative density, specific resistance is now called resistivity. **2.** *Biol. See* species.

specific energy imparted *See* energy imparted.

specific entropy *See* entropy.

specific epithet The second part of the two-part Latin name given to a species of plant (*see* binomial nomenclature). *Compare* specific name.

specific gravity Former name for relative density. *See also* specific.

specific heat capacity *See* heat capacity.

specific humidity *See* humidity.

specific internal energy *See* internal energy.

specific name The second part of the two-part Latin name given to a species of animal (*see* binomial nomenclature). *Compare* specific epithet.

specific resistance Former name for resistivity.

spectral flux density *See* jansky.

spectral types The seven groups into which the majority of stars can be classified according to absorption spectrum characteristics, each denoted by a capital letter. The classes are usually listed in order of decreasing stellar temperature:

O, B, A, F, G, K, M,

and may be remembered by the mnemonic 'Oh be a fine girl kiss me'. There are 10 subdivisions, depending on other factors, denoted by a digit, 0–9, placed immediately after the letter, e.g. B0, A5, G2; there may be further subdivisions, e.g. O9.5

Unusual stellar spectra are indicated by an additional lower-case letter placed immediately after the spectal type, notably Oe, Be, Ae, and Me stars (emission lines present) and Ap stars ('peculiar' A stars showing very strong lines of certain ionized metals).

Fine distinctions between spectra of stars of the same spectral type but different luminosity lead to different luminosity classes, each denoted by a Roman numeral:

Ia	bright supergiants
Ib	supergiants
II	bright giants
III	giants
IV	subgiants
V	main sequence stars
VI	subdwarfs
VII	white dwarfs

(Subdwarfs and white dwarfs are sometimes not included in this classification). The luminosity class is placed after a star's spectral type, separated by a thin space, e.g. B8 Ia, G2 V, K2 III.

spectro- Prefix denoting a spectrum or spectra (e.g. spectrograph, spectrophotometer, spectroscope).

spectroscopic transitions Upper and lower energy levels are indicated by ′ and ″, respectively. A transition is written with the upper energy level first and the lower energy level second, the terms being connected by an en dash, e.g. $^2S_{1/2}-{}^2P_{3/2}$ (electronic transition), $\pi-\pi^*$ (electronic transition), $J'-J''$ (rotational transition), $v'-v''$ (vibrational transition). Absorption and emission transitions can be indicated by the arrows ← and →, respectively, e.g. $^2S_{1/2}\rightarrow{}^2P_{3/2}$, $\pi\rightarrow\pi^*$, $J'\leftarrow J''$.

Radiationless transitions (i.e. internal conversion and intersystem crossing) are often denoted by wavy arrows, e.g. $S_1\rightsquigarrow T_1$.

spectrum (pl. spectra) Adjectival form: **spectral**.

speed *See* velocity.

speed of light in vacuum Symbol: c A fundamental constant equal to

299 792 458 m s^{-1} (exactly).

This value has been recommended since 1975 for universal use. The term is frequently shortened to speed of light. The word 'light' is used here to refer to the whole electromagnetic spectrum. To avoid confusion the term 'speed of electromagnetic waves (or electromagnetic radiation) in vacuum' may be used instead. Use of the word 'velocity' in this context, rather than 'speed', is discouraged.

speed of sound Symbol: c or c_a A physical quantity, the speed of propagation of sound waves in a medium. The *SI unit is the metre per second. The value in dry air at 0 °C is 331.4 m s^{-1}.

Spemann, Hans (1869–1941) German zoologist, embryologist, and histologist.

Spencer, Herbert (1820–1903) British philosopher.

Spencer Jones, Sir Harold (1890–1960) British astronomer.

sperm *See* spermatozoon.

-sperm *Bot.* Noun suffix denoting seeds (e.g. angiosperm, gymnosperm).

spermagonium Use *spermogonium.

spermatheca (not spermotheca; pl. spermathecae) A sac in some female and hermaphrodite invertebrates in which spermatozoa are stored before fertilization.

spermato- (**spermat-** before vowels) Prefix denoting male reproductive cells (e.g. spermatocyte, spermatogenesis, spermatozoon, spermatium).

spermatogenesis The process by which mature spermatozoa are produced in the testis from spermatogonia. It should not be confused with **spermiogenesis**, the final stage of this process, during which spermatids mature to spermatozoa.

spermatogonium (pl. spermatogonia) A primordial spermatozoon. *Compare* spermogonium.

Spermatophyta (cap. S) A division comprising the seed plants, usually subdivided into the *Gymnospermae and *Angiospermae. Also called Embryophyta Siphonogama. In a recent classification seed plants are regarded as a superclass, Spermatophytatinae, of the *Cormatae. Individual name and adjectival form: **spermatophyte** (no cap.).

spermatozoid Another name for antherozoid, a male gamete of lower plants. 'Antherozoid' is usually preferred.

spermatozoon (not spermatozoan; pl. spermatozoa) A mature male gamete in animals and humans. Often shortened to **sperm** (pl. sperms; 'sperm' is often used loosely as the plural form; it should not be used as a synonym for semen). *Compare* spermatozoid.

spermiogenesis The final stage of *spermatogenesis.

spermogonium (preferred to spermagonium; pl. spermogonia) A reproductive structure of certain fungi. Also called pycnidium. *Compare* spermatogonium.

Sperry, Elmer Ambrose (1860–1930) US inventor.

Sperry, Roger Wolcott (1913–) US neurobiologist.

sphaero- *Biol.* Prefix denoting spherical; use *sphero- except in taxonomic names (e.g. Sphaeropsidales).

Sphagnales *See* Sphagnidae.

Sphagnidae A subclass of mosses comprising the bog or peat mosses, contained within the genus *Sphagnum*. In some classifications it is reduced to an order, **Sphagnales**; in others it is elevated to a class, **Sphagnopsida**, within the superclass Bryatinae (*see* Bryopsida). Also called Sphagnobrya.

Sphagnobyra *See* Sphagnidae.

Sphagnopsida *See* Sphagnidae.

Sphagnum (cap. S, ital.) A genus comprising the bog or peat mosses (*see* Sphagnidae). Individual name: **sphagnum** (no cap., not ital.).

S phase (cap. S) A phase of the *cell cycle.

Sphenopsida (cap. S) A class of pteridophytes containing the horsetails (genus *Equisetum*, order Equisetales) and the extinct orders Calamitales, Sphenophyllales, and Pseudoborniales. Individual name and adjectival form: **sphenopsid** (no cap.). In a recent classification sphenopsids are regarded as a superclass, Sphenophyllatinae, in the infradivision Cormatae (vascular plants).

sphere Adjectival form: **spherical** (not spheric). Derived nouns: **sphericity**, **spheroid**.

-sphere Noun suffix denoting a sphere or a shell of a sphere (e.g. bathysphere, chromosphere, lithosphere, stratosphere). Adjectival form: **-spheric**.

sphero- (**spher-** before vowels) Prefix denoting a sphere, spherical or near-spherical, or curvature (e.g. spherometer, spheroplast, spherosome). *See also* sphaero-.

Spiegelman, Sol (1914–83) US microbiologist.

spino- Prefix denoting the spinal cord (e.g. spinobulbar, spinocaudal, spinocerebellar).

spin quantum number Symbol: s (individual entity) or S (whole system); I or J is used for nuclear spin quantum number. A number, either integral or half-integral, that characterizes the intrinsic angular momentum (i.e. spin) of a particle, atom, nucleus, etc. Also called spin angular momentum quantum number.

spiraea (pl. spiraeas) US: **spirea**. Generic name: *Spiraea* (cap. S, ital.).

spirillum (pl. spirilla) Any spirally shaped bacterium. More precisely, the term is used to denote any bacterium belonging to the genera **Aquaspirillum*, **Azospirillum*, **Oceanospirillum*, and **Spirillum*; some authors refer to these genera as the '*Spirillum* group', an assemblage that may also include the genera *Campylobacter* and *Bdellovibrio*. The term 'spirillum' is also used in a more restricted sense to denote individual members of the genus *Spirillum*.

Spirillum (cap. S, ital.) A genus of bacteria. *See* spirillum.

spiro- (**spir-** before vowels) Prefix denoting **1.** spiral (e.g. spirochaete). **2.** respiration (e.g. spirograph).

spirochaete (no cap.) US: **spirochete**. Any bacterium belonging to the order Spirochaetales, which includes the genus *Spirochaeta* (cap. S, ital.) and related genera. When using the term 'spirochaete', avoid possible confusion between members of this genus and those of the order as a whole.

spiroplasma *See* mollicute.

Spitzer, Lyman, Jr (1914–) US astrophysicist.

splanchno- Prefix denoting the viscera (e.g. splanchnocranium, splanchnopleure).

Spm (cap. S, ital.) Abbrev. for suppressor-mutator transposable element. *See* transposon.

sporangium (pl. sporangia) A spore-forming structure in plants: often borne on a **sporangiophore**. Adjectival form: **sporangial**.

Spörer, Gustav Freidrich Wilhelm (1822–95) German astronomer.
Spörer minimum
Spörer's law

Sporichthya (cap. S, ital.) A genus of actinomycete bacteria. Individual name: **sporichthya** (no cap., not ital.; pl. sporichthyae).

sporo- (**spor-** before vowels) Prefix denoting spores (e.g. sporocarp, sporocyst, sporophyll, sporophyte, sporopollenin, sporangium).

sporoactinomycete (one word) Any actinomycete bacterium that develops only as a mycelium. Also called euactinomycete. *Compare* proactinomycete.

Sporobolomycetales An order containing the imperfect or anamorphic yeasts (mirror or shadow yeasts). These are now usually included in the class *Hyphomycetes.

Sporosarcina (cap. S, ital.) A genus of aerobic spore-forming coccoid bacteria. Individual name: **sporosarcina** (no cap., not ital.; pl. sporosarcinae). *See also Sarcina*.

Sporozoa (cap. S) A class of parasitic protozoans. Individual name and adjectival form: **sporozoan** (no cap.); not to be confused with 'sporozoite', the infective stage in some sporozoans.

SPS Abbrev. for Super Proton Synchrotron, at CERN.

spurge Any herbaceous plant of the genus **Euphorbia*.

square brackets *See* brackets.

square metre US: **square meter**. Symbol: m^2 The derived unit of area in *SI units. The SI prefixes permit smaller or larger areas, such as the square millimetre (mm^2) or square kilometre (km^2) to be expressed:
$1\ mm^2 = (10^{-3}\ m)^2 = 10^{-6}\ m^2$
$1\ km^2 = (10^3\ m)^2 = 10^6\ m^2$.
One square metre is equal to 10^{-4} hectare, 1.195 99 square yards, 10.7639 square feet.

square root Denoted by $\sqrt{n}$ or $n^{1/2}$ for a number or quantity n. To indicate the extent of the square root of a more complicated quantity, parentheses can be used, as in $\sqrt{}(2\pi/gk)$, or sometimes a horizontal overline. A cube root is generally denoted by $n^{1/3}$ and so on.

square yard Symbol: yd^2 A traditional UK and US unit of area equal to 0.836 127 square metre.
$1\ yd^2 = 9$ square feet (ft^2)

1 ft^2 = 144 square inches (in^2)
4840 yd^2 = 1 acre.

squid (or **SQUID**) Acronym for superconducting quantum interference device.

sr Symbol for steradian. *See also* SI units.

Sr Symbol for strontium.

Sr (ital.) Symbol for Strouhal number.

SR Abbrev. for special relativity.

SRC Abbrev. for Science Research Council (now superseded by *SERC).

S region (cap. S) Abbrev. and preferred form for **1.** switching region, a recombinant site on a DNA molecule that changes the class of *immunoglobulin heavy chains expressed by the cell. **2.** serum region, the region of the *H–2 locus in mice that codes for *complement proteins.

sRNA Abbrev. for soluble RNA: use tRNA.

SSB Abbrev. for single-sideband (transmission).

SSB protein *Genetics* Abbrev. for single-strand binding protein.

SSE (no stops) Abbrev. and preferred form for south-south-east or south-south-eastern.

SSSI Abbrev. for Site of Special Scientific Interest.

SST Abbrev. for single-sideband transmission.

SSW (no stops) Abbrev. and preferred form for south-south-west or south-south-western.

St Symbol for stokes.

St (ital.) **1.** Abbrev. for stratus. **2.** Symbol for Stanton number.

ST Abbrev. for **1.** (Hubble) Space Telescope. **2.** thermostable toxin. *See* ETEC.

Stahl, Franklin William (1910–) US molecular biologist.
Meselson–Stahl experiment (en dash)

Stahl, Georg Ernst (1660–1734) German chemist and physician.

stalactite *Geol.* Extends downwards from a cave roof (mnemonic: includes c for ceiling).

stalagmite *Geol.* Extends upwards from a cave floor (mnemonic: includes g for ground).

standard atmosphere An internationally established reference for pressure, defined as 101 325 pascals. The atmosphere, symbol atm, was formerly used as a unit of pressure, equal to 101 325 Pa. It should no longer be used as a unit.

standard deviation Symbol: σ (for a distribution) or s (for a sample). Abbrev.: SD (no stops).

standard equilibrium constant *See* equilibrium constant.

standard metabolic rate Abbrev.: SMR (no stops).

standard pressure *See* pressure; stp.

standard temperature *See* stp.

Stanford–Binet test (en dash) Named after A. *Binet and Stanford University.

Stanford Linear Collider Abbrev.: *SLC (no stops).

Stanley, Wendell Meredity (1904–71) US biochemist.

stann- Prefix denoting tin (e.g. stannic, stanniferous, stannous).

stannane SnH_4 The traditional name for tin(IV) hydride.

stannic Denoting compounds in which tin has an oxidation state of +4. The recommended system is to use oxidation numbers, e.g. stannic chloride, $SnCl_2$, has the systematic name tin(IV) chloride.

stannous Denoting compounds in which tin has an oxidation state of +2. The recommended system is to use oxidation numbers, e.g. stannous chloride, $SnCl_2$, has the systematic name tin(II) chloride.

Stanton number Symbol: *St* A dimensionless quantity equal to the ratio of *Nusselt number to *Peclet number. *See also* parameter. Named after Sir Thomas Edward Stanton (1865–1931).

stapes (pl. stapes or stapedes) The stirrup-shaped innermost ear ossicle of mammals.

Staphylococcus (cap. S, ital.) A genus of facultatively anaerobic Gram-positive coccoid bacteria, including some pathogenic species. The generic name of individual species may be abbreviated to *Staph.* (rather than the usual *S.*) to avoid confusion with species of *Streptococcus*. Individual name: **staphylococcus** (no cap., not ital.; pl. staphylococci). Adjectival form: **staphylococcal**.

Strains of *S. aureus*, a common pathogen, produce at least five enterotoxins implicated in food poisoning. These are designated by capital letters, i.e. A, B, C, D, and E; enterotoxin C has three different forms, C_1, C_2, and C_3 (subscript numbers). Several exotoxins are also produced, including the α-, β-, γ-, and δ-haemolysins (lowercase Greek letters, hyphenated). *S. cohnii* has three subspecies, designated by Arabic numerals, e.g. *S. cohnii* (subsp. 2).

star *Astron. See* stellar nomenclature; spectral types. Related adjectives: **stellar, sidereal**.

starburst galaxy (not hyphenated) *Astron.*

Stark, Johannes (1874–1957) German physicist and spectroscopist.
Stark effect
Stark–Einstein equation (en dash)

Starling, Ernest Henry (1866–1927) British physiologist.
Starling's law

-stasis *Med.* Noun suffix denoting lack of movement (e.g. haemostasis, homeostasis). Adjectival form: **-static**.

stat- 1. *See* stato-. **2.** *Obsolete* Prefix denoting an electrostatic unit in the cgs system (e.g. statampere).

-stat Noun suffix denoting a means of keeping something constant or stationary (e.g. cryostat, heliostat, thermostat). Adjectival form: **-static**.

state symbols Symbols used to indicate the physical state of a substance in a chemical equation or formula. State symbols are set in roman type, within parentheses, after the formula, with no intervening space: g for gas, l for liquid, and s for solid. For example,

$$2H_2(g) + O_2(g) \rightarrow 2H_2O(l)$$

is an equation showing the reaction of hydrogen and oxygen in the gaseous states to produce water in the liquid state. Such equations are important in thermodynamics.

-static 1. Adjectival suffix denoting lack of motion (e.g. electrostatic, hydrostatic). **2.** *See* -stasis; -stat; -statics.

-statics Noun suffix denoting a branch of science concerned with lack of motion (e.g. electrostatics). Adjectival form: **-static**.

stationary Not moving or changing. *Compare* stationery.

stationery Paper, etc. *Compare* stationary.

statistics 1. (takes a sing. form of verb) The science itself. **2.** (takes a pl. form of verb) The numerical data involved or the quantities derived from the data. Adjectival form: **statistical**.

stato- (stat- before vowels) Prefix denoting balance or lack of motion (e.g. statocyst, statolith, stator).

Staudinger, Hermann (1881–1965) German chemist.

STD *Telecom.* Abbrev. for subscriber trunk dialling.

stearic acid $CH_3(CH_2)_{16}COOH$ The traditional name for octadecanoic acid.

steato- (steat- before vowels) Prefix denoting fat or fatty (e.g. steatolysis, steatite).

Stebbins, George Ledyard (1906–) US geneticist.

Steenstrup, Johann Japetus Smith (1813–97) Norwegian-born Danish zoologist.

Stefan, Josef (1835–93) Austrian physicist.
***Stefan–Boltzmann constant** (en dash)
Stefan–Boltzmann law (en dash; preferred to Stefan's law)

Stefan–Boltzmann constant (en dash) Symbol: σ (Greek sigma). A fundamental constant equal to

$$5.670\,51 \times 10^{-8}\,\mathrm{W\,m^{-2}\,K^{-4}}.$$

Also called Stefan constant (formerly Stefan's constant).

Stein, William Howard (1911–80) US biochemist.

Steinberger, Jack (1921–) German-born US physicist.

Steiner, Jakob (1796–1863) Swiss geometer.
Steiner's problem
Steiner triplet

Steinheim man Common name for a type of fossil hominid (*see Homo*). Named after Steinheim an der Murr, near Stuttgart, Germany, where the first fossils were found in 1933.

Steinmann, Fritz (1872–1932) Swiss surgeon.
Steinmann's pin

Steinmetz, Charles Proteus (1865–1923) German-born US electrical engineer.
Steinmetz's law *Magnetism*

stelar *Bot.* Relating to steles. *Compare* stellar.

stele (pl. steles; not stelae) The conducting tissue of pteridophytes and seed plants. Different types of stele are denoted by a prefix (e.g. eustele, dictyostele, polystele, protostele, siphonostele). Adjectival form: **stelar**.

stellar Relating to stars. *Compare* stelar.

stellar nomenclature Many bright stars have special names, e.g. Sirius and Canopus. In the more systematic system introduced by Bayer (1603), Greek letters have been allotted in alphabetical order to stars in each constellation in order of brightness as seen from earth. The letter, written out in full with an initial capital, is followed by the genitive form of the constellation name, e.g. Alpha Centauri, Beta Crucis, Gamma Orionis. Usually the abbreviated stellar name is used, in which the letter itself is followed (after a thin space) by the abbreviation of the constellation name, e.g. α Cen, β Cru, γ Ori.
Less bright stars are designated by their number in a particular star catalogue. Modern catalogues generally ignore the constellation and number by *right ascension.
In a binary star or other multiple star system, the component stars are designated A, B, C, in order of decreasing apparent brightness.
A special nomenclature is used for variable stars, i.e. those whose brightness varies with time. Variables in a constellation are denoted by one or two roman capital letters followed by the genitive of the constellation name, or its abbreviation. The letters are allotted (as the variables are discovered) in the order:
R, S, T, . . . , Z, then RR, RS, . . . , RZ, then SS, ST, . . . , SZ, up to ZZ, then AA, . . . , AZ, BB, . . . , BZ, up to QZ (the letter J is not used).
Thereafter the designations V335, V336, etc., are used. Examples of variables are T Tauri, W Virginis, RR Lyrae, RV Tauri, WW Aurigae. In the case of novae, there is an interim designation comprising constellation (genitive form) and year of observation before a permanent variable star designation is made; for example, Nova Herculis 1934 became DQ Herculis.

Steller, Georg Wilhelm (1709–46) German naturalist and explorer.
Steller jay (*Cyanocitta stelleri*)
Steller sealion (*Eumetopias jubatus*)
Steller's eider duck (*Polysticta stelleri*)

STEM Abbrev. for scanning transmission electron microscope.

Steno, Nicolaus (1638–86) Danish anatomist and geologist. Danish name: Niels Stensen.
Steno's law
Stensen's duct

steno- (**sten-** before vowels) Prefix denoting narrow or restricted (e.g. stenohaline, stenopodium, stenosis).

Stensen's duct (not Stenson's duct) *Anat.* Named after N. *Steno.

Stephenson, George (1781–1848) British engineer and inventor.

Stephenson, Robert (1803–59) British civil engineer.

steradian Symbol: sr The *SI unit of solid angle. It is a dimensionless derived unit, i.e. a supplementary unit. The solid angle subtended by the surface of a sphere at its centre is equal to 4π sr. *See also* radian.

stereo- Prefix denoting solid or three-dimensional (e.g. stereochemistry, stereoisomerism, stereoscopic).

sterigma (pl. sterigmata) *Bot.* A projection on a basidium on which a basidiospore is produced. *Compare* stigma.

Stern, Curt (1902–81) German-born US geneticist.

Stern, Otto (1888–1969) German-born US physicist.
Stern–Gerlach experiment (en dash)
Stern layer

sternum (pl. sterna; preferred to sternums) *Anat., zool.* Adjectival form: **sternal**.

Stevenson, Thomas (1818–87) British engineer and meteorologist.
Stevenson screen

Stevin, Simon (1548–1620) Flemish mathematician and engineer. Also called Stevinus.

STH Abbrev. for somatotrophic hormone (*see* somatotrophin).

stib- Prefix denoting antimony (e.g. stibine, stibnite).

stibine SbH_3 The traditional name for antimony(III) hydride.

-stichous Adjectival suffix denoting a row or line (e.g. distichous). Noun form: **-stichy**.

Stieltjes, Thomas Jan (1856–94) Dutch-born French mathematician.
Stieltjes integral

stigma (pl. stigmas; preferred to stigmata) **1.** The receptive surface of a carpel. **2.** An eyespot. Adjectival form: **stigmatic**. *Compare* sterigma.

stilb Symbol: sb A *cgs unit of *luminance equivalent to one candela per square centimetre. In *SI units luminance is measured in candelas per square metre.

stilbene $C_6H_5CH{=}CHC_6H_5$ The traditional name for *trans*-1,2-diphenylethene.

stilboestrol US: **stilbestrol**.

Stillson wrench (not Stilson)

stimulus (pl. stimuli)

Stirling, James (1692–1770) British mathematician.
Stirling numbers
Stirling's formula (or **approximation**)

Stirling, Robert (1790–1878) British engineer.
Stirling cycle
Stirling engine

STLV Abbrev. for simian T-lymphotropic virus.

STM Abbrev. for scanning tunnelling microscope.

stochastic *Stats.*

stoichio- Prefix denoting component parts or proportions (e.g. stoichiometry).

stoichiometric coefficients The relative numbers of atoms, molecules, ions, etc., of reactants and products in a chemical reaction. For a particular substance B, the symbol is ν_B (Greek nu).
It is recommended by IUPAC that the general equation for a chemical reaction be expressed as

$$0 = \Sigma_B \, \nu_B B.$$

This leads to the definition of the **extent of reaction**, symbol ξ (Greek xi):

$$d\xi = (1/\nu_B)dn_B,$$

where n_B is the *amount of substance B; the *SI unit of ξ is the *mole. It follows that the stoichiometric coefficients are negative for reactants and positive for products; this is the reverse of the previous convention.
The rate of change of extent of reaction, $d\xi/dt$, is now called the **rate of conversion of B** and given the symbol $\dot{\xi}$; the SI unit is the mole per second. Because of conflicting usage in different contexts it is strongly recommended that the terms 'rate of reaction' and 'rate of formation' should be avoided in all quantitative work.

stoichiometry (preferred to stoicheiometry) *Chem.* Adjectival form: **stoichiometric**.

stokes (not stoke; pl. stokes) Symbol: St A *cgs unit of viscosity (kinematic viscosity), now discouraged. In *SI units, kinematic viscosity is measured in square metres per second: 1 St = 10^{-6} m^2 s^{-1}. Named after Sir George *Stokes. *See also* poise.

Stokes, Sir George Gabriel (1819–1903) British mathematician and physicist.
***stokes**
Stokes' law
Stokes layer
Stokes' theorem

Stokes, William (1804–78) Irish physician.

Cheyne–Stokes respiration (en dash)
Stokes–Adams attack or **syndrome** (en dash)

stoma 1. (not stomate; pl. stomata) An epidermal pore in plants through which gaseous exchange occurs. **2.** (pl. stomata) A mouth or mouthlike part. Adjectival form (defs. 1. and 2.): **stomal** or **stomatal**. **3.** (pl. stomas) An artificial opening of a tube (e.g. the colon) made during surgery on the surface of the abdomen. Adjectival form: **stomal**.

Stomatococcus (cap. S, ital.) A genus of Gram-positive facultatively anaerobic coccoid bacteria. Individual name: **stomatococcus** (no cap., not ital.; pl. stomatococci). Adjectival form: **stomatococcal**.

-stome Noun suffix denoting a mouth or mouthlike opening (e.g. peristome).

stomodaeum (pl. stomodaea) US: **stomodeum** (pl. stomodea; the US spelling is now often used in British English). The embryonic mouth. Adjectival form: **stomodaeal** (US: **stomodeal**).

stone *See* pound.

stoneworts *See* Charophyta.

stp (or **s.t.p.**) Abbrev. for standard temperature and pressure, the standard conditions for comparing the properties of gases. Standard temperature is 298.15 K (formerly 273.15 K, i.e. 0 °C). It is now recommended by IUPAC that the standard pressure for gases should be 10^5 Pa (formerly 101 325 Pa). *See* pressure.

Strabo (*c.* 63 BC–*c.* AD 23) Greek geographer and historian.

Stradonitz, Friedrich August Kekulé von Usually alphabetized as *Kekulé von Stradonitz.

strain The deformation of a body, temporary or permanent, resulting from an applied *stress. **Linear strain**, symbol ϵ (Greek epsilon) or *e*, is the change in length per unit length; former name tensile strain. **Volume strain**, symbol θ (Greek theta), is the change in volume per unit volume; also called bulk strain. **Shear strain**, symbol γ (Greek gamma), is the angular deformation of a body in radians, the volume remaining constant.

strait *Geog.*

strangeness Symbol: *S* A quantum number associated with subatomic particles. It takes zero or integral values.

Strasburger, Eduard Adolf (1844–1912) German botanist.

Strassmann, Fritz (1902–80) German chemist.

strata Plural of *stratum: always use with a plural form of verb.

strati- Prefix denoting a layer or layers, especially of rock (e.g. stratiform, stratigraphy). *See also* strato-.

strato- Prefix denoting **1.** a layer or layers (e.g. stratomere, stratopause, stratosphere). **2.** stratus clouds (e.g. stratocumulus). *See also* strati-.

stratocumulus (pl. stratocumuli) Abbrev.: *Sc* (cap. S, ital.).

stratum (pl. strata) **1.** Any of the layers of which sedimentary rocks are formed. The word is often used in the plural but this should never be used with the singular form of a verb. **2.** A layer of tissue or cells.

stratum corneum (pl. strata cornea) A layer of the epidermis.

See also strato-. *Compare* stratus.

stratus (pl. strati) Abbrev.: *St* (cap. S, ital.). A type of cloud. *See also* strato-. *Compare* stratum.

strepto- Prefix denoting **1.** twisted (e.g. *Streptococcus*). **2.** *Streptococcus* bacteria (e.g. streptokinase, streptolysin).

Streptobacterium (cap. S, ital.) A subgenus of **Lactobacillus*. Individual name: **streptobacterium** (no cap., not ital.; pl. streptobacteria).

Streptococcus (cap. S, ital.) A genus of Gram-positive lactic acid-producing bacteria, including some pathogenic species. The generic name of individual species may be abbreviated to *Strep.* (rather than the usual *S.*) to avoid confusion with species of *Staphylococcus*. Individual name: **strep-**

tococcus (no cap., not ital.; pl. streptococci). Adjectival form: **streptococcal**.

Various systems of classifying streptococci are in use. One of the most widely used is the Lancefield classification, in which pathogenic strains are assigned to serological groups designated by roman capital letters, according to group-specific antigens in the cell wall, i.e. A, B, C, etc. Some of these groups correspond to a particular species, e.g. group A streptococci correspond to the species *S. pyogenes*, group B is the same as *S. agalactiae*; some groups may include members of more than one species, and others correspond to species that lack a specific epithet, in which case the species is referred to as (for example) group M streptococci, or *Streptococcus* sp. (group M). These groups may be subdivided on the basis of type-specific antigens. Group A (*S. pyogenes*), for example, is subdivided on the basis of the composition of the fimbrial antigenic M protein; hence serovars are designated M1, M2, etc. (no space). Strains may also be described in terms of their haemolytic activity (*see* haemolysis).

Various authors have attempted to subdivide the genus into 'divisions' of a primarily descriptive nature. These groups are given trivial names and have no taxonomic standing. One such scheme comprises the divisions 'pyogenic', 'viridans', 'lactic', and *enterococcus. *See also* pneumococcus.

Streptomyces (cap. S, ital.) A genus of actinomycete bacteria. Individual name and adjectival form: **streptomycete** (no cap., not ital.).

The 378 streptomycete species listed in the Approved Lists of Bacterial Names, published in 1980, are now assigned to so-called species clusters on the basis of shared features. Each cluster takes the name of the earliest validly named member; for example, the *S. anulatus* cluster contains 24 other species and is called a 'major cluster'. 'Minor clusters' contain only two to three species, while others contain only a single species.

Streptophyta A division of plants in the subkingdom *Chlorobionta. It includes the subdivisions Charophytina (stoneworts; *see* Charophyta) and *Embryophytina.

Streptosporangium (cap. S, ital.) A genus of *maduromycete bacteria. Individual name: **streptosporangium** (no cap., not ital.; pl. streptosporangia).

Streptoverticillium (cap. S, ital.) A genus of actinomycete bacteria. Individual name: **streptoverticillium** (no cap., not ital.; pl. streptoverticillia). Adjectival form: **streptoverticilliate**. The species of this genus are now assigned to a number of so-called species clusters, on the basis of shared features. Clusters are named after the earliest validly named member species; for instance, the *S. baldacci* cluster contains 12 other species.

stress A physical quantity, equal to one of a system of opposing forces applied to a body, divided by the area over which it acts. It produces *strain. The *SI unit is the newton per square metre ($N\ m^{-2}$) or the pascal (Pa). **Normal stress**, symbol σ (Greek sigma), tends to stretch or compress in the direction of the applied force; former name: tensile stress. **Shear stress**, symbol τ (Greek tau), is the tangential force per unit area.

stria (pl. striae) *Geol., biol., anat.* Adjectival form: **striated**. Derived noun: **striation**.

strobila (pl. strobilae) **1.** The body of a tapeworm. **2.** A stage in the life cycle of jellyfish. Derived noun: **strobilation**. *Compare* strobilus.

strobilus (pl. strobili) **1.** The reproductive structure of gymnosperms and some pteridophytes: it is the cone of conifers. **2.** A type of dry composite fruit. *Compare* strobila.

Strohmeyer, Friedrich (1776–1835) German chemist.

stroma (pl. stromata) **1.** Tissue forming a matrix, such as the connective tissue of the mammalian ovary or the mass of fungal hyphae in which reproductive structures form. **2.** The matrix of a chloroplast. Adjectival form: **stromatous**.

Strömgren, Bengt Georg Daniel (1908–87) Swedish-born Danish astronomer.

Strömgren sphere

strontium Symbol: Sr *See also* periodic table; nuclide.

Strouhal, Čeněk (1850–1922) Czech scientist.
***Strouhal number**

Strouhal number Symbol: *Sr* A dimensionless quantity equal to lf/v, where l, f, and v are a characteristic length, frequency, and speed. *See also* parameter. Named after C. Strouhal.

Strowger exchange *Telecom.* Named after Almon B. Strowger.

Strutt, John William *See* Rayleigh, Lord.

Struve, Friedrich Georg Wilhelm von (1793–1864) German-born Russian astronomer.

Struve, Otto (1897–1963) Russian-born US astronomer.

STS Abbrev. for space transportation system, formerly used in numbering space shuttle missions.

Student Pseudonym of **William Sealy Gosset** (1876–1937), British statistician.
studentization
Student's distribution (or **Student's *t* distribution**; ital. t)
Student's test (or **Student's *t* test**; ital. t)

Sturgeon, William (1783–1850) British physicist.

Sturm, Jacques-Charles-François (hyphens) (1803–55) French mathematician.
Sturm–Liouville problem (en dash)
Sturm's theorem

Sturtevant, Alfred Henry (1891–1970) US geneticist.

stylus (pl. styluses; not styli) *Acoustics*

-styly Noun suffix denoting **1.** the type or number of styles in a flower (e.g. monostyly). Adjectival form: **-stylous. 2.** the type of jaw suspension in vertebrates (e.g. autostyly, hyostyly). Adjectival form: **-stylic.**

styrene $C_6H_5CH{=}CH_2$ The traditional name for phenylethene.

SU_3 (subscript 3; not SU(3)) *Particle physics, maths.* A type of group.

sub- Prefix denoting **1.** situated beneath (e.g. subarachnoid, subcutaneous, sublittoral, submucosa). **2.** less or smaller than (e.g. subatomic, subclimax, subcritical, subsonic). **3.** part of a whole (e.g. subcontinent, subroutine, subset). **4.** an intermediate rank in plant and animal taxonomy (e.g. subdivision, subkingdom). **5.** *Chem.* a lower than normal amount of an element (e.g. suboxide).

subclass A subdivision of a class in *taxonomy, consisting of a number of similar orders. Names of subclasses are printed in roman (not italic) type with an initial capital letter. Higher plant subclasses typically end in -idae (e.g. Magnoliidae), fungal subclasses in -mycetidae, and algal subclasses in -phycidae; animal subclasses have a variety of endings.

suberic acid $HOOC(CH_2)_6COOH$ The traditional name for hexane-1,6-dicarboxylic acid.

subfamily A subdivision of a family in *taxonomy, consisting of a number of similar genera (sometimes grouped into tribes). Names of subfamilies are usually printed in roman (not italic) type with an initial capital letter; animal subfamilies end in -inae (e.g. Murinae) and plant subfamilies in -oideae (e.g. Rosoideae).

subgiant (not hyphenated) *Astron. See* giant.

suborder A subdivision of an order in *taxonomy, consisting of a number of similar families. Names of suborders are printed in roman (not italic) type with an initial capital letter and typically end in -ineae for plants; animal suborders have various endings (e.g. -morpha for rodent suborders).

subscriber direct dialling Abbrev.: SDD (no stops).

subscriber trunk dialling Abbrev.: STD (no stops).

subset *Maths.* A subset T of a set S is denoted by

$$T \subseteq S,$$

where all members of T are members of S. When there is some member of S that is not a member of T, then T is a proper subset of S, denoted by

$T \subset S.$

See also sets.

subsp. (or **ssp.**; stops; pl. subspp. or sspp.) Abbrev. for *subspecies.

subspecies (pl. subspecies) Abbrev.: subsp. (stop; pl. subspp.). A unit of biological classification (*see* taxonomy) that is a subdivision of a species. Names of subspecies are italicized, being printed in the form *Corvus corone cornix*, *Corvus corone* subsp. *cornix*, or C. *c. cornix* (esp. when discussing several subspecies of the same species). Adjectival form: **subspecific**.

substitute natural gas Abbrev.: SNG (no stops).

succinaldehyde $OHC(CH_2)_2CHO$ The traditional name for butanedial.

succinic acid $HOOC(CH_2)_2COOH$ The traditional name for butanedioic acid.

succinic anhydride The traditional name for *butanedioic anhydride.

Suess, Eduard (1831–1914) Austrian geologist.

sugars In polymers, sequences, tables, etc., individual sugars are denoted by three-letter abbreviations with an initial capital; e.g. Ara (arabinose), Fru (fructose), etc. These abbreviations may be suffixed by an italic lower-case *f* (to indicate furanose) or *p* (to indicate pyranose); e.g. Rib*f* (ribofuranose), Fru*p* (fructofuranose). The most common sugars and their abbreviations are listed alphabetically in the dictionary.

Suipoxvirus (cap. S, ital.) Approved name for a genus of *poxviruses. Vernacular name: swinepox subgroup. Individual name: **suipoxvirus** (no cap., not ital.).

sulcus (not ital.; pl. sulci) *Anat., astron.* In astronomy, the word, with an initial capital, is used in the approved Latin name of an area of furrows and ridges on the surface of a planet or satellite, as in Mashu Sulcus and Philus Sulci, both on Ganymede.

sulfo- (**sulf-** before vowels) US spelling of *sulpho-.

Sulfolobus (cap. S, ital.; not *Sulpholobus*) A genus of extremely thermoacidophilic bacteria.

sulfonyl US spelling of *sulphonyl.

sulfur US spelling of *sulphur.

sulph- *See* sulpho-.

sulphamic acid US: **sulfamic acid**. H_3NSO_3 The traditional name for aminosulphonic acid.

sulphanilic acid $H_2NC_6H_4SO_2OH$ US: **sulfanilic acid**. The traditional name for 4-aminobenzenesulphonic acid.

sulpho- (**sulph-** before vowels) US: **sulfo-** (**sulf-**). Prefix denoting sulphur (e.g. sulphocyanic, sulphamide, sulphonamide).

sulphonic acid US: **sulfonic acid.** Any of a class of organic compounds containing the $-SO_3H$ group joined directly to a carbon atom.

Sulphonic acids are systematically named by prefixing the words sulphonic acid with the name of the substituent hydrocarbon, e.g. methanesulphonic acid, CH_3SO_3H, and propane-2-sulphonic acid, $(CH_3)_2CHSO_3H$.

In nonsystematic nomenclature sulphonic acids are named as in systematic nomenclature but the trivial names of the substituent hydrocarbons are used, e.g. isopropylsulphonic acid, $(CH_3)_2CHSO_3H$.

sulphonyl- US: **sulfonyl-**. *Prefix denoting the group RSO_2-, where R is any alkyl or aryl group (e.g. sulphonyl chloride).

sulphoxide US: **sulfoxide.** Any of a class of organic compounds containing the $=SO$ group joined directly to two carbon atoms.

Sulphoxides are systematically named by adding the word sulphoxide after the names of the hydrocarbon groups joined to the sulphur atom, e.g. ethyl methyl sulphoxide, $CH_3SOCH_2CH_3$, and di-2-propyl sulphoxide, $(CH_3)_2CHSOCH(CH_3)_2$.

In nonsystematic nomenclature sulphoxides are named as in systematic nomenclature but the trivial names of the substituent hydrocarbons are used, e.g. diisopropyl sulphoxide.

sulphur US: **sulfur**. Symbol: S Allotropes may be distinguished as α (rhombic) and β (monoclinic) forms of octasulphur, and polysulphur; *cyclo- or *catena- can be added as prefixes. *See also* periodic table; nuclide.

sulphur dichloride dioxide SO_2Cl_2 US: **sulfur dichloride oxide.** The recommended name for the compound traditionally known as sulphuryl chloride.

sulphur dichloride oxide $SOCl_2$ US: **sulfur dichloride oxide.** The recommended name for the compound traditionally known as thionyl chloride.

sulphuric(IV) acid H_2SO_3 US: **sulfuric(IV) acid.** The recommended name for the compound traditionally known as sulphurous acid.

sulphuric(VI) acid H_2SO_4 US: **sulfuric(VI) acid.** The recommended name for the compound traditionally known as sulphuric acid.

sulphur monochloride S_2Cl_2 US: **sulfur monochloride.** The traditional name for disulphur dichloride.

sulphurous acid H_2SO_3 US: **sulfurous acid.** The traditional name for sulphuric(IV) acid.

sulphur(VI) oxide SO_3 US: **sulfur(VI) oxide.** The recommended name for the compound traditionally known as sulphur trioxide.

sulphur trioxide SO_3 US: **sulfur trioxide.** The traditional name for sulphur(VI) oxide.

sulphuryl chloride SO_2Cl_2 US: **sulfuryl chloride.** The traditional name for sulphur dichloride dioxide.

sumac (not sumach) Any of various trees or shrubs of the genus *Rhus*.

summation Symbol: Σ *See also* limits.

Sumner, James Batcheller (1877–1955) US biochemist.

Sumner, Thomas H. (1807–76) US shipmaster.
Sumner line

sun (preferred to Sun) Terms incorporating the word 'sun' are usually written as one word (e.g. sundial, sunlight, sunrise, sunseeker, sunset, sunshine, sunspot). Two-word terms usually use the related adjective **solar** (e.g. solar cell, solar constant, solar system). *See* solar mass; solar radius; luminosity.

sup (no stop) *Maths* Abbrev. for supremum.

super- Prefix denoting **1.** above or on top of (e.g. superglacial, superposition). **2.** exceeding or excessive in quality, performance, size, value, etc. (e.g. supercomputer, superconductivity, supernormal, supernova, supersaturation). **3.** an intermediate rank in animal *taxonomy (e.g. superfamily). **4.** *Chem.* a greater than normal amount of an element (e.g. superoxide).

superfamily An intermediate category used in animal classification (*see* taxonomy) consisting of a number of similar families; it ranks below an order (or suborder). Names of superfamilies are printed in roman (not italic) type with an initial capital letter and end in -oidea (e.g. Hominoidea).

supergiant (one word) *Astron.* Designated Ia (bright supergiant) or Ib.

superhet *Telecom.* Acronym for supersonic heterodyne.

superhigh frequency Abbrev.: SHF (no stops).

supernova (pl. supernovae; preferred to supernovas) Abbrev.: SN (no stops).
supernova remnant Abbrev.: SNR (no stops).

Super Proton Synchrotron (at *CERN) Abbrev.: SPS (no stops).

supplementary unit *See* SI units.

supra- Prefix denoting above or beyond (e.g. supraorbital, supravital).

supremum (pl. suprema) *Maths.* Abbrev.: sup (no stop).

sur- Prefix denoting above, beyond, or onto (e.g. surjection).

surface acoustic wave Abbrev.: SAW (no stops).

surface density *See* density.

surface density of charge *See* charge density.

surface tension Symbol: σ or γ (Greek sigma, gamma). A physical quantity, the *force in the plane of a surface perpendicular to a line element of the surface, divided by the length of the line element. The *SI unit is the newton per metre (N m^{-1}).

susceptance *See* admittance.

susceptibility (not susceptability) Short for electric susceptibility (*see* permittivity) or magnetic susceptibility (*see* permeability).

Sutherland, Earl (1915–74) US physiologist.

Sv Symbol for sievert. *See also* SI units.

sv. (stop) Abbrev. for serovar.

SV40 (no space) Abbrev. and preferred form for simian virus 40.

S value *See* Svedberg unit.

Svedberg unit Symbol: S A unit for expressing the *sedimentation coefficient of macromolecules and cellular particles analysed in an ultracentrifuge. It is equal to 10^{-13} second. S values indicate the size, shape, and weight of the analysed particles; for example, the prokaryote ribosome has a value of 70 S; a tRNA particle, 4 S; and a lysosome, 9400 S. Particles are often described in terms of their S values, in which case there is no space between the number and the symbol (e.g. a 70S *ribosome). Named after Theodor Svedberg (1884–1971).

Sverdrup, Harald Ulrik (1888–1957) Norwegian oceanographer and meteorologist.
sverdrup (not Sverdrup unit)

SVP Abbrev. for saturated vapour pressure.

SW (no stops) **1.** Abbrev. and preferred form for south-west or south-western. For usage, *see* south. **2.** Abbrev. for short wave.

Swammerdam, Jan (1637–80) Dutch naturalist and microscopist.
Swammerdam's glands

Swan, Sir Joseph Wilson (1828–1914) British inventor and industrialist.

Swanscombe man Common name for a type of fossil hominid (*see Homo*). Named after Swanscombe, Kent, where the first fossil was found in 1935.

sweet william (two words, no initial caps.)

syconium (pl. syconia; preferred to syconus, synconium, and synconus) A type of composite fruit.

Sydenham, Thomas (1624–89) English physician.
Sydenham's chorea

Sykes, William Henry (1790–1872) British soldier and naturalist.
Sykes's monkey (*Cercopithecus albogularis*)

Sylow, P. L. (1832–1918) Norwegian mathematician.
Sylow's theorem
Sylow subgroup

Sylvester, James Joseph (1814–97) British mathematician.

Sylvius, Franciscus (1614–72) Dutch anatomist.
aqueduct of Sylvius Also called aqueduct of the midbrain, cerebral aqueduct.
fissure of Sylvius

sym- (before b, m, or p) Prefix denoting **1.** with or together (e.g. symbiosis, symmetry, sympatric). **2.** fused (e.g. sympetalous, symplast). *See also* syn-.

sym- (ital., always hyphenated) *Chem.* Prefix denoting symmetric (e.g. *sym*-dichloroethane, *sym*-trinitrobenzene). The alternative *s*- is not recommended. *Compare unsym-*.

symbiosis (pl. symbioses) Any close relationship between two individuals of different species (**symbionts**). Usage of the term is sometimes restricted to a relationship in which both participants benefit (i.e. mutualism) but is often extended to include commensalism (one species benefits, the other is unaffected) or parasitism (one species benefits, the other may be harmed). Adjectival form: **symbiotic**.

symmetry Adjectival forms: **symmetric**, **symmetrical**.

Sympetalae Use *Asteridae.

syn- Prefix denoting **1.** with or together (e.g. synapsis, synchronous, syncline, synecology). **2.** fused (e.g. synangium, syncarpous, syndactyly, syngamy). *See also* sym-.

syn- (ital., always hyphenated) *Chem. See cis-*.

synapse The junction between adjacent neurones, across which nerve impulses are transmitted. Adjectival form: **synaptic**. *Compare* synapsis.

synapsis (pl. synapses) The pairing of homologous chromosomes during the first prophase of meiosis. Adjectival form: **synaptic**. *Compare* synapse.

sync (no stop; not synch) Short for synchronization or synchronize.

synchronize Noun forms: **synchronization, synchronism**. Adjectival form: **synchronous**.

syncytium (not syncitium; pl. syncytia) An animal tissue consisting of a mass of protoplasm containing several nuclei. Adjectival form: **syncytial**. *Compare* coenocyte; plasmodium. *See also* acellular.

syndyotaxis *Chem.* Adjectival form: **syndyotactic**.

***Synechocystis* group** (cap. S, ital.) A provisional assemblage of strains of *cyanobacteria previously assigned to five botanical genera, including *Synechocystis*.

Synge, Richard Laurence Millington (1914–) British chemist.

synnema (not synema; pl. synnemata) A compact mass of hyphae produced by certain actinomycete bacteria, particularly the genus *Actinosynnema*.

synodic month *See* month.

synonym *Biol.* One of two or more taxonomic names of the same rank applied to the same group of organisms. Only one can be valid, usually that published first (the senior synonym); the later name (junior synonym) is usually suppressed. *Compare* homonym.

synovia The fluid surrounding a movable joint. Not to be confused with **synovium** (pl. synovia), the membrane that lines the joint cavity and secretes the fluid. Adjectival form: **synovial**.

syntax Adjectival form: **syntactic**.

synthesis (pl. syntheses) Adjectival form: **synthetic**. Verb form: **synthesize**.

synthetic aperture radar (not hyphenated) Abbrev.: SAR (no stops).

syphon Use siphon.

syringa An ornamental shrub of the genus *Philadelphus*. Also called mock orange. *Compare Syringa*.

Syringa (cap. S, ital.) A genus of shrubs and trees that includes lilac (*S. vulgaris*). *Compare* syringa.

systematic 1. Methodical. **2.** Based on or originating from a system. *Compare* systemic.

Système International d'Unités *See* SI units.

systemic *Biol., med.* Relating to or affecting the whole body or all parts of an organism. *Compare* systematic.

systems analysis (for systems in general, hence not 'system') Similarly **systems engineering, systems programmer, systems software**. The word 'system' is often used in the case of a specific system, as in **system specification**.

syzygy *Astron., zool.* Adjectival forms: **syzygial, syzygetic**.

Szent-Györgi, Albert von (hyphen) (1893–1986) Hungarian-born US biochemist.

Szilard, Leo (1898–1964) Hungarian-born US physicist.
Szilard–Chalmers effect (en dash)

T

t Symbol for **1.** tonne. **2.** triplet (in nuclear magnetic resonance spectroscopy). **3.** triton.

t Symbol (light italic type) for **1.** *Celsius temperature (Δt = temperature interval). **2.** *Genetics* *terminator. **3.** *Chem.* transport number.

T Symbol for **1.** ribosylthymine. **2.** tautomeric effect (*see* electron displacement). **3.** *T cell (followed by subscript cap.). **4.** tera-. **5.** tesla. **6.** threonine. **7.** thymine. **8.** true.

T Symbol for **1.** (light ital.) kinetic energy. **2.** (light ital.) period ($T_{1/2}$ = half-life). **3.** (light ital.) thermodynamic temperature (ΔT =

temperature interval, T_C = Curie temperature, T_N = Néel temperature). **4.** (bold ital.) torque. **5.** (light ital.) transmittance. **6.** (light ital.) *Biochem.* *twisting number.

T_m (light ital.) *Biochem.* Symbol for *melting temperature.

***t*-** (ital., always hyphenated) *Chem.* Prefix denoting tertiary (e.g. *t*-butyl alcohol, *t*-butylamine). The alternative *tert*- is not recommended.

2,4,5-T Abbrev. and preferred form for 2,4,5-trichlorophenoxyacetic acid, a herbicide.

T2 *See* phage T2; T-even phage group.

T_3 Abbrev. for triiodothyronine.

T_4 Abbrev. for thyroxine (tetraiodothyronine).

T7 *See* phage T7.

Ta Symbol for tantalum.

tables The title should always appear at the top of a table. Vertical rules should be omitted and horizontal rules kept to a minimum; head and tail rules should be included in most cases. Column headings are usually set in roman (upright) type, either with or without an initial capital, to the left of a column. Where a column contains the values of a physical quantity, the units in which it is expressed should be given in the form:

'physical quantity'/'unit',

as in *graphs.

If the contents of a column are related numbers, the longest item should be ranged left and aligned under the column heading and the other items should be aligned under the decimal point of this number. If the items are unrelated they should all be ranged left and aligned under the column heading.

tachy- Prefix denoting fast or rapid (e.g. tachycardia, tachyon, tachytelic).

TACTAAC box A *consensus sequence of bases (T = thymine, A = adenine, C = cytosine) found in the introns of yeast.

-tactic *See* -taxis.

TAI Abbrev. (French) for International Atomic Time, a scale of time based directly on the atomic radiation defining the *second and maintained by the Bureau International de l'Heure (BIH) in Paris. Legal time in most countries, including the UK, is based on a related scale, that of Coordinated Universal Time (UTC), broadcast internationally. UTC differs from TAI by a whole number of seconds. The difference can be changed in steps of 1 s by the use of a leap second, preferably at the end of December or June.

Takamine, Jokichi (1854–1922) Japanese-born US chemist.

Talbot, William Henry Fox (1800–77) British scientist and photographer.

Talbot's law (preferred to Talbot–Plateau law)

TAME Abbrev. for *N*α-*p*-tosyl-L-arginine methyl ester.

Tamm, Igor Yevgenyevich (1895–1971) Soviet physicist.

tan Symbol for tangent, written with a space or thin space before an angle or variable:

$\tan 30°$ $\tan(-\theta)$ $\tan x$ $\tan(a + b)$.

The symbol tg can also be used.

The inverse function of $y = \tan x$ is denoted by:

$$x = \arctan y$$

or

$$x = \tan^{-1}y \text{ (no space).}$$

tanh Symbol for hyperbolic tangent, written with a space or thin space before a variable as in $\tanh x$, $\tanh(a + b)$. The shortened form th is also permitted.

The inverse function of $y = \tanh x$ is represented using the prefix ar-:

$$x = \operatorname{artanh} y \text{ (or arth } y\text{)}$$

or

$$x = \tanh^{-1}y \text{ (no space).}$$

Tansley, Sir Arthur George (1871–1955) British plant ecologist.

tantalum Symbol: Ta *See also* periodic table; nuclide.

tapeworm (one word) *See* Cestoda.

Tardigrada (cap. T) A phylum of minute aquatic invertebrates formerly regarded as a class of arthropods. Individual name and adjectival form: **tardigrade** (no cap.).

Tarski, Alfred (1902–83) Polish-born US mathematician and logician.

tarsus (pl. tarsi) **1.** The skeleton of the ankle, consisting of a number of small bones (**tarsals** or **tarsal bones**). **2.** The connective tissue of the eyelid. Adjectival form: **tarsal**.

Tartaglia, Niccoló (1500–57) Italian mathematician, topographer, and military scientist.

***d,l*-tartaric acid** HOOCCH(OH)CH(OH)COOH The traditional name for (±)-2,3-dihydroxybutanedioic acid.

TAS-F Abbrev. for tris(dimethylamino)sulphur (trimethylsilyl)difluoride.

TATA box A *consensus sequence of bases (T = thymine, A = adenine) found in eukaryote DNA. Also called Hogness box.

Tatum, Edward Lawrie (1909–75) US biochemist.

tau Greek letter, symbol: τ (lower case), T (cap.).

τ Symbol for **1.** characteristic time interval: $\tau_{1/2}$ = half-life, τ_m = mean life. **2.** chemical shift. **3.** shear stress. **4.** tauon. **5.** transmission coefficient (acoustic). **6.** transmittance.

Tau *Astron.* Abbrev. for Taurus.

Taube, Henry (1915–) US inorganic chemist.

Tauber, Alfred (1866–*c.* 1942) Slovak mathematician.

Tauberian theorem

tauon A negatively charged elementary particle, denoted τ (Greek tau) in nuclear reactions, etc. Its antiparticle, the **positive tauon** or **positive tau**, is denoted by τ^+. Also called tau particle.

tau particle *See* tauon.

taurine $NH_2CH_2CH_2SO_3H$ The traditional name for aminoethylsulphonic acid.

Taurus A constellation. Genitive form: Tauri. Abbrev.: Tau (no stop). *See also* stellar nomenclature.

Taurids (meteor shower)

tauto- Prefix denoting the same or identical (e.g. tautomerism, tautonym).

tautomeric effect *See* electron displacement.

Taxales An order of gymnosperms (class *Coniferopsida) that includes the yews (genus *Taxus*). In some classifications it is regarded as a class, Taxopsida.

taxis (pl. taxes) Directional movement of a whole organism or cell in response to an external stimulus. The word is often used in combination (*see* -taxis). Also called tactic movement. Adjectival form: **tactic**. *Compare* tropism.

-taxis Noun suffix denoting **1.** directional movement of a cell or organism in relation to a stimulus (e.g. chemotaxis, phototaxis, thermotaxis). Adjectival form: **-tactic**. **2.** (or **-taxy**) arrangement or order (e.g. phyllotaxis).

taxon (pl. taxa) A named taxonomic group of any rank. Taxa at the phylum level include Chordata and Echinodermata; Hominidae and Ranunculaceae are examples of taxa at the family level. The word should not be used for the ranks themselves: phyla, classes, orders, etc., are not taxa. *See* taxonomy.

taxonomy The principles and practice of classifying organisms according to their similarities and differences. Plants and animals are classified in a hierarchical series of categories (or ranks). The minimum seven categories required to classify both plants and animals are, in ascending order: *species, *genus, *family, *order, *class, *phylum (for animals) or *division (for plants), *kingdom. Additional intermediate categories may be needed for large complex groups. Most of these take the name of the basic category with a standard prefix. In animal taxonomy the prefixes are (in ascending order): infra-, sub-, and super- (e.g. infraclass, subclass, class, superclass). The names of most botanical intermediate categories are prefixed by sub-. Additional intermediate categories in plant taxonomy include the following, in ascending order (the position of the basic category is indicated in brackets): form, *variety, (species), series, section, (genus), *tribe (the latter is also used in animal taxonomy). In the classification of bacteria and viruses there are few formal categories above the level of family; the term 'group' tends to be used for large assemblages of organisms. There are,

however, a number of additional categories at the subspecific level, e.g. *biovar, *morphovar, *pathovar, *serovar.
A named group of organisms in any category is called a *taxon. Rules for the naming of such groups are specified by the International Code of Zoological Nomenclature (ICZN; for animals), the International Code of Botanical Nomenclature (ICBN; for wild plants, including fungi), and the International Codes of Nomenclature of Cultivated Plants (ICNP), of Bacteria (ICNB), and of Viruses (ICNV; for rules on viral nomenclature, *see* bacteriophage; virus). Taxonomic names are Latinized and printed in roman type except at the genus and species level, when they are italicized (*see* binomial nomenclature). All taxonomic names above the level of species have an initial capital letter. Taxa at intermediate ranks are styled according to the rank immediately above them (e.g. names of plant sections, like genera, are italicized, with an initial capital letter). Some taxonomic names have standardized endings (see entries for individual categories). Ideally, each taxon should have a universally recognized scientific name at a universally accepted level in the taxonomic hierarchy. In practice, however, several systems of classification and nomenclature coexist as the authorities differ in the interpretation of the characters on which taxonomy is based.

Taylor, Brook (1685–1731) British mathematician.
Taylor series
Taylor's theorem

Taylor, Geoffrey Ingram (1886–1975) British physicist.

Taylor, Sir Hugh (Stott) (1890–1974) British chemist.

Tay–Sachs disease (en dash) Also called amaurotic familial idiocy. Named after Warren Tay (1843–1927) and Bernard Sachs (1858–1944).

Tb Symbol for terbium.

t-BOC Abbrev. for *t*-butoxycarbonyl.

TBSV Abbrev. for tomato bushy stunt virus.

tBu, *t*-Bu Use Bu^t (*see* Bu).

Tc Symbol for technetium.

TCA cycle Abbrev. and preferred form for tricarboxylic acid cycle.

TCC Abbrev. for 3,4,4′-trichlorocarbanilide.

T cell (or **T lymphocyte**; not hyphenated unless used adjectivally) A type of lymphocyte responsible for cell-mediated immunity. T cells are subdivided into helper cells, cytotoxic cells, and suppressor cells, designated by subscript capital letters: T_H, T_C, and T_S, respectively. Named after the initial letter of thymus, in which such cells mature. *Compare* B cell.

T-cell differentiation antigens In the past, various systems were devised for naming these molecules, which varied according to species. It is now recommended that the *CD (cluster of differentiation) nomenclature should be adopted for all species. For example, the mouse antigen L3T4 and the human antigen T4 (or Leu 3) are now both known as CD4; the mouse antigen Ly-2,3 (or Lyt-2,3) and the human antigen T8 (or Leu 2) are now called CD8. The presence or absence of specific antigens on the surface of cells is indicated by a plus or minus sign after the name of the antigen; thus cells that carry the CD8 antigen, for example, are denoted 'CD8+ cells'.

T-cell receptor Abbrev. and preferred form: *TCR (no stops).

Tchebyshev Use *Chebyshev.

TCNE Abbrev. for tetracyanoethylene.

TCNQ Abbrev. for 7,7,8,8-tetracyanoquinodimethane.

TCR Abbrev. and preferred form for T-cell (antigen) receptor, a protein on T cells that receives MHC molecules. It consists of two pairs of polypeptide chains, designated $\alpha\beta$ and $\gamma\delta$, respectively, with C, J, D, and V regions like those of *immunoglobulin molecules.

TDA Abbrev. for tetradecen-1-yl acetate.

TDAL Abbrev. for tetradecenal.

TDB Abbrev. (orig. French) for barycentric dynamical time.

TDDA Abbrev. for tetradecadien-1-yl acetate.

TDI Abbrev. for toluene-2,4-diisocyanate.

TDM *Telecom.* Abbrev. for time-division multiplexing.

TDMA *Telecom.* Abbrev. for time-division multiple access.

T-DNA (cap. T, hyphenated) Abbrev. and preferred form for transferred DNA, a segment occupying the *T-region of the Ti plasmid.

TDOL Abbrev. for tetradecen-1-ol.

TDRSS Abbrev. for tracking and data-relay satellite system.

TDT Abbrev. for terrestrial dynamical time.

TDTA Abbrev. for (+)-di-*p*-tolvoyl-D-tartaric acid.

Te Symbol for tellurium.

technetium Symbol: Tc *See also* periodic table; nuclide.

TED Abbrev. for triethylenediamine.

Teepol (cap. T) A trade name for a liquid anionic detergent.

Teflon (cap. T) A trade name for a form of the 'nonstick' plastic, *PTFE.

Teisserenc De Bort, Léon Philippe (1855–1913) French meteorologist.

Tel *Astron.* Abbrev. for Telescopium.

TEL Abbrev. for tetraethyl lead.

tel- *See* telo-.

tele- Prefix denoting at or over a distance (e.g. telecommunications, telemetry, telescope).

Teleostei (cap. T) A large taxon including most extant bony fishes. Individual name and adjectival form: **teleost** (no cap.; not teleostean).

Telescopium A constellation. Genitive form: Telescopii. Abbrev.: Tel (no stop). *See also* stellar nomenclature.

Teletex (cap. T) A trade name for a means of high-speed text transmission using public data networks. *Compare* teletext.

teletext A system, such as Oracle or Ceefax, for transmitting information to domestic TV receivers. *Compare* Teletex.

television Abbrev.: TV or tv (no stops). Verb form: **televise** (not televize).

Telford, Thomas (1757–1834) British architect and engineer.

Teliomycetes A former class of basidiomycete fungi including the rusts and smuts. Also called Hemibasidiomycetes. These groups are now regarded as separate classes, Urediniomycetes and Ustilaginomycetes, respectively.

Teller, Edward (1908–) Hungarian-born US physicist.

Jahn–Teller effect (en dash) Also named after H. A. Jahn (1907–79).

tellurium Symbol: Te *See also* periodic table; nuclide.

telo- (**tel-** before vowels) Prefix denoting **1.** complete or final (e.g. telophase, telencephalon). **2.** at the end (e.g. telocentric, telomere).

TEM Abbrev. for **1.** transmission electron microscope. **2.** transverse electromagnetic.

Temin, Howard Martin (1934–) US molecular biologist.

temp. Abbrev. for temperature.

temperature Abbrev.: temp. (stop). A fundamental physical quantity of a body or region that determines the direction of heat flow (always from a higher to a lower temperature). Scientific measurements are generally made in terms of the *IPTS, which is the practical realization of the *thermodynamic temperature scale. The usual unit is the *kelvin or the *degree Celsius (now defined in terms of the kelvin). The USA commonly employs the IPTS but uses the degree Fahrenheit.

temperature coefficient Symbol: Q_{10} The rate of increase of an enzyme-catalysed reaction for a temperature rise of 10 °C.

temperature interval Symbol: ΔT (for thermodynamic temperatures) or Δt (for Celsius temperatures). A physical quantity relating to a difference in *temperature. The *SI units are the kelvin and the degree Celsius, the two units being identical in size. Also called temperature difference.

Tennant, Charles (1768–1838) British chemical industrialist.

Tennant, Smithson (1761–1815) British chemist.

Tenon, Jacques Réné (1724–1816) French surgeon.
Tenon's capsule

tensile strain Former name for linear strain. *See* strain.

tensile stress Another name for normal stress. *See* stress.

tensor The generalization of a *vector. To distinguish it from a vector, a tensor should be printed in slanted bold sanserif type. If such a type is unavailable, a second-rank tensor (the simplest type) may be indicated by two horizontal arrows, or a double-headed arrow, placed above the italicized symbol. With higher-rank tensors, an index notation should be used uniformly for vectors and tensors, for example:

$$A_i, S_{ij}, R_{ijkl}, R^{ij}_{kl}.$$

tephra Volcanic products, collectively.

TEPP Abbrev. for tetraethyl pyrophosphate.

tera- Symbol: T A prefix to a unit of measurement that indicates 10^{12} times that unit, as in terajoule (TJ). *See also* SI units.

terato- (terat- before vowels) Prefix denoting congenital abnormality (e.g. teratogen, teratogenesis, teratology, teratoma).

terbium Symbol: Tb *See also* periodic table; nuclide.

terephthalic acid $C_6H_4(COOH)_2$ The traditional name for benzene-1,4-dicarboxylic acid.

terminator *Genetics* Symbol: *t* Different terminator sequences are identified by the symbol *t* and additional characters, usually a subscript capital L or R (not ital.), to denote a leftwards or rightwards transcription unit, and a distinguishing number, e.g. t_L2, t_R4, etc.

term symbols *See* orbital angular momentum quantum number.

terra (not ital.) *Astron.* An upland area on the surface of a planet or satellite. The word, with an initial capital, is used in the approved Latin names of such features, as in Ishtar Terra on Venus.

terrestrial (not terrestial)

terrestrial dynamic time Abbrev.: TDT (no stops).

tert- (ital., always hyphenated) *Chem.* Use **t-*.

Tertiary *Geol.* Abbrev.: TT (no stop). **1.** (adjective) Denoting the first period of the Cenozoic era. **2.** (noun; preceded by 'the') The Tertiary period. In certain schemes it is divided into two subperiods, the *Neogene and the *Palaeogene, while in other very recent schemes, it is discarded altogether and the Neogene and Palaeogene are given the status of periods.

Terylene (cap. T) US: **Dacron**. A trade name for a type of polyester fibre and fabric.

TES Abbrev. for 2-[tris(hydroxymethyl)-methylamino-1-ethane]sulphonic acid.

tesla (no cap.; pl. tesla or teslas) Symbol: T The *SI unit of *magnetic flux density.

$$1\ \mathrm{T} = 1\ \mathrm{Wb\ m^{-2}}.$$

Named after Nikola *Tesla.

Tesla, Nikola (1856–1943) Croatian-born US physicist.
***tesla**
Tesla coil

testa (pl. testae) The protective outer layer of a seed.

testis (pl. testes) The reproductive gland of male animals. Use of the synonym **testicle** (and its adjectival form, **testicular**) should be restricted to the mammalian testis.

tetra- 1. (**tetr-** before vowels) Prefix denoting four (e.g. tetrahedron, tetramerous, tetraploid, tetrasporangium, tetravalent). *See also* quadri-. **2.** *Chem.* Prefix denoting the linking of four groups that together form the root of a structure (e.g. tetrachloromethane); **tetrakis-** is used when an expression to be multiplied already contains a multiplicative prefix.

tetraamminecopper(II) sulphate $Cu(NH_3)_4SO_4$ US: **tetraamminecopper(II) sulfate.** The recommended name for the compound traditionally known as cuprammonium sulphate.

1,3,5,7-tetraazaadamantane $(CH_2)_6N_4$ The recommended name for the compound traditionally known as hexamine.

tetrabromomethane CBr_4 The recommended name for the compound traditionally known as carbon tetrabromide.

1,1,2,2-tetrachloroethane $CHCl_2$-$CHCl_2$ The recommended name for the compound traditionally known as acetylene tetrachloride.

tetrachloromethane CCl_4 The recommended name for the compound traditionally known as carbon tetrachloride.

tetracyanoethene $(CN)_2C{:}C(CN)_2$ Abbrev.: TCNE (no stops).

tetraethyl lead(IV) $(C_2H_5)_4Pb$ Abbrev.: TEL (no stops). The recommended name for the compound traditionally known as lead tetraethyl.

tetrahydridoborate(III) A compound containing the ion BH_4^-, e.g. sodium tetrahydridoborate(III), $NaBH_4$. The traditional name is borohydride.

H_2C——CH_2
H_2C N CH_2
H

tetrahydropyrrole
(pyrrolidine)

tetrahydropyrrole The recommended name for the compound traditionally known as pyrrolidine.

tetrakis- *See* tetra-.

1,2,4,5-tetramethylbenzene C_6H_2-$(CH_3)_4$ The recommended name for the compound traditionally known as durene.

tetrapod Any vertebrate animal with four limbs or only secondarily limbless (e.g. snakes, whales), i.e. an amphibian, reptile, bird, or mammal. In some classifications tetrapods are grouped together in the taxon Tetrapoda. *Compare* quadruped.

T-even phage group (cap. T) Vernacular name for a genus of bacteriophages of enterobacteria, of which phage T2 is the type species. The original members were designated T2, T4, T6, etc. (i.e. *T*ype + *even* numbers).

TE wave Short for transverse electric wave.

tex A unit used to express the linear density of textile filaments: 1 tex = 10^{-6} kg m^{-1}.

TF Abbrev. for *transcription factor.

TFA Abbrev. for trifluoroacetic acid.

TFAA Abbrev. for trifluoroacetic anhydride.

TFT Abbrev. for thin-film transistor.

tg Another symbol for tangent. *See* tan.

th Short for tanh.

Th Symbol for thorium.

thalamus (pl. thalami) **1.** Part of the vertebrate forebrain. **2.** The receptacle of a flower. Adjectival form: **thalamic**.

Thales (*c.* 625 BC – *c.* 547 BC) Greek philosopher, geometer, and astronomer.

thallium Symbol: Tl *See also* periodic table; nuclide.

Thallophyta (cap. T) One of two subkingdoms into which plants are divided in certain classifications. It includes all plants except embryophytes (*see* Embryophyta). Formerly, Thallophyta was a division containing all nonanimal organisms without differentiated stems, leaves, and roots (*see* thallus), including algae, fungi, lichens, and bacteria. Individual name and adjectival form: **thallophyte** (no cap.).

thallus (pl. thalli; preferred to thalluses) A plant body not differentiated into roots, stem, and leaves. Adjectival form: **thalloid** (not thallous).

THC Abbrev. for tetrahydrocannabinol.

theca (pl. thecae) *Anat., biol.* A sheathlike or enclosing tissue; the word often occurs in combination (*see* -theca). In botany it is a former name for *ascus.

-theca *Zool.* Noun suffix denoting a case (e.g. ootheca, spermatheca). Adjectival form: **-thecal**. *Compare* -thecium.

-thecium *Bot.* Noun suffix denoting a case (e.g. amphithecium, endothecium). Adjectival form: **-thecial**. *Compare* -theca.

theco- (**thec-** before vowels) Prefix denoting a case or sheath (e.g. thecodont).

Theiler, Max (1899 – 1972) South African-born US virologist.

Thénard, Louis-Jacques (hyphen) (1777–1857) French chemist.
Thénard's blue

Theophrastus (*c.* 372 BC–*c.* 287 BC) Greek botanist and philosopher.

Theorell, Axel Hugo Teodor (1903–82) Swedish biochemist.

theory Verb form: **theorize**. Adjectival form: **theoretical** (preferred to theoretic). Two-word terms, such as game theory, set theory, and wave theory, are not hyphenated unless used adjectivally, when it is usual to use the word 'theory' rather than 'theoretical', e.g. set-theory concept (rather than set-theoretical concept).

Therapsida (cap. T; not Theraspida) An order of extinct mammal-like reptiles. Individual name and adjectival form: **therapsid** (no cap.).

Theria (cap. T) A subclass containing all living mammals except the monotremes, divided into the *Metatheria and *Eutheria, as well as extinct groups. Individual name and adjectival form: **therian** (no cap.).

therm *See* British thermal unit.

thermal conductivity Symbol: λ (Greek lambda). The *heat flow rate per unit area (i.e. the density of heat flow rate) divided by temperature gradient. The *SI unit is usually the watt per metre kelvin.

thermal diffusivity Symbol: a A physical quantity equal to the ratio $\kappa/\rho c_p$, where κ is the *thermal conductivity, ρ the density, and c_p the specific *heat capacity at constant pressure. The *SI unit is the second per square metre.

thermo- (sometimes **therm-** before vowels) Prefix denoting heat or temperature (e.g. thermoammeter, thermocline, thermodynamics, thermoelectric, thermonasty, thermionic).

Thermoactinomyces (cap. T, ital.) A genus of bacteria traditionally classified as *actinomycetes, although this classification has recently been challenged. Individual name and adjectival form: **thermoactinomycete** (no cap., not ital.).

Thermobacterium (cap. T, ital.) A subgenus of **Lactobacillus*. Individual name: **thermobacterium** (no cap., not ital.; pl. thermobacteria).

thermochemical calorie *See* calorie.

thermodynamic temperature Symbol: T A fundamental physical quantity defined in terms of the second law of thermodynamics, independently of any particular substance. The *SI unit is the kelvin (K); the kelvin replaced the degree Kelvin (°K) in 1968. Former name for thermodynamic temperature: absolute temperature. *See also* Celsius temperature.

thermolabile toxin *Microbiol.* Abbrev.: LT (no stops). *See* ETEC.

Thermomonospora (cap. T, ital.; not *Thermonospora*) A genus of actinomycete bacteria. Individual name: **thermomonospora** (no cap., not ital.; pl. thermomonosporas).

thermophilic Denoting microorganisms (known as **thermophiles**) that thrive at high temperatures (55–75 °C).

Thermoplasma (cap. T, ital.) A genus of extremely thermophilic bacteria. Individual name: **thermoplasma** (no cap., not ital.; pl. thermoplasmas).

Thermos (cap. T) A trade name for a type of vacuum flask. The word thermos (no cap.) is used generically, legally so in the USA, for any brand of vacuum flask, but this use should be avoided in the UK.

thermostable toxic *Microbiol.* Abbrev.: ST (no stops). *See* ETEC.

theta Greek letter, symbol: θ (lower case), Θ (cap.).

θ Symbol for **1.** Bragg angle. **2.** *Celsius temperature. **3.** a plane angle. **4.** a polar coordinate. **5.** sidereal time. **6.** volume strain.

Θ Symbol for characteristic temperature (Θ_D = Debye temperature).

theta structure Sometimes written **θ-structure** (Greek theta, hyphenated). The eyelike structure formed by a circular DNA molecule during replication. This type of replication is called theta replication or θ-replication.

Thévenin's theorem *Elec. eng.* Named after Leon Thévenin (1857–1926).

THF Abbrev. for tetrahydrofuran.

thi- *See* thio-.

thia- (**thi-** before vowels) Prefix denoting a heterocyclic compound in which the hetero atom is sulphur (e.g. thiabicycloheptane, thiazole).

thiamin (preferred to thiamine) Permitted synonym: vitamin B_1.

thickness *See* length.

Thiele, Friedrich Karl Johannes (1865–1918) German chemist.

Thiersch, Karl (1822–95) German surgeon.
Thiersch graft

thigmo- Prefix denoting touch (e.g. thigmotaxis). *See also* hapto-.

thin-film transistor (hyphenated) Abbrev.: TFT (no stops).

thin-layer chromatography (hyphenated) Abbrev.: TLC (no stops).

thio- (sometimes **thi-** before vowels) Prefix denoting a substance containing sulphur, especially when a sulphur atom has replaced an oxygen atom (e.g. thioarsenate, thiocyanate, thiosulphate, thiourea, thiamine).

Thiobacillus (cap. T, ital.) A genus of colourless sulphur bacteria. Individual name: **thiobacillus** (no cap., not ital.; pl. thiobacilli).

thiocarbamide $SC(NH_2)_2$ The recommended name for the compound traditionally known as thiourea.

Thiokol (cap. T) A trade name for a synthetic polysulphide rubber.

thiol Any of a class of organic compounds containing the group –SH joined directly to a carbon atom.
Thiols are systematically named by adding the suffix -thiol to the name of the parent hydrocarbon, e.g. methanethiol, CH_3SH, and 2-methylpropanethiol, $(CH_3)_2CHCH_2SH$.
In nonsystematic nomenclature thiols are named by adding the word mercaptan after the name of the hydrocarbon group attached to the sulphur atom, e.g. methyl mercaptan, CH_3SH, and isobutyl mercaptan, $(CH_3)_2CHCH_2SH$.

thionyl- *Prefix denoting the group =SO (e.g. thionyl chloride).

thionyl chloride $SOCl_2$ The traditional name for sulphur dichloride oxide.

Thioploca (cap. T, ital.) A genus of filamentous gliding bacteria. Individual name: **thioploca** (no cap., not ital.).

Thiosphaera (cap. T, ital.; not *Thiosphera*) A genus of colourless sulphur bacteria.

thiourea $SC(NH_2)_2$ The traditional name for thiocarbamide.

third *See* fractions.

THN Abbrev. for 1,2,3,4-tetrahydronaphthalene.

tholus (not ital.; pl. tholi) *Astron.* A hill or dome on the surface of a planet or satellite. The word, with an initial capital, is used in the approved Latin names of such features, e.g. Tharsis Tholus and Jovis Tholus on Mars.

Thomas, Hugh Owen (1834–91) British surgeon.
Thomas splint

Thomas, Sidney Gilchrist (1850–85) British chemist.
Thomas–Gilchrist basic process (en dash) Also named after Percy Gilchrist (1851–1935).

Thompson, Benjamin *See* Rumford, Benjamin Thompson, Count.

Thompson, Sir D'Arcy Wentworth (1860–1948) British biologist.

Thompson, Sir Harold Warris (1908–83) British physical chemist.

Thomsen, Hans Peter Jörgen Julius (1826–1909) Danish chemist.

Thomson, Sir Charles Wyville (1830–82) British marine biologist.

Thomson, Elihu (1853–1937) US electrical engineer.

Thomson, Sir George Paget (1892–1975) British physicist, son of J. J. Thomson.

Thomson, Joseph (1858–94) British explorer.

Thomson's gazelle (*Gazella thomsoni*)

Thomson, Sir Joseph John (1856–1940) British physicist.

Thomson scattering

Thomson, Sir William, Lord Kelvin (1824–1907) British theoretical and experimental physicist. The name Thomson is usually used in preference to *Kelvin in the following:

Joule–Thomson coefficient (en dash) *Thermodynamics* Symbol: μ (Greek mu)

Joule–Thomson effect (en dash) *Thermodynamics*

Thomson effect *Thermoelectricity*

thorax (pl. thoraces) Adjectival form: **thoracic**.

thorium Symbol: Th *See also* periodic table; nuclide.

thou One-thousandth of an inch, i.e. 0.0254 millimetre. Also called mil. The use of both units is discouraged.

Thr (or **thr**; no stop) Abbrev. for threonine. *See* amino acid.

threo- (ital., always hyphenated) *Chem.* Prefix denoting the diastereomer that has no similar substituents on adjacent carbon atoms in the eclipsed configuration (e.g. *threo*-3-phenyl-1-propanol).

threonine Abbrev.: Thr or thr (no stop). Symbol: T *See* amino acid.

threshold (not threshhold)

thrips (pl. thrips) *Zool.*

thrombo- (sometimes **thromb-** before vowels) Prefix denoting the clotting of blood (e.g. thromboembolism, thrombokinase, thromboplastin, thrombosis).

thrombocyte Use *platelet.

throughput (one word)

thulium Symbol: Tm *See also* periodic table; nuclide.

Thunberg, Carl Per (1743–1828) Swedish botanist.

thunbergia or, as generic name (*see* genus), ***Thunbergia***

thymidine Symbol: dT A 2-deoxyribonucleoside (*see* nucleoside) consisting of *thymine combined with D-ribose. *Compare* ribosylthymine.

thymidine 5′-diphosphate Abbrev. and preferred form: dTDP (no stops).

thymidine 5′-phosphate Abbrev. and preferred form: dTMP (no stops).

thymidine 5′-triphosphate Abbrev. and preferred form: dTTP (no stops).

thymine Symbol: T A pyrimidine base. *See also* base pair. *Compare* thymidine.

thyrocalcitonin Use calcitonin.

thyroid-stimulating hormone Abbrev.: TSH (no stops). Also called thyrotrophin.

thyroid-stimulating-hormone-releasing hormone Abbrev. and preferred form: TSH-releasing hormone or TSH-RH (hyphenated, no stops). Also called thyrotrophin-releasing hormone (abbrev.: TRH).

thyrotrophin US: **thyrotropin**. Also called thyroid-stimulating hormone (abbrev.: TSH).

thyrotrophin-releasing hormone Abbrev: TRH (no stops). Also called TSH-releasing hormone (abbrev.: TSH-RH).

thyroxine (preferred to thyroxin) Also called tetraiodothyronine. Abbrev.: T_4 (subscript number).

Ti Symbol for titanium.

tidal wave Use *tsunami.

Tilden, Sir William Augustus (1842–1926) British chemist.

timbre (not timber) *Acoustics*

time Symbol: t A fundamental physical quantity usually indicating duration (a period or interval or time) or a precise moment. The *SI unit is the second (s); the minute (m), hour (h), and day (d) may also be used. 1 d = 24 h = 1440 m = 86 400 s.

The length of the *year depends on how it is defined. *See also* TAI (International Atomic Time).

When giving a time of day, the correct style is 8.30 a.m., 5.15 p.m., etc. The style using a 24-hour clock is 08.30, 17.15.

Normally the word 'time' is used adjectivally in preference to 'temporal'. Two-word terms in which the first word is 'time' (e.g. time base, time constant, time sharing, time switch) are not hyphenated unless used adjectivally (e.g. time-base generator, time-lapse photography).

time-division multiplexing (hyphenated) Abbrev.: TDM (no stops).
time-division multiple access Abbrev.: TDMA (no stops).

timeout (one word) *Computing, etc.*

tin Symbol: Sn *See also* periodic table; nuclide.

Tinbergen, Niko(laas) (1907–88) Dutch-born British zoologist and ethologist.

Ting, Samuel Chao Chung (1936–) US physicist.

tin(IV) hydride SnH_4 The recommended name for the compound traditionally known as stannane.

Ti plasmid (cap. T) Abbrev. and preferred form for tumour-inducing plasmid.

Tiselius, Arne Wilhelm Kaurin (1902–71) Swedish chemist.
Tiselius apparatus

Titan The largest satellite of Saturn. *Compare* Triton.

titanic Designating compounds in which titanium has an oxidation state of +4. The recommended system is to use oxidation numbers, e.g. titanic oxide, TiO_2, has the systematic name titanium(IV) oxide.

titanium Symbol: Ti *See also* periodic table; nuclide.

titanium dioxide TiO_2 A traditional name for titanium(IV) oxide.

titanium(III) oxide Ti_2O_3 The recommended name for the compound traditionally known as titanous oxide.

titanium(IV) oxide TiO_2 The recommended name for the compound traditionally known as titanic oxide or titanium dioxide.

titanous Denoting compounds in which titanium has an oxidation state of +3. The recommended system is to use oxidation numbers, e.g. titanous oxide, Ti_2O_3, has the systematic name titanium(III) oxide.

Titius, Johann Daniel (1729–96) German astronomer.
Titius–Bode law Use Bode's law.

titre US: **titer**. *Chem., immunol.* Derived noun: **titration**.

Tizard, Sir Henry (Thomas) (1885–1959) British chemist and administrator.

Tl Symbol for thallium.

TLC Abbrev. for thin-layer chromatography.

TLR *Photog.* Abbrev. for twin-lens reflex.

TLV Abbrev. for threshold-limit value.

Tm Symbol for thulium.

TMA Abbrev. for trimellitic acid.

TMB Abbrev for 3,3′,5,5′-tetramethylbenzidine.

TMB-8 Abbrev. for 8-(diethylamino)octyl 3,4,5-trimethoxybenzoate.

TMCS Abbrev. for trimethylchlorosilane.

TMEDA Abbrev. for *N*,*N*,*N*′,*N*′-tetramethylethylenediamine.

TML Abbrev. for tetramethyl lead.

TMPTA (or **TMPTMA**) Abbrev. for trimethylolpropane trimethacrylate.

TMS Abbrev. for tetramethylsilane.

TMSDEA Abbrev. for trimethylsilyldiethylamine.

TMTSF Abbrev. for tetramethyltetraselenafulvalene.

TMV Abbrev. for tobacco mosaic virus.

TM wave Short for transverse magnetic wave.

Tn (cap. T) Abbrev. for *transposon.

TNBT Abbrev. for titanium(IV) butoxide.

TNS Abbrev. for 6-(*p*-toluidino)-2-naphthalenesulphonic acid.

TNT Abbrev. for trinitrotoluene.

T-number (hyphenated) *Photog.* T(otal light transmission) number.

toadstool Loosely, any poisonous fungus of the order Agaricales (*see* agaric). The term has no taxonomic significance.

tobacco mosaic virus Abbrev.: TMV (no stops). Type member of the **Tobamovirus* group (vernacular name: tobacco mosaic virus group).

tobacco ringspot virus Abbrev.: TobRV (no stops). Type member of the **Nepovirus* group (vernacular name: tobacco ringspot virus group).

tobacco streak virus Abbrev.: TSV (no stops). Type member of the **Ilarvirus* group (vernacular name: tobacco streak virus group).

Tobamovirus (cap. T, ital.) Approved name for the *tobacco mosaic virus group. Individual name: **tobamovirus** (no cap., not ital.). [from *tob*acco *mo*saic *virus*]

Tobravirus (cap. T, ital.) Approved name for the tobacco rattle virus group. Individual name: **tobravirus** (no cap., not ital.). [from *tob*acco *ra*ttle *virus*]

TobRV Abbrev. for tobacco ringspot virus.

tocopherol The chemical name of vitamin E, existing in several forms, the most important of which is α-tocopherol (others include β-tocopherol and γ-tocopherol).

Todd, Alexander Robertus, Lord (1907–) British biochemist.

TOE *Physics* Abbrev. for theory of everything.

togavirus (one word) Any virus belonging to the family Togaviridae.

Toit, Alexander Logie du Usually alphabetized as *du Toit.

tokamak *Nucl. eng.* Acronym (Russian) for toroidal magnetic chamber.

***o*-tolidine** The traditional name for *3,3′-dimethylbiphenyl-4,4′-diamine.

Tollens, Bernhard Christian Gottfried (1841–1918) German chemist.
Tollens reagent

toluene $C_6H_5CH_3$ The traditional name for methylbenzene.

toluene-4-sulphonyl- (preferred to tosyl-) *Prefix denoting the group *p*-$CH_3C_6H_4S(O)^-{}_2$ (e.g. toluene-4-sulphonyl chloride).

toluic acid $CH_3C_6H_4COOH$ The traditional name for *methylbenzoic acid.

toluidine $CH_3C_6H_4NH_2$ The traditional name for *methylphenylamine.

tomato bushy stunt virus Abbrev.: TBSV (no stops). Type member of the **Tombusvirus* group (vernacular name: tomato bushy stunt virus group).

Tombaugh, Clyde William (1906–) US astronomer.

Tombusvirus (cap. T, ital.) Approved name for the *tomato bushy stunt virus group. Individual name: **tombusvirus** (no cap., not ital.). [from *tom*ato *bu*shy *s*tunt *virus*]

-tome Noun suffix denoting **1.** a cutting instrument (e.g. microtome). **2.** a part or segment (e.g. myotome).

Tomonaga, Sin-Itiro (1906–79) Japanese theoretical physicist.

ton *See* pound. *See also* tonne.

Tonegawa, Susumu (1939–) Japanese immunologist.

tonne Symbol: t A unit of mass equal to 1000 kilograms. 1 tonne = 0.984 207 UK ton or 1.102 31 short (US) tons. Also called metric ton.

top- *See* topo-.

Töpler, August Joseph Ignaz (1836–1912) German physicist.
Töpler pump

topo- (**top-** before vowels) Prefix denoting place or position (e.g. topoclimatology, topography).

topoisomerase (one word) Topo(logical) isomerase: an enzyme that catalyses the conversion of topological isomers. DNA topoisomerases are divided into two classes, designated by Roman numerals: class I and class II.

tornado (pl. tornadoes)

toroid Adjectival form: **toroidal**.

torque Symbol: T A *pseudovector quantity, the (vector) sum of the moments of a system of forces acting on a body; it is equal to the rate of change of *angular momentum. The *SI unit is the newton metre (N m).

torr A unit of pressure defined in terms of the *pascal:
1 torr = 101 325.0/760 Pa ≃ 133.322 Pa.
It is very nearly equal to the millimetre of mercury, mmHg. Use of the torr and the mmHg is now discouraged. Named after *Torricelli.

Torrey, John (1796–1873) US botanist.

Torricelli, Evangelista (1608–47) Italian physicist.
***torr**
Torricellian vacuum

torus (pl. tori) Adjectival form: **toric**.

TosMIC Abbrev. for tosylmethyl isocyanide.

tosyl- Use *toluene-4-sulphonyl-.

total angular momentum quantum number Symbol: j (individual entity) or J (whole system). A number, integral or half-integral, that characterizes the total angular momentum (orbital plus spin angular momentum) of a particle, atom, nucleus, etc.

Townes, Charles Hard (1915–) US physicist.

Townsend, Sir John Sealy Edward (1868–1957) Irish physicist.
Townsend avalanche
Townsend discharge

TP *Surveying* Abbrev. for trigonometric point.

TPB Abbrev. for 1,1,4,4-tetraphenyl-1,3-butadiene.

TΨC arm (cap. Greek psi) The arm of a tRNA molecule characterized by an unusual anticodon triplet containing the nucleosides ribosylthymine (T), pseudouridine (Ψ), and cytidine (C).

TPCK Abbrev. for L-1-*p*-tosylamino-2-phenylethylchloromethyl ketone.

tpi Abbrev. for tracks per inch, a measure of the number of tracks across the radius of a magnetic disk. It is used in specifying the maximum recommended recording density.

TPN Abbrev. for triphosphopyridine nucleotide, a former name for NADP.

TPP Abbrev. for 5,10,15,20-tetraphenyl-21*H*,23*H*-porphine.

Tr (no stop) **1.** *Geol.* Abbrev. for Triassic. **2.** *Chem.* Symbol sometimes used to denote the trityl (triphenylmethyl) group in chemical formulae, e.g. $(C_6H_5)_3CCl$ can be written as TrCl.

TrA *Astron.* Abbrev. for Triangulum Australe.

trabecula (pl. trabeculae) *Anat.* Adjectival forms: **trabecular, trabeculate**.

trachea (pl. tracheae) The windpipe of vertebrates or an analogous structure in invertebrates. Adjectival form: **tracheal**; 'tracheary' is a botanical term applied to water-conducting elements in plants.

tracheid (preferred to tracheide) A water-conducting cell in plants.

Tracheophyta (cap. T) In some plant classification schemes, a division that includes all the vascular plants, i.e. pteridophytes and seed plants. Individual name and adjectival form: **tracheophyte** (no cap.).

trachyte (not trachite) *Geol.*

tracks per inch Abbrev.: tpi (no stops).

Tradescant, John (1570–1633) English gardener and botanist.

Tradescant, John (1608–62) English botanist and plant collector, son of John Tradescant (1570–1633).
tradescantia or, as generic name (*see* genus), ***Tradescantia***

trajectory Adjectival form: **trajectile**.

trans (ital.) Relating to different chromosomes or DNA molecules. A *trans* configuration is one in which loci occur on different chromosomes. A *trans*-acting locus (hyphenated) is one that exercises an effect on other chromosomes. *Compare cis.*

trans- Prefix denoting through, across, or beyond (e.g. transamination, transducer, transfinite, translocation, transpiration).

trans- (ital., always hyphenated) *Chem.* Prefix denoting isomerism in which **1.** two groups are located on either side of the molecular plane due to restricted rotation caused by, for example, a double bond or a cyclic structure (e.g. *trans*-butenedioic acid, *trans*-dichloro-1,2-propadiene). **2.** two substituents on an atom are directly opposite in a square-planar or octahedral inorganic complex (e.g. *trans*-dibromodichlorotitanium(IV)). The alternative *anti-* is not recommended. *Compare cis-*.

transcription factor Any of various proteins that facilitate transcription. They are often designated by the abbreviation TF plus other characters, e.g. HSTF (heat-shock transcription factor, in *Drosophila melanogaster*), CTF (CCAAT-binding transcription factor), and TFIIIA (in *Xenopus*). Some, however, retain a notation that predates the TF notation, e.g. Sp1 (lower-case p, no space).

transference number Symbol: *t* The fraction of the total charge carried by a particular type of ion in electrical conduction in an electrolyte. Also called transport number.

transferred DNA Abbrev. and preferred form: *T-DNA.

transferred region *Genetics* Abbrev. and preferred form: *T-region.

transfer RNA Abbrev.: *tRNA (no stops).

transistor–transistor logic (en dash) Abbrev.: TTL (no stops).

transition state Abbrev.: TS (no stops).

transmission coefficient *See* acoustic absorption coefficient.

transmission electron microscope Abbrev.: TEM (no stops).

transmit–receive switch (en dash) Abbrev.: TR switch.
anti-transmit–receive switch (hyphen, en dash) Abbrev.: ATR switch.

transmittance **1.** *Chem.* Symbol: *T* A dimensionless physical quantity, the ratio I/I_0, where I_0 is the intensity of electromagnetic radiation falling on a body or substance and I the intensity after transmission through it. **2.** *Physics* Symbol: τ (Greek tau). A dimensionless physical quantity, the ratio Φ_{tr}/Φ_o, where Φ_o is the radiant (or luminous) flux falling on a body and Φ_{tr} is the flux transmitted by it.

transonic (not transsonic)

transport number Another name for transference number.

transposable element *See* transposon.

transposon A segment of DNA that can insert at various locations in a chromosome. The term 'transposon' is used mainly in connection with prokaryotes; 'transposable element' is preferred for the equivalent element in eukaryotes. Transposons are designated by the abbreviation Tn (cap. T) plus an identifying Arabic numeral, e.g. Tn10, Tn3, etc. (no intervening space). There are two classes, denoted by Roman numerals. Class II corresponds broadly to the family of transposons known as TnA (cap. A) or Tn3. The simplest transposons are the insertion sequences (*see* IS).
Different notational systems are used for transposable elements in eukaryotes. For example, maize transposable elements (or controlling elements) are allocated to families designated by italicized abbreviations with initial capital letters, e.g. the *Ac-Ds* (*Ac*tivator-*Ds*sociation) family (hyphenated); the *Spm* (*S*u*p*pressor-*m*utator) family, etc. Variants within families may be designated by additional italic letters, e.g. *Spm-s* and *Spm-w* (hyphenated), or by additional italic numbers, e.g. *Ds6*, *Ds9* (not hyphenated).
Bacterial strains carrying transposons have a specific designation to indicate both the type of transposon and the particular host gene involved. This insertion is denoted by a double colon; e.g. *hisA*::Tn10 indicates that the host bacterium is carrying transposon Tn10 affecting the *hisA* gene in the bacterial chromosome. *See* gene; genotype.

transverse electromagnetic Abbrev.: TEM (no stops).

trapezium **1.** (pl. trapeziums; preferred to trapezia) *Maths.* US: **trapezoid**. A quadrilateral with only two sides parallel. **2.** *Anat.* A bone of the wrist. *Compare* trapezius; trapezoid.

trapezius *Anat.* A muscle in the neck. *Compare* trapezium; trapezoid.

trapezoid **1.** *Maths.* US: **trapezium**. A quadrilateral having no parallel sides. **2.** *Anat.* A bone of the wrist. *Compare* trapezium; trapezius.

Traube, Moritz (1826–94) German chemist.

travelling-wave tube (hyphenated) Abbrev.: TWT (no stops).

Travers, Morris William (1872–1961) British chemist.

T-region (cap. T, hyphenated) Abbrev. and preferred form for transferred region, the region of a Ti plasmid that contains T-DNA.

Trematoda (cap. T) A class of parasitic platyhelminths comprising the flukes. Individual name and adjectival form: **trematode** (no cap.).

tren (no stop) Abbrev. for 2,2′,2″-triaminotriethylamine often used in the formulae of coordination compounds, e.g. [CoBr(tren)]Br.

Trendelenburg, Friedrich (not Trendelenberg) (1844–1924) German surgeon.
Trendelenburg position
Trendelenburg test

Treponema (cap. T, ital.) A genus of *spirochaete bacteria. Individual name: **treponeme** (no cap., not ital.).

Trevithick, Richard (1771–1833) British engineer.

TRH Abbrev. for thyrotrophin-releasing hormone.

Tri (no stop) *Astron.* Abbrev. for Triangulum.

tri- (no stop) Prefix denoting **1.** three (e.g. triandrous, triangle, triclinic, tricuspid, trifoliate, tristichous). **2.** *Chem.* the linking of three groups that together form the root of a structure (e.g. triethylamine); **tris-** is used when an expression to be multiplied already contains a multiplicative prefix.

2,2′,2″-triaminotriethylamine *See* tren.

Triangulum A constellation. Genitive form: Trianguli. Abbrev.: Tri (no stop). *See also* stellar nomenclature.

Triangulum Australe A constellation. Genitive form: Trianguli Australis. Abbrev.: TrA (no stops). *See also* stellar nomenclature.

Triassic Abbrev.: Tr (no stop). **1.** (adjective) Denoting the first period in the Mesozoic era. **2.** (noun; preceded by 'the') The Triassic period. Also called Trias. The period is divided into the Lower (or Early) Triassic (or Scythian), Middle Triassic, and Upper (or Late) Triassic epochs (initial cap.); abbrevs. (respectively): Tr_1, Tr_2, and Tr_3 (subscript numerals). *See also* Permo-Triassic.

tribe A category used in biological classification (*see* taxonomy) consisting of a number of similar genera within a very large family or subfamily. Names of tribes are usually printed in roman (not italic) type with an initial capital letter; plant tribes end in -eae (e.g. Astereae) and animal tribes in -ini (e.g. Bovini). Plant tribes may be divided into subtribes, with the ending -inae.

tribo- Prefix denoting friction (e.g. triboelectricity, triboluminescence).

tribromoethanal CBr_3CHO The recommended name for the compound traditionally known as bromal.

tribromoethane $CHBr_3$ The recommended name for the compound traditionally known as bromoform.

tricarbon dioxide OCCCO The recommended name for the compound traditionally known as carbon suboxide.

tricarboxylic acid cycle Abbrev. and preferred form: TCA cycle (no stops). Also called Krebs cycle. Former name: citric acid cycle.

trichloroethanal CCl_3CHO The recommended name for the compound traditionally known as chloral.

trichloroethanoic acid CCl_3COOH. *See* chloroethanoic acid.

trichloromethane $CHCl_3$ The recommended name for the compound traditionally known as chloroform.

(trichloromethyl)benzene $C_6H_5CCl_3$ The recommended name for the compound traditionally known as benzotrichloride.

trichloronitromethane CCl_3NO_2 The recommended name for the compound traditionally known as chloropicrin.

2,4,5-trichlorophenoxyacetic acid Abbrev. and preferred form: 2,4,5-T.

tricho- (**trich-** before vowels) Prefix denoting hair or hairlike structures (e.g. trichogyne, Trichoptera, trichothallic, trichome).

trien (no stop) Abbrev. for trimethylenetetramine often used in the formulae of coordination compounds, e.g. $[Co(trien)(CN)_2]^+$.

triethanolamine $(HOCH_2CH_2)_3N$ The traditional name for tris-(2-hydroxyethyl)amine.

trig (no stop) Abbrev. for trigonometry.

trigonometric point Abbrev.: TP (no stops).

trigonometry Abbrev.: trig (no stop). Adjectival forms: **trigonometric**, **trigonometrical**.

trihalomethane Any trihalogen derivative of methane, e.g. trichloromethane, $CHCl_3$. The traditional name is haloform, e.g. chloroform, iodoform, etc.

3,4,5-trihydroxybenzoic acid $(OH)_3$-C_6H_2COOH The recommended name for the compound traditionally known as gallic acid.

triiodothyronine Abbrev.: T_3 (subscript number).

trillion Originally in the UK, one million million million (10^{18}); in the US, one million million (10^{12}). Avoid use of the word, specifying instead 10^{12} or 10^{18}.

Trilobita (cap. T) A class of extinct marine arthropods common as fossils. Individual name: **trilobite** (no cap.). Adjectival form: **trilobitic**.

1,3,5-trimethylbenzene $C_6H_3(CH_3)_3$ The recommended name for the compound traditionally known as mesitylene.

trimethylenetetramine *See* trien.

Trinitron (cap. T) A trade name for a type of colour TV picture tube.

2,4,6-trinitrotoluene $CH_3C_6H_2(NO_2)_3$ Abbrev.: TNT (no stops). The traditional name for methyl-2,4,6-trinitrobenzene.

triose phosphate Former name for phosphoglyceraldehyde.

trioxoboric(III) acid H_3BO_3 The recommended name for the compound traditionally known as boric acid.

trioxygen O_3 The recommended name in chemistry for the compound traditionally known as ozone. *See* oxygen.

triphenylmethyl- *Prefix denoting the group $(C_6H_5)_3C-$, e.g. triphenylmethyl chloride. Trityl- is not recommended in chemical usage.

triphosphopyridine nucleotide Abbrev.: TPN (no stops). Use NADP.

triplet state *See* electronic states.

TRIS Abbrev. for tris(hydroxymethyl)-aminomethane.

tris- *See* tri-.

tris-(2-hydroxyethyl)amine $(HOCH_2CH_2)_3N$ The recommended name for the compound traditionally known as triethanolamine.

tritium *See* hydrogen.

triton A nucleus of an atom of tritium, $^3H^+$, denoted by t in nuclear reactions, etc.

Triton The largest satellite of Neptune. *Compare* Titan.

trityl- Use *triphenylmethyl-.

tRNA (lower-case t; preferred to sRNA) Abbrev. for transfer RNA. Individual tRNAs are designated by a superscript abbreviation according to the amino acid with which they bind. Thus, the species $tRNA^{Ala}$ binds with alanine, $tRNA^{Leu}$ binds with leucine, and so on. Where more than one tRNA binds with a particular amino acid, the different species are designated by a subscript Arabic numeral, e.g. $tRNA_1^{Glu}$, $tRNA_2^{Glu}$, etc. A tRNA species bound with its amino acid can be given in the form alanyl-$tRNA^{Ala}$ or Ala-$tRNA^{Ala}$ (hyphenated). In prokaryotes the initiator tRNA is designated $tRNA^{fMet}$ (for *N*-formylmethionine). In eukaryotes the initiator species is designated $tRNA_i^{Met}$ (subscript 'i' for initiator).

-tron Noun suffix denoting a particle accelerator or electron tube (e.g. klystron, synchrotron).

-troph Noun suffix denoting an organism with a specified method of nutrition (e.g. autotroph, heterotroph). *See also* -trophic.

-trophic Adjectival suffix denoting **1.** nourishment, nutrition (e.g. heterotrophic, organotrophic). Noun form: **-trophism**. **2.** (US: **-tropic**; the US spelling is widely used in British English) stimulating the development of (e.g. adrenocorticotrophic, gonadotrophic). Derived noun form: **-trophin** (US: **-tropin**).

tropho- (**troph-** before vowels) Prefix denoting nourishment or nutrition (e.g. trophoblast, trophozoite).

-trophy *Med.* Noun suffix denoting development or growth (e.g. dystrophy).

-tropic Adjectival suffix denoting **1.** directional growth of a plant part in relation to a stimulus (e.g. geotropic, orthotropic, plagiotropic). Noun form: **-tropism**. **2.** affinity for (e.g. lymphotropic). **3.** US spelling of *-trophic (def. 2).

tropical year *See* year.

tropism A directional growth movement of a plant part in response to an external stimulus. The word is prefixed to indicate specific responses (e.g. chemotropism, geotropism, phototropism). Also called tropic movement. Adjectival form: **tropic**. *Compare* taxis.

tropo- (**trop-** before vowels) Prefix denoting a change or turning (e.g. tropomyosin, tropophyte, troposphere).

Trousseau, Armand (1801–67) French physician.
Trousseau's sign

Trouton's rule *Chem.* Named after F. T. Trouton (1863–1922).

troy units *See* ounce; pound troy.

Trp (or **trp**; no stop) Abbrev. for tryptophan. *See* amino acid.

TRPGDA Abbrev. for tripropylene glycol diacrylate.

TR switch Short for transmit-receive switch.

Trumpler, Robert Julius (1886–1956) Swiss-born US astronomer.

tryptophan Abbrev.: Trp or trp (no stop). Symbol: W *See* amino acid.

Ts Symbol often used to denote the tosyl (toluene-4-sulphonyl) group in chemical formulae, e.g. $CH_3C_6H_4SO_2Cl$ can be written as TsCl.

TS Abbrev. for transition state.

Tschermak, Gustav (1836–1927) Austrian mineralogist.
tschermakite
Tschermak's molecule

TSH Abbrev. for thyroid-stimulating hormone.

TSH-RH (hyphenated) Abbrev. and preferred form for thyroid-stimulating-hormone-releasing hormone.

Tsiolkovsky, Konstantin Eduardovich (1857–1935) Russian research scientist in aeronautics and astronautics.

tsunami (pl. tsunamis) A seismic ocean wave. Should not be called tidal wave. [from Japanese: port wave]

TSV Abbrev. for tobacco streak virus.

Tsvet, Mikhail Semenovich (1872–1919) Russian botanist.

TT *Geol.* Abbrev. for Tertiary.

TTEGDA Abbrev. for tetraethylene glycol diacrylate.

TTF Abbrev. for tetrathiafulvalene.

TTL Abbrev. for **1.** *Photog.* through the lens. **2.** *Electronics* transistor–transistor logic.

tube Short for electron tube or more specifically cathode-ray tube.

Tuberales A former order containing the truffles. Most of these are now classified as *Pezizales.

Tubulidentata (cap. T) An order of mammals containing only the aardvark. Individual name and adjectival form: **tubulidentate** (no cap.).

Tuc (no stop) *Astron.* Abbrev. for Tucana.

Tucana A constellation. Genitive form: Tucanae. Abbrev.: Tuc (no stop). *See also* stellar nomenclature.

tufa *Geol.* A deposit of calcium carbonate formed by precipitation from water. Adjectival form: **tufaceous**. *Compare* tuff.

tuff *Geol.* A rock composed of consolidated volcanic ash. Adjectival form: **tuffaceous**. *Compare* tufa.

Tufnol (cap. T) A trade name for strong lightweight laminated plastic.

Tull, Jethro (1674–1741) British agriculturalist, writer, and inventor.

tumour US: **tumor**.

tumour-inducing plasmid Abbrev. and preferred form: Ti plasmid (cap. T).

tundra Adjectival form: **tundral**.

tungstate A compound containing the ion WO_4^{2-}, e.g. sodium tungstate, Na_2WO_4. The recommended name is tungstate(VI).

tungsten Symbol: W Former name: wolfram. *See also* periodic table; nuclide.

Tunicata Use *Urochordata.

tunicate Any invertebrate chordate animal belonging to the subphylum *Urochordata (or Tunicata). Although Urochordata is preferred to Tunicata for the taxonomic group, individuals are known as tunicates rather than urochordates. Adjectival form: **tunicate**.

Tupolev, Andrei Niklaievich (1888–1972) Soviet aeronautical engineer.

turbo- Prefix denoting a turbine (e.g. turbocharger, turboelectric, turbogenerator, turbojet).

turgor potential Symbol: Ψ_p Also called pressure potential. *See* water potential.

Turing, Alan Mathison (1912–54) British mathematician.
Turing machine

Turner, David Warren (1927–) British physical chemist.

Turner, Henry Humbert (1892–1970) US physician.
Turner's syndrome

Turner, William (*c.* 1508–68) English physician and botanist.

turnip yellow mosaic virus Abbrev.: TYMV (no stops). Type member of the **Tymovirus* group (vernacular name: turnip yellow mosaic virus group).

Tuve, Merle Antony (1901–82) US geophysicist.

TV Abbrev. for television.

TVO Abbrev. for tractor vaporizing oil.

twin-lens reflex (hyphenated) Abbrev.: TLR (no stops).

twisting number Symbol: *T* (ital.) The number of turns of the B-DNA helix in a *DNA molecule. The value may be nonintegral.

Twort, Frederick William (1877–1950) British bacteriologist.

TWT *Elec. eng.* Abbrev. for travelling-wave tube.

Tycho Brahe Usually alphabetized as *Brahe, Tycho.

Ty element Abbrev. and preferred form for transposon yeast element, one of a family of repeated base sequences occurring in yeasts. They are divided into two classes: Ty1 and Ty917 (no intervening spaces).

Tymovirus (cap. T, ital.) Approved name for the *turnip yellow mosaic virus group. Individual name: **tymovirus** (no cap., not ital.). [from *t*urnip *y*ellow *mo*saic *virus*]

tympanum (pl. tympana or tympanums) The cavity of the middle ear. The word is also used to mean the eardrum, but this should be referred to as the **tympanic membrane** to avoid confusion. Adjectival form: **tympanic**.

TYMV Abbrev. for turnip yellow mosaic virus.

Tyndall, John (1820–93) British physicist.
Tyndall effect

-type Noun suffix denoting form or type (e.g. ecotype, idiotype, somatotype).

Tyr (or **tyr**; no stop) Abbrev. for tyrosine. *See* amino acid.

tyrosine Abbrev.: Tyr or tyr (no stop). Symbol: Y *See* amino acid.

U

u Symbol for **1.** ungerade (often as a subscript to a term symbol). **2.** unified atomic mass unit.

u Symbol (light ital.) for **1.** instantaneous potential difference (*see also* electronics, letter symbols). **2.** *Optics* object distance. **3.** specific internal energy. **4.** a velocity component or speed.

U Symbol for **1.** uracil. **2.** uranium. **3.** uridine.

U Symbol (light ital.) for **1.** internal energy (U_m = molar internal energy). **2.** potential difference (U_m = magnetic potential difference; *see also* electronics, letter symbols).

UARS Abbrev. for upper-atmosphere research satellite.

ubiquinone Former name: coenzyme Q.

UBV *Astron.* Abbrev. for ultraviolet, blue, visual (green-yellow).

UDP Abbrev. and preferred form for uridine 5′-diphosphate.

UEP Abbrev. for *unit evolutionary period.

UHF Abbrev. for ultrahigh frequency.

Uhlenbeck, George Eugene (1900–) Dutch-born US physicist.

UK (no stops) Abbrev. for United Kingdom.

UKAEA Abbrev. for United Kingdom Atomic Energy Authority, a government organization.

UKgal Symbol for UK gallon. *See* gallon.

UKIRT Acronym for UK Infrared Telescope (Mauna Kea, Hawaii).

UKST Abbrev. for UK Schmidt Telescope (Siding Spring, Australia).

ultra- Prefix denoting beyond, surpassing, extreme, or excessive (e.g. ultracentrifuge, ultrasonic, ultrastructure, ultraviolet, ultravirus).

ultrahigh frequency Abbrev.: UHF (no stops).

ultraviolet (not hyphenated) Abbrev.: UV (no stops).

Ulugh Beg (1394–1449) Persian astronomer.

-um Noun suffix denoting a chemical element (e.g. lanthanum, tantalum).

UMa *Astron.* Abbrev. for Ursa Major.

Umbelliferae (cap. U) A large family of dicotyledonous plants, commonly known as the carrot family. Alternative name: Apiaceae. Individual name: **umbellifer** (no cap.). Adjectival form: **umbelliferous**.

umbra (pl. umbras or umbrae) Adjectival form: **umbral**.

UMi *Astron.* Abbrev. for Ursa Minor.

UMIST Abbrev. or acronym for University of Manchester Institute of Science and Technology.

umkehr effect (no cap.) *Meteorol.* [from German: reversal]

umklapp process (no cap.) *Physics* [from German: collapse]

UMP Abbrev. for uridine 5′-phosphate (uridine monophosphate).

UN (no stops) Abbrev. for United Nations. Do not use UNO.

ungerade Abbrev.: u (no stop). The parity of an orbital is described by a symmetry operation: inversion through a centre. An orbital is described as ungerade (German: uneven) if after inversion the sign of the wave function reverses. *See also* gerade.

unguiculate **1.** *Zool.* (Of mammals) having claws or nails, rather than hooves, flippers, etc. **2.** *Bot.* (Of petals) having a clawlike base. *Compare* ungulate.

ungulate Any hoofed herbivorous mammal, formerly regarded as a member of the order Ungulata but now reclassified in either of two separate orders, *Artiodactyla or *Perissodactyla. The term ungulate is still used for descriptive (rather than taxonomic) purposes. *Compare* unguiculate.

Unh Symbol for unnilhexium. *See* element 106.

uni- Prefix denoting one (e.g. uniaxial, unilocular, uninucleate, unisexual). *See also* mono-.

unicellular *Biology* Consisting of a single cell. In describing uninucleate organisms, such as protozoans and certain algae, *acellular is preferred.

unified atomic mass unit *See* atomic mass unit.

unimolecular reaction *See* reaction mechanisms.

union *Maths.* Symbol: ∪ For *sets A and B:

$$A \cup B = \{x \mid x \in A \text{ or } x \in B\}.$$

Compare intersection.

unique region *Genetics* Abbrev. and preferred form: *U region.

United Kingdom Atomic Energy Authority Abbrev.: UKAEA (no stops).

unit evolutionary period Abbrev.: UEP (no stops). *Genetics* The time in millions of years for 1% divergence to occur in the make-up of a particular protein common to two species, or in the base sequences of the corresponding genes.

units *See* SI units.

UNIX (caps.) *Computing* A trade name for a portable operating system.

unnil- *Chem.* Prefix denoting an element with an atomic number exceeding 100, the attached suffix specifying the atomic number (e.g. unnilquadium, unnilpentium, unnilhexium; *see* element 104; element 105; element 106).

University of Manchester Institute of Science and Technology Abbrev. or acronym: UMIST (no stops).

Unp Symbol for unnilpentium. *See* element 105.

Unq Symbol for unnilquadium. *See* element 104.

***unsym*-** (ital., always hyphenated) *Chem.* Prefix denoting unsymmetric (e.g. *unsym*-dichloroethane, *unsym*-trinitrobenzene). The alternative *as*- is not recommended. *Compare sym*-.

upsilon Greek letter, symbol: υ (lower case), Υ (capital).

uracil Symbol: U A pyrimidine base. *See also* base pair. *Compare* uridine.

uranium Symbol: U *See also* periodic table; nuclide.

Uranus A planet. Adjectival form: **Uranian**.

uranyl Denoting a compound containing the ion UO_2^{2+}, e.g. uranyl nitrate $UO_2(NO_3)_2$. The recommended name is uranyl(VI).

Urbain, Georges (1872–1938) French chemist.

urea $(H_2N)_2CO$ The traditional name for carbamide.

ureaplasma *See* mollicute.

Urediniomycetes A class of basidiomycete fungi comprising the rusts. These were formerly regarded as an order, Uredinales, of the class *Teliomycetes.

U region (cap. U) Abbrev. and preferred form for unique region, either of the two regions (U5 and U3) occurring, respectively, at the 5′ and 3′ ends of retroviral RNA and its complementary DNA. They lie adjacent to each terminal *R segment.

ureido- Prefix denoting the group $H_2NCONH-$ derived from urea (e.g. ureidohydantoin).

Urey, Harold Clayton (1893–1981) US physical chemist.

Miller–Urey experiment (en dash)

-urgy Noun suffix denoting a technology (e.g. metallurgy). Adjectival form: **-urgical**.

uridine Symbol: U A ribonucleoside (*see* nucleoside) consisting of *uracil combined with D-ribose.

uridine 5′-diphosphate Abbrev. and preferred form: UDP (no stops).

uridine 5′-phosphate Abbrev. and preferred form: UMP (no stops).

uridine 5′-triphosphate Abbrev. and preferred form: UTP (no stops).

urino- (**urin-** before vowels) Prefix denoting urine or the urinary system (e.g. *urinogenital, uriniferous). *See also* uro-.

urinogenital (one word; the US form, **urogenital**, is widely used in British English) Relating to the reproductive and excretory systems. The term is synonymous with genitourinary, but the two words tend to be used in different contexts, e.g. urinogenital sinus, urinogenital system, genitourinary medicine.

uro- Prefix denoting **1.** a tail (e.g. Urodela, urophysis, urostyle). **2.** urine or the urinary system (e.g. urobilinogen, urochrome, uroporphyrin).

urochord The notochord of an ascidian, usually confined to the tail region. Adjectival form: **urochordal**; not to be confused with urochordate (*see* Urochordata).

Urochordata (cap. U) A subphylum of invertebrate chordate animals including the sea squirts. Also called Tunicata. Individual name: **urochordate** (no cap.), although *tunicate is more widely used. Adjectival form: **urochordate**; not to be confused with urochordal (*see* urochord).

Urodela (cap. U) An order of amphibians comprising the newts and salamanders. Also called Caudata. Individual name and adjectival form: **urodele** (no cap.; not urodelan or urodelous).

urogenital *See* urinogenital.

Ursa Major A constellation. Genitive form: Ursae Majoris. Abbrev.: UMa (no stops). *See also* stellar nomenclature.

Ursa Minor A constellation. Genitive form: Ursae Minoris. Abbrev.: UMi (no stops). *See also* stellar nomenclature.

US Abbrev. for United States (of America), used adjectivally or (preceded by 'the') as a noun.

USA Abbrev. for United States of America.

USgal Symbol for US gallon. *See* gallon.

USSR (no stops) Abbrev. for Union of Soviet Socialist Republics.

Ustilaginomycetes A class of basidiomycete fungi comprising the smuts. These were formerly regarded as an order, Ustilaginales, of the class *Teliomycetes.

UTC Abbrev. (orig. French) for Coordinated Universal Time. *See* TAI.

utero- Prefix denoting the uterus (e.g. uterogestation, utero-ovarian).

uterus (pl. uteri) Adjectival form: **uterine**.

UTP Abbrev. for uridine 5′-triphosphate.

UV Abbrev. for ultraviolet.

UV reactivation Abbrev. for ultraviolet reactivation, another name for *W reactivation.

V

v Symbol for variable absorption, used in infrared spectroscopy.

v Symbol for **1.** (light ital.) instantaneous potential difference (*see also* electronics, letter symbols). **2.** (light ital.) specific volume. **3.** (bold ital.) velocity. **4.** (light ital.) velocity component or speed. **5.** (light ital.) vibrational quantum number.

V Symbol for **1.** guanosine, cytidine, or adenosine (unspecified). **2.** valine. **3.** vanadium. **4.** variable region (of an *immunoglobulin chain). **5.** volt.

V Symbol (light ital.) for **1.** electric potential. **2.** *Optics* image distance. **3.** (or ΔV) potential difference (*see also* electronics, letter symbols). **4.** potential energy. **5.** volume (V_m = molar volume).

vaccinate To render immune by inoculation with a **vaccine**, which stimulates production of specific antibodies. The original vaccine used was the virus for cowpox (**vaccinia**; hence the name). Noun form: **vaccination**. *Compare* inoculate.

vaccinia virus (two words) Type species of the genus **Orthopoxvirus* (vernacular name: vaccinia subgroup). The name was formerly used as a synonym for the cowpox virus, but the two are now considered different entities.

vacuum (pl. vacuums or vacua) A (hypothetical) space free from particulate matter. *Compare* free space. In technology, a vacuum is any region with a pressure below atmospheric pressure.

vadose *Geol.* Denoting ground water occurring above the water table. *Compare* phreatic.

Val (or **val**; no stop) Abbrev. for valine. *See* amino acid.

valence bond *Chem.* Abbrev.: VB (no stops).

valency *Chem.* The US spelling, **valence**, is now acceptable in the UK, especially when used adjectivally (e.g. valence band, valence bond, valence electrons). However, in referring to an element's ability to form bonds it is usual to use 'valency' (e.g. carbon has a valency of four). *See also* -valent.

-valent Suffix denoting valency (e.g. covalent, electrovalent). The series mono-, di-, tri-, tetra-, penta-, hexa-, hepta-, and octavalent is preferred to uni-, bi-, ter-, quadri-, quinque-, sexa-, septa-, and octovalent.

***n*-valeric acid** $CH_3(CH_2)_3COOH$ The traditional name for pentanoic acid.

valine Abbrev.: Val or val (no stop). Symbol: V *See* amino acid.

vallis (not ital.; pl. valles) *Astron.* A valley. The word, with an initial capital, is used in the approved Latin names of such features on the surface of a planet or satellite, as in Kasei Vallis and Valles Marineris on Mars. *See also* chasma.

Vallisneri, Antonio (1661–1730) Italian physician and biologist.
vallisneria or, as generic name (*see* genus), ***Vallisneria*** *Bot.*

valve *Electronics* Now usually called an electron tube, or just tube. Formerly, short for thermionic valve.

vanadate A compound containing the ion VO_3^-, e.g. ammonium vanadate, NH_4VO_3. The recommended name is vanadate(V).

vanadic Denoting compounds in which vanadium has an oxidation state of +3. The recommended system is to use oxidation numbers, e.g. vanadic chloride, VCl_3, has the systematic name vanadium(III) chloride.

vanadic chloride VCl_3 The traditional name for vandium(III) chloride.

vanadium Symbol: V *See also* periodic table; nuclide.

vanadium(IV) dichloride oxide $VOCl_2$ The recommended name for the compound traditionally known as vanadyl chloride.

vanadium(V) oxide V_2O_5 The recommended name for the compound traditionally known as vanadium pentoxide.

vanadium pentoxide V_2O_5 The traditional name for vanadium(V) oxide.

vanadous Denoting compounds in which vanadium has an oxidation state of +2. The recommended system is to use oxidation numbers, e.g. vanadous chloride, VCl_2, has the systematic name vanadium(II) chloride.

vanadous chloride VCl_2 The traditional name for vanadium(II) chloride.

vanadyl Denoting a compound containing the ion VO^{2+} (e.g. vanadyl chloride, $VOCl_2$). The recommended system is to use oxidation numbers, e.g. vanadium(IV) dichloride oxide.

Van Allen, James Alfred (cap. V) (1914–) US physicist.
Van Allen belts Also called radiation belts.

van Alphen effect *See* de Haas–van Alphen effect.

van de Graaff, Robert Jemison (1901–67) US physicist.
van de Graaff accelerator
van de Graaff generator

van de Hulst, Hendrik Christoffell (1918–) Dutch astronomer.

van de Kamp, Peter (1901–) Dutch-born US astronomer.

van der Meer, Simon (1925–) Dutch physicist and engineer.

van der Waals, Johannes Diderik (1837–1923) Dutch physicist.
van der Waals equation
van der Waals forces

V&V (no spaces) Abbrev. for verification and validation.

Vane, Sir John Robert (1927–) British pharmacologist.

van Helmont, Jan Usually alphabetized as *Helmont.

van't Hoff, Jacobus Henricus (1852–1911) Dutch theoretical chemist.
van't Hoff factor
van't Hoff's isochore
van't Hoff's isotherm

Van Vleck, John Hasbrouck (cap. V) (1899–1980) US physicist.
Van Vleck paramagnetism

vapor US spelling of *vapour.

vapour US: **vapor**. Verb form: **vaporize** (not vapourize).
Two-word terms in which the first word is 'vapour' (e.g. vapour density, vapour pressure) are not hyphenated unless used adjectivally (e.g. vapour-phase inhibitor).

vapour-phase chromatography (hyphenated) Abbrev.: VPC (no stops).

var Symbol: var A unit used in the electric power industry to measure reactive *power. It is numerically and dimensionally equal to the *watt.

var. (stop) Abbrev. for *variety (in plant taxonomy).

Varenius, Bernhard (1622–50) German physical geographer.

variable stars, nomenclature *See* stellar nomenclature.

variable surface glycoprotein Abbrev.: VSG (no stops). The major component of the surface coat of trypanosomes, which undergoes antigenic variation and thus eludes the defence mechanisms of the host.

variety Abbrev.: var. (stop). A unit of biological classification (*see* taxonomy) that is a subdivision of a species. Names of varieties are italicized, being printed in the form *Pinus nigra* var. *maritima* (Corsican pine: a variety of the black pine, *P. nigra*). *See also* cultivar.

Varolio, Constanzo (1543–75) Italian anatomist.
***pons Varolii**

vas deferens (pl. vasa deferentia) Either of a pair of ducts conveying spermatozoa from the epididymis to the urethra. *See also* vaso-. *Compare* vas efferens.

vas efferens (pl. vasa efferentia; usually referred to in the pl.) Any of the small tubes that conduct spermatozoa from the testis to the epididymis. *Compare* vas deferens.

Vaseline (cap. V) A trade name for a petroleum jelly.

vaso- Prefix denoting **1.** a vessel, especially a blood vessel (e.g. vasoactive, vasoconstriction, vasomotor). **2.** *Med.* (often **vas-** before vowels) the vas deferens (e.g. vasoligation, vasectomy).

vasoactive intestinal peptide Abbrev. and preferred form: VIP (no stops).

vasodilatation US: **vasodilation.** *See also* dilatation.

vasopressin Originally a trade name (Vasopressin) for antidiuretic hormone (abbrev.: ADH); the word is now used synonymously with ADH.

Vaucouleurs, Gerard Henri De Usually alphabetized as *De Vaucouleurs.

Vauquelin, Louis Nicolas (1763–1829) French chemist.

Vavilov, Nikolai Ivanovitch (1887–*c.* 1942) Soviet plant geneticist.

VAX/VMS (caps., solidus) A trade name for an operating system for Digital Equipment's VAX range of processors, VAX also being a trade name.

VB Abbrev. for valence bond.

VCR Abbrev. for videocassette recorder.

VDU Abbrev. for visual display unit.

Vector analysis symbols

vector	$\boldsymbol{A}$, $\boldsymbol{a}$
absolute value	$\lvert\boldsymbol{A}\rvert$, A
unit coordinate vectors	$\boldsymbol{i}$, $\boldsymbol{j}$, $\boldsymbol{k}$, $\mathbf{e}_x$, $\mathbf{e}_y$, $\mathbf{e}_z$
scalar product of $\boldsymbol{a}$ and $\boldsymbol{b}$	$\boldsymbol{a} \cdot \boldsymbol{b}$
vector product of $\boldsymbol{a}$ and $\boldsymbol{b}$	$\boldsymbol{a} \times \boldsymbol{b}$, $\boldsymbol{a} \wedge \boldsymbol{b}$
dyadic product of $\boldsymbol{a}$ and $\boldsymbol{b}$	$\boldsymbol{a}\,\boldsymbol{b}$
differential vector operator, nabla, del	∇, $\partial/\partial \boldsymbol{r}$
gradient	grad ϕ, $\nabla\phi$
divergence	div $\boldsymbol{A}$, $\nabla \cdot \boldsymbol{A}$
curl	curl $\boldsymbol{A}$, rot $\boldsymbol{A}$, $\nabla \times \boldsymbol{A}$
Laplacian	∇^2

vector A *physical quantity that has magnitude and direction, for example velocity, force, momentum. Symbols of vectors should be printed in bold italic type, e.g. $\boldsymbol{A}$ or $\boldsymbol{a}$; this distinguishes the vector as an entity from its components and from other *scalar quantities, both of which are printed in italic type. If bold italic type is unavailable, a vector can be indicated by a horizontal arrow above the symbol (in italics). Symbols used in vector analysis are shown in the table. *See also* pseudovector.

vector product A quantity involving two *vectors, $\boldsymbol{a}$ and $\boldsymbol{b}$. It is denoted $\boldsymbol{a} \times \boldsymbol{b}$ (bold multiplication sign). It has a magnitude $ab \sin \theta$, where θ is the angle between the vectors $\boldsymbol{a}$ and $\boldsymbol{b}$, and a direction perpendicular to the plane of $\boldsymbol{a}$ and $\boldsymbol{b}$. A vector product is not however a true vector: under a space reflection (a_x becomes $-a_x$, b_x becomes $-b_x$, etc.) the components of $\boldsymbol{a} \times \boldsymbol{b}$ do not change sign. It is therefore often called a *pseudovector. Also called cross product.

Veksler, Vladimir Iosofich (1907–66) Soviet physicist.

Vel (no stop) *Astron.* Abbrev. for Vela.

Vela A constellation. Genitive form: Velorum. Abbrev.: Vel (no stop). *See also* stellar nomenclature.

velocity Symbol: $\boldsymbol{v}$ or $\boldsymbol{c}$; (u, v, w) are the components of velocity $\boldsymbol{c}$. A *vector quantity, the rate of change of *displacement, $\mathrm{d}\boldsymbol{s}/\mathrm{d}t$. The *SI unit is the metre per second or sometimes the kilometre per hour.
The corresponding scalar quantity, which has magnitude but not direction, is speed, symbol v or u; it is the rate of change of distance travelled. Speed and velocity are expressed in the same units. *See also* angular velocity.

velocity of light *See* speed of light in vacuum.

velvet tobacco mottle virus Abbrev.: VTMoV (no stops).

vena cava (pl. venae cavae) *Anat.* Adjectival form: **caval**.

venation 1. The arrangement of veins in an insect's wing. **2.** The arrangement of vascular bundles (veins) in a leaf. *Compare* vernation.

Vening Meinesz, Felix Andries (1887–1966) Dutch geologist.

Venn, John (1834–1923) British logician.
Venn diagram

Venturi, Giovanni Battista (1746–1822) Italian physicist.
venturi (pl. venturis)
venturi meter

Venus A planet. Adjectival form: **Venusian**.

Venus flytrap (not Venus's) A *carnivorous plant, *Dionaea muscipula*.

Verdet constant *Physics* Named after Marcel Emil Verdet (1824–66).

verdigris (not verdegris)

verification and validation *Computing* Abbrev.: V&V (no stops or spaces).

vermi- Prefix denoting worms or wormlike (e.g. vermicide, vermiculite, vermiform, vermifuge).

vernal equinox *Astron.* Former name for dynamic equinox. *See* equinox.

vernation The way in which leaves are folded or rolled in the bud. *Compare* venation.

Vernier, Pierre (*c.* 1580–1637) French mathematician.
vernier
vernier potentiometer
vernier rocket

vers (no stop) Abbrev. for versine (i.e. versed sine), where

$$\text{vers}\,\theta = 1 - \cos\theta.$$

Vertebrata Use *Craniata.

vertebrate Any animal belonging to the subphylum *Craniata (or Vertebrata). Although Craniata is preferred to Vertebrata for the taxonomic group, individuals are known as vertebrates rather than craniates. Adjectival form: **vertebrate**. *Compare* invertebrate.

vertex (pl. vertices; preferred to vertexes)

Verulam, Lord *See* Bacon, Francis.

very high frequency (not hyphenated) Abbrev. and preferred form: VHF (no stops).

Very Large Array *Astron.* Abbrev.: *VLA (no stops).

very large-scale integration (one hyphen) *Electronics* Abbrev. and preferred form: VLSI (no stops).

very long baseline interferometry (not hyphenated) *Astron.* Abbrev. and preferred form: VLBI (no stops).

very low frequency (not hyphenated) Abbrev. and preferred form: VLF (no stops).

Vesalius, Andreas (1514–64) Flemish anatomist.

vesical Relating to or affecting the urinary bladder (vesica). *Compare* vesicle.

vesicle Any small sac or cavity, such as a small membrane-bounded cavity within a cell, a small blister, or a cavity within a rock. Adjectival form: **vesicular**. *Compare* vesical.

VGA Abbrev. for video graphics array.

VHF Abbrev. and preferred form for very high frequency.

VHS Abbrev. for Video Home System.

via **1.** Use 'through', 'by means of', or 'by' as appropriate. **2.** (noun; pl. vias) *Electronics*

vibrational quantum number Symbol: v An integer that governs the vibrational energy of a molecule. *See also* spectroscopic transitions.

vibrio (pl. vibrios) Any comma-shaped bacterium. The term is also used as a name for individuals of the genus **Vibrio*. Adjectival form: **vibrioid**.

Vibrio (cap. V, ital.) A genus of bacteria of the family Vibrionaceae. *See* vibrio.

V. cholerae: the causative agent of cholera.

vic- (ital., always hyphenated) *Chem.* Prefix denoting *vicinal*, indicating that similar substituents are attached to adjacent atoms (e.g. *vic*-dibromoethane, *vic*-dichlorohexafluorotrisilane). *Compare gem-*.

Vidal de La Blache, Paul (1845–1918) French geographer.

video Two-word terms in which the first word is 'video' (e.g. video camera, video mapping, video signal) should not be hyphenated unless used adjectivally (e.g. video-signal information). Many of these terms are now written as one word (*see* video-).

video- Prefix denoting electronically produced visual images (e.g. videocassette, videodisc, videotape).

videocassette recorder (not hyphenated) Abbrev.: VCR (no stops).

video graphics array (not hyphenated) Abbrev.: VGA (no stops).

Video Home System Abbrev.: VHS (no stops).

videotape recorder (not hyphenated) Abbrev.: VTR (no stops).

videotex (not videotext) Generic name for interactive television-based information systems such as Prestel. Former name: viewdata.

Vidicon (cap. V) A trade name for a type of TV camera tube.

Viète, François (1540–1603) French mathematician. Also called Franciscus Vieta.

Viète's (or **Vieta's**) **root theorem**

viewdata Former generic name for videotex and for the UK Prestel service.

viewfinder (one word)

Vigneaud, Vincent Du Usually alphabetized as *Du Vigneaud.

villus (pl. villi; usually referred to in the pl.) A microscopic outgrowth from the surface of some tissues and organs, especially in the intestine or the mammalian placenta. The adjectival forms **villose** and **villous** are most commonly used in botany, to describe a surface covered with soft hairs.

Vincent, Jean Hyacinthe (1862–1950) French physician.

Vincent's angina Use ulcerative gingivitis.

Vine, Frederick John (1939–) British geologist.

Vinogradov, Ivan Matreyevich (1891–1983) Soviet mathematician.

vinyl- *Prefix denoting the $\dot{C}H_2{=}CH-$ group (e.g. vinylbenzene, vinyl chloride). Ethenyl- is recommended in all contexts.

vinyl acetate $CH_2{=}CHOOCCH_3$ The traditional name for ethenyl ethanoate.

vinyl alcohol $CH_2{=}CHOH$ The traditional name for ethenol.

vinyl chloride $CH_2{=}CHCl$ The traditional name for chloroethene.

VIP Abbrev. and preferred form for vasoactive intestinal peptide. A VIP-secreting tumour is known as a **vipoma** (no caps.).

Vir *Astron.* Abbrev. for Virgo.

Virchow, Rudolf Carl (1821–1902) German pathologist.

Virchow–Robin space (en dash)

Virgo A constellation. Genitive form: Virginis. Abbrev.: Vir (no stop). *See also* stellar nomenclature.

Virtanen, Artturi Ilmari (1895–1973) Finnish chemist.
AIV method (from his initials).

virus (pl. viruses) Adjectival form: **viral.**
A system of classifying and naming viruses has been developed by the International Committee on Taxonomy of Viruses (ICTV), and this system is broadly adhered to in this dictionary. For animal, fungal, and bacterial viruses the ranks employed are family, subfamily, genus, and species; there are as yet no formal categories above the level of family. Latinized names are used for the taxa wherever possible; hence names of genera have the ending *-virus*, names of subfamilies end in -virinae, and names of families end in -viridae. Latinized specific epithets are not used, so binomial nomenclature does not obtain. (The ICTV prescribes italicization of all Latinized names; however, this dictionary follows the UK convention of italicizing only names of genera and species.) Many genera and some higher groups do not yet have approved Latinized names, and these are referred to by their English vernacular names. Plant viruses are classified in groups, not families, with the approved group name ending in *-virus* (for consistency, plant virus group names are italicized in this dictionary). The ranks of genus and species are not used in the taxonomy of plant viruses. For the nomenclature of bacterial viruses, *see* bacteriophage.
This move towards a Latinized system of nomenclature has so far found only limited acceptance, in spite of the confusing and cumbersome system of vernacular nomenclature still widely employed. Some have advocated a coding system instead of a traditional nomenclature. Given the special problems associated with classifying and characterizing viruses, a consensus on their nomenclature may be some way off.

viscera (sing. viscus) The organs within a body cavity, especially the abdominal organs. Adjectival form: **visceral.**

visco- Prefix denoting viscosity (e.g. viscoelastic, viscometer).

viscosity Symbol: η (Greek eta). A physical quantity influencing the resistance to flow of a fluid at low speeds. The *SI unit is the newton second per square metre (N s m^{-2}) or pascal second (Pa s). Also called dynamic viscosity.
The kinematic viscosity, symbol ν (Greek nu), is viscosity divided by fluid density, η/ρ. The SI unit is the square metre per second (m^2 s^{-1}).

viscous 1. Describing fluids that are thick, with a high resistance to flow; not to be confused with viscus (*see* viscera). **2.** Relating to viscosity.

viscus The singular of *viscera. *Compare* viscous.

visual display unit (not hyphenated) *Computing* Abbrev.: VDU (no stops).

vitamin A Chemical name: retinol.

vitamin B complex A group of water-soluble vitamins that, although chemically unrelated, are obtained from similar sources and function as coenzymes. Individual vitamins within this group are described as 'a B vitamin' or 'a vitamin of the B complex'; there is no such entity as 'vitamin B', although this term is incorporated in the trivial names of some of these vitamins, followed by a subscript Arabic numeral:
vitamin B_1 (thiamin)
vitamin B_2 (riboflavin)
vitamin B_6 (pyridoxine)
vitamin B_{12} (cyanocobalamin)
Other B vitamins include biotin, folic acid, nicotinic acid, and pantothenic acid.

vitamin C Chemical name: ascorbic acid.

vitamin D A fat-soluble vitamin occurring in two forms, designated vitamin D_2 (ergocalciferol) and vitamin D_3 (cholecalciferol).

vitamin E Chemical name: *tocopherol.

vitamin K A fat-soluble vitamin, occurring in three forms, designated vitamin K_1, vitamin K_2, and vitamin K_3. Chemically, these forms comprise three groups of quinones.

Viton (cap. V) A trade name for a chemically resistant synthetic rubber.

Vitreoscilla (cap. V, ital.) A genus of filamentous gliding bacteria. Individual name: **vitreoscilla** (no cap., not ital.).

vitrify (**vitrifies, vitrifying, vitrified**) Noun form: **vitrification**. Related adjective: **vitreous**.

vitro- (**vitr-** before vowels) Prefix denoting glass or glasslike (e.g. vitroclastic).

viviparity A type of animal reproduction in which the embryo develops within and obtains nourishment from the mother. Adjectival form: **viviparous**. *Compare* vivipary.

vivipary The development of young plants or bulbils on a parent plant in place of flowers. *Compare* viviparity.

VLA *Astron.* Abbrev. for Very Large Array (New Mexico).

VLBI *Astron.* Abbrev. for very long baseline interferometry.

VLDL Abbrev. for very low-density lipoprotein.

Vleck, John Hasbrouck Van Usually alphabetized as *Van Vleck.

VLF Abbrev. for very low frequency.

VLSI *Electronics* Abbrev. and preferred form for very large-scale integration.

VMS *See* VAX/VMS.

vocal cords (not vocal chords)

Vogel, Hermann Karl (1842–1907) German astronomer.

Vogt–Russell theorem (en dash) *See* Russell, Henry Norris.

Voigt, Woldemar (1850–1919) German physicist.
Voigt effect

Vol (no stop) *Astron.* Abbrev. for Volans.

vol. Abbrev. for volume.

Volans A constellation. Genitive form: Volantis. Abbrev.: Vol (no stop). *See also* stellar nomenclature.

volcano (pl. volcanoes) Adjectival form: **volcanic**. Derived noun: **volcanism** (preferred to vulcanism).

Volkmann, Alfred Wilhelm (1800–77) German anatomist.
Volkmann canal

volt Symbol: V The *SI unit of electric potential, potential difference, and electromotive force.

$$1\,V = 1\,W\,A^{-1} = 1\,J\,C^{-1}.$$

Named after Count Alessandro *Volta.

Volta, Count Alessandro Giuseppe Antonio Anastasio (1745–1827) Italian physicist.
***volt**
voltage
voltaic cell
voltaic pile

voltage An *electric potential, *potential difference, or *electromotive force expressed in volts. The word is probably best avoided in technical writing.

voltameter Former name for coulombmeter. *Compare* voltmeter.

volt ampere (two words) Symbol V A A unit used in the electric power industry to measure apparent *power. It is numerically equal to the watt.

Volterra, Vito (1860–1940) Italian physicist.
Lotka–Volterra equations (en dash)
Volterra's integral equations

voltmeter An instrument for measuring potential difference. *Compare* voltameter.

volume Symbol: V or v A physical quantity indicating extent in three dimensions. The *SI unit is the cubic metre or the litre.

volume density of charge *See* charge density.

volume strain *See* strain.

volume unit Abbrev.: VU (no stops).

volva (pl. volvae) A cuplike membrane encircling the base of the stalk of many basidiomycete fungi. *Compare* vulva.

von Baer Usually alphabetized as *Baer.

von Baeyer, Johann Usually alphabetized as *Baeyer.

von Braun, Wernher Magnus Maximilian (1912–77) German-born US rocket engineer.

von Buch, Christian Leopold (1774–1853) German geographer and geologist.

von Eötvös, Baron Usually alphabetized as *Eötvös.

von Euler, Ulf Svante (1905–83) Swedish physiologist.

von Fehling, Hermann Usually alphabetized as *Fehling.

von Fraunhofer, Josef Usually alphabetized as *Fraunhofer.

von Frisch, Karl Usually alphabetized as *Frisch.

von Helmholtz, Hermann Usually alphabetized as *Helmholtz.

von Hevesy, George Usually alphabetized as *Hevesy.

von Hofmann, August Usually alphabetized as *Hofmann.

von Humboldt, Baron Usually alphabetized as *Humboldt.

von Klitzing, Klaus (1943–) German physicist.

von Kupffer, Karl Usually alphabetized as *Kupffer.

von Laue, Max Usually alphabetized as *Laue.

von Leydig, Franz Usually alphabetized as *Leydig.

von Liebig, Justus Usually alphabetized as *Liebig.

von Linde, Karl Usually alphabetized as *Linde.

von Lindemann, Carl Usually alphabetized as *Lindemann.

von Mayer, Julius Usually alphabetized as *Mayer.

von Miller, Oskar Usually alphabetized as *Miller.

von Naegeli, Karl Usually alphabetized as *Naegeli.

von Neumann, John (originally Johann) (1903–57) Hungarian-born US mathematician.

non von Neumann architecture (not hyphenated)

von Neumann machine

von Purbach, Georg Usually alphabetized as *Purbach.

von Recklinghausen, Friedrich (1833–1910) German pathologist.

von Recklinghausen's disease 1. A disease of nerves. **2.** A bone disease.

von Richthofen, Ferdinand (1833–1905) German geographer.

von Sachs, Julius Usually alphabetized as *Sachs.

von Siebold, Karl Usually alphabetized as *Siebold.

von Stradonitz, Friedrich August Kekulé Usually alphabetized as *Kekulé von Stradonitz.

von Szent-Györgi, Albert Usually alphabetized as *Szent-Györgi.

von Waldeyer-Hartz, Heinrich Usually alphabetized as *Waldeyer-Hartz.

von Weizsäcker, Baron Carl Friedrich Usually alphabetized as *Weizsäcker.

von Welsbach *See* Auer, Karl.

von Zeppelin, Graf Usually alphabetized as *Zeppelin.

-vorous Adjectival suffix denoting feeding on (e.g. carnivorous, herbivorous, omnivorous). Noun form: **-vore**.

vortex (pl. vortices; preferred to vortexes) Derived noun: **vorticity**.

VPC Abbrev. for vapour-phase chromatography.

Vries, Hugo de Usually alphabetized as *de Vries.

VSEPR Abbrev. for valence-shell/electron-pair repulsion.

VSG Abbrev. for *variable surface glycoprotein.

VTMoV (lower-case o) Abbrev. for velvet tobacco mottle virus.

VTR Abbrev. for videotape recorder.

VU *Acoustics* Abbrev. for volume unit.

Vul (no stop) *Astron.* Abbrev. for Vulpecula.

vulcanism *Geol.* Use volcanism.

vulcanite *Chem.* Derived noun: **vulcanization**.

Vulpecula A constellation. Genitive form: Vulpeculae. Abbrev.: Vul (no stop). *See also* stellar nomenclature.

vulva (pl. vulvae or vulvas) The female external genital organs, collectively. *Compare* volva.

v/v (no stops) Abbrev. for volume in volume.

W

w Symbol for weak absorption, used in infrared spectroscopy.

w Symbol (light ital.) for **1.** mass fraction (w_B = mass fraction of substance B). **2.** a velocity component.

W 1. Symbol for **i.** adenosine or thymidine (or uridine) (unspecified). **ii.** tryptophan. **iii.** tungsten. **iv.** watt. **v.** W particle (W^+, W^-). **2.** Abbrev. for *west or western.

W Symbol (light ital.) for **1.** *Eng.* load. **2.** weight. **3.** work. **4.** *Biochem.* *writhing number.

Waage, Peter (1833–1900) Norwegian chemist.
Guldberg–Waage theory (en dash)

Waals, Johannes Diderik van der Usually alphabetized as *van der Waals.

Wacker process *Organic chem.*

Waddington, Conrad Hal (1905–75) British embryologist and geneticist.

wadi (not wady; pl. wadis)

Wadsworth mounting *Spectroscopy*

Wagner–Meerwein rearrangement (en dash) *Organic chem.*

Waksman, Selman Abraham (1888–1973) Russian-born US biochemist.

Wald, George (1906–) US biochemist.

Walden, Paul (1863–1957) Russian-born German chemist.
Walden inversion
Walden's rule

Waldenström, Jan (1906–) Swedish biochemist.
Waldenström's disease.

Waldeyer-Hartz, Heinrich Wilhelm Gottfried von (hyphen) (1836–1921) German anatomist and physiologist.
Waldeyer's ring

Walker, Sir James (1863–1935) British chemist.

Wallace, Alfred Russel (not Russell) (1823–1913) British naturalist.
Wallace effect
Wallace's line

Wallach, Otto (1847–1931) German chemist.

Wallis, Sir Barnes Neville (1887–1979) British engineer.

Wallis, John (1616–1703) English mathematician and theologian.
Wallis's product

Walton, Ernest Thomas Sinton (1903–) Irish physicist.
Cockcroft–Walton generator (en dash)

WAN *Computing* Acronym for wide-area network.

Wankel, Felix (1902–88) German engineer.
Wankel engine

Warburg, Otto Heinrich (1883–1970) German physiologist.
Warburg effect
Warburg manometer

Ward, Joshua (1685–1761) English physician and chemist.

Ward-Leonard system (hyphen) *Elec. eng.* Named after Harry Ward Leonard (not hyphenated) (1861–1915), US electrical engineer.

Waring, Edward (1734–98) British mathematician.
Waring's problem

Wassermann, August von (not Wasserman) (1866–1925) German bacteriologist.
Wassermann reaction (or **test)**

water Two-word terms in which the first word is 'water' (e.g. water balance, water lily, water potential, water softening, water table, water vapour) should not be hyphenated unless used adjectivally (e.g. water-balance components). *See also* hydrates.

watercourse (one word)

watercress (one word) *See also Nasturtium.*

waterfall (one word)

water lily (two words) Any of various aquatic plants of the family Nymphaeaceae. Note that they are not true *lilies.

water potential Symbol: Ψ (Greek cap. psi) Water conduction in plants was previously measured in terms of *osmotic pressure; since this does not take account of capillary forces, water relations in plants are now expressed in terms of water potential, which has three components: osmotic potential, pressure potential, and matric potential:

$$\Psi_w = \Psi_o + \Psi_p + \Psi_m.$$

watershed (one word)

waterspout (one word)

Watson, David Meredith Seares (1886–1973) British palaeontologist.

Watson, James Dewey (1928–) US biochemist.
Watson–Crick model (en dash)

Watson, Sir William (1715–87) British physicist, physician, and botanist.

watt (no cap.) Symbol: W The *SI unit of power.

$$1\,\mathrm{W} = 1\,\mathrm{J\,s^{-1}}.$$

In electric power technology, active power is expressed in watts, apparent power in volt amperes, and reactive power in vars. Named after James *Watt.

Watt, James (1736–1819) British instrument maker and inventor.
***watt**
Watt engine
wattmeter

watt second Symbol: W s or W·s A unit of work or energy given the name joule in SI units. *See also* kilowatt hour.

wave Two-word terms in which the first word is 'wave' (e.g. wave equation, wave mechanics, wave motion, wave theory, wave velocity) are not hyphenated unless used adjectivally (e.g. wave-equation solutions). Some two-word terms are now usually written as one word (e.g. waveband, wavebase, waveform, wavefront, waveguide, *wavelength, *wavenumber, wavetrain, wavetrap).

wave function Symbol: Ψ (Greek cap. psi). A mathematical function representing the amplitude of a wave associated with a particle. The value $|\Psi|^2\mathrm{d}T$ is proportional to the probability of finding the particle in an element of space dT.
The wave function for the hydrogen atom can be represented by:

$$\Psi(r,\theta,\phi) = R(r)\Theta(\theta)\Phi(\phi),$$

where r, θ, and ϕ are polar coordinates, $R(r)$ is the radial wave function, and $\Theta(\theta)\Phi(\phi)$ is the angular wave function.

wavelength (one word) Symbol: λ (Greek lambda). A physical property of a periodic wave, such as a light or sound wave. It is the distance between two points of equal phase (e.g. two maxima) in the direction of propagation. The *SI unit is the metre; the angstrom may also be used for electromagnetic radiation.

wavenumber (one word) Symbol: σ (Greek sigma). A physical quantity, the reciprocal of *wavelength, i.e. $1/\lambda$. The *SI unit is the reciprocal metre (m^{-1}). Also called repetency.
The **circular wavenumber** (or **angular wavenumber**), symbol k, is equal to $2\pi\sigma$.

Wb Symbol for weber. *See also* SI units.

WBC Abbrev. for white blood cell. *See* leucocyte.

W chromosome (not hyphenated) The smaller of the *sex chromosomes in birds and lepidopterans, equivalent to the Y chromosome in other organisms.

We (ital.) Symbol for Weber number.

weber (no cap.) Symbol: Wb The *SI unit of *magnetic flux.

$$1\,\mathrm{Wb} = 1\,\mathrm{V\,s}.$$

Named after W. E. *Weber.

Weber, Ernst Heinrich (1795–1878) German physiologist and psychologist.
Weber–Fechner law (en dash) Use Weber's law.
Weberian ossicles *Zool.*

Weber, Wilhelm Eduard (1804–91) German physicist.
***weber**

Weber number Symbol: *We* A dimensionless quantity equal to $\rho v^2 l/\sigma$, where ρ is the density, v and l a characteristic speed and length, and σ is *surface tension. *See also* parameter. Named after Moritz Weber (1871–1951).

Weddell, James (1787–1834) British navigator.
Weddell Sea
Weddell seal (*Leptonychotes weddelli*)

Wegener, Alfred Lothar (1880–1930) German meteorologist and geologist.

Weierstrass, Karl Wilhelm Theodor (1815–97) German mathematician.
Weierstrass (or **Weierstrassian**) **functions**
Weierstrass test (or **M test**)

weight Symbol: *W* Abbrev.: wt (no stop). A scalar physical quantity, the gravitational force exerted on a body at the surface of the earth (or another planet or a satellite), giving it an acceleration equal to the local *acceleration of free fall, *g*. The weight of a body must not be confused with its *mass, *m*: $W = mg$, therefore weight varies as *g* varies. The *SI unit is the newton but in everyday usage weight is measured in units of mass.

Weigle reactivation Abbrev. and preferred form: *W reactivation.

Weil, Adolf (1848–1916) German physician.
Weil's disease

Weil, André (1906–) French mathematician.

Weil, Edmund (1880–1922) Austrian physician.
Weil–Felix reaction (en dash)

Weinberg, Steven (1933–) US physicist.

Weinberg, Wilhelm (1862–1937) German physician.
Hardy–Weinberg equation (en dash)
Hardy–Weinberg equilibrium (en dash)
Hardy–Weinberg law (en dash)

Weismann, August Friedrich Leopold (not Weisman) (1834–1914) German biologist.
Weismannism

Weiss, Pierre Ernst (1865–1940) French physicist.
Curie–Weiss law (en dash)
Weiss constant
Weiss magneton

Weissenberg, Karl (1893–) Austrian-born physicist.
Weissenberg camera
Weissenberg effect

Weizmann, Chaim Azriel (1874–1952) Russian-born Israeli chemist.

Weizsäcker, Baron Carl Friedrich von (1912–) German physicist.
Bethe–Weizsäcker cycle (en dash)

Welch, William Henry (1850–1934) US pathologist and bacteriologist.
Welch's bacillus Old name for *Clostridium perfringens*.

Weldon, Walter (1832–85) British industrial chemist.
Weldon process

Weller, Thomas Huckle (1915–) US microbiologist.

Welwitschia (cap. W, ital.) A genus of unusual gymnosperms in the class *Gnetopsida. In some classifications it is separated into its own order, Welwitschiales (*see* Gneticae). Individual name: **welwitschia** (no cap., not ital.; pl. welwitschias). Named after F. M. J. Welwitsch (1807–72).

Went, Friedrich August Ferdinand Christian (1863–1935) Dutch botanist.

Werner, Abraham Gottlob (1750–1817) German mineralogist and geologist.

Werner, Alfred (1866–1919) French-born Swiss chemist.
Werner complexes
Werner's theory

west Adjectival forms: **west, western**. Abbrev.: W (no stop). Use the abbrev. only descriptively, not in place names or concepts (e.g. W London, W Canada, but Western Isles, West Indies, West Wind Drift, western hemlock).

Western blotting (cap. W) A chromatographic technique used in the analysis of polypeptides and proteins. It is named by analogy to the similar technique of *Southern blotting.

Westinghouse, George (1846–1914) US inventor.
Westinghouse brake

west-north-west (hyphenated) Adjectival form: **west-north-western**. Abbrev. (for both): WNW (no stops).

Weston, Edward (1850–1936) British-born US electrical engineer.
Weston cell

Weston number

west-south-west (hyphenated) Adjectival form: **west-south-western**. Abbrev. (for both): WSW (no stops).

Weyl, Hermann (1885–1955) German mathematician.

Wharton, Thomas (1614–73) English physician.
Wharton's duct
Wharton's jelly

Wheatstone, Sir Charles (1802–75) British physicist.
Wheatstone bridge

Wheeler, John Archibald (1911–) US theoretical physicist.

Whewell, William (1794–1866) British physicist and philosopher.

Whipple, Fred Lawrence (1906–) US astronomer.

White, Gilbert (1720–93) British naturalist.
White's thrush (*Zoothera dauma*)

white blood cell (preferred to white blood corpuscle) Abbrev.: WBC (no stops). *See* leucocyte.

white dwarf *Astron. See* dwarf.

Whitehead, Alfred North (1861–1947) British-born US mathematician and philosopher.

white lead $Pb_3(OH)_2(CO_3)_2$ The traditional name for lead(II) carbonate hydroxide.

whiteout (one word) *Meteorol.*

white vitriol $ZnSO_4$ A former name for zinc sulphate; no longer in scientific usage.

Whittaker, Sir Edmund Taylor (1873–1956) British mathematician and physicist.

Whittle, Sir Frank (1907–) British aeronautical engineer.

Whitworth, Sir Joseph (1803–87) British engineer.
Whitworth metal
Whitworth thread

WHO Abbrev. for World Health Organization.

WHT Abbrev. for William Herschel Telescope (La Palma).

wide-area network (hyphenated) *Computing* Acronym: WAN (no stops).

Widmanstätten structure *Astron., metallurgy* Named after A. J. Widmanstätten (1754–1849).

Wiedemann–Franz law (en dash) *Physics* Named after G. H. Wiedemann (1826–99) and R. Franz (1827–1902).

Wieland, Heinrich Otto (1877–1957) German chemist.

Wien, Max C. (1866–1938) German physicist.
Wien-bridge oscillator
Wien effect

Wien, Wilhelm Carl Werner Otto Fritz Franz (1864–1928) German physicist.
Wien's displacement law
Wien's formula

Wiener, Norbert (1894–1964) US mathematician.

Wigglesworth, Sir Vincent Brian (1899–) British entomologist.

Wigner, Eugene Paul (1902–) Hungarian-born US physicist.
Breit–Wigner formula (en dash)
Wigner effect
Wigner energy
Wigner nuclides
Wigner supermultiplets

Wilcke, Johan Carl (1732–96) German-born Swedish physicist.

Wild, John Paul (1923–) British-born Australian radio astronomer.

Wildt, Rupert (1905–76) German-born US astronomer.

wild type (hyphenated when used adjectivally) The normal form of a phenotype, genotype, or allele, as is most commonly found in wild populations. In the simplest systems of genotypic notation (*see* genotype) the wild-type allele of a gene is represented simply by a plus sign; for instance, + + or +/+ for a homozygote, $+a$ or $+/a$ for a heterozygote. In other cases a superscript plus sign is attached to the symbol for the allele; for example, w^+

designates the wild-type allele at the white-eye locus in the fruit fly, *Drosophila melanogaster* (*see* gene). The same notation can be used for alleles of bacterial genes (e.g. *lacY*$^{+}$) and also for phenotypic designations (*see* phenotype).

Wilkes, Maurice Vincent (1913–) British computer engineer.

Wilkins, Maurice Hugh Frederick (1916–) New Zealand biophysicist.

Wilkinson, Sir Denys Haigh (1922–) British physicist.

Wilkinson, Sir Geoffrey (1921–) British inorganic chemist.
Wilkinson's catalyst

William Hershel Telescope (in La Palma) Abbrev.: WHT (no stops).

Williams, Robert R. (1886–1965) US chemist.

Williams, Robley Cook (1908–) US biophysicist.

Williamson, Alexander William (1824–1904) British chemist.
Williamson ether synthesis

Williams tube *Computer eng.* Named after F. C. Williams (1911–77).

Willis, Thomas (1621–75) English anatomist and physician.
circle of Willis

Willstätter, Richard (1872–1942) German chemist.

Wilms, Marx (1867–1918) German pathologist.
Wilms' tumour

Wilson, Alexander (1766–1813) British-born US ornithologist.

Wilson, Charles Thomson Rees (1869–1959) British physicist.
Wilson cloud chamber

Wilson, Edmund Beecher (1856–1939) US biologist.

Wilson, Edward Osborne (1929–) US entomologist, ecologist, and sociobiologist.

Wilson, John Tuzo (1908–) Canadian geophysicist.

Wilson, Kenneth G. (1937–) US theoretical physicist.

Wilson, (Samuel Alexander) Kinnier (1878–1937) US-born British neurologist.
Wilson's disease

Wilson, Robert Woodrow (1936–) US astrophysicist.

wimp (or **WIMP**) *Computing* Acronym for windows, icons, menus, and pointers.

Wimshurst, James (1832–1903) British engineer.
Wimshurst machine

winchester *Chem.* A large cylindrical glass bottle. *Compare* Winchester.

Winchester (cap. W) *Computing* A small high-capacity hard disk. Also called Winchester disk. *Compare* winchester.

Windaus, Adolf Otto Reinhold (1876–1959) German chemist.

Winkler, Clemens Alexander (1838–1904) German chemist.

Winslow, Jakob Benignus (1669–1760) Danish anatomist.
foramen of Winslow

Wistar, Caspar (1761–1818) US anatomist.
Wistar rats (from the Wistar Institute in Philadelphia founded by Caspar's grand-nephew)
wisteria** or, as generic name, ***Wisteria

wisteria (not wistaria) Generic name: *Wisteria* (cap. W, ital.). Named after Caspar *Wistar, hence the common variant spelling 'wistaria', but the official spelling for the generic name is *Wisteria* and this is recommended for the common name.

Wiswesser line notation *Chem.* Named after W. J. Wiswesser.

Wittig, Georg Friedrich Karl (1897–1987) German organic chemist.
Wittig reaction
Wittig rearrangement

WLN Abbrev. for Wiswesser line notation.

WMO Abbrev. for World Meteorological Organization.

WNW Abbrev. and preferred form for west-north-west or west-north-western.

Wöhler, Friedrich (1800–82) German chemist.

Wolf, Charles J. E. (1827–1918) French astronomer.
Wolf–Rayet star (en dash) Also named after G. A. P. Rayet (1839–1906).

Wolf, Johann Rudolf (1816–93) Swiss astronomer.
Wolf number Also called Zurich relative sunspot number.

Wolf, Maximilian Franz Joseph Cornelius (1863–1932) German astronomer.
Wolf diagram

Wolff, Kaspar Friedrich (1734–94) German anatomist and physiologist.
Wolffian (US wolffian) body Another name for mesonephros.
Wolffian (US wolffian) duct Another name for mesonephric duct.

wolfram Former name for tungsten.

Wollaston, William Hyde (1766–1828) British chemist and physicist.
Wollaston prism

Wood, Robert Williams (1868–1955) US physicist.
Wood's glass

Woodward, Sir Arthur Smith (1864–1944) British palaeontologist.

Woodward, John (1665–1728) British geologist.

Woodward, Robert Burns (1917–79) US chemist.
Woodward–Hoffmann rules (en dash)

word A fixed number of *bits treated as a single unit by the hardware of a computer. In present-day computers there are usually 16 or 32 bits in a word. The main store in a computer is divided into either *byte or word storage locations. A byte is shorter than a word.

word processing (preferred to word-processing) Abbrev.: WP or wp (no stops).
word processor (preferred to word-processor) Abbrev.: WP or wp (no stops).

work Symbol: W A physical quantity expressed as the product $F{\cdot}s$ of a *force and the distance through which its point of application moves in the direction of the force. The *SI unit is the joule. *See also* energy.

work function Symbol: ϕ or Φ (Greek lower-case or cap. phi), where $\Phi = e\phi$ and e is the electronic charge. A physical quantity used in solid-state physics, the energy difference between an electron at rest at infinity and an electron at the Fermi level within a substance. The *SI unit is the joule or sometimes the electronvolt.

workstation (one word) *Computing*

World Health Organization Abbrev.: WHO (no stops).

World Meteorological Organization Abbrev.: WMO (no stops).

WORM (or **worm**) *Computing* Acronym for write once read many (times).

WP (or **wp**) Abbrev. for word processing or word processor.

W particles *See* gauge bosons.

W reactivation Abbrev. and preferred form for Weigle reactivation, the phenomenon in which UV-irradiated lambda phages show greater infectivity in irradiated host *E. coli* than in nonirradiated *E. coli*. Named after Jean Weigle. Also called UV reactivation.

Wright, Sewall (1889–1988) US statistician and geneticist.
***Sewall Wright effect**

writhing number Symbol: W (ital.) The number of turns of superhelix in a DNA molecule. The value may be nonintegral.

Wronskian *Maths.* Named after Jósef Wroński (real surname Hoene; 1776–1853).

W–R stars (en dash; or **WR stars**) Short for Wolf–Rayet stars.

WSW Abbrev. and preferred form for west-south-west or west-south-western.

wt (no stop) Abbrev. for weight.

Wu, Chien-Shiung (1912–) Chinese-born US physicist.

Wurtz, Charles Adolphe (1817–84) French chemist.
Wurtz reaction (or **synthesis)**

w/v (no stops) Abbrev. for weight in volume of solution.

w/w (no stops) Abbrev. for weight in weight.

Wyckoff, Ralph Walter Graystone (1897–) US crystallographer and electron microscopist.

Wynne-Edwards, Vero Copner (hyphen) (1906–) British zoologist.

wysiwyg (or **WYSIWYG**) *Computing* Acronym for what you see (on the screen) is what you get (from the printer).

X

x Symbol (light ital.) for **1.** a Cartesian coordinate (usually horizontal). **2.** mole fraction (x_B = mole fraction of substance B).

X Symbol for **1.** xanthosine. **2.** a halogen in chemical formulae, e.g. MgX. **3.** an unknown amino acid (in a sequence).

X Symbol (light ital.) for **1.** exposure. **2.** reactance.

Xaa Abbrev. for an unknown amino acid (in a sequence).

xantho- (**xanth-** before vowels) Prefix denoting yellow (e.g. xanthophyll).

Xanthomonas (cap. X, ital.) A genus of *pseudomonad bacteria. Individual name: **xanthomonad** (no cap., not ital.).

Xanthophyta A division comprising the yellow-green algae. Former name: Heterokontae (*see* heterokont). In a recent classification it is reduced to the status of a class, Xanthophyceae, in the division *Chromophyta, subkingdom Chromobionta.

xanthosine Symbol: X *See* nucleoside.
xanthosine 5′-diphosphate Abbrev. and preferred form: XDP (no stops).
xanthosine 5′-phosphate Abbrev. and preferred form: XMP (no stops).
xanthosine 5′-triphosphate Abbrev. and preferred form: XTP (no stops).

x-axis (ital. x, hyphenated)

X chromosome (not hyphenated) *See* sex chromosome.

XDP Abbrev. for xanthosine 5′-diphosphate.

Xe Symbol for xenon.

xeno- (**xen-** before vowels) Prefix denoting strange or foreign (e.g. xenoblastic, xenocryst, xenolith).

xenon Symbol: Xe *See also* periodic table; nuclide.

xero- (**xer-** before vowels) Prefix denoting dryness (e.g. xerography, xerophyte, xerosere).

Xerox (cap. X) A trade name for a type of xerographic process. The word should not be used to describe the equipment involved nor the photocopy produced nor should it be used as a verb. The word xerox (no cap.) is sometimes used generically, but not correctly, for any photocopy or photocopying process.

xi Greek letter, symbol: ξ (lower case), Ξ (cap.).
ξ *Chem.* Symbol for extent of reaction.
Ξ Symbol for Xi particle (Ξ^0 or Ξ^-).

xiphi- (or **xipho-**) Prefix denoting swordlike (e.g. xiphisternum, Xiphosura).

XMP Abbrev. for xanthosine 5′-phosphate (xanthosine monophosphate).

XOR (or **xor**) Short for exclusive-OR. Also written EXOR or exor. *See* logic symbols.

X-PES Abbrev. for X-ray photoelectron spectroscopy.

X-ray (cap. X; hyphenated in noun, verb, and adjectival forms) US: **x-ray** or **X-ray**. Do not use 'X-ray' (noun) as a synonym for *radiograph.

XTP Abbrev. for xanthosine 5′-triphosphate.

x-unit Symbol: xu An X-ray standard value equal to
$1.002\ 077\ 89 \times 10^{-13}$ metre, xu(Cu)
$1.002\ 099\ 38 \times 10^{-13}$ metre, xu(Mo).

XUV (or **EUV**) Abbrev. for extreme ultraviolet.

Xyl (no stop) Abbrev. for xylose.

xylene $C_6H_4(CH_3)_2$ The traditional name for *dimethylbenzene.

xylic acid $C_6H_3(CH_3)_2COOH$ The traditional name for 2,4-dimethylbenzoic acid.

xylo- (**xyl-** before vowels) Prefix denoting wood (e.g. xylocarp, xylophilous).

xylose Abbrev.: Xyl (no stop). *See* sugars.

Y

y Symbol (light ital.) for a Cartesian coordinate (usually vertical).

Y Symbol for **1.** thymidine or cytidine (unspecified). **2.** tyrosine. **3.** yttrium.

Y Symbol (light ital.) for **1.** admittance. **2.** hypercharge.

Yagi, Hidetsuga (1886–1976) Japanese electrical engineer.
Yagi aerial

Yalow, Rosalyn Sussman (1921–) US physicist.

Yang, Chenning (1922–) Chinese-born US physicist.
Yang–Mills theory (en dash) Also named after R. L. Mills (1927–).

Yanofsky, Charles (1925–) US geneticist.

yard Symbol: yd The traditional UK and US unit of length, now defined as equal to exactly 0.9144 metre. It has been superseded in scientific measurements by the metre (*see* SI units).
1 yard = 3 feet = 36 inches
1 furlong = 10 chains = 220 yards
1 mile = 8 furlongs = 1760 yards.

***y*-axis** (ital. y, hyphenated)

Yb Symbol for ytterbium.

Y chromosome (not hyphenated) *See* sex chromosome.

yd Symbol for yard.

year 1. General symbol: *a* The period of the earth's revolution around the sun, measured relative to a particular frame of reference in the sky. The length of the period depends on the frame of reference. For example, the **sidereal year** (defined by the sun's passage among the stars) has 365.256 36 days; the **tropical year** (defined by the sun's passage through the solstices and equinoxes) has 365.242 19 days.
2. A time interval of 365 days or, in the case of a leap year, of 366 days. Leap years are those years that are divisible by 4, with the exception of century years that are not divisible by 400. Thus 1988, 1992, and 2000 are leap years but 1989, 1990, 1991, and 2100 are not. Over a cycle of 400 years the average length of the year is 365.2425 days, which is close to the tropical year.

yeasts *See* Endomycetales; Hyphomycetes. For genetic nomenclature of yeasts, *see* gene.

yellow-green algae *See* Xanthophyta.

Yersin, Alexandre Emile John (1863–1943) Swiss bacteriologist.
****Yersinia***

Yersinia (cap. Y, ital.) A genus of bacteria of the family *Enterobacteriaceae.
Y. pestis: the agent responsible for bubonic plague, formerly called *Pasteurella pestis*.
Individual name: **yersinia** (no cap., not ital.; pl. yersinias).

YIG *Electronics* Acronym for yttrium iron garnet.

-yl Noun suffix denoting a radical, especially a hydrocarbon radical with one free valency (e.g. ethyl, phenyl).

-ylidene Noun suffix denoting a hydrocarbon radical with two free valencies on the same carbon atom (e.g. ethylidene, benzylidene).

Young, James (1811–83) British industrial chemist.

Young, Thomas (1773–1829) British physicist, physician, and Egyptologist.
Young's fringes
***Young modulus**
Young's slits

Young modulus Symbol: *E* or sometimes *Y* A physical quantity equal to the ratio of normal *stress to the resulting linear *strain, σ/ϵ, for stresses smaller than the yield stress. The *SI unit is the newton per square metre. Also called Young's modulus, modulus of elasticity.

-yse Preferred to -yze in British English as the suffix for verbs derived from *-lysis* (e.g. analyse, catalyse, electrolyse, paralyse); the preferred US spelling is -yze. *Compare* -ize.

ytterbium Symbol: Yb *See also* periodic table; nuclide.

yttrium Symbol: Y *See also* periodic table; nuclide.

yttrium iron garnet Acronym: YIG (no stops).

Yukawa, Hideki (1907–81) Japanese physicist.
Yukawa potential

Z

z Symbol (light ital.) for **1.** a Cartesian coordinate. **2.** charge number (for cell reaction; z_B = charge number of an ion B). **3.** specific energy imparted.

Z 1. *Computing* A formal specification notation. **2.** Symbol for **i.** glutamic acid or glutamine (unspecified). **ii.** Z particle (Z^0).

Z Symbol (light ital.) for **1.** impedance. **2.** proton number.

(*Z*)- (ital., parentheses, always hyphenated) *Chem.* Prefix denoting a geometric isomer in which the highest priority substituent groups are located on the same side of a double bond (e.g. (*Z*)-3-methyl-2-pentenoic acid). The priority of the substituent groups is obtained using the Cahn–Ingold–Prelog sequence rules. *Compare* (*E*)-.

***z*-axis** (ital. z, hyphenated)

Z chromosome (not hyphenated) The larger of the *sex chromosomes in birds and lepidopterans, equivalent to the X chromosome in other organisms.

Zeeman, Pieter (1865–1943) Dutch physicist.
Zeeman effect

Zeiss, Carl (1816–88) German optical-instrument maker.

Zelenchukskaya Observatory The Special Astrophysical Observatory of the Academy of Sciences of the USSR, sited near Zelenchukskaya in the N Caucasus.

Zener, C. M. (1905–) US physicist.
Zener breakdown
Zener diode
Zener effect

zenithal hourly rate (not hyphenated) *Astron.* Abbrev.: ZHR (no stops).

Zeno of Elea (*c.* 490 BC–*c.* 430 BC) Greek philosopher.
Zeno's paradoxes

zeolite *Min., chem.* Adjectival form: **zeolitic.**

Zeppelin, Graf Ferdinand von (1838–1917) German airship designer.

Zermelo–Fraenkel system (en dash) *Maths.* Named after Ernst Zermelo (1871–1953) and A. Fraenkel.

Zernike, Frits (1888–1966) Dutch physicist.

Zerodur (cap. Z) A trade name for a type of glass–ceramic material little affected by temperature changes.

zero-point energy *See* internal energy.

zeta Greek letter, symbol: ζ (lower case), Z (cap.).

ZHR *Astron.* Abbrev. for zenithal hourly rate.

Ziegler, Karl (1898–1973) German chemist.
Ziegler process
Ziegler–Natta catalysts (en dash) or **Ziegler catalysts**

Ziehl, F. (1857–1926) German neurologist.
Ziehl–Neelsen method (en dash)
Ziehl's stain

zinc Symbol: Zn *See also* periodic table; nuclide.

zinc sulphate $ZnSO_4$ US: **zinc sulfate.** The recommended name for white vitriol.

Zinder, Norton David (1928–) US geneticist.

Zinjanthropus (cap. Z, ital.) The generic name originally given to fossil remains of hominids now classified as *Australopithecus boisei* (*see Australopithecus*).

Zinn, Walter Henry (1906–) Canadian-born US physicist.

zirconium Symbol: Zr *See also* periodic table; nuclide.

Zn Symbol for zinc.

-zoa 1. Noun suffix denoting animals or the animal kingdom in taxonomic names (e.g. Protozoa). Adjectival form: **-zoan** (no initial cap.). **2.** *See* -zoon.

zodiac Adjectival form: **zodiacal**.

-zoid Noun suffix denoting a motile entity derived from a plant (e.g. antherozoid). *Compare* -zoon.

Zöllner, Johann Karl Friedrich (1834–82) German astronomer and physicist.
Zöllner illusion

zoo- Prefix denoting animals (e.g. zoogeography, zoogloea, zoonosis, zooplankton, zoospore, zooid).

zoogloea (pl. zoogloeae) US: **zooglea** (pl. zoogleas). A structure formed by colonies of *pseudomonad bacteria of the genus *Zoogloea* (cap. Z, ital.). It consists of a gelatinous matrix containing many bacterial cells. Adjectival form: **zoogloeal** (US: **zoogleal**).

-zoon (pl. **-zoa**) Noun suffix denoting a motile entity derived from an animal (e.g. spermatozoon). *Compare* -zoid.

Z particles *See* gauge bosons.

Zr Symbol for zirconium.

Zsigmondy, Richard Adolf (1865–1929) Austrian chemist.

Zuckerkandl, Emil (1849–1910) Austrian anatomist.
Zuckerkandl's bodies Use para-aortic bodies

Zuckerman, Solly, Lord (1904–84) British zoologist and educationalist.

Zweig, George (1937–) Soviet-born US physicist.

Zwicky, Fritz (1898–1974) Swiss-born US astronomer and physicist.
Zwicky catalogue

zwitterion *Chem.* Adjectival form: **zwitterionic**. [from German: hermaphrodite ion]

Zworykin, Vladimir Kosma (1889–1982) Soviet-born US electrical engineer.

zygo- (**zyg-** before vowels) Prefix denoting a union or pair (e.g. zygomorphy, zygospore, zygotene).

Zygomycetes (cap. Z) A class of fungi of the subdivision *Zygomycotina that includes the pin moulds (genus *Mucor*). Individual name and adjectival form: **zygomycete** (no cap.).

Zygomycotina A subdivision of true fungi (Eumycota) containing two classes, *Zygomycetes and Trichomycetes. In a recent classification these fungi are regarded as a division, Zygomycota.

-zyme Noun suffix denoting an enzyme (e.g. lysozyme).

zymo- (**zym-** before vowels) Prefix denoting enzymes or fermentation (e.g. zymogen, zymosan, zymase).

Zymomonas (cap. Z, ital.) A genus of bacteria. Individual name: **zymomonad** (no cap., not ital.).

Appendix 1 The Electromagnetic spectrum

Name of radiation	Frequency/hertz	Wavelength/metre
gamma rays	$3 \times 10^{19} - 3 \times 10^{23}$	$10^{-15} - 10^{-10}$
X-rays	$2.3 \times 10^{16} - 3 \times 10^{19}$	$10^{-11} - 13 \times 10^{-9}$
ultraviolet	$7.5 \times 10^{14} - 2.3 \times 10^{16}$	$(13 - 400) \times 10^{-9}$
visible light	$(4.1 - 7.5) \times 10^{14}$	$(400 - 730) \times 10^{-9}$
infrared	$3 \times 10^{11} - 4.1 \times 10^{14}$	$730 \times 10^{-9} - 1 \times 10^{-3}$
radio waves		
EHF	$3 \times 10^{10} - 3 \times 10^{11}$	$10^{-3} - 10^{-2}$
SHF	$3 \times 10^{9} - 3 \times 10^{10}$	$10^{-2} - 10^{-1}$
UHF	$3 \times 10^{8} - 3 \times 10^{9}$	$10^{-1} - 1$
VHF	$3 \times 10^{7} - 3 \times 10^{8}$	$1 - 10$
HF	$3 \times 10^{6} - 3 \times 10^{7}$	$10 - 10^{2}$
MF	$3 \times 10^{5} - 3 \times 10^{6}$	$10^{2} - 10^{3}$
LF	$3 \times 10^{4} - 3 \times 10^{5}$	$10^{3} - 10^{4}$
VLF	$3 \times 10^{3} - 3 \times 10^{4}$	$10^{4} - 10^{5}$

Appendix 2 Graphical symbols used in electronics

Table 2.1 Qualifying graphical symbols

Symbol	Meaning
∼	alternating current
↗	variability (noninherent)
	variability in steps
	thermal effect
	electromagnetic effect
	radiation, electromagnetic nonionizing
	coherent radiation
	ionizing radiation
	positive-going pulse
	negative-going pulse
	pulse of a.c.
	positive-going step function
	negative-going step function
	fault

Table 2.2 Graphical symbols

Symbol	Meaning
●	connection of conductors
○	terminal (circle may be filled in)
	junction of conductors
	plug & socket (male & female)
	earth
	primary cell or accumulator (longer line represents +ve pole)
	battery of accumulators or primary cells
	switch, general symbol; make contact
	resistor, general symbol (first form preferred)
	variable resistor
	resistor with sliding contact
	capacitor, general symbol (first form preferred)
	inductor, coil, winding, choke, general symbol
	inductor with magnetic core
	transformer, 2 windings
	piezoelectric crystal, 2 electrodes
	semiconductor diode, general symbol
	light-emitting diode, general symbol
	photodiode
	pnp transistor
	npn transistor
	JFET, n-type channel
	JFET, p-type channel

Table 2.2 Graphical symbols (continued)

Symbol	Meaning
	IGFET, enhancement type, single gate, p-type channel without substrate connection
	amplifier, general symbol
&	AND gate, general symbol
≥1	OR gate, general symbol
1	inverter (NOT gate)
&	NAND gate (negated AND)
≥1	NOR gate (negated OR)
=1	exclusive-OR gate
*	indicating instrument (first form) & recording instrument; asterisk is replaced by symbol of unit of quantity being measured (e.g. V for voltmeter, A for ammeter, or by some other appropriate symbol)
	antenna, general symbol

Appendix 3 Letter symbols used in electronics

Table 3.1 Use of letter and subscript symbols

<table>
<tr><td rowspan="2" colspan="2"></td><td colspan="2">Basic letter</td></tr>
<tr><td>Lower-case</td><td>Capital</td></tr>
<tr><td rowspan="2">Subscript(s)</td><td>Lower-case</td><td>instantaneous value of the varying component</td><td>Without special sign or subscript
r.m.s. value of the varying component
With special sign or subscript
maximum (peak) value of the varying component
average value of the varying component</td></tr>
<tr><td>Capital</td><td>instantaneous total value</td><td>Without special sign or subscript
continuous (d.c.) value without signal
With special sign or subscript
total maximum (peak) value
total average value</td></tr>
</table>

Table 3.2 Recommended general subscripts

AV, av	average
F, f	forward
M, m	maximum (peak) value
MIN, min	minimum value
O, o	open ciruit
R, r	reverse or, as a 2nd subscript, repetitive
S, s	short circuit or, as a 2nd subscript, surge and/or nonrepetitive
(BR)	breakdown
(OV)	overload
tot	total

Table 3.3 Recommended general subscripts for parameters

F, F	forward; forward transfer
I, i	input
O, o	output
R, r	reverse; reverse transfer
T	depletion laer
1	input
2	output

Appendix 4 The Geological time scale

Eon	Era	Period	Subperiod	Epoch	Million years ago
Phanerozoic	Cenozoic	Quaternary		Holocene	
					0.01
				Pleistocene	
					2
		Tertiary	Neogene	Pliocene	
					5.1
				Miocene	
					24.6
			Palaeogene	Oligocene	
					38
				Eocene	
					54.9
				Palaeocene	
					65
	Mezozoic	Cretaceous		Upper	
					97.5
				Lower	
					144
		Jurassic		Upper	
					163
				Middle	
					188
				Lower	
					213
		Triassic		Upper	
					231
				Middle	
					243
				Lower	
					248
	Palaeozoic	Permian		Upper	
					258
				Lower	
					286
		Carboniferous	Upper (Silesian)		
					320
			Lower (Dinantian)		
					360
		Devonian		Upper	
					374
				Middle	
					387
				Lower	
					408
		Silurian			
					438
		Ordovician			
					505
		Cambrian			
					590

Appendix 5 Mathematical symbols

Table 5.1 General symbols

ratio of circumference of circle to its diameter	π
base of natural logarithms	e
imaginary unit: $i^2 = -1$	i, j
infinity	∞
equal to	$=$
not equal to	$\neq$
identically equal to	$\equiv$
corresponds to	$\triangleq$
approximately equal to	$\approx$
asymptotically equal to	$\simeq$
proportional to	$\propto$
approaches	$\to$
greater than	$>$
smaller than	$<$
much greater than	$\gg$
much less than	$\ll$
greater than or equal to	$\geq$
less than or equal to	$\leq$
plus	$+$
minus	$-$
plus or minus	$\pm$
a multiplied by b	ab, $a \cdot b$, $a \times b$
a divided by b	a/b, $\frac{a}{b}$, ab^{-1}
a raised to the power n	a^n
magnitude of a	$\|a\|$
square root of a	$\sqrt{a}$, $\sqrt{}a$, $a^{\frac{1}{2}}$
mean value of a	$\bar{a}$, $\langle a \rangle$
factorial p	$p!$
binomial coefficient: $n!/[p!(n-p)!]$	$\binom{n}{p}$

Table 5.2 Symbols for functions

exponential of x	$\exp x$, e^x
logarithm to base a of x	$\log_a x$
natural logarithm of x	$\ln x$, $\log_e x$
common logarithm of x	$\lg x$, $\log_{10} x$
binary logarithm of x	$\mathrm{lb}\, x$, $\log_2 x$
sine of x	$\sin x$
cosine of x	$\cos x$
tangent of x	$\tan x$, $\mathrm{tg}\, x$
cotangent of x	$\cot x$, $\mathrm{ctg}\, x$
secant of x	$\sec x$
cosecant of x	$\mathrm{cosec}\, x$, $\csc x$
inverse sine x	$\sin^{-1} x$, $\arcsin x$
inverse cosine x	$\cos^{-1} x$, $\arccos x$
inverse tangent x	$\tan^{-1} x$, $\arctan x$
integral	$\int$
summation	Σ
product	Π
finite increase of x	Δx
variation of x	δx
total differential of x	$\mathrm{d}x$
function of x	$f(x)$
composite function of f and g	$f \circ g$
convolution of f and g	$f * g$
limit of $f(x)$	$\lim_{x \to a} f(x)$, $\lim\limits_{x \to a} f(x)$
derivative of f	$\frac{\mathrm{d}f}{\mathrm{d}x}$, $\mathrm{d}f/\mathrm{d}x$, f'
time derivative of f	$\dot{f}$
partial derivative of f	$\frac{\partial f}{\partial x}$, $\partial f/\partial x$, $\partial_x f$, f_x
total differential of f	$\mathrm{d}f$
variation of f	δf

Appendix 6 The Periodic table

IA	IIA	IIIA	IVA	VA	VIA	VIIA	VIII			IB	IIB	IIIB	IVB	VB	VIB	VIIB	0
				1 H													2 He
3 Li	4 Be											5 B	6 C	7 N	8 O	9 F	10 Ne
11 Na	12 Mg											13 Al	14 Si	15 P	16 S	17 Cl	18 Ar
19 K	20 Ca	21 Sc	22 Ti	23 V	24 Cr	25 Mn	26 Fe	27 Co	28 Ni	29 Cu	30 Zn	31 Ga	32 Ge	33 As	34 Se	35 Br	36 Kr
37 Rb	38 Sr	39 Y	40 Zr	41 Nb	42 Mo	43 Tc	44 Ru	45 Rh	46 Pd	47 Ag	48 Cd	49 In	50 Sn	51 Sb	52 Te	53 I	54 Xe
55 Cs	56 Ba	57* La	72 Hf	73 Ta	74 W	75 Re	76 Os	77 Ir	78 Pt	79 Au	80 Hg	81 Ti	82 Pb	83 Bi	84 Po	85 At	86 Rn
87 Fr	88 Ra	89† Ac															

s–block (IA–IIA); Transition elements (IIIA–IIB); d–block (IVA–IIB); p–block (IIIB–0)

*Lanthanides	57 La	58 Ce	59 Pr	60 Nd	61 Pm	62 Sm	63 Eu	64 Gd	65 Tb	66 Dy	67 Ho	68 Er	69 Tm	70 Yb	71 Lu
†Actinides	89 Ac	90 Th	91 Pa	92 U	93 Np	94 Pu	95 Am	96 Cm	97 Bk	98 Cf	99 Es	100 Fm	101 Md	102 No	103 Lr

f–block (58–71, 90–103)

Appendix 7 SI units

Table 7.1 Base SI units

Quantity	Base SI unit	Symbol
length	metre	m
mass	kilogram	kg
time	second	s
electric current	ampere	A
thermodynamic temperature	kelvin	K
amount of substance	mole	mol
luminous intensity	candela	cd

Table 7.2 Prefixes used with SI units

Factor	Prefix	Symbol
10^{18}	exa–	E
10^{15}	peta-	P
10^{12}	tera-	T
10^{9}	giga-	G
10^{6}	mega-	M
10^{3}	kilo-	k
10^{2}	hecto-	h
10	deca-	da
10^{-1}	deci-	d
10^{-2}	centi-	c
10^{-3}	milli-	m
10^{-6}	micro-	μ
10^{-9}	nano-	n
10^{-12}	pico-	p
10^{-15}	femto-	f
10^{-18}	atto-	a

Table 7.3 Derived SI units with special names

Quantity	Name of derived SI unit	Symbol	Expressed in terms of base or derived SI units
frequency	hertz	Hz	1 Hz = 1 s^{-1}
force	newton	N	1 N = 1 kg m s^{-2}
pressure, stress	pascal	Pa	1 Pa = 1 N m^{-2}
energy, work, quantity of heat	joule	J	1 J = 1 N m
power	watt	W	1 W = 1 J s^{-1}
electric charge, quantity of electricity	coulomb	C	1 C = 1 A s
electric potential, potential difference, tension, electromotive force	volt	V	1 V = 1 J C^{-1}
electric capacitance	farad	F	1 F = 1 C V^{-1}
electric resistance	ohm	Ω	1 Ω = 1 V A^{-1}
electric conductance	siemens	S	1 S = 1 Ω^{-1}
flux of magnetic induction, magnetic flux	weber	Wb	1 Wb = 1 V s
magnetic flux density, magnetic induction	tesla	T	1 T = 1 Wb m^{-2}
inductance	henry	H	1 H = 1 Wb A^{-1}
Celsius temperature	degree Celsius	°C	1 °C = 1 K
luminous flux	lumen	lm	1 lm = 1 cd sr
illuminance	lux	lx	1 lx = 1 lm m^{-2}
activity (of a radionuclide)	becquerel	Bq	1 Bq = 1 s^{-1}
absorbed dose, specific energy	gray	Gy	1 Gy = 1 J kg^{-1}
dose equivalent	sievert	Sv	1 Sv = 1 J kg^{-1}
plane angle	radian	rad	
solid angle	steradian	sr	

Appendix 8 The Greek alphabet

Letters		Name
Capital	Lower case	
Α	α	alpha
Β	β	beta
Γ	γ	gamma
Δ	δ	delta
Ε	ϵ	epsilon
Ζ	ζ	zeta
Η	η	eta
Θ	θ	theta
Ι	ι	iota
Κ	κ	kappa
Λ	λ	lambda
Μ	μ	mu
Ν	ν	nu
Ξ	ξ	xi
Ο	ο	omicron
Π	π	pi
Ρ	ρ	rho
Σ	σ	sigma
Τ	τ	tau
Υ	υ	upsilon
Φ	φ	phi
Χ	χ	chi
Ψ	ψ	psi
Ω	ω	omega

Appendix 9 References

The Harvard (Author-Date) System

This is the version of the Harvard system preferred by Oxford University Press for its science books.

TEXT REFERENCES

Give the author's name and the year in parentheses: ' . . . (Smith 1991)'.

Where the author's name occurs naturally in the sentence, only the year is added in parentheses: 'Finch (1990) found . . . '.

When the same author has published more than one cited work in the same year, add '*a*', '*b*', etc. (ital.) to the year of publication (in order of appearance in the text): 'As Jones (1990*a*) has stated . . . He went on to show (Jones 1990*b*) . . . '.

If there are two authors, the surnames of both should be given before the date: ' . . . (Harris and Brown 1989).'.

If there are more than two authors, the name of the first author only should be given, followed by *et al.* (ital.): 'Three investigations (Denker *et al.* 1987; Ferris and Lindberg 1989; Lloyd *et al.* 1991) found . . . '. Note semicolons separating references.

For unpublished theses use the form: 'The work of Rogers (unpublished thesis, 1985) has led to . . . In her unpublished thesis (1991) Pauling investigated . . . '. Include details in the list of references.

For personal communications use the form: 'There is some evidence (Moore, personal communication) . . . '. Do not give further details in the list of references.

For works in the process of being published, use the form '(Richards, in press)'. Give as many details as possible in the list of references and remember to update if necessary at proof stage.

When citing an anonymous editorial in a journal use the name of the journal and the date: ' (*Lancet* 1989)', and list this under 'L' in the list of references.

When citing a work produced by an organization (e.g. the World Health Organization) for which the author's name is not given, the name of the organization (abbreviated if possible) may be used in place of the author's name: '(WHO 1989)'. List organizations alphabetically by the abbreviated form in the list of references; the full name should also be given in parentheses immediately after the abbreviation: 'WHO (World Health Organization) (1989)'.

When citing a string of references in the text, it is best to list them in chronological order: 'There were three experiments (Bonnington 1983; Jacobs 1985; Callow 1986 . . . '. Alphabetical order is also acceptable, however, but adopt a consistent policy throughout the text.

LIST OF REFERENCES

Order of entries

References are listed in alphabetical order of author. Single-author works are listed first, arranged chronologically; two-author works are listed second, in alphabetical order of second author, then chronologically; multiauthor works are listed third, arranged chronologically (not in alphabetical order of co-author, since '*et al.*' (not the co-authors' names) is used in the text):

Jones, F. (1980)
Jones, F. (1981)
Jones, F. and Smith, P. (1991)

Jones, F. and Thomas, G. (1990)
Jones, F. and Veevers, L. (1984*a*)
Jones, F. and Veevers, L. (1984*b*)
Jones, F. and Veevers, L. (1985)
Jones, F., Watts, D., Smith, G., and Brown, M. (1989*a*)
Jones, F., Barlow, D., Low, S., and King, N. (1989*b*)
Jones, F., Gore, P., and Davies, G. (1991)

References to articles in journals or periodicals

Items should be included in the following order:
(1) Author's name
- Use the form: 'Smith, D. G., Brown, A., and Eliot, F.'
- For multiauthor works, give the names of *all* the authors. However, if there are more than six authors, just name the first six authors, followed by '*et al.*'.

(2) Year of publication
- Put this in parentheses.
- Follow this by '*a*', '*b*' (inside the parentheses) if you are (i) citing more than one item published by the same author in the same year, or (ii) citing multiauthor works with the same first author, published in the same year (even if the co-authors are different).
- Put a full stop after parentheses.
- If the paper is still in the process of publication, put '(In press.)' at the end of the whole entry.

(3) Title of the paper
- If you prefer to leave out paper titles, do so consistently throughout the typescript.
- This should not be in quotation marks.
- Use lower case for all words in the title except the first word and proper names.
- Treat the spelling in paper titles as you would a quotation, i.e. do not change it to house style.
- The title is followed by a full stop.

(4) Title of the journal
- Give this in full.
- Underline the journal title (to indicate italic type).
- Use capitals for the main words, e.g. *American Journal of Physical Anthropology*.

(5) Volume number
- This will appear in bold type (e.g. **16**; denote this by a wavy underline).
- If letters form part of the volume number these will be in bold too, and set without a space between them and the volume number, e.g. **A135**.

(6) Issue number (only if essential)
- Use this only if the pagination of the journal is by issue rather than by volume. Leave it out unless it is essential.
- Use the form: *Scientific American*, **47**, (3), 63–4.

(7) Page numbers
- If you are giving only one citation, give the page numbers here.
- If you are making several citations of different parts of a work, give the page numbers in your text reference, e.g. '(Jones 1986, p. 114)', and give the page numbers for the whole article here.
- Give the spread of page numbers, e.g. '49–50', '111–25', not just the first page.
- Omit 'p.' or 'pp.' before the page numbers.

EXAMPLES

Haile, N. S. (1958). The snakes of Borneo, with a key to their species. *Sarawak Museums Journal*, **8**, 743–71.

Hooijer, D. A. (1957*a*). A *Stegodon* from Flores. *Treubia*, **24**, 119–29.

Hooijer, D. A. (1957*b*). Three new giant prehistoric rats from Flores, Lesser Sunda Islands. *Zoologisch Mededeelingen, Leiden*, **35**, 229–314.

Sprent, P. (1982). Discussion of Dr Atkinson's paper. *Journal of the Royal Statistical Society*, **344**, 22–4.

References to books, pamphlets, etc.

Items should appear in the following order:
(1) Author's name
- Use the form: 'Smith, D. G., Brown, A., and Eliot, F.' If the 'author' is an editor or editors, follow this by '(ed.)'.
- For multiauthor works, give the names of *all* authors. However, if there are more than six authors, just name the first six authors followed by '*et al.*'.

(2) Year of publication
- Put this in parentheses.
- Follow this by '*a*', '*b*', if you are (i) citing more than one item published by the same author in the same year, or (ii) citing multiauthor works with the same first author, published in the same year.
- If the book is still in the process of publication, add '(In press.)' as the last item (following the place of publication).

(3) Title of the chapter
- If you prefer to leave out chapter titles, do so consistently throughout the typescript.
- This should not be in quotation marks.
- Lower case for all words in title, except first word/proper names.
- Treat spelling of these as a quotation – do not change to house style.

(4) 'In'
- Needed only for references to multiauthor works.
- Use cap. 'I'.

(5) Title of book
- Underline this to denote italic type.
- Lower case for all words in title, except first word/proper names.

(6) Volume
- Use the form 'Vol. 1'. (Note: cap. 'V', full stop.)

(7) Edition (if not 1st)
- Use the form '(2nd edn)' '(revised edn)'.

(8) Translator's name (if any)
- Use the form '(trans. G. Fraser)'.

(9) Name of the editor(s) (if any)
- Use the form '(ed. P. Hordern and L. Thomas)'.
- Lower-case 'e' for 'ed.'.
- Use 'ed.' whether singular or plural, not 'eds'.
- Initial precedes the surname.

(10) Page numbers

- If you are referring to a particular chapter or paper or section in a book give the page numbers here.
- Give spread of page numbers, e.g. '49–50', not just the first page.
- Insert 'pp.' before the page numbers, e.g. 'pp. 9–10', 'p. 33'.

(11) Publisher

- Use the simplest form, e.g. 'Wiley' not 'John Wiley & Sons'.

(12) Place of publication

- Found on the title page of the book.
- If the title page gives, e.g. 'London–New York–Toronto', it is the city mentioned first in this list that is the place of publication (the rest should not be cited in the reference).
- No need to give this if it is obvious from the publisher's name, e.g. most university presses (but watch for titles published by branches abroad – e.g. OUP, New York – this will be clear from the title page).

EXAMPLES

Benneson, A. (ed.) (1980). *Control of communicable diseases*, (13th edn). American Public Health Association, Washington.

Bliss, C. I. (1970). *Statistics in biology*, Vol. 2. McGraw Hill, New York.

Britter, R. E. (1985). Diffusion and decay in stably-stratified turbulent flows. In *Turbulence and diffusion in stable environments*, The Institute of Mathematics and its Applications Conference Series, No. 4, (ed. J. C. R. Hunt), pp. 1–13. Oxford University Press.

Cochrane, S. H., O'Hara, J., and Leslie, J. (1980). *The effects of education on health*, World Bank Staff Working Paper, No. 405. World Bank, Washington.

Goodall, G. (1980). Stimulus and response learning in signalled punishment. Unpublished D.Phil. thesis. University of Sussex.

Gorman, M. (1979). *Island ecology*. Chapman and Hall, London.

Hader, R. J. and Grandage, A. H. E. (1958). Simple and multiple regression analyses. In *Experimental designs in industry*, (ed. V. Chew), pp. 108–37. Wiley, New York.

Health Education Council. *Cystitis*. Leaflet, obtainable from: Health Education Council, 78 New Oxford Street, London W1LA 1AH.

Kimble, J., Sulston, J., and White, J. (1979). In *Cell lineage, stem cells and determination*, INSERM Symposium, No. 10, (ed. N. Le Douarin), pp. 73–9. Elsevier, Amsterdam.

Sanders, D. (1983). Premenstrual tension. In *Women's problems in general practice*, (ed. A. McPherson and A. Anderson), pp. 42–62. Oxford University Press.

Takhtajan, A. O. (1969). *Flowering plants: origin and dispersal*, (trans. C. Jeffrey). Oliver & Boyd, Edinburgh.

The Numbered System

In this system, each reference is numbered in order of appearance in each chapter. References are listed at the ends of chapters (or, by chapter, at the end of the book).

There are some disadvantages to this system. Each time a reference is added (or removed) at typescript stage the numbers of all the other references in the chapter and the list of references will need to be altered (unless a sophisticated word-processing system is used). The same applies if references are added or removed at proof stage, but at this stage such corrections are very expensive. It is also difficult for the reader to find at a later date any particular reference in the list of references, as they will not normally be listed alphabetically. Numbers in the text are less immediately informative to the reader, as they are not accompanied by the writer's name.

However, there are some advantages in using this system for certain types of book. For example, for some literary works – scientific biographies, collections of lectures, etc. – the numbered system works well, as the references do not impede the flow of the text. With this system the list of references can be expanded to include some comment and extra information as well. This can dispense with the need for footnotes or unnecessary detail in the text itself.

TEXT REFERENCES

In the version of the numbered system used by Oxford University Press superscript numbers are used; note that these numbers go outside any punctuation: 'This was first discovered by Jones,[1] and was later confirmed by Smith.[2]'

When referring to several references, separate these numerals by commas: 'Jenkins[1,5] did further research on this subject.'

If there is a long string of references, give a range of numbers: 'The early accounts of the treatment[17–20] emphasize the side-effects.'

In mathematical works, the superscript numbers are given in square brackets so that they do not get confused with other notation.

LIST OF REFERENCES

Each reference is preceded by a number (not superscript) followed by a full stop. The order of items within a reference and the typographical style are the same as for the Harvard system (see above):

1. Grimes, C. M. (1974). Cost effectiveness in diagnostic X-ray departments. *Radiography*, **40**, 285–6.